introduction to vector and tensor analysis

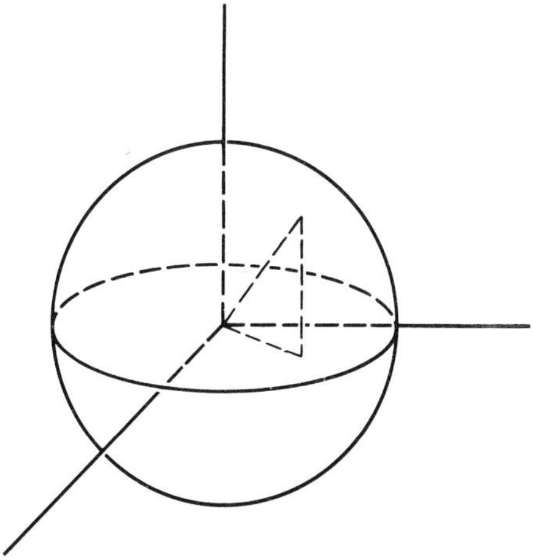

introduction to vector and tensor analysis

ROBERT C. WREDE, PROFESSOR OF MATHEMATICS, SAN JOSE STATE COLLEGE, SAN JOSE, CALIFORNIA

Dover Publications, Inc., New York

Published in Canada by General Publishing Company, Ltd., 30 Lesmill Road, Don Mills, Toronto, Ontario.
Published in the United Kingdom by Constable and Company, Ltd.

This Dover edition, first published in 1972, is an unabridged and corrected republication of the work originally published by John Wiley and Sons, Inc., in 1963.

International Standard Book Number: 0-486-61879-X
Library of Congress Catalog Card Number: 72-79300

Manufactured in the United States of America
Dover Publications, Inc.
180 Varick Street
New York, N. Y. 10014

To my Mother
and the
Memory of my Father

preface

Vector and tensor analysis are mathematical tools appropriate for the development of concepts in many areas, including geometry, analysis, and linear algebra. Even though vector analysis is a proper subset of tensor analysis, it has grown up separately. For this reason two distinct notations have developed with many variations of each. The vector notation, introduced primarily by Gibbs, has found wide use in differential geometry, especially curve theory, and in classical mechanics. The vector approach is rapidly finding its way into calculus and various engineering texts and into writings on aspects of biology, economics, and other sciences which had little use for such a viewpoint a few years ago.

The tensor notation and concepts grew up with the development of differential geometry, particularly surface theory and the extension of geometric ideas to spaces of n dimensions. For some years the theoretical physicist and the applied mathematician have found tensor analysis to be a valuable tool.

In emphasizing the utilitarian aspects of vector and tensor analysis, we must not overlook the fact that each theory involves the development of a complete mathematical system; that is, with respect to vector analysis, basic entities, called vectors, are introduced and the algebra and calculus of these objects are developed. The situation for tensor analysis is similar but more complex.

In this text I have tried to point out the role played in the development of vector analysis by concepts that are a part of linear algebra. There is an emphasis on the interrelationship of geometric and algebraic modes of expression. The classical notation of vector analysis is used, but notation more appropriate to tensor analysis is gradually introduced and correlated with the common vector notation. This is done in part to familiarize the student with tensor symbols and in part to facilitate the statement of proofs. However, the major distinction between this text and others of an introductory nature is the emphasis on transformation theory and the ramifications of that emphasis. In the first four chapters it is assumed that the space is Euclidean, and orthogonal Cartesian, general Cartesian, and general coordinate transformations, respectively, are presented. Simultaneously n-tuples of real numbers are associated with orthogonal Cartesian systems, collections of such n-tuples (a collection consisting of one n-tuple from each orthogonal Cartesian system) are defined as Cartesian vectors, and the vector concept is extended to the more general transformation groups. Transformation laws relating vector components in different coordinate systems are carefully stated. The significance of this mode of introduction of the vector form is indicated by various examples. In particular, the importance of the invariance of form, accomplished by the procedure described, is demonstrated in Chapter 2, Section 6. This development of special relativity requires no great depth of mathematical knowledge and serves to illustrate the importance of transformation theory. I believe that many of my students found it to be a bright spot in the courses from which this book is derived.

I have applied my notes (or I should say series of notes) in two ways. The first four chapters with the starred sections deleted were used for a one-semester, three-unit vector analysis course, which had a differential equations prerequisite but not an advanced calculus prerequisite. The students were mostly mathematics, physics, and engineering majors. I have also used the starred sections, with Chapter 5, as the basis for a one-semester, three-unit course in tensor analysis for seniors and beginning graduate students. The section on general relativity was not included. I have used the material on general relativity in an expanded form, as well as material on special relativity, in a three-unit course on applications of tensor analysis. I suggest also that this book could serve the purposes of an integrated two-semester course in vector and tensor analysis.

It is my belief that it is particularly suitable as preparation for differential geometry, applied mathematics, and theoretical physics. I think it will also provide a bridge between elementary aspects of linear algebra, geometry, and analysis, and I have tried to add some perspective by the inclusion of a significant amount of historical information.

San Jose, California
August 1963

Robert C. Wrede

acknowledgments

A textbook is necessarily a compilation of historical information concerning the subject at hand. My choice of topics and procedures has been influenced by many sources. Of these I would like particularly to acknowledge the influence of my former teacher Professor Vaclav Hlavaty (Indiana University).

I am indebted to Professors Erwin Kreyszig (Graz, Austria) and Bernard Friedman (University of California, Berkeley), both of whom carefully read the manuscript and made many valuable suggestions and corrections, and also to Dr. Dale Ruggles (San Jose State College), who eliminated numerous errors from the manuscript and discussed the point of view of the text with me.

I wish to express my thanks to my wife Jeanne S. Wrede, who drafted the initial set of illustrations for the book and performed many other tasks that contributed to its completion.

I am grateful to Mrs. Gerry Dunckel for typing the manuscript and for doing an excellent job in the tedious task of compiling the index.

Finally I would like to extend my thanks to the staff of John Wiley & Sons for their consideration and care in the preparation of this text.

R. C. W.

contents

chapter 1 the algebra of vectors, 1

Historical summary, 1
1. *Introductory concepts*, 7
2. *Linear dependence or independence of a set of number n-tuples*, 23
3. *Transformation equations relating rectangular Cartesian coordinate systems*, 35
4. *Definitions of Cartesian scalar and vector*, 51
5. *The inner product*, 56
5*. *General Cartesian coordinates*, 65
6. *ℰ systems and determinants*, 85
7. *The cross product*, 96
7*. *ℰ systems and the cross product in general Cartesian systems*, 111
8. *The algebra of matrices*, 114

chapter 2 the differentiation of vectors, 123

1. *The differentiation of vectors*, 123
2. *Geometry of space curves*, 139
3. *Kinematics*, 147
4. *Moving frames of reference*, 155
4*. *A tensor formulation of the theory of rotating frames of reference*, 166
5. *Newtonian orbits*, 171
6. *An introduction to Einstein's special theory of relativity*, 176

chapter 3 partial differentiation and associated concepts, 196

1. *Surface representations*, 196
2. *Vector concepts associated with partial differentiation*, 205
3. *Identities involving* ∇, 218
4. *Bases in general coordinate systems*, 220
5. *Vector concepts in curvilinear orthogonal coordinate systems*, 237
6. *Maxima and minima of functions of two variables*, 245

chapter 4 integration of vectors, 254

1. *Line integrals*, 255
2. *Surface integrals*, 270

2*. *An introduction to surface tensors and surface invariants*, 282

3. *Volume integrals*, 288
4. *Integral theorems*, 291

chapter 5 tensor algebra and analysis, 304

1. *Fundamental notions in n-space*, 305
2. *Transformations and tensors*, 315
3. *Riemannian geometry*, 324
4. *Tensor processes of differentiation*, 331
5. *Geodesics*, 341
6. *The parallelism of Levi-Civita*, 348
7. *The curvature tensor*, 354
8. *Algebraic properties of the curvature tensor*, 361
9. *An introduction to the general theory of relativity*, 371

answers to odd-numbered problems, 383

index, 411

chapter 1 the algebra of vectors

Historical Summary

The nineteenth century was one of great achievement. Darwin's theory of evolution and geological discoveries concerning the formation of the earth influenced man's philosophical outlook. Outstanding accomplishments in electricity and magnetism were followed by technological conquests that made the fruits of science available to the many. These developments stimulated discovery in mathematics and attracted more individuals to its study. The mathematician's station in life changed. Previously his existence had often depended on the benevolence of some royal personage, but in the nineteenth century more and more mathematicians were employed, primarily as teachers. The accompanying greater freedom of thought and increased motivation produced an outstanding era in mathematics. France and Germany were the centers of activity, and toward the end of the century Italian mathematics began to flourish. On the other hand Britain was just emerging from the shadow of Newton's brilliance and made only a few fresh contributions.

Vector analysis was born in the middle of the nineteenth century. Its underlying concepts are due primarily to William Rowan Hamilton (1805–1865, Irish) and Herman Grassmann (1809–1877, German). These ideas were developed on a foundation of the physical and mathematical thought of many centuries.

The process of addition in vector analysis is derived from, or at least correlates with, the parallelogram law of composition of forces. Aristotle (384–322 B.C., Greek) was aware of this law for the special case of a rectangle. Simon Stevin (1548–1620, Belgian) employed the principle in developing ideas in static mechanics. The parallelogram law of forces was formally set forth by Galileo Galilei (1564–1642, Italian); however, it is not clear that even he recognized its full scope. In any case, the need for a mathematical theory embracing an operation satisfying the fundamental laws of real number addition, yet different in character, was unmistakable.

From the mathematical point of view, vector and tensor analysis is a study of geometric entities and algebraic forms not dependent on coordinate system. Since this statement is meaningless without the existence of coordinate systems and transformation equations relating them, the quest for historical perspective carries us back to their inception.

René Descartes (1596–1650, French) and Pierre Fermat (1601–1665, French) share in the credit for the discovery of the utility of coordinate systems in relating geometry and algebra. Today the concept of coordinate system seems almost trivial; hence it is difficult to realize the stimulus that was given to the field of mathematics when a method was devised that made possible the amalgamation of geometry and algebra. Actually we must be careful in assigning credit for the introduction of the coordinate system concept, for, although Descartes' *La Géometrie* (1637) gives impetus to the idea, works by Fermat and John Wallis (1616–1703, English) correspond much more closely to modern day analytical geometry. And indeed Apollonius had a characterization of conic sections in terms of what we now call coordinates.

With the idea of coordinate system established in the first half of the seventeenth century, the second half saw strides taken in terms of the geometric representation of complex numbers $a + bi$[1] (a, b real numbers, $i = \sqrt{-1}$). John Wallis in 1673 published his *Algebra* in which he represented the real part of $a + bi$ along what we would call the axis of the reals, or horizontal axis, and then measured vertically a distance corresponding to b. Although this representation appears to be similar to that used today, it was not until 1798 that the idea of an axis of imaginaries was introduced. This concept, put forth by Caspar Wessel (1745–1818, Norwegian) in a paper entitled "On the Analytic Representation of Direction: an Attempt," completed the development of our present-day geometric image of the complex numbers. Wessel's name is not always associated with the diagrammatic representation of complex numbers. In French and English literature we find the term Argand[2] diagram, whereas in German writings it is often called the Gaussian plane. In the northern European countries Wessel is usually honored. Complex numbers are

[1] The term "complex numbers" was introduced by Carl Fredrich Gauss (1777–1855, German).

[2] Argand (1768–1822, French) arrived at a similar conclusion in 1806, several years after Wessel.

important in the historical background of vectors for several reasons. Just as the law of composition of forces served as an impetus to the development of vector addition, complex numbers provide a mathematical analogy to it. Vector addition is algebraically of the same form as complex number addition. Furthermore, the representation of a complex number by means of a line segment from the origin to a point in a plane associated with a number pair (a, b) treads on the toes of the vector concept. In fact, for some purposes it is quite convenient to represent a complex number by means of a line segment with a direction, sense, and magnitude. (We shall see that such a segment is the geometric representative of a vector.) Multiplication in terms of complex numbers introduces a process more general than that associated with real numbers. In particular, multiplication by $\sqrt{-1}$ has a geometric interpretation as a 90° rotation. (For example, $\sqrt{-1}(2) = 2i$.) It is only a small step forward to ask how a type of multiplication can be introduced in three-space such that the geometric interpretation corresponds to rotation. The answer to this question assumes fundamental importance in the development of vector analysis.

As we look back on the nineteenth century it is apparent that a mathematical theory in terms of which physical laws could be described and their universality checked was needed. The prerequisites for the development of such a theory were available. Figuratively speaking, two men stepped forward to do the job.

Hamilton and Grassman introduced the fundamental concepts of vector analysis from quite divergent modes of thought and in significantly different frameworks. Hamilton[3] seems to have been inspired mainly by a necessity for appropriate mathematical tools with which he could apply Newtonian mechanics to various aspects of astronomy and physics. From the strictly mathematical standpoint he was perhaps stimulated by the desire to introduce a binary operation that could be interpreted physically by means of a rotation in space. On the other hand, Grassmann's motivations were of a more philosophic nature. His chief desire seems to have been that of developing a theoretical algebraic structure on which geometry of any number of dimensions could be based.

Both Hamilton and Grassmann succeeded in one sense but at least partially failed in another. Hamilton developed an algebra based on four fundamental units $(1, i, j, k)$ with which many aspects of mechanics could be handled. In particular, his so-called theory of quaternions introduced

[3] The term "vector" is due to Hamilton. See Felix Klein, *Elementary Mathematics from an Advanced Viewpoint—Geometry*, translated by E. R. Hedrick and C. A. Noble, Dover 1939, p. 47.

a noncommutative binary operation which is known today under the name "cross product." Hamilton's quaternion ($a + bi + cj + dk$, where $i^2 = j^2 = k^2 = ijk = -1$) was not completely the product of his own imagination. August Ferdinand Möbius (1790–1868, German), a pupil of Gauss, had done much work along this line in his *Barycentric Calculus* (1827).[4] For this reason, and also because of its revolutionary nature, the introduction of a noncommutative operation must be considered as Hamilton's outstanding success. His failure, from the viewpoint of having his original work live on in the mathematical literature, existed in the fact that the theory of quaternions was too complicated in structure. Despite the numerous applications pointed out by Hamilton and the devoted advocacy of quaternions by P. G. Tait (1831–1900, Scottish) and others, the theory of quaternions was not able to survive in its original form.

Tait, an ardent disciple of Hamilton's quaternions, spent many years of his life in a valiant struggle to make the theory a fundamental language of mathematics. His success seems to have been limited to the admission by James C. Maxwell (1831–1879, Scottish) that these ideas, as distinguished from the algebraic structure of quaternions, could become valuable in the theoretical development of electricity and magnetism.

While Hamilton labored with quaternions, the German gymnasium teacher Grassmann (roughly equivalent to a high school teacher in the United States) was working on a highly theoretical and philosophic work *Lineale Ausdehnungslehre* (theory of linear extension), published in 1844. By defining an n-dimensional manifold as the set of all real number n-tuples $(X^1 \cdots X^n)$ and introducing a set of n fundamental units $e_1 \cdots e_n$ Grassmann was able to construct the hypercomplex numbers $X^1e_1 + \cdots + X^ne_n$. In terms of these numbers, he defined two binary operations, one of which is an addition of n-tuples that you will encounter in the first section of this work. The other operation introduced by Grassmann was no doubt motivated by real-number multiplication. The basic idea can be arrived at by naïvely multiplying

$$(X^1e_1 + \cdots + X^ne_n)(Y^1e_1 + \cdots + Y^ne_n).$$

By commuting the space variables with the fundamental units, we obtain the form

$$X^1Y^1e_1e_1 + X^1Y^2e_1e_2 + X^2Y^1e_2e_1 + \cdots.$$

A variety of results can be derived from this expression by defining the binary operation with respect to pairs of unit elements in different ways.

[4] The title was a bit long. *Der barycentrische Calcül, ein neues Huelfsmittel zur analytischen Behandlung der Geometrie darstellt und insbesondere auf die Bildung neuer Classen von Aufgaben und die Entwickelung mehrerer Eigenschaften der Kegelschnitte angewendet*, Leipzig, Verlag von Johann Ambrosius Barth, pp. 1–454.

As the reader proceeds with his examination of the development of the text, he will be able to observe that the so-called scalar and vector products as well as the tensor of second order are special cases of Grassmann's generalized multiplication.

In spite of the great merit of Grassmann's work, it made little impression on the scientific world. Even a revised edition of his theory of extension, published in 1862, which was to some degree simplified and which had many amplifications of the original, failed to be clear enough or of a style interesting enough to attract the attention of more than a few of his co-workers. Neither Hamilton nor Grassmann was able to stand back and take the objective look at his work needed to bring about a clarity and simplicity of form that focuses attention on the important contributions. This task was left to others.

In particular, John Willard Gibbs (1839–1903, United States) one of the outstanding mathematical physicists of the nineteenth century, was instrumental in developing the form of vector analysis found in present-day American texts. In lecturing to students at Yale, Gibbs felt a need for a simpler mathematical framework for the theoretical aspects of such subjects as electromagnetics and thermodynamics. His familiarity with the work of both Hamilton and Grassmann enabled him to pick out those of its aspects that seemed to apply best to the needs of theoretical physics. Apparently Gibbs did not think of his development of vector analysis as a contribution original enough to warrant publication; thus his notes on the subject were circulated among only a limited number of interested people. It was some twenty years after the development of his original notes that Gibbs reluctantly consented to the presentation of his work in book form. The book by E. B. Wilson, published in 1901, proved valuable in advancing the cause of vector methods. In fact, most of the texts on vector analysis printed to date have adhered to the form set down by Gibbs.

With the support of the use of vector methods in the development of theoretical physics by such outstanding men as O. Heaviside (1850–1925, English) in England and A. Föppel (1854–1924, German) in Germany, numerous texts on the subject began to appear in the United States, Germany, Britain, France, and Italy. By the beginning of the twentieth century vector analysis had become firmly entrenched as a tool for the development of geometry and theoretical physics.

As previously stated, Grassmann's *Ausdehnungslehre* contained much more than a basis for elementary vector analysis. Implicit in the work are many of the basic ideas of modern-day tensor analysis.[5] Possibly

[5] The term "tensor" was originated in the study of elasticity. See Dirk J. Struîk, *A Concise History of Mathematics II*, Dover, 1948, p. 283.

because of the more complicated notation, surely because of a lack of pressing need, the tensor theory was slower in coming into formal being. A realization of the need for the tensor ideas can be closely tied to the development of the intrinsic geometry of surfaces by Karl Friedrich Gauss. This significant contribution to mathematics was followed up by Riemann's generalization of the concepts to a space of n dimensions. Bernhard Riemann (1826–1866, German) a pupil of Gauss, based the metric properties of an n-dimensional space on a fundamental quadratic form.

$$ds^2 = \sum_{\alpha=1}^{n} \sum_{\beta=1}^{n} g_{\alpha\beta}\, dx^\alpha\, dx^\beta,$$

which had the same character as Gauss's first fundamental form on a surface. He generalized the concept of curvature on a surface to n-dimensional space and in general set up a form of geometry that was a prime target for the algebraic methods existing in the works of Grassmann and others. Riemann's doctoral thesis "On the Hypotheses Which Lie at the Foundations of Geometry," read before the Philosophical Faculty of the University of Göttingen in 1854 and published in 1868, after his death, stirred much interest in the mathematical world. Such men as E. Beltrami (1835–1900, Italian), E. B. Christoffel (1829–1900, German), and R. Lipschitz (1831–1904, German) made contributions that laid a further foundation for the development of the algebra and calculus of n-dimensional manifolds.

At the close of the nineteenth century the ideas were compiled and developed by Gregorio Ricci Curbastro (1853–1925, Italian) into what is now known as the algebra and calculus of tensors. Ricci's researches are enumerated in a memoir published in collaboration with his pupil Tullio Levi-Civita, "Méthodes de calcul différentiel et leurs applications," *Math. Ann.*, **54**, 1901. Tullio Levi-Civita (1873–1945, Italian), himself, made major contributions to geometry. In particular, his generalization of the concept of parallelism to Riemannian spaces was a significant addition to the absolute calculus of Ricci.

In spite of the fact that Ricci and Levi-Civita pointed out many applications for tensor calculus in both mathematics and physics, the subject was at the beginning of the twentieth century little more than the plaything of a small group of mathematicians. No great impetus had yet come about to enliven the interest of the scientific world. However, fortunately for both tensor analysis and physical theory, the elements of the subject were passed into the hands of Albert Einstein (1878–1955, Switzerland) by M. Grossmann (1878–1936, Switzerland) of Zurich.

Einstein had revolutionized scientific and philosophic thought with the introduction of his special theory of relativity around 1905. With the tool of tensor analysis and the n-dimensional space concepts of Riemann at hand, he was able to make significant generalizations of the previous theory. Einstein's general theory of relativity served to focus attention on tensor calculus and its merits. Since 1916 wide areas of application in theoretical physics, applied mathematics, and differential geometry have been found. Vector and tensor notation is making its way into elementary books in analysis, linear algebra, and other areas of mathematics. In a significant measure vector and tensor analysis is serving as the universal language Hamilton and Grassmann envisioned in their original theories.

1. *Introductory Concepts*

A course in vector algebra and analysis must presuppose a foundation of mathematical knowledge and a certain maturity of mind. The original intent was to state this desired foundation of knowledge in outline form. But it soon became evident that to put forth in an axiomatic way the needed fundamental ideas of geometry, algebra, and analysis would lead the discussion far afield. Therefore this introductory section is concerned only with a few basic ideas which are sometimes not given enough emphasis.

The concepts of point, line, plane, and space are used without definition, a customary viewpoint in the development of geometry since the time of David Hilbert.[6] These basic ideas have been abstracted from man's experience, along with that which he perceives as reality, and made a part of the foundation of geometry. With respect to Euclidean geometry, which is to serve as a setting for this development, you can get a firm intuitive feeling for point, line, and plane in the pictorial representation that you are no doubt familiar with from your high school geometry.

Depending on previous training, the use of undefined terms may or may not seem strange. Indeed in his original collection of mathematical works Euclid "defined" point, line, and plane. However, with the evolution of logical thought over the centuries, it became evident that these definitions were meaningless. In fact, a significant development in man's mode of thinking is the realization that certain concepts must be taken as undefined; otherwise a process of circular reasoning necessarily results.

By making use of the idea of coordinate systems, algebraic characterizations for point, line, and plane can be obtained. For example, given a coordinate system, a point in a plane can be represented by an ordered pair of numbers and then a line in a plane can be characterized as the set of all points whose coordinates satisfy a linear algebraic equation. Such

---

[6] David Hilbert (1862–1943, German). His *Grundlagen der Geometrie*, 1900, has become a model for the modern axiomatic approach to geometry.

equations are no doubt familiar to the reader. It must be cautioned that to construe this as a definition of the Euclidean[7] line would only lead to circular reasoning, for without line the coordinate system (and Euclidean measurements) cannot be conceived and without coordinate system the algebraic characterization fades into an animal of a different sort; that is, without a coordinate system in the back of one's mind, a linear equation has no well-established geometric meaning. For example, interpret $v = u$ in terms of rectangular Cartesian and then polar coordinates in the plane.

Many of the considerations of the text are made with respect to a three-dimensional Euclidean space or a two- or one-dimensional subspace. However, these ideas are expressed in a notational fashion that is immediately extendable to n-dimensional space.

In some instances this procedure may result in notational devices that seem a little more complex than necessary. However, many aspects of modern-day science cannot be adequately described in terms of a three-dimensional Euclidean model. Hence it seems appropriate that this introductory mathematical material should pave the way for that which lies ahead. The mathematical necessity of the notational devices is illustrated by the development of tensor algebra; physical usage is amply demonstrated in the development of special and general relativity theory.

By means of a coordinate system the triples of real numbers can be put into one-to-one correspondence with the points of a three-dimensional Euclidean space. Probably the simplest type of coordinate system, hence the one most appropriate for use at the start of this development, is the so-called rectangular Cartesian coordinate system.

It is assumed that the reader is somewhat familiar with this system. Therefore, the discussion that follows is brief, its chief purpose being that of introducing the notation and point of view that is used throughout the book.

A rectangular Cartesian coordinate system is constructed by means of three concurrent and mutually perpendicular lines called axes.

The elements of the set of all real numbers are put in one-to-one correspondence with the points of each axis in such a way that the point of concurrence corresponds to 0 in each case; a common unit of measurement is employed. The variables of these one-to-one correspondences are designated by X^1, X^2, X^3. Hence it is convenient to denote the lines as the X^1, X^2, and X^3 axes, respectively. Then, to each point of the three-dimensional space there corresponds an ordered triple of real numbers

---

[7] Of course, we can define in a purely algebraic sense a linear equation as a number line. However, having done so, the Euclidean line is only one of many possible geometric interpretations.

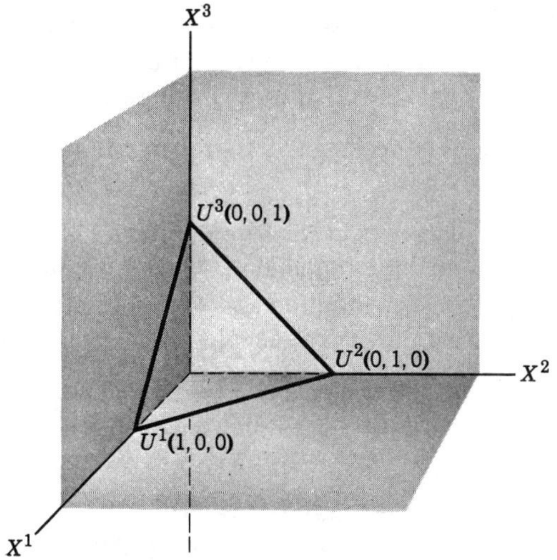

Fig. 1-1.1

(X^1, X^2, X^3) and conversely. Note that the word "ordered" is significant. For example, given the triple (2, 1, 5), 2, 1, 5, respectively, replace X^1, X^2, X^3, and no other ordering of replacement is possible. The reader is likely to be more familiar with the symbols X, Y, Z than with X^1, X^2, X^3. However, the usage indicated will have notational advantages that will become evident as we read further into the book. For example, a linear form $a_1X^1 + a_2X^2 + a_3X^3$ can be represented by the compact expression a_jX^j.[8]

A rectangular Cartesian coordinate system is said to be left-handed or right-handed. In order to make the distinction between these two types, consider the triangle of Fig. 1-1.1 determined by the unit points U^1, U^2, U^3.

Definition 1-1.1. If, as observed from the origin (0, 0, 0), the triangle orientation $U^1 \to U^2 \to U^3$ is clockwise, the associated rectangular Cartesian coordinate system is said to be right-handed. Similarly, if the orientation $U^1 \to U^2 \to U^3$ is counterclockwise, the coordinate system is said to be left-handed.

Comments on the relation between left- and right-hand coordinate systems are deferred until Section 7. By that stage of the game appropriate information concerning transformations and elements of volume will be

---

[8] The summation convention which leads to this representation is introduced in Chapter 1, Section 3.

available to facilitate the discussion. (See Example 1-7.7.) For present purposes it suffices to say that either type of coordinate system may be used, but consistency demands that no switching back and forth be done.

Convention 1-1.1. The rectangular Cartesian coordinate systems used in this text are right-handed.

One of the major objectives of the subject at hand is to place geometry on an algebraic and analytic foundation. Such an approach leads to a healthy interchange between intuition and rigor. The following introduction of an n-tuple space provides a stepping stone in this direction.

Definition 1-1.2. Let the elements of S be ordered n-tuples $(A^1, A^2, \cdots, A^n)$ of real numbers that satisfy the following properties:

(a) (*Addition*)

$$(A^1, A^2, \cdots, A^n) + (B^1, B^2, \cdots, B^n)$$
$$= (A^1 + B^1, A^2 + B^2, \cdots, A^n + B^n).$$

(b) (*Multiplication of an ordered n-tuple of real numbers by a real number*)

$$\beta(A^1, A^2, \cdots, A^n) = (\beta A^1, \beta A^2, \cdots, \beta A^n),$$

where β is any real number.

The set S will be called an n-tuple space.[9] An ordered n-tuple of real numbers $(A^1, \cdots, A^n)$ is also represented by a bold face **A** and the numbers $(A^1, \cdots, A^n)$ are called the components of **A**.

Much of the desired algebraic foundation follows immediately from the fact that the n-tuples of the set S are defined in terms of real numbers. After setting forth the consequences of this fact, it will be my purpose to correlate these algebraic properties with the appropriate geometric facts.

Definition 1-1.3. Two n-tuples **A** and **B** are equal if and only if $A^j = B^j$, $j = 1, 2, \cdots n$

Definition 1-1.4. The n-tuple with components $(0, 0, \cdots, 0)$ is called the zero n-tuple and denoted by **0**

The fundamental laws of n-tuples are established in the next two theorems.

---

[9] The reader familiar with the concept of a vector space, as introduced in an algebra course or text, will recognize the space introduced above as such. The reason for the present use of the term n-tuple space is that, in differential geometry and in areas of physics that deal with transformation theory, there is a further restriction on the vector concept. In this book the word "vector" is used in the more limited sense. Birkhoff and MacLane, *A Survey of Modern Algebra*, seventh printing, Macmillan, 1958, p. 159–163.

Theorem 1-1.1. Elements of the n-tuple set S satisfy the following laws:

(1-1.1)
(a) (*Closure law*) If $\mathbf{A}$ and $\mathbf{B}$ are elements of S, then so is $\mathbf{A} + \mathbf{B}$.
(b) (*Commutative law of addition*) $\mathbf{A} + \mathbf{B} = \mathbf{B} + \mathbf{A}$.
(c) (*Associative law of addition*) $(\mathbf{A} + \mathbf{B}) + \mathbf{C} = \mathbf{A} + (\mathbf{B} + \mathbf{C})$.
(d) (*Identity element for addition*) The n-tuples $\mathbf{0}$ is the unique element of S with the property $\mathbf{A} + \mathbf{0} = \mathbf{0} + \mathbf{A} = \mathbf{A}$.
(e) (*Additive inverse*) Corresponding to each $\mathbf{A}$ in S is a unique element $(-\mathbf{A})$ such that $\mathbf{A} + (-\mathbf{A}) = \mathbf{0}$.

PROOF. These laws (1-1.1a–e) are consequences of the corresponding real-number laws and the fundamental properties of n-tuples, as described in Definition 1-1.2. The proofs of (1-1.1b) and (1-1.1d) are set forth in order to illustrate the method of approach. The other proofs are left to the student.

PROOF OF (1-1.1b). According to Definition 1-1.2, the n-tuple $\mathbf{A} + \mathbf{B}$ has components $(A^1 + B^1, \cdots, A^n + B^n)$. Because the real numbers are commutative,

$$(A^1 + B^1, \cdots, A^n + B^n) = (B^1 + A^1, \cdots, B^n + A^n).$$

Since the right-hand member of this relation represents $\mathbf{B} + \mathbf{A}$, the proof is complete.

PROOF OF (1-1.1d). Let $\mathbf{X}$ be an n-tuple such that

$$\mathbf{A} + \mathbf{X} = \mathbf{A}.$$

There are two questions to be answered. Is $\mathbf{X}$ unique? What are the components of $\mathbf{X}$? Again the answer is supplied by making use of the associated real-number properties. According to definitions (1-1.2) and (1-1.3),

$$A^j + X^j = A^j, \qquad j = 1, \cdots, n.$$

Because the A^j and X^j are real numbers, we obtain the unique solutions

$$X^j = 0, \qquad j = 1, \cdots, n.$$

Therefore $\mathbf{X}$ is the n-tuple with components $(0, \cdots, 0)$, as was to be shown.

The concept of additive inverse as introduced in (1-1.1e) makes possible the following definition of subtraction.

Definition 1-1.5. For every pair of n-tuples $\mathbf{A}$ and $\mathbf{B}$ let

(1-1.2) $$\mathbf{A} - \mathbf{B} = \mathbf{A} + (-\mathbf{B}).$$

We say that $\mathbf{B}$ is subtracted from $\mathbf{A}$.

According to this definition, subtraction is the inverse operation to that of addition.

Theorem 1-1.2. Let α and β represent arbitrary elements from the set of real numbers, and **A** and **B** represent elements of the n-tuple set S.

(1-1.3)

(a) (*Closure under real number multiplication*) $\alpha\mathbf{A}$ is an element of S.

(b) (*Real number identity element*) $1\mathbf{A} = \mathbf{A}$.

(c) (*Law of distribution of an n-tuple with real numbers*) $(\alpha + \beta)\mathbf{A} = \alpha\mathbf{A} + \beta\mathbf{A}$.

(d) (*Law of distribution of a real number with n-tuples*) $\alpha(\mathbf{A} + \mathbf{B}) = \alpha\mathbf{A} + \alpha\mathbf{B}$.

(e) (*Law of association of real numbers with an n-tuple*) $\alpha(\beta\mathbf{A}) = (\alpha\beta)\mathbf{A}$.

PROOF. As in Theorem 1-1.1, the proofs result from analogous properties of real numbers. The proof of (1-1.3a) is presented as an illustration of the method employed.

PROOF OF (1-1.3a). **A** has components $(A^1, \cdots, A^n)$. According to Definition 1-1.2b,

$$\alpha(A^1, \cdots, A^n) = (\alpha A^1, \cdots, \alpha A^n).$$

Because the set of real numbers is closed under multiplication, $(\alpha A^1, \cdots, \alpha A^n)$ is an ordered n-tuple of real numbers and therefore an element of S. This completes the proof.

Let us now turn our attention to the geometric significance of n-tuples and their properties. Most of the considerations of this book pertain to three-dimensional Euclidean space. On occasion examples are presented in terms of two- and even one-dimensional Euclidean space. For this reason, and because nobody can visualize a space of dimension greater than three anyway, the geometric illustrations are three-, two-, or one-dimensional. However, it is worth emphasizing that the corresponding algebraic structure is set up so that it is easily extendable to integer values of $n > 3$. Since it is not difficult to find physical examples in which $n > 3$, this approach can be of significant future value. (See Chapter 2, Section 6.)

Suppose that a rectangular Cartesian coordinate system is given. (See Fig. 1-1.2.) Then an n-tuple **A** can be represented graphically by an arrow[10] with its initial point at the origin and its terminal point at the position with coordinates (A^1, A^2, A^3). However, this is not the only possible arrow representation of the n-tuple. As a matter of fact, any arrow with

[10] The term "arrow" is being used in an intuitive sense. We are to visualize a line segment with one end designated as the initial point and the other as the terminal point.

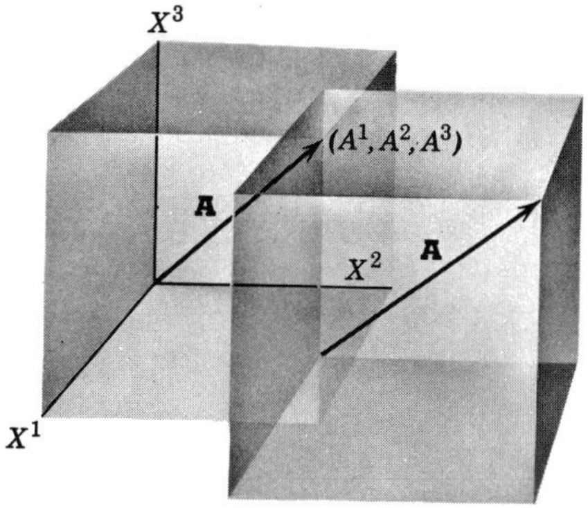

Fig. 1-1.2

initial and terminal points $P_0(X_0^1, X_0^2, X_0^3)$ and $P_1(X_1^1, X_1^2, X_1^3)$ respectively, such that

$$\text{(1-1.4)}\quad A^1 = X_1^1 - X_0^1, \qquad A^2 = X_1^2 - X_0^2, \qquad A^3 = X_1^3 - X_0^3$$

can be considered a representative of the n-tuple.
Therefore an n-tuple has as its geometric analogue a family of arrows, each of which has a common magnitude, direction, and sense. According to (1-1.4), the components of **A** can be interpreted as measurements of the perpendicular (or parallel) projections of any geometric representative onto the coordinate axes. (Again see Fig. 1-1.2) The geometric interpretation of n-tuples gives added meaning to their representation by a single boldfaced letter such as **A**.

The terms magnitude, direction, and sense are used in an intuitive manner. The meaning of each is made precise as follows: the concept of distance as based on the Pythagorean theorem is fundamental to Euclidean geometry. This development presupposes knowledge of the theorem. Hence, if $P_0(X_0^1, X_0^2, X_0^3)$ and $P_1(X_1^1, X_1^2, X_1^3)$ are any two points in Euclidean three-space and the coordinate system is rectangular Cartesian, the distance $|P_0 P_1|$ is given by

$$\left[(X_1^1 - X_0^1)^2 + (X_1^2 - X_0^2)^2 + (X_1^3 - X_0^3)^2\right]^{1/2}.$$

If we think of the components of **A** as orthogonal projections on the coordinate axes [see (1-1.4)], we have

$$\left[(A^1)^2 + (A^2)^2 + (A^3)^2\right]^{1/2} = \left[[(X_1^1 - X_0^1)^2 + (X_1^2 - X_0^2)^2 + (X_1^3 - X_0^3)^2\right]^{1/2}.$$
(1-1.5a)

The concept of magnitude of an n-tuple is then correlated to that of distance by means of the following definition.

Definition 1-1.6. Let $|\mathbf{A}|$ represent the magnitude of $\mathbf{A}$; then

$$|\mathbf{A}| = [(A^1)^2 + (A^2)^2 + (A^3)^2]^{1/2}. \tag{1-1.5b}$$

The word direction is used in an analogous way to the word parallel, as is common in solid analytic geometry when we talk about direction numbers.

Definition 1-1.7. Two n-tuples $\mathbf{A}$ and $\mathbf{B}$ are said to have the same direction (are parallel) if and only if their components are proportional; that is, if and only if

$$(B^1, B^2, B^3) = \beta(A^1, A^2, A^3) \tag{1-1.6}$$

for some nonzero real number β.

The geometric terms "same direction" or "parallel" take on intuitive significance when associated with the families of arrows representing $\mathbf{A}$ and $\mathbf{B}$. Representatives of these families are shown in Fig. 1-1.3.

The word "sense" is used as a refinement of the term direction; that is, if two n-tuples are parallel, we can further indicate an orientation by specifying that they have the same sense or are oppositely sensed. Again, the geometric interpretation is clearly expressed by the arrow representatives. (See Fig. 1-1.3.)

Definition 1-1.8. Suppose two n-tuples, $\mathbf{A}$ and $\mathbf{B}$, had the same direction [i.e., to satisfy (1-1.6)]. They will have the same sense if $\beta > 0$ but opposite senses if $\beta < 0$.

The algebra of n-tuples evolves from the properties of addition and multiplication of an n-tuple by a real number. (See Definition 1-1.2.) We can supplement our intuition significantly by examining the geometric interpretation of these operations. The law of composition, as demonstrated in Fig. 1-1.4, is a so-called parallelogram law.

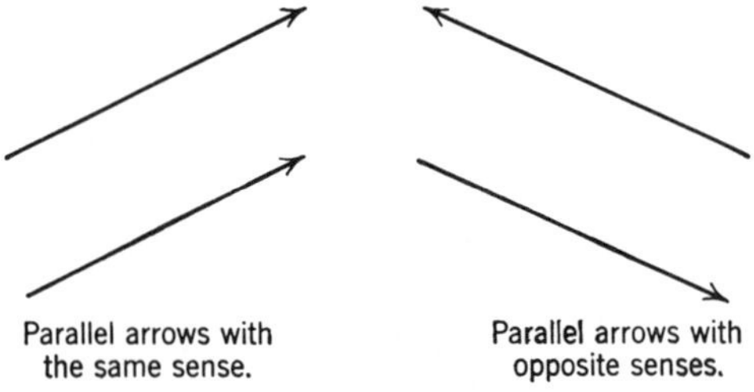

Fig. 1-1.3

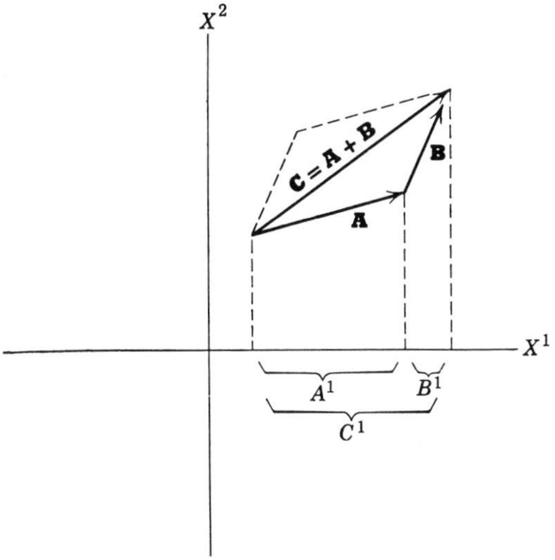

Fig. 1-1.4

The geometric justification for this name is obtained by constructing an arrow representative of **A**, forming that representative of **B** which has its initial point at the terminal point of **A**, and then constructing the arrow that joins the initial point of **A** to the terminal point of **B**. This join, which is the diagonal of a parallelogram determined by **A** and **B**, is a geometric representative of **A** + **B**. Since forces, linear velocities, and linear accelerations are known from experimentation to compound by means of a parallelogram law, possibly the reader can get a first glimpse of the way in which the mathematical structure presently being developed can be used in building models for certain physical problems.

In the last paragraph n-tuple addition was geometrically interpreted. Also, the process of multiplication of an n-tuple by a real number can be represented simply. According to Definition 1-1.2, $\beta\mathbf{A}$ has components $(\beta A^1, \beta A^2, \beta A^3)$. Hence the magnitude as determined by (1-1.5b) is

(1-1.7) $$|\beta\mathbf{A}| = [(\beta A^1)^2 + (\beta A^2)^2 + (\beta A^3)^2]^{1/2} = |\beta|\,|\mathbf{A}|.$$

Therefore $\beta\mathbf{A}$ is geometrically represented by an arrow parallel to a representation of **A**, with the same or opposite sense to that of **A**, depending on whether $\beta \gtrless 0$, and with magnitude differing by the factor $|\beta|$.

Construction of geometric interpretations associated with the laws of Theorems 1-1.1 and 1-1.2 is left to the reader.

Most of the physical and geometric usages of the structure presently being introduced depend on further developments. However, the following simple examples are available at this stage.

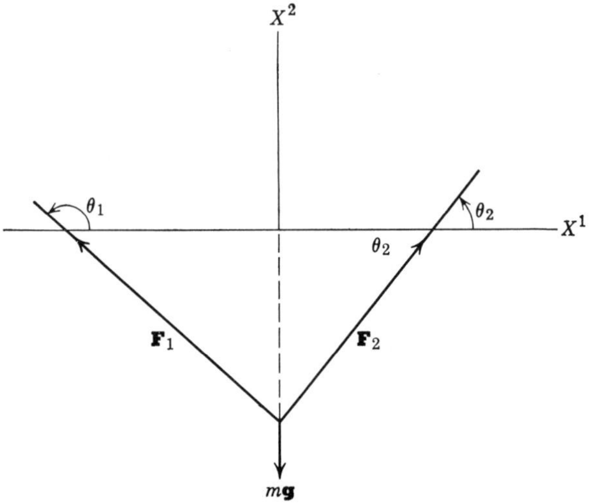

Fig. 1-1.5

Example 1-1.1. The study of a particle at rest is often called statics. Since classical mechanics is primarily concerned with problems involving motion of a particle, it might be thought that statics could have no real interest. However, as soon as we bring to mind the possibility of several forces acting on a given particle in such a way that the particle remains in a state of equilibrium, the problems of statics take their rightful place as a subset of the problems of motion study. As a particular illustration, consider the following. A particle of mass m is suspended by two weightless strings as in Fig. 1-1.5. The idealization to weightless strings enables us to work with a model in which the strings are replaced by straight lines. There is a gravitational force $m\mathbf{g}$ acting on the particle. Furthermore, if forces (i.e., tensions) $\mathbf{F}_1$ and $\mathbf{F}_2$ are associated with the strings,

$$\mathbf{F}_1 + \mathbf{F}_2 + m\mathbf{g} = \mathbf{0}.$$

The components of $\mathbf{F}_1$ are $|\mathbf{F}_1| \cos \theta_1$, $|\mathbf{F}_1| \sin \theta_1$, whereas those of $\mathbf{F}_2$ are $|\mathbf{F}_2| \cos \theta_2$ and $|\mathbf{F}_2| \sin \theta_2$. Therefore

$$|\mathbf{F}_1| \cos \theta_1 + |\mathbf{F}_2| \cos \theta_2 = 0$$
$$|\mathbf{F}_1| \sin \theta_1 + |\mathbf{F}_2| \sin \theta_2 - m\,|\mathbf{g}| = 0.$$

Upon solving for $|\mathbf{F}_1|$ and $|\mathbf{F}_2|$, we obtain

$$|\mathbf{F}_1| = \frac{-m\,|\mathbf{g}| \cos \theta_2}{\sin (\theta_2 - \theta_1)} \qquad |\mathbf{F}_2| = \frac{m\,|\mathbf{g}| \cos \theta_1}{\sin (\theta_2 - \theta_1)}.$$

Thus, by identifying a state of equilibrium with the resultant **0** and using the basic laws of n-tuples, we are able to obtain the magnitudes of the string tensions.

Example 1-1.2. The fact that a line is determined by a point and a direction leads at once to an algebraic representation in terms of n-tuples. Suppose that the given point is denoted by P_0 and the direction is determined by an n-tuple **B**. As in Fig. 1-1.6, let $\mathbf{r}_0$ be an arrow with the initial point at the origin and the end point at P_0. For convenience of illustration, use that arrow representative of **B** which has its initial point at P_0. Then an arrow **r**, with its initial point at the origin and its end point at an arbitrary position on the line, is determined according to the parallelogram law of addition by

$$\mathbf{r} = \mathbf{r}_0 + \mathbf{B}f(t), \tag{1-1.8}$$

where $f(t)$ is a continuous function[11] with range values $-\infty < f(t) < \infty$. Often it is convenient to choose $f(t)$ such that $f(t) = 0$ for $t = 0$. The requirement that the range values $f(t)$ include the complete set of real numbers ensures that the line rather than a segment of it is represented. Choices of f such that (1-1.8) represents a line segment are also of interest and are used later in the discussion.

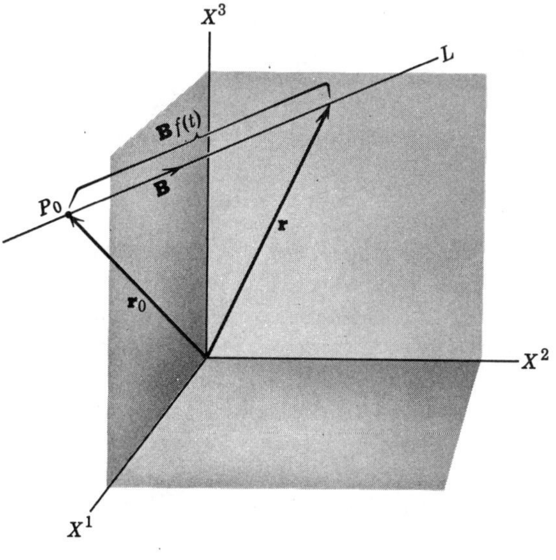

Fig. 1-1.6

[11] For a discussion of function, see Chapter 2, Section 1.

It is left to the reader to show that the vector representation (1-1.8) is equivalent either to the set of parametric equations

$$\begin{aligned} X^1 &= X_0^1 + B^1 f(t) \\ X^2 &= X_0^2 + B^2 f(t) \\ X^3 &= X_0^3 + B^3 f(t) \end{aligned} \tag{1-1.9a}$$

or, when $B^j \neq 0$, $j = 1, 2, 3$, to the symmetric form

$$\frac{X^1 - X_0^1}{B^1} = \frac{X^2 - X_0^2}{B^2} = \frac{X^3 - X_0^3}{B^3}. \tag{1-1.9b}$$

If one of the components B^j is equal to zero, say, for example, $B^1 = 0$, then the symmetric form must be replaced by

$$\frac{X^2 - X_0^2}{B^2} = \frac{X^3 - X_0^3}{B^3}, \qquad X^1 = X_0^1.$$

Therefore the symmetric form is not generally valid but rather subject to a variety of exceptions. On the other hand, the parametric form (1-1.9a) and the equivalent arrow or n-tuple equation (1-1.8) have general validity. The universal character of these representations is a significant point in favor of their usage.

Example 1-1.3. Suppose that the function of the preceding example were

$$f(t) = t,$$

where t is defined on the set of all real numbers. On the one hand, the relation

$$\mathbf{r} = \mathbf{r}_0 + \mathbf{B}t$$

represents a line. On the other hand, this relation can be interpreted as the algebraic analogue of Newton's first law of motion (i.e., every body tends to remain in a state of rest or of uniform rectilinear motion unless compelled to change its state through the action of an impressed force). This analogy is obtained by interpreting the values of the parameter t as measures of time and by giving the line the kinematical interpretation of a path generated by a moving particle. Such physical interpretations of geometric models, as alluded to in this example, are dealt with in the discussions of kinematics and dynamics in a later chapter.

Example 1-1.4. Suppose the function f of Example 1-1.2 is given by the rule

$$f(t) = \sin t,$$

where t is defined on some subset of real numbers. The set of range values of f lies in the closed interval

$$-1 \leq \sin t \leq 1.$$

Therefore, when this function is used in the form (1-1.8), only a segment of a line is represented.

A fundamental problem of theoretical physics is that of formulating universally valid laws relating natural phenomena. Often the problem is restricted to that of finding relations valid with respect to a certain set of frames of reference. Newton's second law of motion (i.e., the rate of change of momentum is proportional to the impressed force and occurs in the direction of the applied force) provides such an example. The law, the simplest form of which is

$$\mathbf{F} = \frac{dm\mathbf{v}}{dt},$$

where $\mathbf{F}$ represents force and $m\mathbf{v}$ represents momentum, is valid only in frames of reference that are in an unaccelerated state of motion. Such frames are called inertial systems and probably exist only ideally. From the point of view of mathematics we associate a coordinate system with each frame of reference and then ask how the entities in the algebraic expression of the physical law transform in order that the form of the law may remain the same (i.e., invariant). Because the transformation idea is of such importance, the development of vector analysis[12] in this book is built around it. As stepping stones to the introduction of the vector concept, we make use of the set of orthogonal Cartesian coordinate systems and the geometric interpretation of n-tuples.

Having constructed one rectangular Cartesian coordinate system in a Euclidean space,[13] it is clear that any number of systems can be constructed, each distinct from any other. If an n-tuple in three-space (A^1, A^2, A^3) is given with respect to the initially constructed coordinate system, then by its arrow representation, and by means of orthogonal projection, we can determine triples in each of the other systems. (See Fig. 1-1.7.) This thought leads us to the concept of a Cartesian vector.

Definition 1-1.9. The set $\{A^1, A^2, A^3\}$ of all triples (A^1, A^2, A^3), $(\bar{A}^1, \bar{A}^2, \bar{A}^3)$, etc., determined by orthogonal projections of a common arrow representation on the axes of the associated rectangular Cartesian

[12] The term "vector" was introduced by Hamilton.

[13] The existence of a rectangular Cartesian coordinate system is equivalent to the space being Euclidean. As an example of a non-Euclidean space, hence one in which a Cartesian coordinate system cannot be constructed, consider the surface of a sphere.

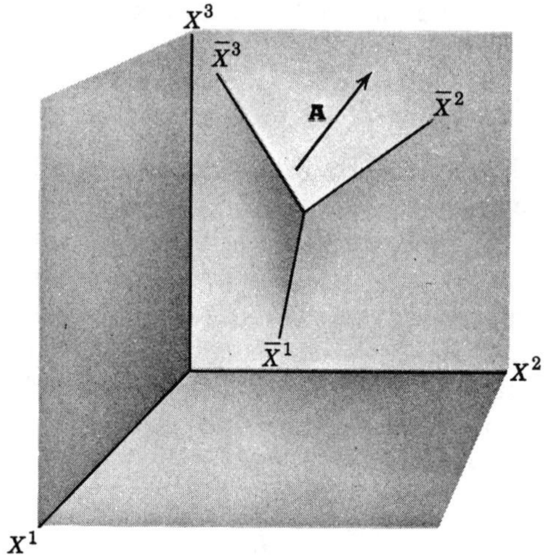

Fig. 1-1.7

coordinate system is said to be a Cartesian vector. Any two triples of this set are related by a particular law of transformation. (This law has yet to be specified.)

A Cartesian vector has a family of arrows as its geometric representative; with respect to a coordinate system it is characterized by a magnitude, a direction, and a sense, and its components in any coordinate system satisfy the algebraic laws pertaining to n-tuples.[14] In most elementary books on vector analysis a vector is identified as an arrow. We take the more complicated viewpoint of the preceding definition because by defining a Cartesian vector as a collection of n-tuples, any two of which are related by a specified law of transformation, the concept is given an absolute meaning from the point of view of algebra. On the one hand, this viewpoint is consistent with the absolute geometric meaning for a Euclidean space of dimension $n \leq 3$. On the other hand, the algebraic framework opens the door to a generalized definition of the vector concept expressed in terms of a set of transformations but not dependent on the geometric interpretation as an arrow. This general approach has significance for Euclidean spaces of dimensions $n > 3$ and to considerations in non-Euclidean spaces.

[14] There are two facts that must not be confused. On the one hand, the term "family of arrows" alludes to the fact that there is an arrow at each point of space. On the other hand, the expression "sets of triples" refers to the fact that an arrow can be projected on an infinite set of orthogonal axes.

The law of transformation alluded to in Definition 1-1.9 is introduced for Cartesian vectors in Section 3 and in more generality in later sections.

Because there is an identification of the geometric representative of a Cartesian vector and a 3-tuple, and in order to avoid involvement in a complexity of notation, the same boldfaced symbols are used for Cartesian vectors that have been used for n-tuples.

Problems

1. Graph a right-hand rectangular Cartesian coordinate system for three dimensions.

(a) Construct an arrow with initial point at (1, 2, 0) and end point (2, 5, 0). What are the magnitudes of the perpendicular projections of this arrow on the three coordinates axes?

(b) Construct the arrow of the same family as the arrow of part (a) with initial point at the origin.

(c) Repeat (a) and (b) for an arrow with initial point (1, 2, 3) and terminal point (2, 5, 6).

2. The initial and terminal points, respectively, of several arrows are given. Find the n-tuple corresponding to each arrow.

(a) (5, 1, 3) and (4, 3, 6).

(b) (0, 0, 0) and $(-1, 4, 7)$.

(c) $(-3, -5, -1)$ and $(-2, 4, 6)$.

(d) (2, 5, 1) and (2, 5, 3).

3. The components of an n-tuple are 2, 5, -1. What are the components of an n-tuple of opposite sense and of the same magnitude?

4. n-tuples **A** and **B** have components 1, 3, -2 and 5, 1, 7, respectively. What are the components of $\mathbf{C} = \mathbf{A} + \mathbf{B}$?

5. What is the magnitude of each of the n-tuples **A**, **B**, and **C** of Problem 4?

6. (a) Use (1-1.5b) to find the magnitude of an n-tuple with components 6, 9, $6\sqrt{3}$.

(b) Find the components of a unit n-tuple with the same direction and sense as the n-tuple of (a).

7. It is found experimentally that forces compound according to the parallelogram law and that they can be represented geometrically by arrows. If the forces on a joint of a bridge have components $[0, 3, -5]$, $[0, -3, -5]$ and $[0, 0, 10]$ what is the total force? (Find components and magnitude.)

8. Velocities also fall into the arrow category. In the accompanying diagram the two arrows represent a plane's velocity in still air (PV) and wind velocity (WV). If the components are $400 \cos 35°$, $400 \sin 35°$, and $15 \cos 320°$, $15 \sin 320°$, respectively, find the components of the resultant velocity. (Actually, angular measurements are made clockwise from the

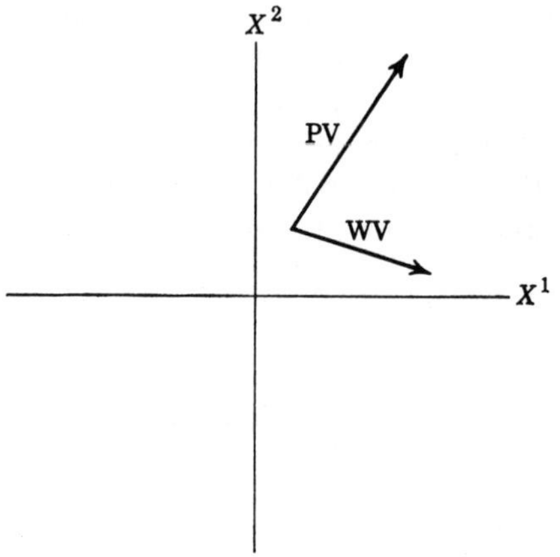

north in navigation problems rather than counterclockwise from the positive X^1 axis as expected here.)

9. (a) Complete the proof of Theorem 1-1.1.
(b) Complete the proof of Theorem 1-1.2.

10. Construct a diagrammatic representation for the associative law of addition (1-1.1c).

11. Prove $\beta\mathbf{A} = \mathbf{0}$ and $\beta \neq 0$ implies that $\mathbf{A} = \mathbf{0}$.

12. Solve the n-tuple equation $3\mathbf{R} + 5\mathbf{B} = \mathbf{C}$ for $\mathbf{R}$ in each of the following cases:
(a) **B** and **C**, respectively, have components (2, 4, 0), (1, 5, 0).
(b) **B** and **C**, respectively, have components (1, −4, 6), (2, 1, 3).

13. Prove for nonzero n-tuples that if **A** is parallel to **C**, **B** is parallel to **D**, and **A** is parallel to **B** then **C** is parallel to **D**.

14. (a) Suppose that $|\mathbf{A}| = |\mathbf{B}|$; give an example such that $\mathbf{A} \neq \mathbf{B}$.
(b) Prove that $\mathbf{A} - \mathbf{B} = \mathbf{0}$ implies that $\mathbf{A} = \mathbf{B}$.

15. Show that the joint of the midpoints of two sides of a triangle is parallel to the third side and has a magnitude of one half that of the third side.

16. Arrows are drawn from the center of a regular pentagon to the vertices. Show that the sum of these arrows is zero.

17. (a) Suppose that in Example 1-1.1 the angles θ_1 and θ_2 were 135 and 30°, respectively. If $m = 10$ and $|\mathbf{g}| = 32$, determine the magnitudes of $\mathbf{F}_1$ and $\mathbf{F}_2$.
(b) For what values of θ_1 and θ_2 is the solution not valid?

18. Suppose that a particle of mass m is suspended by three weightless strings. Set up the n-tuple equation for this problem. If $\mathbf{F}_1, \mathbf{F}_2, \mathbf{F}_3$ have components $(F_{x_1}, F_{y_1}, F_{z_1})$, $(F_{x_2}, F_{y_2}, F_{z_2})$, $(F_{x_3}, F_{y_3}, F_{z_3})$, respectively, write out the set of equations. Does this set of equations provide sufficient information to bring about a solution for $|\mathbf{F}_1|, |\mathbf{F}_2|, |\mathbf{F}_3|$?

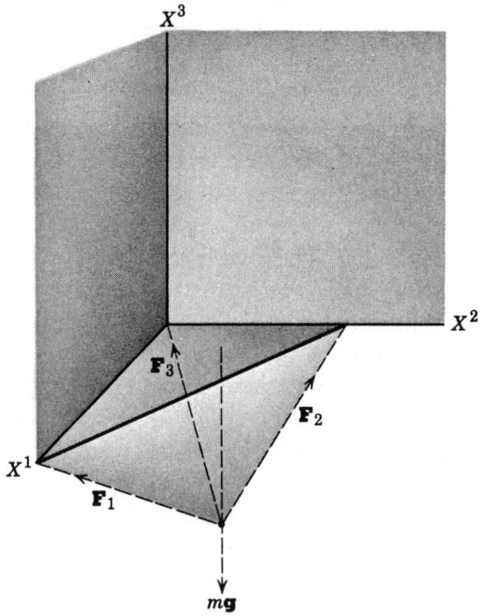

19. Find the parametric and symmetric representations of a line determined by
(a) $P(1, 5, 3)$ and $\mathbf{B} = (-1, 4, 6)$,
(b) $P(-1, 3, 2)$ and $\mathbf{B} = (2, -5, 3)$.

20. Show that the lines

$$\begin{aligned} X^1 &= 2 + 2t, & X^1 &= 6s, \\ X^2 &= 1 - 4t, & X^2 &= 5 + 3s, \\ X^3 &= 0, & X^3 &= 0 \end{aligned}$$

are perpendicular.

21. Consider the line $\mathbf{r} = \mathbf{r}_0 + \mathbf{B}t$ of Example 1-1.3 as representative of the path of a particle in motion and interpret the parameter t as a measure of time. Show that the distance traversed in a fixed interval of time, $\Delta t = t_2 - t_1$, is independent of t_2 and t_1.

2. *Linear Dependence or Independence of a Set of n-tuples*

The concepts of linear dependence and linear independence permeate many areas of mathematics. The reader, knowingly or unknowingly, was exposed to the ideas while evaluating integrals by partial fractions and while considering exact differential equations of the form $M\,dx + N\,dy = 0$.

If we were to examine some of those books that pioneered the study of vector analysis,[15] we would not find discussions of linear dependence or

[15] See E. B. Wilson, *Vector Analysis—Gibbs*, Yale University Press, 1901. Kelland, Tait, and Knott, *Introduction to Quaternions*, Macmillan, first edition, 1873.

independence. The properties associated with these concepts are assumed on a purely geometrical basis and are expressed in terms of the arrow[16] representations of the vectors. For example, it is said that two arrows are collinear (or parallel) or not collinear, whereas three arrows are coplanar or not coplanar. We can obtain a geometric insight into the concepts of this section by visualizing the arrow representatives of a pair of dependent n-tuples as collinear and a dependent triple as coplanar. On the other hand, an independent pair determines a plane, whereas three independent arrows cannot be in a plane. These statements provide a correct intuitional background. However, they are not really operational; that is, given three sets of components (say 2, 5, 3; 1, 4, 7; 5, 14, 13), we cannot easily test for linear dependence or independence.

By resorting to algebraic statements in defining the concepts of linear dependence and independence precision is gained and no element of geometric intuition need be lost. This is therefore the approach that is taken.

This section is restricted to considerations of linear dependence and independence of n-tuples. Corresponding statements concerning Cartesian vectors follow in an intuitive sense from the geometric representations and can be validated in a precise algebraic sense as soon as appropriate transformation laws are established. (See Sections 3 and 6 of Chapter 1.)

Definition 1-2.1a. A set of pn-tuples $\mathbf{A}_1, \cdots, \mathbf{A}_p$ is said to be linearly dependent if and only if there are p real numbers $\beta_1, \cdots, \beta_p$, not all zero, such that

$$\sum_{j=1}^{p} \beta_j \mathbf{A}_j \equiv \beta_1 \mathbf{A}_1 + \cdots + \beta_p \mathbf{A}_p = \mathbf{0}. \tag{1-2.1}$$

Definition 1-2.1b. A set of n-tuples is linearly independent if and only if it is not a linearly dependent set.

An alternative way of expressing the idea of linear independence is stated by the following theorem. The symbol $\leftrightarrow$ is to be read "is equivalent to the statement" and thought of as a logical implication in both directions.

Theorem 1-2.1

$$\left\{\begin{matrix}\text{The set of } n\text{-tuples} \\ \mathbf{A}_1, \cdots, \mathbf{A}_p \text{ is linearly} \\ \text{independent.}\end{matrix}\right\} \leftrightarrow \left\{\begin{matrix}\text{the only condition under} \\ \text{which (1-2.1) holds is that} \\ \text{all } \beta_i = 0.\end{matrix}\right\}$$

[16] The term "arrow" is not specifically used.

PROOF. If the set $\mathbf{A}_1, \cdots, \mathbf{A}_p$ is an independent set of n-tuples, it is not dependent. Hence, according to Definition 1-2.1a, all $\beta_i = 0$. The implication in the other direction is also a direct consequence of the basic definitions.

Theorem 1-2.2. Any nonempty subset of a set of linearly independent n-tuples is itself linearly independent.

PROOF. Let $\mathbf{A}_1, \cdots, \mathbf{A}_p$ be a linearly independent set of n-tuples. Suppose that some nonempty subset $\mathbf{A}_{j_1}, \cdots, \mathbf{A}_{j_k}$, $k < p$, were not linearly independent. This means that there are real numbers $\beta_{j_1}, \cdots, \beta_{j_k}$, not all zero, such that

(1-2.2a)
$$\sum_{i=1}^{k} \beta_{j_i}\mathbf{A}_{j_i} = \mathbf{0}.$$

By simply choosing each of $\beta_{j_{k+1}}, \cdots, \beta_{j_p}$ as zero, the relation (1-2.2a) can be extended to

(1-2.2b)
$$\sum_{i=1}^{p} \beta_{j_i}\mathbf{A}_{j_i} = \mathbf{0},$$

where not all β_{j_i} are zero. But this contradicts the linear independence of the set $\mathbf{A}_1, \cdots, \mathbf{A}_p$. Therefore the supposition that some subset $\mathbf{A}_{j_1}, \cdots, \mathbf{A}_{j_k}$ is linearly dependent is not valid and the proof is complete.

It is the purpose of this section to present the ideas of linear dependence and independence as simply as possible and with no undue amount of generality. Therefore the next two theorems come right down to earth, so to speak, and deal with the problem of determining how many n-tuples can be independent in a three-dimensional space.

Theorem 1-2.3. In a three-dimensional Euclidean space, referred to a rectangular Cartesian coordinate system, the n-tuples (1, 0, 0), (0, 1, 0), and (0, 0, 1) determine a linearly independent set.

PROOF. Let (1, 0, 0), (0, 1, 0) and (0, 0, 1) be represented by $\boldsymbol{\iota}_1$, $\boldsymbol{\iota}_2$, and $\boldsymbol{\iota}_3$, respectively. Then, according to Theorem 1-2.1, we test for linear independence by examining the possible sets of values $\beta_1, \beta_2, \beta_3$ which can satisfy

(1-2.3a)
$$\beta_1\boldsymbol{\iota}_1 + \beta_2\boldsymbol{\iota}_2 + \beta_3\boldsymbol{\iota}_3 = \mathbf{0}.$$

This equation can be rewritten in the form

(1-2.3b)
$$\beta_1(1, 0, 0) + \beta_2(0, 1, 0) + \beta_3(0, 0, 1) = (0, 0, 0),$$

which, according to the defining properties of n-tuples, reduces to

(1-2.3c)
$$(\beta_1, \beta_2, \beta_3) = (0, 0, 0).$$

Since (1-2.3c) constitutes a unique set of values for $\beta_1, \beta_2, \beta_3$, the n-tuples ι_1, ι_2, and ι_3 form a linearly independent set.

The preceding theorem indicates that in Euclidean three-space a linearly independent set of n-tuples can have at least three members. The next theorem establishes the fact that this number is maximum.

Theorem 1-2.4. In a three-dimensional Euclidean space any set of four n-tuples is linearly dependent.

PROOF. Let the n-tuples be represented by the symbols $\mathbf{A}_1, \cdots, \mathbf{A}_4$. In order to test for linear dependence or independence, we can assume that the n-tuples (i.e., their components) are known and determine the possible sets of $(\beta_1, \cdots, \beta_4)$ such that (1-2.1) is satisfied. The object is to show that at least one set, the elements of which are not all zero, exists. The form [see (1-2.1)]

$$\text{(1-2.4a)} \quad \beta_1(A_1{}^1, A_1{}^2, A_1{}^3) + \beta_2(A_2{}^1, A_2{}^2, A_2{}^3) + \beta_3(A_3{}^1, A_3{}^2, A_3{}^3) + \beta_4(A_4{}^1, A_4{}^2, A_4{}^3) = (0, 0, 0)$$

is equivalent to the set of linear equations

$$\text{(1-2.4b)} \quad \begin{aligned} \beta_1 A_1{}^1 + \beta_2 A_2{}^1 + \beta_3 A_3{}^1 + \beta_4 A_4{}^1 &= 0, \\ \beta_1 A_1{}^2 + \beta_2 A_2{}^2 + \beta_3 A_3{}^2 + \beta_4 A_4{}^2 &= 0, \\ \beta_1 A_1{}^3 + \beta_2 A_2{}^3 + \beta_3 A_3{}^3 + \beta_4 A_4{}^3 &= 0. \end{aligned}$$

We employ an inductive process to show that this set of equations has nontrivial solutions. Consider one linear equation in two unknowns:

$$\text{(1-2.4c)} \quad \beta_1 A_1{}^1 + \beta_2 A_2{}^1 = 0.$$

If $A_1{}^1$ and $A_2{}^1$ are different from zero, we may choose an arbitrary nonzero value for β_2 and then solve for β_1. If one of $A_1{}^1$ and $A_2{}^1$ is zero but not both, say $A_2{}^1 = 0$ for the sake of argument, then we can choose $\beta_1 = 0$ and β_2 arbitrarily. If $A_1{}^1 = A_2{}^1 = 0$, then both β_1 and β_2 can be chosen arbitrarily; therefore one equation in two unknowns always has nontrivial solutions. Next consider two equations in three unknowns:

$$\text{(1-2.4d)} \quad \begin{aligned} \beta_1 A_1{}^1 + \beta_2 A_2{}^1 + \beta_3 A_3{}^1 &= 0, \\ \beta_1 A_2{}^1 + \beta_2 A_2{}^2 + \beta_3 A_2{}^3 &= 0. \end{aligned}$$

If $A_1{}^1 = A_2{}^1 = A_3{}^1 = 0$, then β_1, β_2, and β_3 can be chosen to satisfy the second equation and the values will automatically satisfy the first equation. Assume that one of these components is nonzero, say $A_3{}^1$; then

$$\text{(1-2.4e)} \quad \beta_3 = -\frac{1}{A_3{}^1}(\beta_1 A_1{}^1 + \beta_2 A_2{}^1).$$

When β_3 is replaced in the second equation by this expression, we obtain the single equation in two unknowns,

$$(A_3{}^1A_1{}^2 - A_1{}^1A_3{}^2)\beta_1 + (A_3{}^1A_2{}^2 - A_2{}^1A_3{}^2)\beta_2 = 0. \tag{1-2.4f}$$

We have seen that this expression has nontrivial solutions. Any solution (β_1, β_2) of this equation determines a value for β_3 according to (1-2.4e). Therefore we have established that a set of two equations in three unknowns has nontrivial solutions. The argument must be carried one step further to show that the equations in (1-2.4b) have nonzero solutions. We leave this task to the reader.[17]

The thoughtful student will find it of interest to prove the following two statements which are obvious generalizations of Theorems 1-2.3 and 1-2.4, respectively.

(1-2.5a) There are at least n linearly independent n-tuples in an n-dimensional Euclidean space.

(1-2.5b) In an n-dimensional Euclidean space any set of $n + 1$ n-tuples is dependent.

The set of n-tuples ι_1, ι_2, ι_3, with components (1, 0, 0), (0, 1, 0), (0, 0, 1), respectively, will play an important part in the sequel. A typical geometric representation of the triple has each initial point at the origin. The components are also coordinates of the end points. For future applications it is important to realize that the triad may be placed at any other point of space. (See Fig. 1-2.1.)

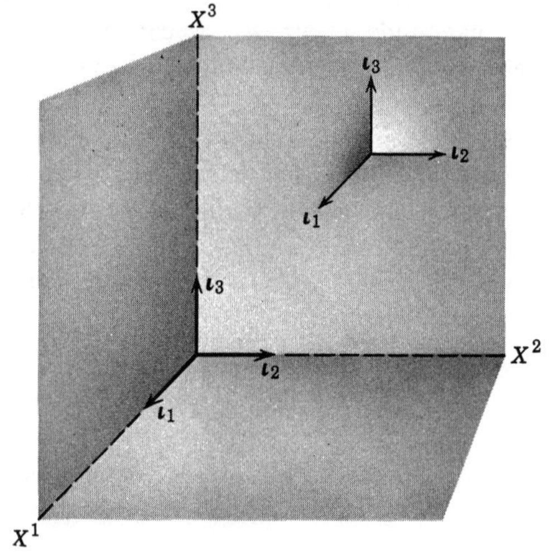

Fig. 1-2.1

---

[17] This theorem also may be proved by the method of proof used in Theorem 1-2.6.

Since ι_1, ι_2, and ι_3 form a linearly independent set of n-tuples, it is impossible to represent one of them as a linear combination of the other two. However, according to Theorem 1-2.4, ι_1, ι_2, ι_3, along with any fourth n-tuple of the space, determine a linearly dependent set; and therein lies their merit.

Example 1-2.1. Let **A** be an arbitrary n-tuple in Euclidean three-space. Consider

$$\mathbf{A} = \alpha_1\iota_1 + \alpha_2\iota_2 + \alpha_3\iota_3, \tag{1-2.6a}$$

where α_1, α_2, α_3 are real numbers yet to be determined. By putting (1-2.6a) in the equivalent form,

$$(A^1, A^2, A^3) = \alpha_1(1, 0, 0) + \alpha_2(0, 1, 0) + \alpha_3(0, 0, 1), \tag{1-2.6b}$$

it is immediately observed that

$$\alpha_1 = A^1, \qquad \alpha_2 = A^2, \qquad \alpha_3 = A^3;$$

that is, when an n-tuple **A** is expressed in terms of the set ι_1, ι_2, and ι_3, the coefficients are the components of **A**. Therefore

$$\boxed{\mathbf{A} = A^1\iota_1 + A^2\iota_2 + A^3\iota_3.} \tag{1-2.6c}$$

Furthermore, the n-tuples ι_1, ι_2, and ι_3 are represented by mutually perpendicular unit (magnitude 1) arrows, as can be seen from geometric inspection. These properties further enhance their value for the expression of concepts of n-tuple algebra.

So far in this section no methods other than employment of basic definition have been introduced for determining whether a given set of n-tuples is dependent. The next objective is the development of these methods.

The statement of the next theorem can be made more conveniently if the concept of the rank of a matrix

$$\begin{pmatrix} A^1 & A^2 & A^3 \\ B^1 & B^2 & B^3 \end{pmatrix}$$

of numbers is first introduced.[18] This is done by means of the following definition.

[18] A matrix is simply a rectangular array of numbers. A limited discussion of the algebra of matrices and the correlation of matrix concepts with vector concepts appears in Section 8.

Definition 1-2.2. Consider the three second-order determinants,

$$\begin{vmatrix} A^1 & A^2 \\ B^1 & B^2 \end{vmatrix}, \qquad \begin{vmatrix} A^1 & A^3 \\ B^1 & B^3 \end{vmatrix}, \qquad \begin{vmatrix} A^2 & A^3 \\ B^2 & B^3 \end{vmatrix},$$

associated with the matrix

$$\begin{pmatrix} A^1 & A^2 & A^3 \\ B^1 & B^2 & B^3 \end{pmatrix}.$$

The rank of the matrix is

(a) 2 if at least one of the second-order determinants is different from zero,

(b) 1 if at least one element of the array has a value different from zero but each second-order determinant has value zero,

(c) 0 if every element of the matrix has value zero.

Theorem 1-2.5. In a Euclidean three-space a set of two n-tuples **A**, **B** is linearly dependent if and only if

$$\operatorname{rank}\begin{pmatrix} A^1 & A^2 & A^3 \\ B^1 & B^2 & B^3 \end{pmatrix} < 2.$$

PROOF. If **A** and **B** are dependent,

$$\alpha(A^1, A^2, A^3) + \beta(B^1, B^2, B^3) = (0, 0, 0), \tag{1-2.7a}$$

where not both α and β are zero; (1-2.7a) is equivalent to

$$\begin{aligned} \alpha A^1 + \beta B^1 &= 0, \\ \alpha A^2 + \beta B^2 &= 0, \\ \alpha A^3 + \beta B^3 &= 0. \end{aligned} \tag{1-2.7b}$$

If any second-order determinant were different from zero, say, for the purpose of illustration,

$$\begin{vmatrix} A^1 & B^1 \\ A^2 & B^2 \end{vmatrix} \neq 0,$$

the pair of equations (1-2.7b) with these coefficients could be satisfied only by $\alpha = \beta = 0$. But this is contrary to hypothesis.

Conversely, if the rank of the matrix is less than 2, the fact that the second-order determinants have value zero leads to

(1-2.7c)

$$A^1B^2 - A^2B^1 = 0, \qquad A^1B^3 - A^3B^1 = 0, \qquad A^2B^3 - A^3B^2 = 0$$

or

$$\frac{A^1}{B^1} = \frac{A^2}{B^2} = \frac{A^3}{B^3} = \lambda, \quad \text{if } B^1B^2B^3 \neq 0; \tag{1-2.7d}$$

that is

$$A^1 = \lambda B^1, \qquad A^2 = \lambda B^2, \qquad A^3 = \lambda B^3. \tag{1-2.7e}$$

Hence **A** and **B** are dependent. It is left to the reader to comment on the situation in which $B^1B^2B^3 = 0$.

Example 1-2.2. Let

$$\mathbf{A} = 2\iota_1 + 3\iota_2,$$
$$\mathbf{B} = \iota_1 - \iota_2;$$

then

$$\begin{vmatrix} 2 & 3 \\ 1 & -1 \end{vmatrix} = -5 \neq 0.$$

Therefore the arrows **A** and **B** are linearly independent.

Theorem 1-2.6. In Euclidean three-space a set of three *n*-tuples **A**, **B**, **C** is linearly dependent if and only if

$$\begin{vmatrix} A^1 & A^2 & A^3 \\ B^1 & B^2 & B^3 \\ C^1 & C^2 & C^3 \end{vmatrix} = 0. \tag{1-2.8}$$

PROOF. If **A**, **B**, and **C** are linearly dependent, then, according to Definition 1-2.1a, there are α, β, γ, not all zero, such that $\alpha\mathbf{A} + \beta\mathbf{B} + \gamma\mathbf{C} = \mathbf{0}$. For the purpose of illustration assume that $\alpha \neq 0$; then

$$A^1 = -\frac{1}{\alpha}(\beta B^1 + \gamma C^1),$$
$$A^2 = -\frac{1}{\alpha}(\beta B^2 + \gamma C^2), \tag{1.2-9a}$$
$$A^3 = -\frac{1}{\alpha}(\beta B^3 + \gamma C^3).$$

According to (1-2.9a), one row of the determinant of components is a linear combination of the other two; therefore

$$\begin{vmatrix} A^1 & A^2 & A^3 \\ B^1 & B^2 & B^3 \\ C^1 & C^2 & C^3 \end{vmatrix} = 0, \tag{1-2.9b}$$

as was to be shown.

Conversely, if (1-2.9b) holds the set of equations

$$\alpha(A^1, A^2, A^3) + \beta(B^1, B^2, B^3) + \gamma(C^1, C^2, C^3) = (0, 0, 0) \tag{1-2.9c}$$

is satisfied by

$$\alpha = \begin{vmatrix} B^2 & B^3 \\ C^2 & C^3 \end{vmatrix}, \qquad \beta = -\begin{vmatrix} A^2 & A^3 \\ C^2 & C^3 \end{vmatrix}, \qquad \gamma = \begin{vmatrix} A^2 & A^3 \\ B^2 & B^3 \end{vmatrix}. \tag{1-2.9d}$$

This fact is easily verified by direct substitution. The determinants of (1-2.9d) are cofactors[19] of the elements of the first column of (1-2.9b). The cofactors of elements of any other column might have been chosen, and indeed might have to be chosen, if all of those in (1-2.9d) turn out to be of value zero. If all cofactors are zero, the n-tuples must be checked for dependence in pairs. (See Theorem 1-2.5.)

Example 1-2.3. Let

$$\begin{aligned} \mathbf{A} &= 2\iota_1 - \iota_2 + 3\iota_3, \\ \mathbf{B} &= \iota_1 - 5\iota_2 + 2\iota_3, \\ \mathbf{C} &= 5\iota_1 - 16\iota_2 + 9\iota_3; \end{aligned}$$

then

$$\begin{vmatrix} 2 & -1 & 3 \\ 1 & -5 & 2 \\ 5 & -16 & 9 \end{vmatrix} = 0.$$

Therefore the given n-tuples are linearly dependent. Geometrically, this means that all three arrow representatives are parallel to a common plane. If the n-tuples are dependent in pairs, the arrow representatives are parallel to a common line.

Example 1-2.4. It is not too difficult to choose a triple of linearly independent n-tuples. For example,

$$\begin{aligned} \mathbf{A} &= \iota_1 + 3\iota_2, \\ \mathbf{B} &= 2\iota_1 + 5\iota_2, \\ \mathbf{C} &= \iota_3 \end{aligned}$$

satisfy this criterion, since

$$\begin{vmatrix} 1 & 3 & 0 \\ 2 & 5 & 0 \\ 0 & 0 & 1 \end{vmatrix} = -1 \neq 0.$$

Since any four n-tuples are dependent, an arbitrary n-tuple $\mathbf{B}$ can be expressed in terms of any set of three linearly independent n-tuples,

[19] A cofactor differs from a minor at most by sign.

$\mathbf{A}_1, \mathbf{A}_2, \mathbf{A}_3$. This statement is easily verified by writing

(1-2.10)

$$(B^1, B^2, B^3) = \beta_1(A_1^{\ 1}, A_1^{\ 2}, A_1^{\ 3}) + \beta_2(A_2^{\ 1}, A_2^{\ 2}, A_2^{\ 3}) + \beta_3(A_3^{\ 1}, A_3^{\ 2}, A_3^{\ 3}).$$

Since $\mathbf{A}_1, \mathbf{A}_2, \mathbf{A}_3$ are linearly independent,

$$\begin{vmatrix} A_1^{\ 1} & A_1^{\ 2} & A_1^{\ 3} \\ A_2^{\ 1} & A_2^{\ 2} & A_2^{\ 3} \\ A_3^{\ 1} & A_3^{\ 2} & A_3^{\ 3} \end{vmatrix} \neq 0.$$

Therefore the nonhomogeneous system of three equations in three unknowns $\beta_1, \beta_2, \beta_3$, implied by (1-2.10), has a unique set of solutions.

The fact that an arbitrary *n*-tuple in three-dimensional space can be represented in terms of any linearly independent set of three *n*-tuples leads to the following definition.

Definition 1-2.3. An *n*-tuple basis in a three-dimensional Euclidean space is constituted by any set of three linearly independent *n*-tuples.

A particularly valuable basis is the set $\boldsymbol{\iota}_1, \boldsymbol{\iota}_2, \boldsymbol{\iota}_3$. Its usefulness will be demonstrated in many future situations.

Example 1-2.5. The representation

$$\mathbf{r} = \mathbf{r}_0 + \mathbf{B}t$$

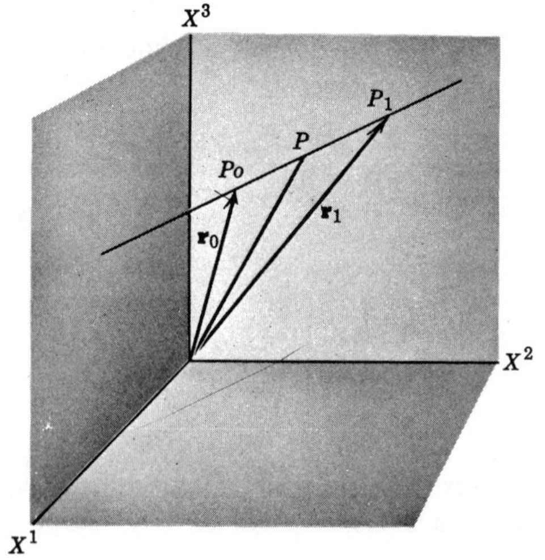

Fig. 1-2.2

of a line was introduced in Example 1-1.3. Let $\mathbf{r}_1$ be an arrow with its initial point at the origin and its end point at a position P_1 on the line. (See Fig. 1-2.2.) Specifically, take $\mathbf{B} = \mathbf{r}_1 - \mathbf{r}_0$. The preceding representation then has the form

$$\mathbf{r} = (1 - t)\mathbf{r}_0 + t\mathbf{r}_1. \tag{1-2.11}$$

If $\mathbf{r}_0$ and $\mathbf{r}_1$ are linearly independent n-tuples, they determine a two-dimensional basis. The fact that the end point of $\mathbf{r}$ generates a line and not a two-space is a consequence of the condition on the coefficients; that is, if p and q represent the coefficients of $\mathbf{r}_0$ and $\mathbf{r}_1$, then $p + q = 1$ for all t. (See Problem 8d.)

Problems

1. Show that the n-tuples $\mathbf{A} = 2\iota_1 + \iota_2 - 5\iota_3$, $\mathbf{B} = 3\iota_1 + 2\iota_2 + \iota_3$, and $\mathbf{C} = \iota_1 - 11\iota_3$ are linearly dependent.
2. Prove that the three n-tuples $2\iota_1 + 3\iota_2$, $5\iota_1$, and $2\iota_3$ are linearly independent.
3. Show that the n-tuples $\mathbf{A} = \iota_1 + 3\iota_2 + 3\iota_3$, $\mathbf{B} = 2\iota_1 + \iota_2 - 3\iota_3$, and $\mathbf{C} = \iota_1 + 5\iota_2 - \iota_3$ are linearly independent.
4. Show that
   (a) $(1, 5, 7)$ and $(-3, -15, -21)$ are linearly dependent n-tuples;
   (b) $(2, -3, 6)$ and $(1, 4, 7)$ are linearly independent n-tuples.
5. Suppose that each of the arrows $\mathbf{A}(1, -3, 5)$, $\mathbf{B}(2, 4, 7)$, and $\mathbf{C}(1, 6, 4)$ has its initial point at the origin. Are they coplanar?
6. (a) Arrow representatives of combinations of the linearly independent n-tuples $\mathbf{A}$ and $\mathbf{B}$ satisfy the triangle relationship indicated by the accompanying diagram. Find x and y.

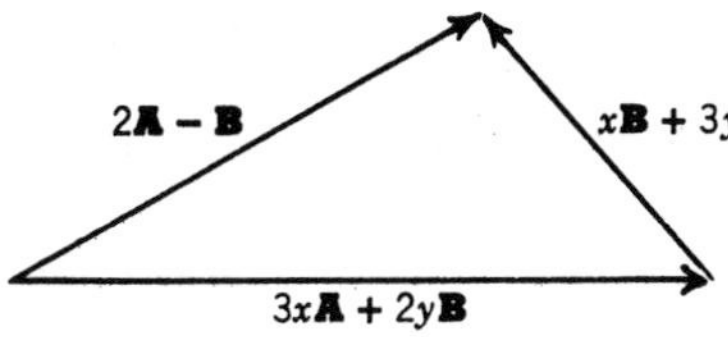

   (b) Three forces acting at a point of a bridge are expressed in terms of linearly independent n-tuples $\mathbf{A}$ and $\mathbf{B}$ as follows (no other forces act at this point):

$$2p\mathbf{A} + q\mathbf{B}, \quad q\mathbf{A} - 5p\mathbf{B}, \quad 3\mathbf{A} - 2\mathbf{B}.$$

   What must the values of p and q be in order to put this point of the bridge in equilibrium?

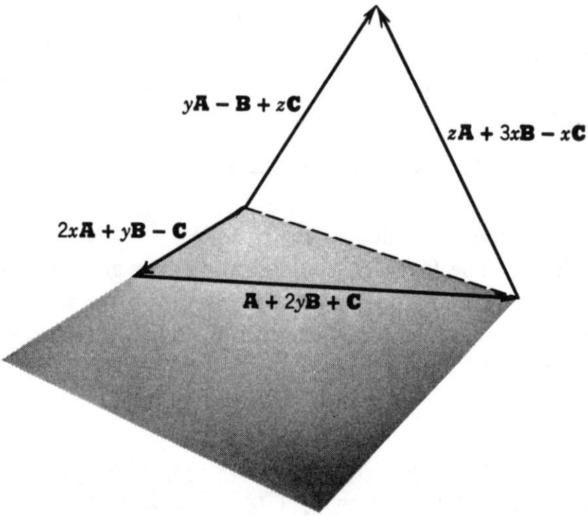

(c) Arrow representatives of combinations of the linearly independent n-tuples **A**, **B**, and **C** satisfy the relation indicated in the accompanying diagram. Find x, y, and z.

7. (a) An n-tuple **C** has two representations in terms of the linearly independent pair **A**, **B**.

$$\mathbf{C} = (X - 3)\mathbf{A} - Y\mathbf{B}, \qquad \mathbf{C} = -Y\mathbf{A} - (X + 2)\mathbf{B}$$

Find X and Y.

(b) Let a triangle be determined by the end points of the linearly independent arrows $\mathbf{r}_1$, $\mathbf{r}_2$, and $\mathbf{r}_3$, each of which has its initial point at the origin. Show that the medians of this triangle intersect at the end point of $\frac{1}{3}(\mathbf{r}_1 + \mathbf{r}_2 + \mathbf{r}_3)$ if the initial point of this arrow is taken to be the origin.

(c) Prove that the diagonals of a parallelogram bisect one another.

8. (a) Suppose the line of Example 1-2.5 contains the origin; what can be said about the n-tuples $\mathbf{r}_1$ and $\mathbf{r}_0$?

(b) Assume that the representation in (1-2.11) corresponds to Fig. 1-2.2. Denote the end point of $\mathbf{r}$ by P. For what values of t is P to the left of P_0; between P_0 and P_1; to the right of P_1?

(c) Obtain the midpoint formulas

$$X_p^{\ 1} = \tfrac{1}{2}(X_0^{\ 1} + X_1^{\ 1})$$
$$X_p^{\ 2} = \tfrac{1}{2}(X_0^{\ 2} + X_1^{\ 2})$$
$$X_p^{\ 3} = \tfrac{1}{2}(X_0^{\ 3} + X_1^{\ 3})$$

from (1-2.11).

(d) Prove: A necessary and sufficient condition that three points P_0, P_1, P_2 be collinear is that

$$\mathbf{r}_0 = p\mathbf{r}_1 + q\mathbf{r}_2, \qquad p + q = 1,$$

where $\mathbf{r}_0$, $\mathbf{r}_1$, $\mathbf{r}_2$ are arrows with initial points fixed at the origin and end points at P_0, P_1, P_2, respectively.

9. Prove: A necessary and sufficient condition that four points be coplanar is that

$$\mathbf{r}_1 = p\mathbf{r}_2 + q\mathbf{r}_3 + s\mathbf{r}_4,$$
$$p + q + s = 1,$$

where $\mathbf{r}_1$, $\mathbf{r}_2$, $\mathbf{r}_3$, and $\mathbf{r}_4$ are arrows with their initial points at the origin.

10. Complete the proof of Theorem 1-2.1.

11. Express **A** as a linear combination of **B**, **C**, and **D** if the components of the n-tuples are (1, 3, 5), (0, 2, 1), (−2, 1, 4) and (1, 3, 2), respectively.

12. (a) Is the set consisting of one n-tuple, $\mathbf{A} \neq \mathbf{0}$, linearly dependent or independent?
(b) Show that any set of n-tuples containing the zero n-tuple is a linearly dependent set.

13. A variable X has all but a finite number of reals as its domain. How is linear independence involved in expressing $(3X + 2)/(X - 2)(X - 5)$ as a sum of fractions $A/(X - 2) + B/(X - 5)$?

14. Prove statements 1-2.5a and 1-2.5b. *Hint:* Use the method of mathematical induction to prove 1-2.5b.

3. *Transformation Equations Relating Rectangular Cartesian Coordinate Systems*

A rectangular Cartesian coordinate system imposes a one-to-one correspondence between the points of Euclidean three-space and the set of all ordered triples of real numbers. A second rectangular Cartesian system brings about another correspondence. The purpose of this section is to investigate the nature of those transformations that relate such coordinate representations of the three-space.

We shall find that the desired transformations are special types of linear transformations; that is, they are expressed by means of first-degree equations. Linear transformations have an origin that predates the formal development of coordinate systems. Francois Viète (1540–1603, French) solved quadric, cubic, and quartic equations by a process that made use of linear transformations. Felix Klein's (1849–1925) famous "Erlanger Programm" of 1872 firmly established their importance in the area of geometry, for the essence of his program was to classify geometries by means of their invariants with respect to given groups of linear transformations.

The transformation idea has more than historical interest. It plays a major role in the present-day study of physical laws. In fact, the use of vector analysis as a descriptive language for physical sciences is largely based on the invariant properties of vector relations under certain types of transformations.

The specific transformations studied in this section are called translations and rotations. It is of some philosophic interest to note that although the names connote motion there is, in fact, none. This imposition made by the English language on our thought processes probably does no harm and has certain intuitional advantages.

Recall that in Section 1 a Cartesian vector was defined as a collection of n-tuples (one n-tuple associated with each rectangular Cartesian coordinate system) with any two of its elements related by transformation equations. These transformation equations have not yet been specified. However, the results of this section make it possible to introduce them.

Since more than one coordinate system is involved in the considerations that follow, we shall begin to employ the term "Cartesian vector," but it is convenient and also appropriate to use the word "arrow" to point out a geometric property.

Certain algebraic preliminaries are necessary. The notations and conventions, here introduced, are fundamental throughout the book.

Definition 1-3.1. An arrow is called a position arrow if and only if the initial point is fixed at the origin of coordinates. The symbols $\mathbf{r}$, $\mathbf{r}_0$, $\mathbf{r}_1$, etc., are used to denote these arrows.[20]

Since the initial point of $\mathbf{r}$ is at the origin of coordinates, the coordinates (X^1, X^2, X^3) of the end point are simultaneously components of the arrow (i.e., of the n-tuple which the arrow represents). Therefore

$$\mathbf{r} = X^1\boldsymbol{\iota}_1 + X^2\boldsymbol{\iota}_2 + X^3\boldsymbol{\iota}_3 = \sum_{j=1}^{3} X^j\boldsymbol{\iota}_j. \tag{1-3.1}$$

The summation sign used in (1-3.1) is not really essential; that is, we might just as well use the repetition of the index j as a means of indicating summation. Such is the viewpoint of the so-called Einstein convention traditionally used in tensor analysis. This notational concept is precisely stated here and is used throughout the book.

Convention 1-3.1. Whenever the same index symbol appears in a term of an algebraic expression both as a subscript and a superscript, the expression is to be summed over the range of that index, called a dummy index.

---

[20] It will be seen that the property of the initial point of an arrow, being at the origin of coordinates, is not maintained under translations. Since the property is not independent of the coordinate system, there is no Cartesian vector, that is, class of triples, by means of which it can be expressed. For this reason we refrain from using the term "position vector" found in most elementary texts on vector analysis.

With Convention 1-3.1 in mind, the relation (1-3.1) can be expressed in the form

$$\boxed{\mathbf{r} = X^j \boldsymbol{\iota}_j.} \tag{1-3.2}$$

The so-called Kronecker delta, $\delta_j^{\,k}$, named after the German mathematician L. Kronecker (1823–1891), is another notational device that is both simple and useful.

Definition 1-3.2. The Kronecker delta is denoted by the symbol $\delta_j^{\,k}$ and has numerical values

$$\delta_j^{\,k} = \begin{cases} 0 \\ 1 \end{cases} \text{if} \begin{cases} j \neq k \\ j = k \end{cases}. \tag{1-3.3}$$

Both symbols, δ_{jk} and δ^{jk}, are used with the same meaning as the Kronecker delta; that is, if $j = k$, the value of either symbol is one, whereas, if $j \neq k$, the value of either symbol is zero.

Example 1-3.1. The Kronecker delta is a set of nine numbers (in three-space). When written out as a square matrix,

$$\begin{pmatrix} \delta_1^{\,1} & \delta_1^{\,2} & \delta_1^{\,3} \\ \delta_2^{\,1} & \delta_2^{\,2} & \delta_2^{\,3} \\ \delta_3^{\,1} & \delta_3^{\,2} & \delta_3^{\,3} \end{pmatrix} = \begin{pmatrix} 1 & 0 & 0 \\ 0 & 1 & 0 \\ 0 & 0 & 1 \end{pmatrix}. \tag{1-3.4}$$

The relation (1-3.4) is understood by equating corresponding elements. A shorthand for the left side of (1-3.4) is $(\delta_j^{\,k})$, where the lower index represents the row of the matrix and the upper index represents the column.

Example 1-3.2. If i ranges over the values 1, 2, 3,

$$\delta_i^{\,i} = \delta_1^{\,1} + \delta_2^{\,2} + \delta_3^{\,3} = 3.$$

Example 1-3.3. If i and j range over the values 1, 2, 3,

$$X^j = a_i^{\,j} \bar{X}^i \tag{1-3.5a}$$

represents a set of three equations with three terms in each right-hand member; that is, the set

$$\begin{aligned} X^1 &= a_1^{\,1}\bar{X}^1 + a_2^{\,1}\bar{X}^2 + a_3^{\,1}\bar{X}^3, \\ X^2 &= a_1^{\,2}\bar{X}^1 + a_2^{\,2}\bar{X}^2 + a_3^{\,2}\bar{X}^3, \\ X^3 &= a_1^{\,3}\bar{X}^1 + a_2^{\,3}\bar{X}^2 + a_3^{\,3}\bar{X}^3. \end{aligned} \tag{1-3.5b}$$

As previously stated, the index i in (1-3.5a) is a dummy index and could be replaced by any other letter except j. Also note that j can be replaced by any other letter except i. In other words,

$$X^p = a_i^{\,p} \bar{X}^i \tag{1-3.5c}$$

is equivalent to (1-3.5a). An index such as j in (1-3.5a), which distinguishes between the equations of the set, is called a free index. Sets of equations expressed in the index notation have a consistency with respect to the position of an index which is valuable for purposes of checking the accuracy of the set. For example, as in (1-3.5c), a free index appears in the same position in every term of the relation.

Although the summation convention is simply stated, it can give some concern to the beginning student. The following examples may be of assistance in assimilating the idea.

Example 1-3.4. If i and j have the range 1, 2,

$$a_{ij}b^{ij} = a_{1j}b^{1j} + a_{2j}b^{2j} = a_{11}b^{11} + a_{12}b^{12} + a_{21}b^{21} + a_{22}b^{22}.$$

Note that whether the summation on i or j is carried out first does not matter. It is important to realize that i and j are dummy indices and may be replaced by any other two distinct letters; that is,

$$a_{ij}b^{ij} = a_{pq}b^{pq} = a_{rs}b^{rs}, \text{ etc.}$$

Example 1-3.5. If p and q range over the values 1, 2, 3, the expression $a_{pq}b^{pq}$ represents a sum of nine terms.

$$a_{pq}b^{pq} = a_{11}b^{11} + a_{12}b^{12} + a_{13}b^{13} + a_{21}b^{21} + a_{22}b^{22}$$
$$+ a_{23}b^{23} + a_{31}b^{31} + a_{32}b^{32} + a_{33}b^{33}.$$

Note that we would never write $a^{jj}b_{jj}$ because the nature of the summations in such an expression is not well defined; that is, we must denote different index summations by different letters.

The first set of transformation equations to be introduced relates the coordinates (X^1, X^2, X^3) and $(\bar{X}^1, \bar{X}^2, \bar{X}^3)$ of two rectangular Cartesian systems whose corresponding axes are parallel. (See Fig. 1-3.1.) These are called equations of translation. The Cartesian vector concept is employed in obtaining them.

Theorem 1-3.1. Let (X^1, X^2, X^3) and $(\bar{X}^1, \bar{X}^2, \bar{X}^3)$, respectively, represent coordinates of rectangular Cartesian coordinate systems, the axes of which are parallel. The sets of coordinates are related by means of the transformation equations

$$X^j = \bar{X}^j + X_0^{\ j}, \tag{1-3.6}$$

where $(X_0^{\ 1}, X_0^{\ 2}, X_0^{\ 3})$ represent the unbarred coordinates of the origin of the barred system.

PROOF. In Fig. 1-3.1 the arrows $\mathbf{r}$, $\mathbf{r}_0$, and $\bar{\mathbf{r}}$ can be associated with the unbarred system. According to the parallelogram law of addition

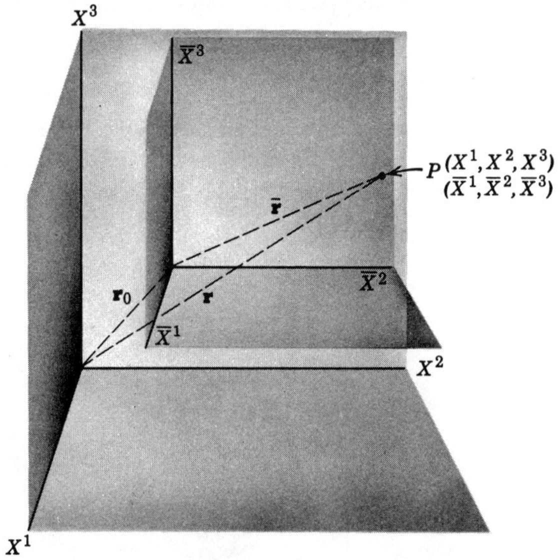

Fig. 1-3.1

(i.e., the law of addition of n-tuples),

(1-3.7a) $$\bar{\mathbf{r}} = \mathbf{r} - \mathbf{r}_0.$$

On the one hand, $\bar{\mathbf{r}}$, thought of as a Cartesian vector, has an n-tuple representation in the barred system (i.e., $\bar{X}^1, \bar{X}^2, \bar{X}^3$). Because the axes of the two systems are parallel, these same symbols represent the components of $\bar{\mathbf{r}}$ in the unbarred system. Therefore, either directly from (1-3.7a) or from the equivalent n-tuple form,

(1-3.7b) $$(\bar{X}^1, \bar{X}^2, \bar{X}^3) = (X^1, X^2, X^3) - (X_0^{\ 1}, X_0^{\ 2}, X_0^{\ 3}),$$

we obtain the transformation equations (1-3.6).

Example 1-3.6. A sphere of radius 2 and with center at (1, 2, 3), when referred to a coordinate system X^1, X^2, X^3, is algebraically represented by the equation

(1-3.8a) $$(X^1 - 1)^2 + (X^2 - 2)^2 + (X^3 - 3)^2 = 4.$$

If a second rectangular Cartesian coordinate system, denoted by a bar, is introduced such that

(1-3.8b) $$\bar{X}^1 = X^1 - 1, \qquad \bar{X}^2 = X^2 - 2, \qquad \bar{X}^3 = X^3 - 3,$$

then, with respect to this system, the equation of the sphere is

$$(\overline{X}^1)^2 + (\overline{X}^2)^2 + (\overline{X}^3)^2 = 4.$$

(See Fig. 1-3.2.)

The equations of translation are a simple sort, yet they hold an important place in the history of classical mechanics and of relativistic mechanics. More information on this point appears in Chapter 2.

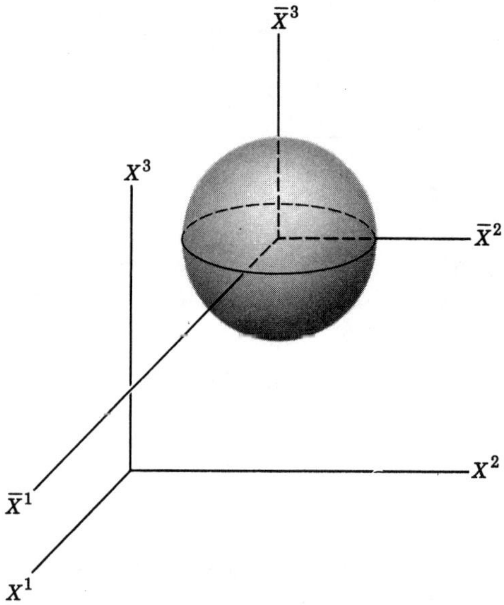

Fig. 1-3.2

Let us turn our attention to a discussion of transformations relating coordinates X^j and $\overline{X}^j$ of rectangular Cartesian systems with a common origin. Such transformations are called rotations.

Theorem 1-3.2. The ordered triples (X^1, X^2, X^3) and $(\overline{X}^1, \overline{X}^2, \overline{X}^3)$ associated with rectangular Cartesian coordinate systems having a common origin and such that there is no change of unit distance along coordinate axes are related by the transformation equations

(1-3.9a) $$\boxed{X^j = a_k^{\ j}\overline{X}^k,}$$

where the coefficients of transformation $a_k^{\ j}$ are direction cosines satisfying the conditions

(1-3.9b) $$\sum_{j=1}^{3} a_k^{\ j} a_p^{\ j} = \delta_{kp}.$$

PROOF. Let $\bar{\iota}_j$ represent a set of unit orthogonal basis arrows, associated with the barred system in the same way as the set ι_j is associated with the unbarred system. We interpret $\mathbf{r}$ and $\bar{\mathbf{r}}$ as symbols for n-tuples in the respective systems, but as representatives of the same Cartesian vector. Furthermore, the barred basis arrows $\bar{\iota}_1$, $\bar{\iota}_2$, $\bar{\iota}_3$, when considered as Cartesian vectors, have representations in terms of the unbarred basis ι_1, ι_2, ι_3.

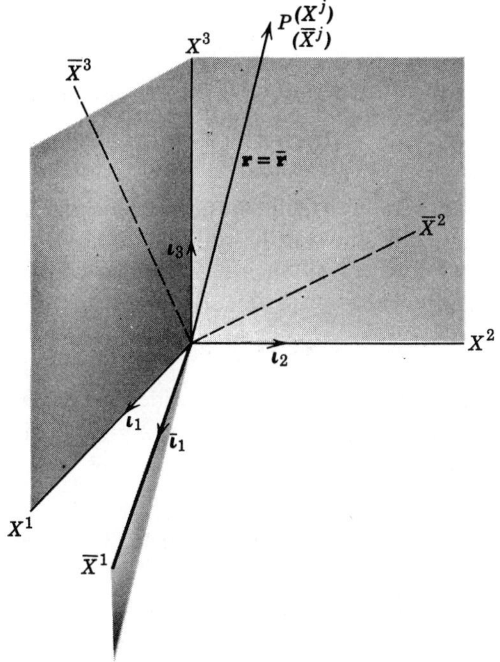

Fig. 1-3.3

Therefore the relation $\mathbf{r} = \bar{\mathbf{r}}$ or

$$X^j\iota_j = \bar{X}^k\bar{\iota}_k \tag{1-3.10}$$

has a significance in a single coordinate system. This fact enables us to employ the algebraic properties of n-tuples. Of course, a double interpretation of the symbols $\bar{X}^k$ is involved. On the one hand they are thought of as coordinates in the barred system, and on the other they are considered as numbers somehow associated with the unbarred system.

The next step toward obtaining the relations (1-3.9a) is to express the $\bar{\iota}_k$ in terms of the ι_j. In particular, consider $\bar{\iota}_1$. Since $\bar{\iota}_1$ is of length one, its projections in the directions of ι_1, ι_2, ι_3 can be represented by the cosines of the angles made by $\bar{\iota}_1$ with ι_1, ι_2, ι_3, respectively. Let $\alpha_1{}^1$, $\alpha_1{}^2$, $\alpha_1{}^3$

represent the angles made by $\bar{\iota}_1$ with ι_1, ι_2, and ι_3, respectively. Furthermore, let

$$a_1^{\ j} = \cos \alpha_1^{\ j}.$$

Then

$$\bar{\iota}_1 = a_1^{\ 1}\iota_1 + a_1^{\ 2}\iota_2 + a_1^{\ 3}\iota_3 = a_1^{\ j}\iota_j.$$

If, in general, the angle[21] from ι_j to $\bar{\iota}_k$ is designated by $\alpha_k^{\ j}$ and

(1-3.11a) $$a_k^{\ j} = \cos \alpha_k^{\ j},$$

then

(1-3.11b) $$\bar{\iota}_k = a_k^{\ j}\iota_j.$$

When (1-3.11b) is substituted into (1-3.10), we obtain

$$X^j\iota_j = \bar{X}^k a_k^{\ j}\iota_j,$$

$$(X^j - a_k^{\ j}\bar{X}^k)\iota_j = \mathbf{0}.$$

Since the arrow set ι_j is linearly independent, the result is (1-3.9a).

The relation (1-3.9b) follows from the solid analytic geometry formula for the angle between two directions. The formula, as expressed in terms of direction cosines, is

(1-3.12) $$\cos \theta = \sum_{j=1}^{3} a_k^{\ j} a_p^{\ j}.$$

If $k = p$, then, of course, $\bar{\iota}_k$ and $\bar{\iota}_p$ coincide and $\cos 0 = 1$. If $k \neq p$, then $\bar{\iota}_k$ and $\bar{\iota}_p$ are orthogonal and $\cos \pi/2 = 0$. This completes the proof.

Those students who are not familiar with (1-3.12) may defer its consideration until Section 5 has been studied. The formula is actually developed there.

The significance of the conditions (1-3.9b) can be made more distinct by writing out the coefficients of transformation in the form of a square matrix.

(1-3.13) $$(a_k^{\ j}) = \begin{pmatrix} a_1^{\ 1} & a_1^{\ 2} & a_1^{\ 3} \\ a_2^{\ 1} & a_2^{\ 2} & a_2^{\ 3} \\ a_3^{\ 1} & a_3^{\ 2} & a_3^{\ 3} \end{pmatrix}.$$

According to (1-3.9b), the sum of the squares of the elements of any row is one, whereas the sum of products of corresponding elements of different rows is zero.

With a square matrix of numbers, such as (1-3.13), we can associate a determinant. For notational convenience we denote the value of the determinant of $(a_k^{\ j})$ by a. The next theorem evaluates this determinant.

---

[21] The angle $\alpha_k^{\ j}$ should be thought of as formed by the positive senses of $\bar{\iota}_k$ and ι_j and measured from ι_j to $\bar{\iota}_k$. However, it should be noted that cosine is an even function, and therefore the direction of measurement of the angle is not algebraically significant.

Theorem 1-3.3. We have

$$a^2 = 1. \tag{1-3.14}$$

PROOF. Consider (1-3.9b). The determinant of (δ_{kp}) is one. The determinant of the left-hand side of (1-3.9b) is a^2. This proves the theorem.

No doubt some of the readers of this book have not been exposed to determinant multiplication, hence do not realize that the determinant of the left-hand side of (1-3.9b) is as indicated. I refer them to Section 6 in which the fundamental ideas concerning determinants are reviewed and elaborated.

The determinant theory to be developed in Section 6 will enable us to see that our previous restriction to a single type of coordinate system (right-handed) requires that

$$a = 1. \tag{1-3.15}$$

It will also be determined that $a = -1$ if and only if $\bar{\iota}_1$, $\bar{\iota}_2$, $\bar{\iota}_3$ have an orientation corresponding to an odd permutation of ι_1, ι_2, ι_3.

Since $a \neq 0$, the linear equations in (1-3.9a) can be solved; that is, inverse transformation equations can be obtained by a straightforward application of determinants (Cramer's rule). The procedure is facilitated by means of the following definition.

Definition 1-3.3. Let

$$A_j^k = \frac{\text{cofactor of } a_k^j \text{ in } \det(a_k^j)}{a}. \tag{1-3.16}$$ [22]

As an immediate consequence of definition (1-3.3), we have

$$\text{(a)} \quad A_j^k a_k^p = \delta_j^p, \qquad \text{(b)} \quad A_j^k a_q^j = \delta_q^k. \tag{1-3.17}$$

It need be observed only that (1-3.17a, b) represent expansions of the determinant of (a_j^k) around a given column (row) so that we obtain the determinant divided by itself (i.e., 1) or a determinant with repeated

[22] The cofactors of a_1^1, a_2^1, a_3^1 in

$$\begin{vmatrix} a_1^1 & a_1^2 & a_1^3 \\ a_2^1 & a_2^2 & a_2^3 \\ a_3^1 & a_3^2 & a_3^3 \end{vmatrix}$$

are

$$\begin{vmatrix} a_2^2 & a_2^3 \\ a_3^2 & a_3^3 \end{vmatrix}, \qquad \begin{vmatrix} a_1^3 & a_1^2 \\ a_3^3 & a_3^2 \end{vmatrix}, \qquad \begin{vmatrix} a_1^2 & a_1^3 \\ a_2^2 & a_2^3 \end{vmatrix}$$

The cofactor of a_j^k is the second-order determinant, obtained by striking out the jth row and kth column, multiplied by $(-1)^{j+k}$.

columns (rows). This statement is illustrated in detail in the following example.

Example 1-3.7. Choose $j = p = 1$ in (1-3.17a). Then

$$A_1^{\ k} a_k^{\ 1} = A_1^{\ 1} a_1^{\ 1} + A_1^{\ 2} a_2^{\ 1} + A_1^{\ 3} a_3^{\ 1}$$

$$= \frac{\begin{vmatrix} a_2^{\ 2} & a_2^{\ 3} \\ a_3^{\ 2} & a_3^{\ 3} \end{vmatrix}}{a} a_1^{\ 1} + \frac{\begin{vmatrix} a_1^{\ 3} & a_1^{\ 2} \\ a_3^{\ 3} & a_3^{\ 2} \end{vmatrix}}{a} a_2^{\ 1} + \frac{\begin{vmatrix} a_1^{\ 2} & a_1^{\ 3} \\ a_2^{\ 2} & a_2^{\ 3} \end{vmatrix}}{a} a_3^{\ 1}$$

$$= \frac{\begin{vmatrix} a_1^{\ 1} & a_1^{\ 2} & a_1^{\ 3} \\ a_2^{\ 1} & a_2^{\ 2} & a_2^{\ 3} \\ a_3^{\ 1} & a_3^{\ 2} & a_3^{\ 3} \end{vmatrix}}{a} = \frac{a}{a} = 1.$$

On the other hand, if we choose $j = 1, p = 2$,

$$A_1^{\ k} a_k^{\ 2} = A_1^{\ 1} a_1^{\ 2} + A_1^{\ 2} a_2^{\ 2} + A_1^{\ 3} a_3^{\ 2}$$

$$= \frac{\begin{vmatrix} a_2^{\ 2} & a_2^{\ 3} \\ a_3^{\ 2} & a_3^{\ 3} \end{vmatrix}}{a} a_1^{\ 2} + \frac{\begin{vmatrix} a_1^{\ 3} & a_1^{\ 2} \\ a_3^{\ 3} & a_3^{\ 2} \end{vmatrix}}{a} a_2^{\ 2} + \frac{\begin{vmatrix} a_1^{\ 2} & a_1^{\ 3} \\ a_2^{\ 2} & a_2^{\ 3} \end{vmatrix}}{a} a_3^{\ 2}$$

$$= \frac{\begin{vmatrix} a_1^{\ 2} & a_1^{\ 2} & a_1^{\ 3} \\ a_2^{\ 2} & a_2^{\ 2} & a_2^{\ 3} \\ a_3^{\ 2} & a_3^{\ 2} & a_3^{\ 3} \end{vmatrix}}{a} = \frac{0}{a}.$$

The reader not familiar with the relation (1-3.17a) should experiment with other examples.

The algebraic facility that evolves from the relation (1-3.17b) is demonstrated in the next theorem in which transformation equations inverse to those of (1-3.9a) are obtained.

Theorem 1-3.4. The set of transformation equations inverse to (1-3.9a) is

(1-3.18) $$\bar{X}^p = A_j^{\ p} X^j.$$

PROOF. According to (1-3.9a),

$$X^j = a_k^{\ j} \bar{X}^k.$$

If the three equations of this set are multiplied by $A_1{}^p$, $A_2{}^p$, $A_3{}^p$ and the resultant expressions are added [for the sake of brevity we say that (1-3.9a) is multiplied and summed with $A_j{}^p$], then

$$A_j{}^p X^j = A_j{}^p a_k{}^j \bar{X}^k = \delta_k{}^p \bar{X}^k = \bar{X}^p.$$

The last two members of this equation were obtained by making use of relation (1-3.17b) and the definition of the Kronecker delta. This completes the proof.

Before turning attention to examples of rotational transformations, it is worthwhile to establish the following algebraic properties of the coefficients of transformation.

Theorem 1-3.5. The set of inverse transformation coefficients $A_j{}^k$ is obtained from the set $a_p{}^q$ by interchanging rows and columns of the matrix; that is,

(1-3.19) $$\boxed{A_k{}^j = a_j{}^k,}$$

PROOF. According to (1-3.17a),

$$A_j{}^k a_k{}^p = \delta_j{}^p.$$

If this set of relations is multiplied and summed with $a_q{}^p$, then

$$\sum_{p=1}^{3} A_j{}^k a_k{}^p a_q{}^p = \sum_{p=1}^{3} \delta_j{}^p a_q{}^p.$$

By making use of (1-3.9b) on the left and the definition of the Kronecker delta on the right, we obtain the result

$$A_j{}^k \, \delta_{kq} = a_q{}^j.$$

or

$$A_j{}^q = a_q{}^j,$$

as was to be proved.

The orthogonality conditions (1-3.9b) hold for columns of the matrix of coefficients as well as for rows. This fact is demonstrated in the next theorem.

Theorem 1-3.6. We have

(1-3.20) $$\sum_{j=1}^{3} a_j{}^q a_j{}^k = \delta^{qk}.$$

PROOF. According to (1-3.19),

$$A_k{}^j = a_j{}^k.$$

Multiply and sum this relation with $a_j^{\ q}$. Then

$$a_j^{\ q} A_k^{\ j} = \sum_{j=1}^{3} a_j^{\ q} a_j^{\ k}.$$

From (1-3.17a) it follows that $a_j^{\ q} A_k^{\ j} = \delta_k^{\ q}$; therefore the proof is complete.

The reader should observe that the relation (1-3.19) implies that the $A_j^{\ k}$ satisfy the same orthogonality conditions as the $a_j^{\ k}$. The algebraic properties of the coefficients of transformation may be summed up as follows:

(1-3.21)

$$\text{(a)}\quad \sum_{p=1}^{3} a_j^{\ p} a_k^{\ p} = \delta_{jk}, \qquad \text{(b)}\quad \sum_{p=1}^{3} a_p^{\ j} a_p^{\ k} = \delta^{jk},$$

$$\text{(c)}\quad a = 1, \qquad \text{(d)}\quad A_j^{\ k} = a_k^{\ j},$$

$$\text{(e)}\quad \sum_{p=1}^{3} A_j^{\ p} A_k^{\ p} = \delta_{jk}, \qquad \text{(f)}\quad \sum_{p=1}^{3} A_p^{\ j} A_p^{\ k} = \delta^{jk},$$

$$\text{(g)}\quad A = 1.$$

(1-3.21h) $$A_j^{\ k} a_q^{\ j} = \delta_q^{\ k}, \qquad A_k^{\ j} a_j^{\ q} = \delta_k^{\ q}.$$

The orthogonality of the transformations (as well as preservation of the unit of length) can be characterized by (1-3.21a). As indicated previously, (1-3.21c) is an assumed condition that restricts the set of coordinate systems to the right-handed type (or to the left-handed type). The other relations which are not purely algebraic in nature follow from these. Note that relations (1-3.21h) are purely algebraic; that is, they are a consequence of the definition of the $A_j^{\ k}$ and have nothing to do with the fact that transformations of rotation are under discussion.

The significance of the conditions (1-3.21) is further illustrated by the following example.

Example 1-3.8. The matrix of coefficients

(1-3.22a) $$(a_j^{\ k}) = \begin{pmatrix} \frac{\sqrt{3}}{2} & \frac{1}{2} & 0 \\ \frac{\sqrt{3}}{4} & \frac{-3}{4} & \frac{1}{2} \\ \frac{1}{4} & \frac{-\sqrt{3}}{4} & \frac{-\sqrt{3}}{2} \end{pmatrix}$$

satisfy the conditions in (1-3.21a, b, c), hence are valid coefficients of transformation relating two right-handed rectangular Cartesian coordinate systems. [Note that (1-3.21b) is a consequence of (1-3.21a).] To see that these conditions are met, the reader must notice that the sum of squares of any row (column) is one; the sum of products of corresponding elements of different rows (columns) is zero, and the value of the determinant of coefficients is one. Similar remarks hold for the inverse transformation coefficients:

$$(1\text{-}3.22\text{b}) \qquad (A_j^{\,k}) = \begin{pmatrix} \frac{\sqrt{3}}{2} & \frac{\sqrt{3}}{4} & \frac{1}{4} \\ \frac{1}{2} & \frac{-3}{4} & \frac{-\sqrt{3}}{4} \\ 0 & \frac{1}{2} & \frac{-\sqrt{3}}{2} \end{pmatrix}$$

Quite legitimately we might wonder how the set of coefficients (1-3.22a) is obtained. This question is considered in Section 8, where a simple method for generating sets of three-space transformation coefficients is introduced. The following example is pertinent to that method.

Example 1-3.9. Rotations in a plane, say the X^1, X^2 plane, are special types of three-space rotations. Therefore the appropriate equations of transformation should be contained in our general results. The equations of rotation result from (1-3.9a) by noting that $\mathbf{\iota}_1$, $\mathbf{\iota}_2$, and $\mathbf{\iota}_3$, respectively, make angles of $\pi/2$, $\pi/2$, and 0 with $\bar{\mathbf{\iota}}_3$. Since $\cos \pi/2 = 0$, $\cos 0 = 1$, (1-3.9a) reduces to

$$(1\text{-}3.23\text{a}) \qquad \begin{aligned} X^1 &= a_1^{\,1}\bar{X}^1 + a_2^{\,1}\bar{X}^2 \\ X^2 &= a_1^{\,2}\bar{X}^1 + a_2^{\,2}\bar{X}^2 \\ X^3 &= \bar{X}^3. \end{aligned}$$

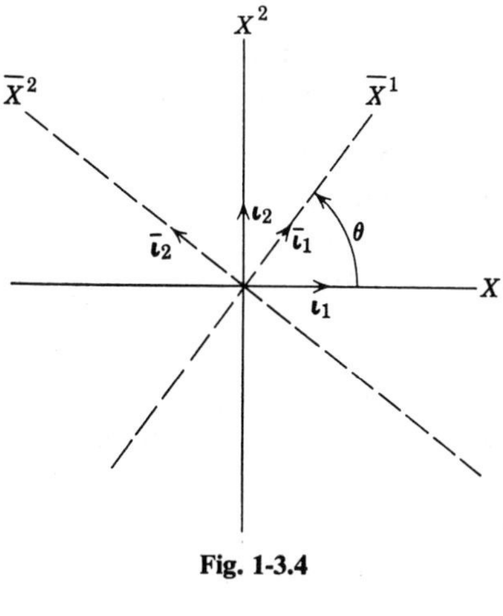

Fig. 1-3.4

If the angle of rotation is denoted by θ, where θ is measured from X^1 to $\bar{X}^1$, then

$$
\begin{aligned}
&a_1^{\ 1} = \cos(\bar{\iota}_1, \iota_1) = \cos\theta, \\
&a_1^{\ 2} = \cos(\bar{\iota}_1, \iota_2) = \cos\left(\frac{\pi}{2} - \theta\right) = \sin\theta, \\
&a_2^{\ 1} = \cos(\bar{\iota}_2, \iota_1) = \cos\left(\frac{\pi}{2} + \theta\right) = -\sin\theta, \\
&a_2^{\ 2} = \cos(\bar{\iota}_2, \iota_2) = \cos\theta.
\end{aligned} \tag{1-3.23b}
$$

Therefore (1-3.23a) can be put in the form

$$
\begin{aligned}
X^1 &= \cos\theta\,\bar{X}^1 - \sin\theta\,\bar{X}^2, \\
X^2 &= \sin\theta\,\bar{X}^1 + \cos\theta\,\bar{X}^2.
\end{aligned} \tag{1-3.23c}
$$

Intuitively, we can think of a counterclockwise rotation from X^j to $\bar{X}^j$.

Using the relation $A_j^{\ k} = a_k^{\ j}$, we determine that the inverse transformation equations are

$$
\begin{aligned}
\bar{X}^1 &= \cos\theta X^1 + \sin\theta X^2, \\
\bar{X}^2 &= -\sin\theta X^1 + \cos\theta X^2.
\end{aligned} \tag{1-3.23d}
$$

This rotation can be considered as clockwise from X^j to $\bar{X}^j$.

Example 1-3.10. As a simple illustration of a transformation of rotation in the X^1,X^2 plane, let $\theta = \pi/6$. Then (1-3.23c) takes the form

$$
\begin{aligned}
X^1 &= \frac{\sqrt{3}}{2}\bar{X}^1 - \frac{1}{2}\bar{X}^2, \\
X^2 &= \frac{1}{2}\bar{X}^1 + \frac{\sqrt{3}}{2}\bar{X}^2.
\end{aligned}
$$

Often the algebraic representation of a geometric configuration is much simpler when referred to a particular coordinate system. Consider a hyperbola which when referred to the coordinates X^1, X^2 has the equation

$$(X^1)^2 + 2\sqrt{3}X^1X^2 - (X^2)^2 = 2.$$

When expressed in terms of the barred system of coordinates, the equation of the hyperbola is of the simpler form,

$$(\bar{X}^1)^2 - (\bar{X}^2)^2 = 1.$$

Problems

1. This problem deals with notational concepts. We define

$$\delta_{js}^{kp} = \delta_j^{\ k}\delta_s^{\ p}, \qquad \delta_{js}^{[kp]} = \tfrac{1}{2}[\delta_{js}^{kp} - \delta_{js}^{pk}].$$

The first expression is a generalization of the Kronecker δ, whereas the second symbolizes a particular combination of such generalized entities represented by means of a bracket notation.

(a) If the indices j, s, k, p are placeholders for 1, 2, 3, how many elements are in the set δ_{js}^{kp}?
(b) What numerical values do these elements take on?
(c) Evaluate $\delta_{js}^{[kp]}$ for all possible sets of values j, s, k, p.
(d) Show that $\delta_{js}^{[kp]} = -\delta_{js}^{[pk]}$.
(e) Show that $\delta_{js}^{[kp]} = \delta_{[js]}^{kp}$, where the brackets about j, s imply

$$\delta_{[js]}^{kp} = \tfrac{1}{2}[\delta_{js}^{kp} - \delta_{sj}^{kp}].$$

2. (a) Evaluate $\delta_j{}^k\, \delta_k{}^p\, \delta_p{}^j$ where j, k, p take on values 1, 2, 3.
(b) Write out the equation $\delta_{ij}X^iX^j = 0. \qquad i, j = 1, 2, 3.$

3. The equation of a sphere in a given rectangular Cartesian coordinate system is

$$(X^1)^2 + (X^2)^2 + (X^3)^2 - 6X^1 + 2X^2 - 8X^3 + 25 = 0.$$

How is this coordinate system related to one with corresponding axes parallel and origin at the center of the sphere?

4. Give an example of a transformation relating a left-hand and a right-hand rectangular Cartesian system (i.e., an inversion) and therefore such that $a = -1$.

5. Illustrate the validity of (1-3.17b); that is, $A_j{}^k a_q{}^j = \delta_q{}^k$, by writing out the details for two or three choices of q and k.

6. Show that the determinant of coefficients $a_k{}^j$ of the transformation equations (1-3.9a) can be expressed in terms of partial derivatives as follows:

$$\left|\frac{\partial X^j}{\partial \bar{X}^k}\right| = \begin{vmatrix} \dfrac{\partial X^1}{\partial \bar{X}^1} & \dfrac{\partial X^2}{\partial \bar{X}^1} & \dfrac{\partial X^3}{\partial \bar{X}^1} \\[2ex] \dfrac{\partial X^1}{\partial \bar{X}^2} & \dfrac{\partial X^2}{\partial \bar{X}^2} & \dfrac{\partial X^3}{\partial \bar{X}^2} \\[2ex] \dfrac{\partial X^1}{\partial \bar{X}^3} & \dfrac{\partial X^2}{\partial \bar{X}^3} & \dfrac{\partial X^3}{\partial \bar{X}^3} \end{vmatrix}$$

7. Consider a transformation of rotation in the X^1, X^2 plane. Using the equations in (1-3.23c), find the X^j coordinates of a point whose $\bar{X}^j$ coordinates are (2, 3, 0) when (a) $\theta = \pi/4$, (b) $\theta = \pi/3$.

8. (a) Show that the following matrix of numbers satisfies the conditions in (1-3.21a, b, c).

$$\begin{pmatrix} \dfrac{-1}{2} & \dfrac{-1}{2} & \dfrac{-1}{\sqrt{2}} \\[2ex] \dfrac{1}{2} & \dfrac{1}{2} & \dfrac{-1}{\sqrt{2}} \\[2ex] \dfrac{1}{\sqrt{2}} & \dfrac{-1}{\sqrt{2}} & 0 \end{pmatrix}$$

(b) If the matrix in (a) is considered as a set of coefficients of a rectangular Cartesian transformation, find the set of inverse transformation coefficients.

(c) If the $\bar{X}^j$ coordinates of a point are (1, −2, 3), find the X^j coordinates, given that the transformation coefficients $a_j^{\ k}$ are those of (a).

9. Suppose that a set of transformation coefficients is given by

$$(a_j^{\ k}) = \begin{pmatrix} \frac{1}{2} & \frac{-\sqrt{3}}{2} & 0 \\ \frac{\sqrt{3}}{2} & \frac{1}{2} & 0 \\ 0 & 0 & 1 \end{pmatrix}$$

(a) Show that $a = 1$.

(b) Find $(A_j^{\ k})$.

(c) Describe the transformation.

10. (a) Show that the algebraic form

$$\sum_{j=1}^{3} (X^j - X_0^{\ j})^2,$$

representing the square of the distance determined by two points P_0 and P_1, is invariant (i.e., remains unchanged) under rectangular Cartesian transformations.

(b) Suppose that $X^1 = 4\bar{X}^1$, $X^2 = 4\bar{X}^2$, $X^3 = 4\bar{X}^3$; then show that the forms

$$\sum_{j=1}^{3} (X_1^{\ j} - X_0^{\ j})^2 \quad \text{and} \quad \sum_{j=1}^{3} (\bar{X}_1^{\ j} - \bar{X}_0^{\ j})^2$$

are not equal. Suppose that the X^j coordinate system is rectangular Cartesian. Is the $\bar{X}^j$ system rectangular Cartesian? Why is such a transformation excluded from the set of orthogonal Cartesian rotations?

(c) Assume that $X^1 = 4\bar{X}^1$, $X^2 = 3\bar{X}^2$, $X^3 = 2\bar{X}^3$. Is the form

$$\sum_{j=1}^{3} (X_1^{\ j} - X_0^{\ j})^2$$

invariant with respect to this transformation?

11. Parametric equations of a line are of the form (see Example 1-1.3)

$$X^j = X_1^{\ j} + B^j t.$$

Suppose that $B^j = X_2^{\ j} - X_1^{\ j}$ ($X_1^{\ j}$, $X_2^{\ j}$ are coordinates of points on the line) and t is unchanged by a coordinate transformation. Show that the algebraic form of the parametric equations of a line is unchanged by an orthogonal Cartesian transformation (i.e., a translation or rotation).

4. *Definition of Cartesian Scalar and Vector*

A Cartesian vector was initially defined in terms of the set of all rectangular Cartesian coordinate systems. A law of transformation relating the vector components in any pair of systems was alluded to but not specifically stated. It is the purpose of this section to supply a precise mode of transformation for the components of a Cartesian vector as well as to introduce the concept of a scalar, which is also dependent on a set of transformations.

The terms vector and scalar are both due to W. R. Hamilton. Each constituted a part of his so-called quaternion, with neither having the stature of an entity in its own right. A significant measure of the credit for dissociating the vector and scalar concepts from the quaternion ideas and thereby putting them within a structure convenient to geometry and physics belongs to W. Gibbs.

The definitions that follow specify the meanings of vector and scalar with respect to the set of transformations relating rectangular Cartesian coordinate systems.† These statements are formulated so that they will correlate with the usages of the terms vector and scalar when more general sets of transformations are involved. (See Section 5*.)

In the following definition the symbol $\{U^1, U^2, U^3\}$ denotes the collection of ordered-number triples (U^1, U^2, U^3), $(\bar{U}^1, \bar{U}^2, \bar{U}^3)$, etc; one triple is associated with each possible rectangular Cartesian coordinate system.

Definition 1-4.1. A Cartesian vector is a collection $\{U^1, U^2, U^3\}$ of ordered triples, each associated with a rectangular Cartesian coordinate system and such that any two satisfy the transformation law

$$U^j = \frac{\partial X^j}{\partial \bar{X}^k} \bar{U}^k. \tag{1-4.1}$$

The triples (U^1, U^2, U^3) etc., are called components of the Cartesian vector in the respective coordinate systems.

As stated in Section 1, it is notationally convenient to use the same boldface symbol **U** that previously has characterized an n-tuple (and the family of arrows geometrically representing the n-tuple).

For a transformation of rotation, that is, $X^j = a_k^{\ j} \bar{X}^k$, the relation in (1-4.1) reduces to

$$U^j = a_k^{\ j} \bar{U}^k, \tag{1-4.2a}$$

whereas for a translation, $X^j = \bar{X}^j + \bar{X}_0^{\ j}$, it takes the form

$$U^j = \delta_k^{\ j} \bar{U}^k = \bar{U}^j. \tag{1-4.2b}$$

† For clarity of presentation the definitions are phrased for the case $n = 3$.

It is worth noting that the components of a Cartesian vector transform under rotations as the coordinates transform. Furthermore, the collection of coordinate differences $\{X_1^{\,j} - X_0^{\,j}\}$ is a Cartesian vector, for when the relation in (1-3.9a), that is,

$$X^j = a_k^{\,j}\bar{X}^k,$$

is applied to the coordinates of P_0 and P_1, we have

$$X_1^{\,j} - X_0^{\,j} = a_k^{\,j}\bar{X}_1^{\,k} - a_k^{\,j}\bar{X}_0^{\,k} = a_k^{\,j}(\bar{X}_1^{\,k} - \bar{X}_0^{\,k}). \tag{1-4.3}$$

The coordinate differences clearly satisfy (1-4.2a). A corresponding verification of the statement for translations is left to the reader. We already know that a Cartesian vector can be represented geometrically by an arrow (any one of a family of arrows). The preceding statement establishes the fact that an arrow determined by a pair of points, P_0 and P_1, gives rise to a Cartesian vector.

There is some need for caution, as pointed out by the following pair of theorems.

Theorem 1-4.1a. The concept of position arrow is not a vector concept with respect to the set of translations; that is, the components of two position arrows $\mathbf{r}$ and $\bar{\mathbf{r}}$ of systems X^j and $\bar{X}^j$, one translated from the other, are not related by the transformation law (1-4.2b).

PROOF. The proof consists of nothing more than writing down the transformation equations

$$X^j = \bar{X}^j + X_0^{\,j}.$$

Rather than being equal, the components of $\mathbf{r}$ and $\bar{\mathbf{r}}$ differ by the terms $X_0^{\,j}$. .

Theorem 1-4.1b. Suppose that the X^j and $\bar{X}^j$ systems are related by means of a rotation. The components (1, 0, 0) of the basis arrow $\boldsymbol{\iota}_1$ are not related to the components (1, 0, 0) of the basis arrow $\bar{\boldsymbol{\iota}}_1$ by means of the law (1-4.2a).

PROOF. We have

$$\bar{\boldsymbol{\iota}}_j = a_j^{\,k}\boldsymbol{\iota}_k.$$

Therefore the components of $\bar{\boldsymbol{\iota}}_1$ in the X^j system are $(a_1^{\,1}, a_1^{\,2}, a_1^{\,3})$. These coefficients of transformation in general do not take on values (1, 0, 0).

If the components of a Cartesian vector are known in one coordinate system, then they can be found in any other coordinate system of the set. The procedure for doing so is illustrated in the next example.

Example 1-4.1. Suppose that two coordinate systems X^j and $\bar{X}^j$ are related by the transformation equations of Example 1-3.10. If the

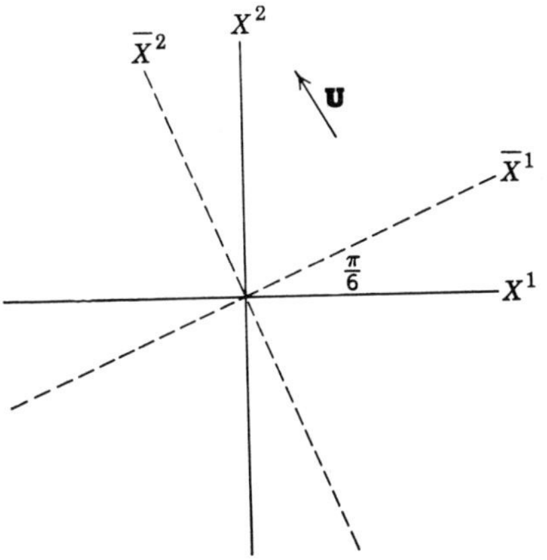

Fig. 1-4.1

components of a vector **U** are $(0, \frac{1}{2})$ in the barred system, the components in the unbarred system are determined as follows (see Fig. 1-4.1):

$$U^1 = \frac{\sqrt{3}}{2}(0) - \frac{1}{2}\left(\frac{1}{2}\right) = -\frac{1}{4},$$

$$U^2 = \frac{1}{2}(0) + \frac{\sqrt{3}}{2}\left(\frac{1}{2}\right) = \frac{\sqrt{3}}{4}.$$

The vector concept received much of its early impetus from the theoretical physicist. Today it fittingly plays a fundamental role in many aspects of physics. Linear velocities, accelerations, and forces fit nicely into the vector pattern, as do electric- and magnetic-field strengths. Also, much of the vector terminology and structure has wandered far afield and is becoming more and more important in biological and social sciences. For example, hereditary patterns are studied with the aid of vectors and matrices, as are linear programming problems.[23]

A second concept which has evolved in the development of theoretical physics is that of the scalar. The definition of "scalar," which has come down from Hamilton,[24] states that it is a quantity possessing magnitude but no direction. Such entities as mass, time, density, and temperature are given as examples. However, the prize example is a real number. This definition will not satisfy our needs, but it is worth mentioning for two reasons. First of all, it is vague, as illustrated by the incorrect prize example. A real number does not have to be associated with magnitude.

---

[23] An introduction to these ideas can be found in Kemeny, Snell, Thompson, *Finite Mathematics*, Prentice Hall, 1957.

[24] See Gibbs's interpretation, *Vector Analysis*, Gibbs, Yale University Press, 1947.

It can be given no interpretation whatsoever, it can denote position, or it can distinguish one object from another as in an index. Secondly, the use of the word "scalar" to denote a real number, or more generally an element of a field underlying a vector space, has become prevalent in the area of algebra.

It fits our purpose to use a definition of the term that is common to modern-day physics, differential geometry, and other fields in which transformation theory plays a prominent part. From a historical point of view this approach, which specifies a scalar as any quantity invariant under all transformations of coordinates, can be found in the texts of Felix Klein.[25] The definition is now stated formally. The symbol $\{\Phi\}$ denotes a collection of functions, one associated with each rectangular Cartesian coordinate system.

Definition 1-4.2. If the elements of $\{\Phi\}$ satisfy the relation

$$\Phi(X^1, X^2, X^3) = \bar{\Phi}(\bar{X}^1, \bar{X}^2, \bar{X}^3), \tag{1-4.4}$$

then the collection $\{\Phi\}$ is called a Cartesian scalar with respect to the rectangular Cartesian set of transformations. The functions Φ, $\bar{\Phi}$, etc., represent the scalar in their respective coordinate systems.

Under this definition such a phenomenon as temperature, which is a function of position and not dependent on the coordinate representation of that position, can be thought of as a numerical scalar. Other entities, such as the magnitude of a vector, have an additional property in that the form of their algebraic representation is the same in all coordinate systems of a given collection. These entities are said to be invariants.

Example 1-4.2. Suppose that a function f has a domain consisting of all triples (X^1, X^2, X^3) that satisfy the relation $(X^1)^2 + (X^2)^2 + (X^3)^2 = 1$; that is, the ordered triples of the domain correspond to the points of the unit sphere. If the rule for the function is $f(X^1\ X^2\ X^3) = \ln r$ (where $r = [(X^1)^2 + (X^2)^2 + (X^3)^2]^{1/2}$), then this function is a scalar representative with respect to the group of orthogonal Cartesian rotations. Furthermore, the function has an invariant algebraic form, since

$$(X^1)^2 + (X^2)^2 + (X^3)^2 = (\bar{X}^1)^2 + (\bar{X}^2)^2 + (\bar{X}^3)^2$$

under all rotations.

There are many physical ideas that can be represented by the scalar concept. For example, electrostatic or magnetostatic potential has scalar

---

[25] Klein, *op. cit.*, p. 47.

representations in terms of integrals containing functions that are proportional to $1/r$. The very useful innovation of the physicist called "work" is also of scalar character.

In the next section more information of a positive nature concerning scalars is discussed, but, before going on, there are a few negative bits of information that have some interest.

Example 1-4.3. The components of a Cartesian vector do not represent Cartesian scalars with respect to the rotation transformations in (1-3.9a). This statement follows at once from the fact that the vector components satisfy the transformation law

$$U^j = a_k{}^j \bar{U}^k.$$

This law is not of the form of (1-4.4). Problem 3b at the end of this section deals with the nature of vector components with respect to the set of translations.

Example 1-4.4. The distance formula

$$(X_1{}^1 - X_0{}^1)^2 + (X_1{}^2 - X_0{}^2)^2 + (X_1{}^3 - X_0{}^3)^2$$

is an invariant with respect to the transformations (1-3.9a, b).

Problems

1. The components $\bar{U}^k$ of a vector in an $\bar{X}^k$ coordinate system are $(\frac{1}{2}, 0, 0)$. The coefficients of an orthogonal Cartesian rotation relating the $\bar{X}^k$ and X^k systems are

$$(a_j{}^k) = \begin{pmatrix} 0 & -1 & 0 \\ 1 & 0 & 0 \\ 0 & 0 & 1 \end{pmatrix}.$$

   (a) Compute the components U^k.
   (b) Construct the two coordinate systems and a representation of the vector.

2. The components $\bar{U}^k$ of a vector with respect to an $\bar{X}^k$ coordinate system are (0, 1, 3). Coefficients of an orthogonal Cartesian rotation relating the $\bar{X}^k$ and X^k systems are

$$(a_j{}^k) = \begin{pmatrix} \dfrac{\sqrt{3}}{2} & \dfrac{1}{2} & 0 \\ -\dfrac{\sqrt{3}}{4} & \dfrac{3}{4} & \dfrac{1}{2} \\ \dfrac{1}{4} & -\dfrac{\sqrt{3}}{4} & \dfrac{\sqrt{3}}{2} \end{pmatrix}.$$

   Compute the U^k components of the vector.

3. (a) Show that distance is a scalar quantity with respect to translations.
   (b) Show that the components of a vector are scalars with respect to translation.
4. Suppose the n-tuple representatives P^j and Q^j of vectors **P** and **Q** are linearly dependent; then show that the representatives $\bar{P}^k$ and $\bar{Q}^k$ in a second rectangular Cartesian coordinate system are dependent. Note that this result justifies speaking of linearly dependent Cartesian vectors and associating the results of Section 2 with them.

5. *The Inner Product*

Many of the concepts studied in a course concerning vector and tensor algebra have resulted from fundamental ideas of Hamilton and Grassmann. As we pointed out in the historical introduction, these ideas have been subject to so much sifting, sorting, and adding that it is difficult to assign individual credit for them. However, one concept which specifically resulted from the ingenuity of Hermann Grassmann is used to introduce a second type of vector binary operation. (Addition constituted the first type.) This process has intimate ties with real-number multiplication.

In analogy to the procedure of Grassmann, consider two three-space Cartesian vectors, each expressed in terms of a basis $\boldsymbol{\iota}_1$, $\boldsymbol{\iota}_2$, $\boldsymbol{\iota}_3$; that is, $\mathbf{P} = P^1\boldsymbol{\iota}_1 + P^2\boldsymbol{\iota}_2 + P^3\boldsymbol{\iota}_3$ and $\mathbf{Q} = Q^1\boldsymbol{\iota}_1 + Q^2\boldsymbol{\iota}_2 + Q^3\boldsymbol{\iota}_3$. When naïvely multiplied together,

$$\text{(1-5.1a)} \quad \mathbf{PQ} = P^1Q^1\boldsymbol{\iota}_1\boldsymbol{\iota}_1 + P^1Q^2\boldsymbol{\iota}_1\boldsymbol{\iota}_2 + P^1Q^3\boldsymbol{\iota}_1\boldsymbol{\iota}_3 + P^2Q^1\boldsymbol{\iota}_2\boldsymbol{\iota}_1 + P^2Q^2\boldsymbol{\iota}_2\boldsymbol{\iota}_2 + P^2Q^3\boldsymbol{\iota}_2\boldsymbol{\iota}_3 + P^3Q^1\boldsymbol{\iota}_3\boldsymbol{\iota}_1 + P^3Q^2\boldsymbol{\iota}_3\boldsymbol{\iota}_2 + P^3Q^3\boldsymbol{\iota}_3\boldsymbol{\iota}_3.$$

The expression (1-5.1a) has no preassigned significance. In fact, a variety of binary operations can be specified, each depending on the meaning assigned to the nine quantities $\boldsymbol{\iota}_j\boldsymbol{\iota}_k$, $j, k = 1, 2, 3$.

We introduce a binary operation for a pair of Cartesian vectors **P** and **Q**, symbolized by

$$\text{(1-5.1b)} \qquad \mathbf{P} \cdot \mathbf{Q} = P^jQ^k\boldsymbol{\iota}_j \cdot \boldsymbol{\iota}_k,$$

in the following definition.

Definition 1-5.1. Let

$$\text{(1-5.2)} \qquad \boldsymbol{\iota}_j \cdot \boldsymbol{\iota}_k = \cos\theta_{jk} = \delta_{jk}.$$

We specify that θ_{jk} has either the value $\pi/2$ or 0 and is determined by the positive senses of $\boldsymbol{\iota}_j$ and $\boldsymbol{\iota}_k$.

As a consequence of this definition, relation (1-5.1b) takes the form

$$\text{(1-5.3)} \qquad \mathbf{P} \cdot \mathbf{Q} = P^jQ^k\delta_{jk} = P^1Q^1 + P^2Q^2 + P^3Q^3.$$

Because of the notation employed, $\mathbf{P} \cdot \mathbf{Q}$ is often called the "dot" product of $\mathbf{P}$ and $\mathbf{Q}$. The term "inner product" is also commonly associated with the expression.

A common practice, when building a mathematical structure, is to look toward the set of real numbers for comparison and analogy; that is, we ask whether a newly introduced operation satisfies the same laws as a corresponding real-number operation. The following theorem points out that the inner product conforms to several basic rules of real numbers.

Theorem 1-5.1. The fundamental properties of the inner product are the following:

(1-5.4a) (*Commutative law*) $\mathbf{P} \cdot \mathbf{Q} = \mathbf{Q} \cdot \mathbf{P}.$

(1-5.4b) (*Associative law with respect to a real number*)

$$(\alpha\mathbf{P}) \cdot \mathbf{Q} = \alpha(\mathbf{P} \cdot \mathbf{Q}).$$

(1-5.4c) (*Distributive law*) $\mathbf{P} \cdot (\mathbf{Q} + \mathbf{R}) = \mathbf{P} \cdot \mathbf{Q} + \mathbf{P} \cdot \mathbf{R}.$

(1-5.4d) (*Positive definite property*) $\mathbf{P} \cdot \mathbf{P} \geq 0;\quad \mathbf{P} \cdot \mathbf{P} = 0 \leftrightarrow \mathbf{P} = \mathbf{0}.$

PROOF. The proofs are left to the reader. The first three follow from the definitions of inner product, vector or n-tuple addition, and corresponding properties of real numbers. The proof of (1-5.4d) is trivial with respect to the form (1-5.3), since a sum of real-number squares can never be negative.

Note that the inner product of two vectors is not a vector. This is described by saying that the operation is not closed.

The next matter to come under discussion is the behavior of the inner product when subjected to transformations of the orthogonal Cartesian set. Of course, the symbol $\mathbf{P} \cdot \mathbf{Q}$ is only a notation for the inner product of two vectors $\mathbf{P}$ and $\mathbf{Q}$. To investigate the way in which this product transforms, we must concentrate attention on the right-hand side of (1-5.3).

Theorem 1-5.2. The form $\sum_{j=1}^{3} P^j Q^j$ is an invariant Cartesian scalar. [See (1-3.9a) and (1-3.9b).]

PROOF. By making use of (1-3.9a) and (1-3.9b), the transformation can be carried out as follows:

$$\begin{aligned}\sum_{j=1}^{3} P^j Q^j &= \sum_{j=1}^{3} (a_r{}^j \bar{P}^r)(a_s{}^j \bar{Q}^s) = \sum_{j=1}^{3} a_r{}^j a_s{}^j \bar{P}^r \bar{Q}^s \\ &= \delta_{rs} \bar{P}^r \bar{Q}^s = \sum_{k=1}^{3} \bar{P}^k \bar{Q}^k.\end{aligned}$$

The last member of the string of equalities results when the meaning of the delta symbol is invoked. The proof for translations is trivial.

In the exercises at the end of the section the reader is asked to show that the form (1-5.4b) is not preserved under linear transformations whose coefficients do not satisfy the orthogonality conditions in (1-3.9b). This illustrates the fact that whether a given algebraic form is invariant depends on the group of transformations under consideration.

The availability of the dot product greatly enhances the procedure for exhibiting various geometric concepts in Euclidean three-space. (This statement is even more significant for $n > 3$.) Magnitude and angle are fundamental to the metric structure of Euclidean space. These concepts are dealt with in the following theorems.

Theorem 1-5.3. The magnitude of a Cartesian vector **P** is represented by the Cartesian scalar $\{|\mathbf{P}|\}$, where

$$|\mathbf{P}| = (\mathbf{P} \cdot \mathbf{P})^{1/2} = [(P^1)^2 + (P^2)^2 + (P^3)^2]^{1/2} = (\delta_{jk}P^jP^k)^{1/2}. \tag{1-5.5}$$

PROOF. The magnitude of a Cartesian vector is expressed in any given coordinate system by the representation (1-1.5b), that is $\sqrt{\Sigma P^jP^j}$. In Theorem 1-5.2 it was shown that $\sum_{j=1}^{3} P^jP^j$ was a scalar representative; hence $(\Sigma P^jP^j)^{1/2}$ is a scalar representative. The expression $\delta_{jk}P^jP^k$ is just another notation for $\sum_{j=1}^{3} P^jP^j$.

If $|\mathbf{P}| = 1$, then **P** is said to be a unit Cartesian vector.

The foregoing has been introduced in a manner analogous to the procedure of Grassmann in order to emphasize that the inner product is only one of a possible variety of binary operations. Indeed, another binary operation is presented in Section 7. In Gibbs's development of vector analysis a geometric form of the scalar product is given. This form,

$$\mathbf{P} \cdot \mathbf{Q} = |\mathbf{P}|\,|\mathbf{Q}| \cos\theta, \tag{1-5.6a}$$

was suggested by physical applications, such as the concept of work, and geometric considerations, for example, that of the angle between two directions. In the next theorem we verify that the representation (1-5.6a) is consistent.

Theorem 1-5.4. $\mathbf{P} \cdot \mathbf{Q}$ and $|\mathbf{P}|\,|\mathbf{Q}| \cos\theta$ are equivalent relations; that is,

$$\sum_{j=1}^{3} P^jQ^j = |\mathbf{P}|\,|\mathbf{Q}| \cos\theta, \tag{1-5.6b}$$

where θ is the angle measured from **P** to **Q** and such that $0 \le \theta \le \pi$.

PROOF. As a matter of convenience, choose geometric representatives of **P** and **Q** with their initial points at the origin of coordinates. (See

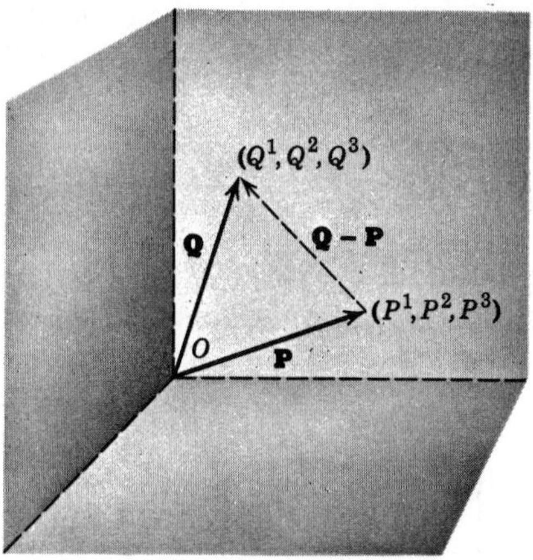

Fig. 1-5.1

Fig. 1-5.1.) According to the law of cosines,

$$|\mathbf{Q} - \mathbf{P}|^2 = |\mathbf{P}|^2 + |\mathbf{Q}|^2 - 2\,|\mathbf{P}|\,|\mathbf{Q}| \cos \theta. \tag{1-5.7a}$$

This relation can also be expressed in the form

$$\delta_{jk}(Q^j - P^j)(Q^k - P^k) = \delta_{jk}P^jP^k + \delta_{jk}Q^jQ^k - 2\,|\mathbf{P}|\,|\mathbf{Q}| \cos \theta. \tag{1-5.7b}$$

By writing out the summation in the left member of (1-5.7b) the reader can verify that a usual multiplication of the parenthetic expressions, followed by a distribution of δ_{jk}, is valid. Therefore

(1-5.7c)

$$\delta_{jk}Q^jQ^k - 2\delta_{jk}P^jQ^k + \delta_{jk}P^jP^k = \delta_{jk}P^jP^k + \delta_{jk}Q^jQ^k - 2\,|\mathbf{P}|\,|\mathbf{Q}| \cos \theta.$$

Subtracting the appropriate terms from each member and then dividing by -2, we obtain

$$\delta_{jk}P^jQ^k = |\mathbf{P}|\,|\mathbf{Q}| \cos \theta. \tag{1-5.7d}$$

Since (1-5.7d) is equivalent to (1-5.6b), the proof is complete.

By making the association $|\mathbf{P}|(|\mathbf{Q}| \cos \theta)$ in the right member of (1-5.6) we are able to interpret the inner product geometrically as the magnitude of **P** multiplied by the perpendicular projection of **Q** on **P** taken with the appropriate sign. Of course, **P** and **Q** can be interchanged in this statement. The interchange would correspond to the association $|\mathbf{Q}|(|\mathbf{P}| \cos \theta)$.

It is worthy of notice that the relation (1-5.6) should not be new to the reader. When written in terms of the components of **P** and **Q**,

$$\cos\theta = \frac{P^1Q^1 + P^2Q^2 + P^3Q^3}{[(P^1)^2 + (P^2)^2 + (P^3)^2]^{1/2}[(Q^1)^2 + (Q^2)^2 + (Q^3)^2]^{1/2}} \tag{1-5.8}$$

This expression is recognizable as the formula for the angle made by two directions, usually presented in a development of solid analytic geometry.

Other representations of geometric concepts and entities in Euclidean three-space depend in a large measure on the relationship between the inner product and the ideas of distance and angle. The following theorems and examples draw attention to some of the elementary ideas and configurations.

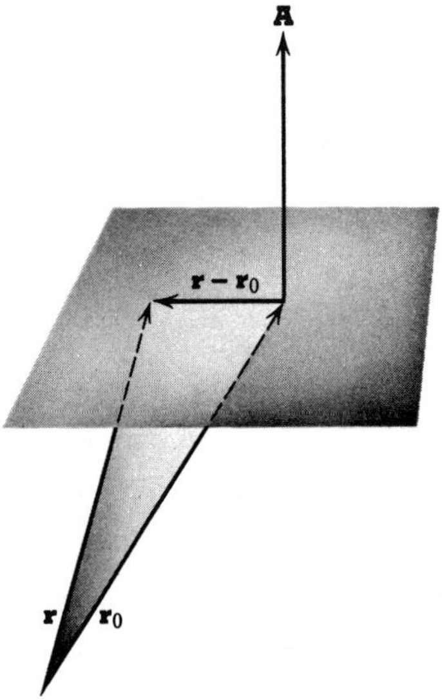

Fig. 1-5.2

Theorem 1-5.5. If **P** and **Q** are nonzero vectors,

(1-5.9) $\mathbf{P}\cdot\mathbf{Q} = 0 \leftrightarrow$ **P** and **Q** are orthogonal.

PROOF. From (1-5.6) it follows that, if $\theta = \pi/2$, then $\mathbf{P}\cdot\mathbf{Q} = 0$, and conversely.

Example 1-5.1a. An equation of a plane can be obtained by making use of the characterization of orthogonality stated in Theorem 1-5.5. Take (as in Fig. (1-5.2) the geometric representation of a fixed vector **A** whose initial point has coordinates $X_0^{\,j}$. Then the end point of a position vector **r**, subject to the condition,

$$\mathbf{A}\cdot(\mathbf{r} - \mathbf{r}_0) = 0, \tag{1-5.10a}$$

generates a configuration in space, which we call a plane. Note that the equation of a plane can be put in the expanded form

$$A^1(X^1 - X_0^{\,1}) + A^2(X^2 - X_0^{\,2}) + A^3(X^3 - X_0^{\,3}) = 0. \tag{1-5.10b}$$

Another notational form for the equation of a plane, which is sometimes useful, is

$$\delta_{jk}A^j(X^k - X_0^{\,k}) = 0. \tag{1-5.10c}$$

Example 1-5.1b. The plane through (2, 1, 5) and perpendicular to the vector $\mathbf{A} = 3\iota_1 + 5\iota_2 - \iota_3$ has the equation

$$3(X^1 - 2) + 5(X^2 - 1) - 1(X^3 - 5) = 0.$$

Of course, this equation can also be put in the form

$$3X^1 + 5X^2 - X^3 = 6.$$

An interesting formula for the perpendicular distance between a plane and a point follows from writing the equation of the plane in "Hesse's normal form";[26] that is, in the form

$$\frac{\mathbf{A}}{|\mathbf{A}|} \cdot (\mathbf{r} - \mathbf{r}_0) = 0. \tag{1-5.11}$$

The word "normalized" implies, as indicated in (1-5.11), that the direction perpendicular to the plane has been expressed in terms of a vector of magnitude 1, that is, $\mathbf{A}/|\mathbf{A}|$.

Theorem 1-5.6. The perpendicular distance D, determined by a point with coordinates $X_1{}^j$ and a plane with the equation (1-5.11), is

$$D = \left| \frac{\mathbf{A}}{|\mathbf{A}|} \cdot (\mathbf{r}_1 - \mathbf{r}_0) \right|. \tag{1-5.12}$$

PROOF. Let the components of **C** be given by the coordinate differences

$$C^j = X_1{}^j - X_0{}^j. \tag{1-5.13}$$

From the geometric interpretation of the inner product it follows that

$$D = \left| \text{proj. } \mathbf{C} \text{ on } \frac{\mathbf{A}}{|\mathbf{A}|} \right| = \big|\, |\mathbf{C}| \cos\theta \big|.$$

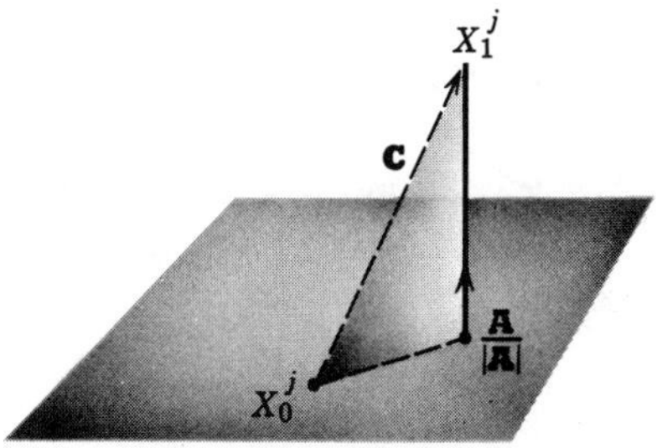

Fig. 1-5.3

[26] The form was named after the German mathematician Otto Hesse (1811–1874).

Since

$$\cos\theta = \frac{\mathbf{A}}{|\mathbf{A}|} \cdot \frac{C}{|\mathbf{C}|},$$

it follows that

$$D = \left| \frac{\mathbf{A}}{|\mathbf{A}|} \cdot \mathbf{C} \right|.$$

The relation (1-5.1) is obtained when $\mathbf{C}$ is replaced by $\mathbf{r}_1 - \mathbf{r}_0$, a substitution justified by (1-5.13).

Example 1-5.2. If the point with coordinates (2, 5, 1) and the plane whose equation is $3(X^1 - 2) + 2(X^2 - 1) - 5(X^3 + 2) = 0$ are given, the perpendicular distance (1-5.12) is

$$D = \left| \frac{3(2-2) + 2(5-1) - 5(1+2)}{\sqrt{9+4+25}} \right| = \left| \frac{-7}{\sqrt{38}} \right| = \frac{7}{\sqrt{38}}.$$

Example 1-5.3. Another common geometric configuration with a simple vector representation is the sphere. A sphere is geometrically defined as the set of all points equidistant from a fixed point. If, as in Fig. 1-5.4, the fixed point has coordinates X_0^j and the radius of the sphere is denoted by a, then $\mathbf{r} - \mathbf{r}_0$ is a vector of constant magnitude a. The algebraic representation of the sphere is

$$(\mathbf{r} - \mathbf{r}_0) \cdot (\mathbf{r} - \mathbf{r}_0) = a^2. \tag{1-5.14a}$$

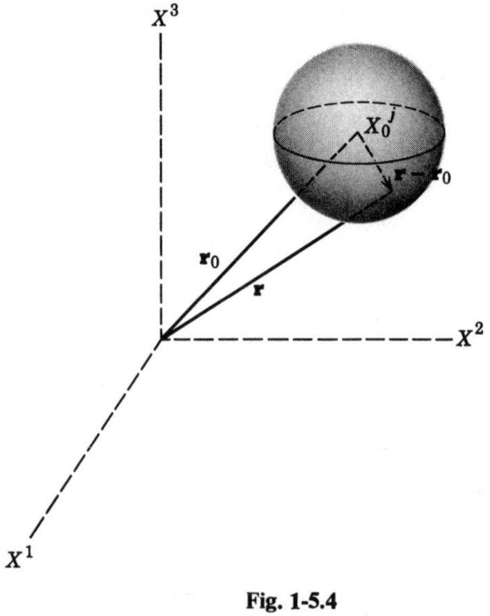

Fig. 1-5.4

This representation can also be put in the form

$$|\mathbf{r} - \mathbf{r}_0|^2 = a^2. \tag{1-5.14b}$$

Problems

1. Suppose $\mathbf{A} = 3\iota_1 + 2\iota_2 - 5\iota_3$, $\mathbf{B} = 2\iota_1 - 7\iota_2 + 4\iota_3$.
   (a) Compute $\mathbf{A} \cdot \mathbf{B}$, $\mathbf{A} \cdot \mathbf{A}$, and $\mathbf{B} \cdot \mathbf{B}$.
   (b) Determine a unit vector with the sense and direction of **A**.
   (c) Determine the angle $0 \leq \theta \leq \pi$ made by **A** and **B**.
2. Does the relation (1-5.3) in any way define $\mathbf{A} \cdot \mathbf{B} \cdot \mathbf{C}$? Explain.
3. Suppose that the coordinates X^j and $\bar{X}^j$ of two rectangular Cartesian systems are related by means of the transformation equations $X^j = a_k{}^j \bar{X}^k$, where

$$(a_k{}^j) = \begin{pmatrix} \frac{\sqrt{3}}{2} & \frac{1}{2} & 0 \\ \frac{-1}{2} & \frac{\sqrt{3}}{2} & 0 \\ 0 & 0 & 1 \end{pmatrix}.$$

   (a) Describe the rotation.
   (b) If $\bar{\mathbf{A}} = 3\bar{\iota}_1 + 5\bar{\iota}_2$, $\bar{\mathbf{B}} = 2\bar{\iota}_1 - 4\bar{\iota}_2$, find $\bar{\mathbf{A}} \cdot \bar{\mathbf{B}}$.
   (c) Find **A** and **B**.
   (d) Compute $\mathbf{A} \cdot \mathbf{B}$.

   Remark. Since $\sum_{j=1}^{3} A^j B^j$ is an invariant with respect to orthogonal Cartesian transformations, the result of (d) is known from (b). The purpose of the problem is to illustrate this scalar relation through numerical calculation.

4. Perform the tasks analogous to (b), (c), and (d) of Problem 3 if **A** and **B** have components 1, 3, 5 and −2, 3, 4, respectively, and

$$(A_j{}^k) = \begin{pmatrix} \frac{\sqrt{3}}{2} & \frac{\sqrt{3}}{4} & \frac{1}{4} \\ \frac{1}{2} & \frac{-3}{4} & \frac{-\sqrt{3}}{4} \\ 0 & \frac{1}{2} & \frac{-\sqrt{3}}{2} \end{pmatrix}.$$

5. Show that $\mathbf{A} = 2\iota_1 - 3\iota_2 + \iota_3$ and $\mathbf{B} = 5\iota_1 + 2\iota_2 - 4\iota_3$ are orthogonal.
6. Suppose that an object is moving along a straight line a distance $|\mathbf{r}_1 - \mathbf{r}_0|$ with a direction and sense corresponding to the vector $\mathbf{r}_1 - \mathbf{r}_0$. Let **F** represent a constant force acting on the object. If the "work" done by

this force is defined as

$$W = \mathbf{F} \cdot (\mathbf{r}_1 - \mathbf{r}_0),$$

(a) when is there no work done?

(b) Make an analysis of the way in which the work done varies with the direction and sense of **F**.

7. (a) Show that the diagonals of a square are perpendicular by showing that $\mathbf{E}_1 \cdot \mathbf{E}_2 = 0$. (See the accompanying figure.)

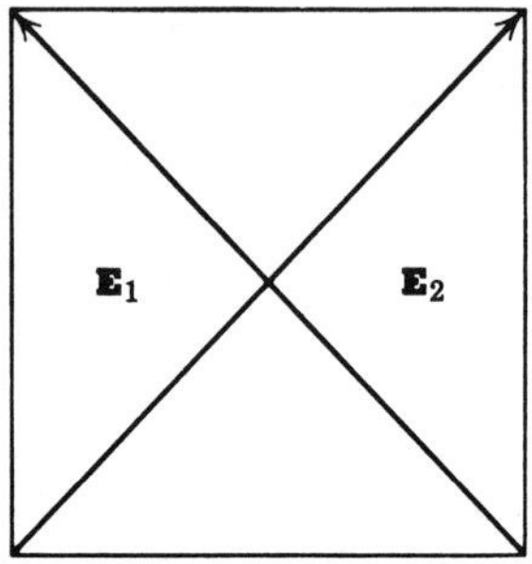

(b) Show that the diagonals of a rhombus are perpendicular. (A rhombus is a four-sided polygon, all sides of which are equal in length.)

8. (a) A line passes through the point $(1, -3, 2)$ and has the direction of $\mathbf{B} = 2\iota_1 - 5\iota_2 - 4\iota_3$. What are the parametric and symmetric representations of the line?

(b) Find the parametric representation of the line in an $\bar{X}^j$ coordinate system related to the X^j system of (a) by means of the transformation equations

$$\bar{X}^j = A_k{}^j X^k,$$

where

$$(A_k{}^j) = \begin{pmatrix} \dfrac{-1}{2} & \dfrac{-1}{2} & \dfrac{-1}{\sqrt{2}} \\ \dfrac{1}{2} & \dfrac{1}{2} & \dfrac{-1}{\sqrt{2}} \\ \dfrac{1}{\sqrt{2}} & \dfrac{-1}{\sqrt{2}} & 0 \end{pmatrix}.$$

9. Show that the two lines with parametric representations

$$\begin{aligned} X^1 &= 5 + 2t & \qquad X^1 &= 2 - 6s \\ X^2 &= 3 + 7t \quad \text{and} & X^2 &= 4 + 2s \\ X^3 &= 2 - 4t & X^3 &= 6 + \tfrac{1}{2}s, \end{aligned}$$

respectively, are perpendicular.

10. What is a necessary and sufficient condition that two lines with vector representations $\mathbf{r} = \mathbf{r}_1 + \mathbf{B}_1 t$ and $\mathbf{r} = \mathbf{r}_2 + \mathbf{B}_2 s$, respectively, be (a) parallel, (b) perpendicular?

11. (a) What is an equation of the plane through $(-2, 4, 1)$ and perpendicular to $\mathbf{A} = 3\iota_1 + \iota_2 - 5\iota_3$?
(b) Find the equation of this plane in the $\bar{X}^j$ system of Problem 8b.

12. Find a necessary and sufficient condition that two planes be parallel.

13. What is the meaning of the statement: two planes are perpendicular? (Refer to a text on solid analytic geometry.)

14. Prove that $\mathbf{A}/|\mathbf{A}|$ is a unit vector. $\mathbf{A} \neq \mathbf{0}$.

15. Find the perpendicular distance between the plane with equation $2(X^1 - 3) - 5(X^2 + 1) + 3(X^3 - 4) = 0$ and the point $(1, 2, -1)$.

16. Is the algebraic representation of a line dependent on a coordinate system?

17. Show that the angle θ determined by two vectors $\mathbf{P}$ and $\mathbf{Q}$ satisfies

$$\sin^2\theta = \frac{(\mathbf{P}\cdot\mathbf{P})(\mathbf{Q}\cdot\mathbf{Q}) - (\mathbf{P}\cdot\mathbf{Q})^2}{(\mathbf{P}\cdot\mathbf{P})(\mathbf{Q}\cdot\mathbf{Q})}.$$

18. Find components of a vector $\mathbf{C}$ perpendicular to linearly independent vectors $\mathbf{P}$ and $\mathbf{Q}$.
Hint: Solve the equations $\mathbf{P}\cdot\mathbf{C} = 0$ and $\mathbf{Q}\cdot\mathbf{C} = 0$.

5*. *General Cartesian Coordinates*

Historical perspective on the significance of linear transformations in geometry can be obtained from an examination of Felix Klein's "Erlanger Programm" of 1872, mentioned in Section 3. This program,[27] which was inspired by the work of Arthur Cayley (1821–1895, English), defines a geometry as a theory of the invariants of a transformation group.[28] From this viewpoint, projective geometry, which was suggested by Cayley as the universal geometry, is characterized by the set of transformations

$$\bar{X}^1 = \frac{a_1X^1 + a_2X^2 + a_3X^3 + a_4}{d_1X^1 + d_2X^2 + d_3X^3 + d_4},$$

$$\bar{X}^2 = \frac{b_1X^1 + b_2X^2 + b_3X^3 + b_4}{e_1X^1 + e_2X^2 + e_3X^3 + e_4},$$

$$\bar{X}^3 = \frac{c_1X^1 + c_2X^2 + c_3X^3 + c_4}{f_1X^1 + f_2X^2 + f_3X^3 + f_4}.$$

Incidence relations are among the invariants of this group of transformations.

---

[27] Klein, *op. cit.*, 130.
[28] See Definition 1-5*.3.

Affine geometry[29] consists of the study of the invariants of a subset of the foregoing transformation relations; that is,

$$\bar{X}^1 = a_1 X^1 + a_2 X^2 + a_3 X^3 + a_4,$$
$$\bar{X}^2 = b_1 X^1 + b_2 X^2 + b_3 X^3 + b_4,$$
$$\bar{X}^3 = c_1 X^1 + c_2 X^2 + c_3 X^3 + c_4.$$

A significant aspect of affine geometry is its inclusion of the parallel concept which plays no part in the more general projective geometry.

When orthogonality conditions (1-3.21) are imposed on the coefficients

$$\begin{pmatrix} a_1 & a_2 & a_3 \\ b_1 & b_2 & b_3 \\ c_1 & c_2 & c_3 \end{pmatrix},$$

the transformations characterize Euclidean metric geometry. Sometimes the orthogonality conditions are slightly relaxed to enable inclusion of the transformations that change size but do not destroy proportionalities; that is,

$$\bar{X}^1 = \lambda X^1, \qquad \bar{X}^2 = \lambda X^2, \qquad \bar{X}^3 = \lambda X^3.$$

Under this circumstance the transformations characterizing Euclidean metric geometry are called similarity transformations.

If $a_4 = b_4 = c_4 = 0$, the affine transformations leave the origin of coordinates fixed and are called centered affine transformations. This is the set with which we shall be concerned.

The centered affine transformations form a transformation group. The meaning of the term group is indicated by the next definition.

Definition 1-5*.1. A set of transformations $R, S, T, \cdots$ is said to form a group if the following properties hold:

(a) (*Closure*) Two successive transformations of the set result in a transformation of the set.

(b) (*Associative law*) $(RS)Ty^j = R(ST)y^j$.

(c) (*Existence of an identity element*). The transformation I with coefficients $\delta_j^{\ k}$ is an element of the set of transformations.

(d) (*Existence of a unique inverse to each transformation of the set*). To each transformation T of the set there is a unique transformation T^{-1} such that the coefficients of T followed by T^{-1} or T^{-1} followed by T are $(\delta_j^{\ k})$.

---

[29] The term affine apparently goes back to Euler and Möbius.

Theorem 1-5*.1. The centered affine transformations form a group.

PROOF. Consider successive transformations

$$\bar{y}^k = B_j^{\ k} y^j, \qquad \bar{\bar{y}}^q = C_k^{\ q} \bar{y}^k. \tag{1-5*.1a}$$

We have

$$\bar{\bar{y}}^q = C_k^{\ q} B_j^{\ k} y^j. \tag{1-5*.1b}$$

Denote the set of nine numbers represented by $C_k^{\ q} B_j^{\ k}$ by

$$D_j^{\ q} = C_k^{\ q} B_j^{\ k}. \tag{1-5*.1c}$$

We have

$$\bar{\bar{y}}^q = D_j^{\ q} y^j. \tag{1-5*.1d}$$

If the determinant $(D_j^{\ q}) \neq 0$, the property of closure has been exhibited. The fact that $D \neq 0$ follows from the knowledge that B and C are nonzero and that D represents the product of determinants B and C. (The student not familiar with determinant multiplication should see Section 6.)

To consider the associative property, make use of the relations in (1-5*.1a) along with a third transformation

$$\bar{\bar{\bar{y}}}^r = D_q^{\ r} \bar{\bar{y}}^q.$$

Then

$$\bar{\bar{\bar{y}}}^r = (D_q^{\ r} C_k^{\ q}) B_j^{\ k} y^j = D_q^{\ r}(C_k^{\ q} B_j^{\ k}) y^j$$

represents a statement of the associative law. That the law holds is fundamentally a consequence of the associative and distributive laws of real numbers. This can be verified in the straightforward but tedious manner of writing out the summations.

Since $\bar{y}^j = \delta_k^{\ j} y^k$ is a centered affine transformation, (c) holds. Finally, to each transformation $\bar{y}^j = B_k^{\ j} y^k$ of the set there corresponds $y^k = b_j^{\ k} \bar{y}^j$ such that

$$b_j^{\ k} B_q^{\ j} = \delta_q^{\ k}, \ \text{det.}\ (B_q^{\ j}) \neq 0.$$

Therefore (d) holds and the theorem is proved.

The important ideas to be developed in this text could be stated without formally introducing the group concept. We might ask, then why introduce it? There are at least two reasons. First of all the existence of a well-defined meaning of the term in the mind of the reader adds measurably to the fluency of communication. Second, the group concept has played such a fundamental role in mathematics, especially the ideas developed in the nineteenth century, that the reader should be aware of its presence in this discussion.

The seed of the group idea was planted by Lagrange (1736–1813, French) around 1770–1771. By critically examining solutions of equations of second, third, and fourth degree (with the purpose of extending his examination to equations of higher degree), he made a beginning in the theory of permutation groups. The term "group" is actually due to E. Galois (1811–1832, French), who was stimulated by interest in showing that equations of a degree higher than four could not in general be solved by radicals. The group concept entered into geometry principally through the writings of Sophus Lie (1842–1899, Norwegian) on continuous groups. It culminated in Klein's "Erlanger Programm," in which it was used to classify the geometries of that day.

The properties (a)–(d), which characterize linear transformations as a group, have been pointed out. It is left to the reader to show that the orthogonal linear transformations, that is, those relating rectangular Cartesian coordinate systems, form a group contained in the linear group.

The "Erlanger Programm," or the definition of a geometry as the set of invariants of a transformation group, is to be understood from the viewpoint of space mappings; that is, if we think of a transformation as a replacement of a space by a new one, the question of finding those entities that have an absolute meaning is of prime significance. In this book we are interpreting the transformations differently. From our viewpoint the space is fixed (i.e., it is a Euclidean metric space), and the transformations simply bring about a change of coordinate system. From this standpoint we can employ the affine transformation equations when considering metric concepts. Of course, representations for distance or angle will not have an invariant algebraic form.

In this section it is shown that the affine transformations relate Cartesian coordinate systems (i.e., systems determined by straight-line axes). In particular, the axes of these systems need not be perpendicular to one another. Representation of various concepts in the Cartesian systems will give us some direction as to which of the concepts considered in the framework of rectangular Cartesian systems should be generalized as well as how to bring about the generalizations. Moreover, the availability of Cartesian systems of reference will be valuable when considering the special theory of relativity in a later chapter.

Perhaps the simplest way of assimilating new ideas is to look at special cases. The following example affords this opportunity.

Example 1-5*.1. Let (X^1, X^2, X^3) represent rectangular Cartesian coordinates while the triple (Y^1, Y^2, Y^3) represents an arbitrary set of

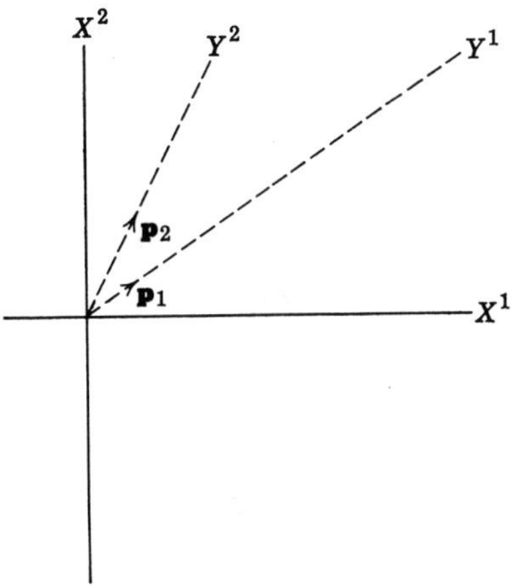

Fig. 1-5*.1

Cartesian coordinates. (See Fig. 1-5*.1.) Consider the linear transformation equations

$$\begin{aligned} Y^1 &= 2\sqrt{3}X^1 - 2X^2, \\ Y^2 &= -X^1 + \sqrt{3}X^2, \\ Y^3 &= X^3. \end{aligned} \tag{1-5*.2a}$$

It follows from these transformation equations that

$$\begin{aligned} Y^1 = 0 &\quad \text{if and only if} \quad 2\sqrt{3}X^1 - 2X^2 = 0, \\ Y^2 = 0 &\quad \text{if and only if} \quad -X^1 + \sqrt{3}X^2 = 0, \\ Y^3 = 0 &\quad \text{if and only if} \quad X^3 = 0. \end{aligned}$$

Therefore $Y^1 = 0$, $Y^2 = 0$, $Y^3 = 0$ can be identified with the coordinate planes $2\sqrt{3}X^1 - 2X^2 = 0$, $-X^1 + \sqrt{3}X^2 = 0$, and $X^3 = 0$, respectively. The axes of the Y^j system correspond to the lines of intersection of these planes. The fact that the plane $Y^3 = 0$ corresponds to the plane $X^3 = 0$ simplifies matters. We are able to interpret the axes $Y^1 = 0$ and $Y^2 = 0$ as lines in this common plane. Specifically $Y^2 = 0$, that is, the Y^1 axis, corresponds to the line $-X^1 + \sqrt{3}X^2 = 0$ and the Y^2 axis is incident with $2\sqrt{3}X^1 - 2X^2 = 0$.

Now let us investigate other aspects of this transformation. The matrix of coefficients of the transformation is

$$(B_j^{\,k}) = \begin{pmatrix} 2\sqrt{3} & -1 & 0 \\ -2 & \sqrt{3} & 0 \\ 0 & 0 & 1 \end{pmatrix} \tag{1-5*.2b}$$

where as usual the lower index represents row and the upper index represents column. Note that the equations in (1-5*.2a) correspond to the form

(1-5*.2c) $$Y^k = B_j^k X^j.$$

The determinant of the transformation has the value

(1-5*.2d) $$B = 4.$$

Since this value is different from zero, we can obtain the inverse transformation by use of Cramer's rule. More simply, the inverse transformation coefficients can be obtained from the definition

(1-5*.3a) $$b_j^k = \frac{\text{cofactor of } B_k^j \text{ in det } (B_k^j)}{B}.$$

As in Section 3, the definition implies the algebraic conditions

(1-5*.3b) $$b_j^q B_q^k = \delta_j^k, \qquad b_q^k B_j^q = \delta_j^k.$$

These relations are used freely throughout this section.

(1-5*.3c) $$(b_j^k) = \begin{pmatrix} \frac{\sqrt{3}}{4} & \frac{1}{4} & 0 \\ \frac{1}{2} & \frac{\sqrt{3}}{2} & 0 \\ 0 & 0 & 1 \end{pmatrix}$$

Note that

(1-5*.3d) $$b = \frac{1}{4}$$

and therefore

(1-5*.3e) $$Bb = 1.$$

It will be seen that the fact that the product of the values of the determinant of transformation and its inverse equals one is a general property of transformations.

It has been assumed that the space under consideration is a Euclidean metric space. We have defined a Cartesian vector as a collection of n-tuples, one n-tuple associated with each rectangular Cartesian system, and we have taken a family of arrows as the geometric representative of the collection. The question that we pose for ourselves is that of extending the algebraic formulation of the vector concept to nonorthogonal Cartesian coordinate systems. We attack the problem by the consideration of bases that can be associated with these systems. A basis that occurs naturally is

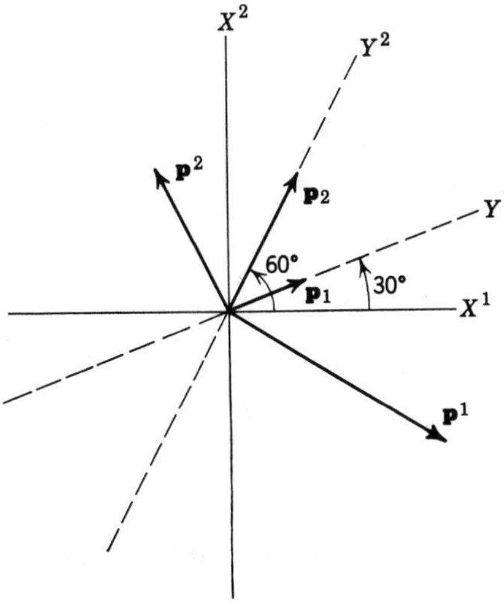

Fig. 1-5*.2

geometrically represented by arrows of fixed magnitude, direction, and sense along the coordinate axes. Let us examine the set of n-tuples $\mathbf{p}_1$, $\mathbf{p}_2$, $\mathbf{p}_3$, given rise to by the relations

(1-5*.4) $$\mathbf{p}_j = b_j^{\ k}\boldsymbol{\iota}_k.$$

With respect to the example under consideration [see (1-5*.3b)], we have

(1-5*.5) $$\mathbf{p}_1 = \frac{\sqrt{3}}{4}\boldsymbol{\iota}_1 + \frac{1}{4}\boldsymbol{\iota}_2$$
$$\mathbf{p}_2 = \frac{1}{2}\boldsymbol{\iota}_1 + \frac{\sqrt{3}}{2}\boldsymbol{\iota}_2$$

It is clear that these n-tuples can be represented along the Y^1 and Y^2 axes, since the lines corresponding to these axes have direction numbers $(\sqrt{3}, 1, 0)$ and $(2, 2\sqrt{3}, 0)$, respectively. (See Fig. 1-5*.2.) It is worth noting that the basis $\mathbf{p}_j$, introduced by (1-5*.5), is not composed of unit n-tuples. In particular,

$$|\mathbf{p}_1| = \tfrac{1}{2}.$$

In this example it does happen that $\mathbf{p}_2$ and $\mathbf{p}_3$ are of unit length.

Before generalizing our observations and specifying the significance of introducing a basis as we have just done, it would be convenient to illustrate a second way of determining a basis. Suppose that the n-tuples $\mathbf{p}^1$, $\mathbf{p}^2$, $\mathbf{p}^3$

are given by the relations

$$\mathbf{p}^j = \sum_{k=1}^{3} B_k{}^j \boldsymbol{\iota}_k. \tag{1-5*.6}$$

(The significance of the superscript notation is discussed presently.) Again, in terms of our example, we have

$$\begin{aligned} \mathbf{p}^1 &= 2\sqrt{3}\boldsymbol{\iota}_1 - 2\boldsymbol{\iota}_2 \\ \mathbf{p}^2 &= -\boldsymbol{\iota}_1 + \sqrt{3}\boldsymbol{\iota}_2. \end{aligned} \tag{1-5*.7}$$

By straightforward computation, we find that the bases (1-5*.5) and (1-5*.7) satisfy the conditions (see Fig. 1-5*.2)

$$\mathbf{p}_j \cdot \mathbf{p}^k = \delta_j{}^k; \tag{1-5*.8}$$

that is,

$\mathbf{p}^1$ is perpendicular to $\mathbf{p}_2$ and $\mathbf{p}_3$,
$\mathbf{p}^2$ is perpendicular to $\mathbf{p}_3$ and $\mathbf{p}_1$,
$\mathbf{p}^3$ is perpendicular to $\mathbf{p}_1$ and $\mathbf{p}_2$.

This completes the example.

The preceding example can be used as a guide to intuition as we proceed with a general development of vector concepts in Cartesian coordinates. Note that the linear transformations

$$\begin{aligned} &\text{(a)} \quad Y^j = B_k{}^j X^k, \\ &\text{(b)} \quad X^k = b_j{}^k Y^j, \qquad B \neq 0, \end{aligned} \tag{1-5*.9}$$

where the X^j are rectangular Cartesian coordinates, at once establish the fact that the Y^j system is Cartesian (i.e., referred to straight-line axes). For example, if $Y^1 = Y^2 = 0$, the conditions

$$\begin{aligned} B_1{}^1 X^1 + B_2{}^1 X^2 + B_3{}^1 X^3 &= 0 \\ B_1{}^2 X^1 + B_2{}^2 X^2 + B_3{}^2 X^3 &= 0. \end{aligned} \tag{1-5*.9c}$$

determine a set of points for which only Y^3 varies. Since each of the equations represents a plane, the curve of intersection is a straight line.

The next theorem establishes the fact that an arbitrary pair of Cartesian coordinate systems is related by a linear transformation.

Theorem 1-5*.2. The coordinates, Y^j and $\bar{Y}^j$, of a pair of three-space Cartesian coordinate systems are related by linear transformations.

PROOF. Let X^j represent rectangular Cartesian coordinates such that

$$Y^j = B_k{}^j X^k, \qquad X^k = d_q{}^k \bar{Y}^q, \qquad B \neq 0, \qquad d \neq 0.$$

By a straightforward substitution

$$Y^j = B_k^{\ j} d_q^{\ k} \bar{Y}^q.$$

The products $B_k^{\ j} d_q^{\ k}$ give rise to nine constant coefficients of transformation which are denoted by $c_q^{\ j}$; that is,

(1-5*.10a) $$c_q^{\ j} = B_k^{\ j} d_q^{\ k}.$$

When this notational substitution is made in the preceding relation, we have

(1-5*.10b) $$Y^j = c_q^{\ j} \bar{Y}^q.$$

Since $d \neq 0$ and $B \neq 0$, the inverse transformation equations

(1-5*.10c) $$\bar{Y}^s = C_j^{\ s} Y^j, \qquad C_j^{\ s} = D_r^{\ s} b_j^{\ r}$$

can be uniquely determined. Therefore $c \neq 0$, which completes the proof.

We are able to introduce a basis into a Cartesian coordinate system by means of the relation (1-5*.4); that is,

(1-5*.10d) $$\mathbf{p}_j = b_j^{\ k} \boldsymbol{\iota}_k \qquad (\text{or } \bar{\mathbf{p}}_j = d_j^{\ k} \boldsymbol{\iota}_k).$$

For purposes of identification the basis introduced in this way is called a covariant basis. Incidentally, the reader should establish the fact that $\mathbf{p}_1$, $\mathbf{p}_2$, $\mathbf{p}_3$ are linearly independent (see Problem 9); otherwise the term basis would be inappropriate.

Theorem 1-5*.3. The covariant basis n-tuples $\mathbf{p}_j$ can be geometrically represented by arrows along the corresponding coordinate axes.

PROOF. We have

$$Y^j = B_k^{\ j} X^k.$$

If $Y^1 = Y^2 = 0$ and the conditions in (1-5*.9c) determining the Y^3 axis are multiplied by $B_3^{\ 2}$ and $B_3^{\ 1}$, respectively, then subtracted, we have

$$\frac{X^1}{B_3^{\ 1} B_2^{\ 2} - B_3^{\ 2} B_2^{\ 1}} = \frac{X^2}{B_3^{\ 1} B_1^{\ 1} - B_3^{\ 1} B_1^{\ 2}}.$$

A corresponding elimination of X^2 when combined with the preceding result leads to the symmetric form of the rectangular Cartesian representation of the Y^3 axis;

(1-5*.11)

$$\frac{X^1}{-(B_3^{\ 1} B_2^{\ 2} - B_3^{\ 2} B_2^{\ 1})} = \frac{X^2}{-(B_3^{\ 2} B_1^{\ 1} - B_3^{\ 1} B_1^{\ 2})} = \frac{X^3}{-(B_2^{\ 1} B_1^{\ 2} - B_2^{\ 2} B_1^{\ 1})}.$$

The direction numbers exhibited in (1-5*.11) are cofactors in the determinant $(B_j^{\ k})$ and therefore proportional to $b_3^{\ 1}$, $b_3^{\ 2}$, $b_3^{\ 3}$. (See 1-5*.3a.)

Therefore

$$\mathbf{p}_3 = b_3^{\ j}\boldsymbol{\iota}_j$$

can be represented along the Y^3 axis. The corresponding considerations for $\mathbf{p}_1$ and $\mathbf{p}_2$ are left to the reader.

Theorem 1-5*.4. Covariant bases $\mathbf{p}_j$ and $\bar{\mathbf{p}}_j$, associated with Cartesian coordinate systems satisfying (1-5*.10b, c), are related by

$$\text{(1-5*.12)} \qquad \begin{aligned} &\text{(a)} \quad \mathbf{p}_j = C_j^{\ q}\bar{\mathbf{p}}_q, \\ &\text{(b)} \quad \bar{\mathbf{p}}_q = c_q^{\ j}\mathbf{p}_j. \end{aligned}$$

PROOF. We have (see 1-5*.10a and 1-5*.10d)

$$\mathbf{p}_j = b_j^{\ k}\boldsymbol{\iota}_k = b_j^{\ k}D_k^{\ q}\bar{\mathbf{p}}_q = C_j^{\ q}\bar{\mathbf{p}}_q.$$

This completes the proof of (a); (b) is left to the reader.

Following the procedure of the example, we introduce a second basis in a Cartesian system Y^j by means of (1-5*.6); that is,

$$\mathbf{p}^j = \sum_{k=1}^{3} B_k^{\ j}\boldsymbol{\iota}_k.$$

This basis is called a contravariant basis. Consideration of the determinant of the components $B_k^{\ j}$ enables us to establish the linear independence of $\mathbf{p}^1$, $\mathbf{p}^2$, $\mathbf{p}^3$; therefore the term "basis" is justified. A procedure similar to that employed in the proof of Theorem 1-5*.3 leads to the relations

$$\text{(1-5*.13)} \qquad \begin{aligned} &\text{(a)} \quad \mathbf{p}^j = c_q^{\ j}\bar{\mathbf{p}}^q, \\ &\text{(b)} \quad \bar{\mathbf{p}}^q = C_j^{\ q}\mathbf{p}^j. \end{aligned}$$

In the introductory example we were able to determine that the bases $\mathbf{p}_j$ and $\mathbf{p}^j$ were reciprocal by calculating the appropriate dot products. The next theorem exhibits the generality of this reciprocity.

Theorem 1-5*.5. The covariant and contravariant bases, $\mathbf{p}_j$ and $\mathbf{p}^j$, are reciprocal; that is,

$$\text{(1-5*.14)} \qquad \mathbf{p}_j \cdot \mathbf{p}^k = \delta_j^{\ k}.$$

PROOF. The proof follows by employment of (1-5*.4) and (1-5*.6). We have

$$\mathbf{p}_j \cdot \mathbf{p}^k = b_j^{\ q}\boldsymbol{\iota}_q \cdot \sum_{r=1}^{3} B_r^{\ k}\boldsymbol{\iota}_r = \sum_{r=1}^{3} b_j^{\ q}B_r^{\ k}\delta_{qr} = b_j^{\ r}B_r^{\ k} = \delta_j^{\ k}.$$

If a basis $\boldsymbol{\iota}_1$, $\boldsymbol{\iota}_2$, $\boldsymbol{\iota}_3$ associated with a rectangular Cartesian system is considered, we find that it is self-reciprocal. In other words, a basis $\boldsymbol{\iota}^j$, introduced according to (1-5*.14), that is,

$$\iota_j \cdot \iota^k = \delta_j^k,$$

does not differ from the original basis ι_j.

The determination of covariant and contravariant bases for general Cartesian systems makes possible a natural interpretation of the vector concept for those systems.

We start with the definitions that follow. In the first it is assumed that triples of numbers (G^1, G^2, G^3), $(\bar{G}^1, \bar{G}^2, \bar{G}^3)$, etc., are specified in each Cartesian coordinate system. As before, the symbol $\{G^1, G^2, G^3\}$ denotes the collection of all such triples.

Definition 1-5*.2a. A contravariant vector is a collection $\{G^1, G^2, G^3\}$ of ordered triples, one triple associated with each Cartesian system, the elements of which satisfy the transformation law

$$G^j = \frac{\partial Y^j}{\partial \bar{Y}^k} \bar{G}^k. \tag{1-5*.15a}$$

The triples (G^1, G^2, G^3) etc., are called the contravariant components of a contravariant vector in the respective coordinate systems.

It is also assumed that ordered triples of functions (G_1, G_2, G_3), $(\bar{G}_1, \bar{G}_2, \bar{G}_3)$ can be associated with the respective Cartesian systems.

Definition 1-5*.2b. A covariant vector[30] is a collection $\{G_1, G_2, G_3\}$ of ordered triples, one triple associated with each Cartesian system, the elements of which satisfy the transformation law

$$G_j = \frac{\partial \bar{Y}^k}{\partial Y^j} \bar{G}_k. \tag{1-5*.15b}$$

The triples (G_1, G_2, G_3), etc., are called the covariant components of a covariant vector in the respective coordinate systems.

The use of subscripts to denote covariant components is adhered to throughout the text.

Since the laws of transformation of the coordinates are linear, that is,

$$Y^j = b_k^{\ j} \bar{Y}^k, \qquad \bar{Y}^k = B_j^{\ k} Y^j,$$

the sets of partial derivatives $\partial Y^j/\partial \bar{Y}^k$ and $\partial \bar{Y}^k/\partial Y^j$ are nothing more than the sets of constant coefficients $b_k^{\ j}$ and $B_j^{\ k}$, respectively. It should also be noted that the orthogonal Cartesian rotations are included as a subgroup of the more general group of transformations now under consideration.

---

[30] The terms contravariant, covariant, and invariant were introduced by the English mathematician James Joseph Sylvester (1814–1897).

The next two examples stipulate geometric interpretations of contravariant and covariant vectors and their components.

Example 1-5*.2a. Let G^1, G^2, G^3 be the components of a contravariant vector in a Cartesian coordinate system Y^j. Consider the form $G^j\mathbf{p}_j$. Transformation rules for the components G^j and the basis elements $\mathbf{p}_j$ have been formulated so that the form is invariant; that is, by employing (1-5*.12) and (1-5*.15a) we obtain the relation

$$G^j\mathbf{p}_j = c_k{}^j\bar{G}^k C_j{}^q\bar{\mathbf{p}}_q = \delta_k{}^q\bar{G}^k\bar{\mathbf{p}}_q = \bar{G}^k\bar{\mathbf{p}}_k. \tag{1-5*.16a}$$

In particular, with respect to a rectangular Cartesian system $\bar{X}^j$, we have

$$G^j\mathbf{p}_j = \bar{G}^j\bar{\boldsymbol{\iota}}_j = \bar{G}. \tag{1-5*.16b}$$

Therefore in a Euclidean metric space referred to affine coordinates the contravariant vector components can be interpreted as components of a Cartesian vector. This means that a geometric representation by means of an arrow (or family of arrows) is valid. We use this fact to find the geometric significance of the contravariant components with respect to a general Cartesian system. Suppose, for simplicity of geometric presentation, we consider a contravariant vector with components $(G^1, G^2, 0)$; then

$$\mathbf{G} = G^1\mathbf{p}_1 + G^2\mathbf{p}_2. \tag{1-5*.17a}$$

The basis n-tuples $\mathbf{p}_1$ and $\mathbf{p}_2$ are not necessarily of unit magnitude. Therefore we replace (1-5*.17a) by

$$\mathbf{G} = G^1\,|\mathbf{p}_1|\,\hat{\mathbf{p}}_1 + G^2|\mathbf{p}_2|\,\hat{\mathbf{p}}_2, \tag{1-5*.17b}$$ [31]

where $\hat{\mathbf{p}}_1$ and $\hat{\mathbf{p}}_2$ are of unit length.

The fact that $G^1\,|\mathbf{p}_1|$ and $G^2\,|\mathbf{p}_2|$ are measures of the parallel projections of the arrow representative of **G** on the Y^1 and Y^2 axes, respectively, follows immediately from the parallelogram law of addition. (See Fig. 1-5*.3a.) If the units of measurement along each axis are specified by $|\mathbf{p}_1|$ and $|\mathbf{p}_2|$, the components themselves, G^1 and G^2, represent parallel projections onto the Y^1 and Y^2 axes, respectively.

The preceding geometric interpretation of contravariant vector components is not dependent on two dimensions in any way; we let $G^3 = 0$ for simplicity of pictorial representation.

[31] A basis of unit n-tuples $\hat{\mathbf{p}}_1, \hat{\mathbf{p}}_2, \hat{\mathbf{p}}_3$ is called a physical basis. If we make the identifications $g_1 = |\mathbf{p}_1|\,G^1$, $g_2 = |\mathbf{p}_2|\,G^2$, $g_3 = |\mathbf{p}_3|\,G^3$, then g_1, g_2, and g_3 are called physical components of the vector **G**. These physical components are geometrically represented by parallel projections on the coordinate axis and are often used in making physical interpretations.

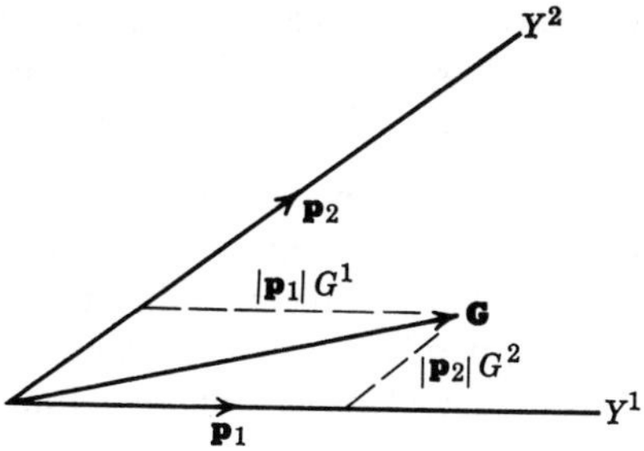

Fig. 1-5*.3a

Example 1-5*.2b. Suppose that G_1, G_2, G_3 are the components of a covariant vector. As a consequence of the transformation rules for G_j and $\mathbf{p}^j$, we obtain the invariance of the form $G_j\mathbf{p}^j$. The details are left to the reader. With respect to a rectangular Cartesian coordinate system, $\bar{X}^j$.

$$G_j\mathbf{p}^j = \bar{G}_j\bar{\mathbf{\iota}}^j = \sum_{j=1}^{3} \bar{G}_j\bar{\mathbf{\iota}}_j = \bar{\mathbf{G}};$$

that is, the contravariant basis in the rectangular system does not differ from the covariant basis. Therefore the components $\bar{G}^j$ are Cartesian vector components.

The components of a covariant vector with respect to a general Cartesian coordinate system are readily interpreted by the arrow representation of the Cartesian vector $\mathbf{G}$. Consider the form

$$\mathbf{G} = G_j\mathbf{p}^j.$$

If we take a dot product of the relation with $\mathbf{p}_k$,

(1-5*.18) $$\mathbf{p}_k \cdot \mathbf{G} = G_j\mathbf{p}_k \cdot \mathbf{p}^j = G_k.$$

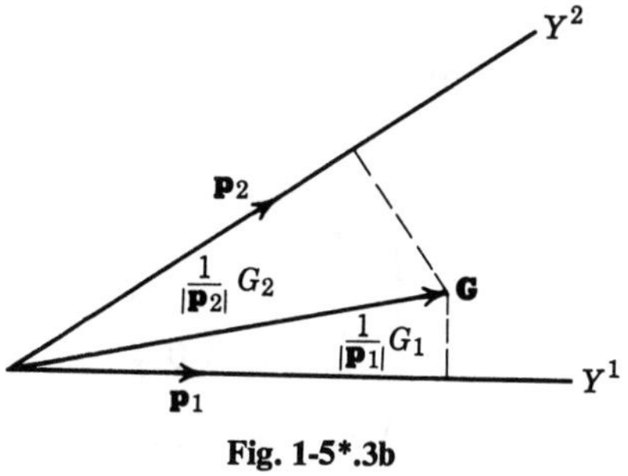

Fig. 1-5*.3b

In other words, the covariant components G_k are measures of the orthogonal projections of the arrow representative of **G** onto the coordinate axes when we measure in units of $1/|\mathbf{p}_1|$, $1/|\mathbf{p}_2|$ and $1/|\mathbf{p}_3|$, respectively.

Geometric interpretations of contravariant and covariant vectors are usually given in terms of projections on the covariant-basis arrows, since the coordinate axes correspond to them in direction. However, a fully analogous interpretation could be given with respect to the contravariant basis by an interchange of the association of parallel and orthogonal projections.

In a metric space (i.e., a space in which provision is made for measurement of distance and angle) the concepts of contravariant vector and covariant vector are not completely distinct.

A more detailed investigation of the relationship between contravariant and covariant entities is facilitated by the introduction of a matrix of numbers (h_{jk}). Later it will be found that the set h_{jk} has profound geometric significance.

Definition 1-5*.3. Let

(1-5*.19) $$h_{jk} = \mathbf{p}_j \cdot \mathbf{p}_k.$$

The numbers h_{jk} are easily expressed in terms of transformation coefficients relating the Cartesian coordinate system Y^j to a rectangular Cartesian system X^j. If

$$X^j = b_k^{\ j} Y^k,$$

we have

$$h_{jk} = \mathbf{p}_j \cdot \mathbf{p}_k = b_j^{\ q}\boldsymbol{\iota}_q \cdot b_k^{\ r}\boldsymbol{\iota}_r = \delta_{qr} b_j^{\ q} b_k^{\ r} = \sum_{q=1}^{3} b_j^{\ q} b_k^{\ q};$$

that is,

(1-5*.20a) $$h_{jk} = \sum b_j^{\ q} b_k^{\ q}.$$

Either from (1-5*.10) or (1-5*.20a) we can observe that the set h_{jk} is symmetric. In other words,

(1-5*.20b) $$h_{jk} = h_{kj}.$$

Example 1-5*.3. In Example 1-5*.1 the matrix of coefficients $(b_j^{\ k})$ is specifically

$$(b_j^{\ k}) = \begin{pmatrix} \frac{\sqrt{3}}{4} & \frac{1}{4} & 0 \\ \frac{1}{2} & \frac{\sqrt{3}}{2} & 0 \\ 0 & 0 & 1 \end{pmatrix}.$$

Therefore the matrix of h_{jk} is

$$(h_{jk}) = \begin{pmatrix} \frac{1}{4} & \frac{\sqrt{3}}{4} & 0 \\ \frac{\sqrt{3}}{4} & 1 & 0 \\ 0 & 0 & 1 \end{pmatrix}.$$

The numbers h_{jk} provide the algebraic link between the covariant and contravariant bases. The linear relation between the two is stipulated in the next theorem.

Theorem 1-5*.6. We have

$$\mathbf{p}_q = h_{qj}\mathbf{p}^j. \tag{1-5*.21}$$

PROOF. According to (1-5*.14),

$$\mathbf{p}^j \cdot \mathbf{p}_k = \delta_k^{\ j}.$$

Multiplying and summing each side of this expression with h_{qj} leads to

$$h_{qj}\mathbf{p}^j \cdot \mathbf{p}_k = h_{qj}\,\delta_k^{\ j} = h_{qk} = \mathbf{p}_q \cdot \mathbf{p}_k \tag{1-5*.22a}$$

or

$$(h_{qj}\mathbf{p}^j - \mathbf{p}_q) \cdot \mathbf{p}_k = 0. \tag{1-5*.22b}$$

The set of Cartesian vectors $\mathbf{p}_k$ is a linearly independent set. Therefore not all three of $\mathbf{p}_1$, $\mathbf{p}_2$, $\mathbf{p}_3$ can be perpendicular to the expression in parentheses. We must conclude that

$$h_{qj}\mathbf{p}^j - \mathbf{p}_q = \mathbf{0}.$$

This completes the proof.

By introducing a set of numbers inverse to the set h_{jk} we obtain a complete reciprocity.

Definition 1-5*.4. The numbers h^{jk} satisfy the relation

$$h^{jk}h_{kq} = \delta_q^{\ j}. \tag{1-5*.23}$$

Theorem 1-5*.7. We have

$$\mathbf{p}^s = h^{sq}\mathbf{p}_q. \tag{1-5*.24}$$

PROOF. Multiply and sum both members of (1-5*.21) with h^{sq}; then

$$h^{sq}\mathbf{p}_q = h^{sq}h_{qj}\mathbf{p}^j = \delta_j^{\ s}\mathbf{p}^j = \mathbf{p}^s.$$

This completes the proof.

In a Euclidean metric space covariant and contravariant vectors are not completely independent of one another. Every covariant vector gives rise to a contravariant vector and conversely. Corresponding covariant and contravariant vectors are equivalent to the same Cartesian vector. Suppose that G^1, G^2, G^3 are the components of a contravariant vector. Then

$$\mathbf{G} = G^j\mathbf{p}_j$$

represents a Cartesian vector referred to the covariant basis $\mathbf{p}_j$. The reciprocal or contravariant basis $\mathbf{p}^k$ can be introduced as indicated by (1-5*.21). We then have

$$\mathbf{G} = G^j h_{jk}\mathbf{p}^k.$$

This relation points out that linear combinations, $G^j h_{jk}$, of the components of the contravariant vector serve as components of a covariant vector. We formalize this fact by denoting the combinations with the same symbol G; that is,

$$G_k = h_{kj}G^j. \tag{1-5*.25a}$$

Note that it does not matter whether we write h_{jk} or h_{kj}, since the set is symmetric.

A process of multiplying and summing (1-5*.25a) along with the use of (1-5*.23) produces the form

$$G^j = h^{jk}G_k. \tag{1-5*.25b}$$

From this form it is inferred that each covariant vector gives rise to a contravariant vector. In the terminology usually employed in tensor analysis the components h_{jk} and h^{jk} are said to lower and raise indices, respectively. The precise meaning of this terminology, for example, with respect to (1-5*.25a), is that linear combinations of the contravariant components G^j are replaced by the covariant symbols G_k and that the coefficients of the linear combinations are the h_{kj}.

A Cartesian vector $\mathbf{G}$ can be written in either of the forms

$$\mathbf{G} = G^j\mathbf{p}_j = G_j\mathbf{p}^j.$$

It is this fact that initiated the statement that covariant and contravariant vectors are not entirely independent. This statement actually has meaning, not only with respect to a Euclidean metric space but whenever a metric is introduced into a space and the algebraic manipulations of raising and lowering indices by means of a set of numbers resulting from that metric are possible.

Example 1-5*.4. The identity of covariant and contravariant bases, $\boldsymbol{\iota}_j$ and $\boldsymbol{\iota}^j$, with respect to a rectangular Cartesian coordinate system, has already been pointed out. This fact can be symbolically expressed by writing

$$\boldsymbol{\iota}_j = \delta_{jk}\boldsymbol{\iota}^k.$$

Therefore we have

$$\mathbf{G} = G^j\boldsymbol{\iota}_j = G^j\,\delta_{jk}\boldsymbol{\iota}^k = G_k\boldsymbol{\iota}^k,$$

where

$$G_k = \delta_{jk}G^j = G^k.$$

In other words, the covariant and contravariant components do not differ in a rectangular Cartesian system.

Let us turn our attention to the sets of numbers (h_{jk}) and (h^{jk}). These matrices have played a part in our previous considerations. The following development points up their fundamental importance.

Theorem 1-5*.8a. With respect to transformations (1-5*.10a,b), the numbers h_{jk} transform according to the law

(1-5*.26) $$h_{jk} = C_j{}^pC_k{}^q\bar{h}_{pq}.$$

The proof, which follows from the definition of h_{jk} and (1-5*.12a), is left to the reader.

Theorem 1-5*.8b. With respect to transformations (1-5*.10a), the Kronecker delta, $\delta_j{}^k$, transforms according to the law

(1-5*.27) $$\delta_j{}^k = C_j{}^pc_q{}^k\,\bar{\delta}_p{}^q.$$

PROOF. Because of the inverse relation of the sets of transformation coefficients, the expression is merely an identity.

Theorem 1-5*.9. With respect to the transformations in (1-5*.10a), the numbers h^{jk} transform according to the law

(1-5*.28) $$h^{jk} = c_p{}^jc_q{}^k\bar{h}^{pq}.$$

PROOF. By definition we have

$$h_{jk}h^{jp} = \delta_k{}^p.$$

By using (1-5*.26) and (1-5*.27) this relation can be put in the form

(1-5*.29a) $$C_j{}^qC_k{}^r\bar{h}_{qr}h^{jp} = C_k{}^sc_t{}^p\,\bar{\delta}_s{}^t.$$

If (1-5*.29a) is multiplied with $c_v{}^k$ and summed,

$$C_j{}^q\,\bar{\delta}_v{}^r\bar{h}_{qr}h^{jp} = \bar{\delta}_v{}^sc_t{}^p\,\bar{\delta}_s{}^t$$

or

(1-5*.29b) $$C_j{}^q\bar{h}_{qv}h^{jp} = \bar{\delta}_v{}^sc_s{}^p.$$

Multiplying with $\bar{h}^{vk}$ and summing leads to

$$C_j^{\ k} h^{jp} = \bar{h}^{sk} c_s^{\ p}. \tag{1-5*.29c}$$

Application of the coefficients $c_k^{\ q}$ produces the result

$$h^{qp} = c_s^{\ p} c_k^{\ q} \bar{h}^{sk}.$$

This completes the proof.

Now that the transformation rules satisfied by the numbers h_{jk} and h^{jk} have been specified, we turn to the role that they play in describing the basic structure of the space.

Theorem 1-5*.10. The square of the magnitude of a Cartesian vector **G** is represented by any of the forms

$$\mathbf{G} \cdot \mathbf{G} = h_{jk} G^j G^k = h^{jk} G_j G_k = G_j G^j. \tag{1-5*.30}$$

The proof is left to the reader.

Theorem 1-5*.11. The cosine of the angle θ, determined by the positive senses of two Cartesian vectors **V** and **W**, is represented by any of the forms

$$\cos\theta = \frac{\mathbf{V} \cdot \mathbf{W}}{\sqrt{\mathbf{V} \cdot \mathbf{V}}\sqrt{\mathbf{W} \cdot \mathbf{W}}} = \frac{h_{jk} V^j W^k}{|\mathbf{V}|\,|\mathbf{W}|} = \frac{h^{jk} V_j W_k}{|\mathbf{V}|\,|\mathbf{W}|} = \frac{V^j W_j}{|\mathbf{V}|\,|\mathbf{W}|}. \tag{1-5*.31}$$

The proof is left to the reader.

The matrices (h_{jk}) and (h^{jk}) are instrumental in expressing the metric concepts of magnitude and angle. When considerations of non-Euclidean spaces are made, the metric ideas are defined in terms of these matrices. They provide the first examples of tensors of the second order.

Definition 1-5*.5a. The collection $\{h_{jk}\}$, the elements of which satisfy the transformation law (1-5*.26), is called the fundamental metric tensor with respect to the centered affine group. It is said to be of covariant order 2.

Definition 1-5*.5b. The collection $\{h^{jk}\}$, the elements of which satisfy the transformation law (1-5*.28), is called the associated metric tensor with respect to the centered affine group. It is said to be a tensor of contravariant order 2.

Definition 1-5*.5c. The collection $\{\delta_j^{\ k}\}$, the elements of which satisfy the transformation law (1-5*.27), is called a mixed tensor, with respect to the centered affine group, of contravariant order 1 and of covariant order 1.

Definitions 1-5*.5a,b,c serve as a model for the general tensor definition. Although it is not the present purpose to pursue the idea in any greater depth, the general definition is presented for completeness.

It is assumed that the Cartesian coordinates y^j and $\bar{y}^j$ are related by means of the transformation equations

$$y^j = c_k{}^j \bar{y}^k.$$

Furthermore, suppose that a set of numbers, $T^{k_1 \cdots k_q}_{j_1 \cdots j_p}$, $\bar{T}^{k_1 \cdots k_q}_{j_1 \cdots j_p}$, respectively, is associated with each Cartesian coordinate system.

Definition 1-5*.6. The collection $\{T^{k_1 \cdots k_q}_{j_1 \cdots j_p}\}$, the elements of which satisfy the transformation law

$$T^{k_1 \cdots k_q}_{j_1 \cdots j_p} = c^{k_1}_{r_1} \cdots c^{k_q}_{r_q} C^{s_1}_{j_1} \cdots C^{s_p}_{j_p} \bar{T}^{r_1 \cdots r_q}_{s_1 \cdots s_p} \tag{1-5*.32}$$

is said to be a mixed tensor with respect to the centered affine group of contravariant order q and covariant order p.

Problems

1. Prove (b) of (1-5*.12).
2. Suppose that a Cartesian vector $\mathbf{G}$ is represented in terms of the basis $\mathbf{p}_1$ and $\mathbf{p}_2$ of Example 1-5*.1. Show that $|\mathbf{G}| = 1$ if $G^1 = 2$ and $G^2 = 0$.
3. (a) Show that the set of n-tuples $\mathbf{p}_1$, $\mathbf{p}_2$, $\mathbf{p}_3$ is linearly independent.
   (b) Show that the set of n-tuples $\mathbf{p}^1$, $\mathbf{p}^2$, $\mathbf{p}^3$ is linearly independent.
4. (a) Construct a matrix of coefficients $(B_j{}^k)$ such that $B = 1$ but the transformation is not rectangular Cartesian.
   (b) Find $(b_j{}^k)$ and b with respect to the matrix of (a).
5. (a) Suppose that the X^j coordinates are rectangular Cartesian and that the Y^j coordinates corresponded to a general Cartesian system. (See the accompanying diagram.) Let $Y^3 = X^3$, thereby reducing the problem to one in two dimensions. Show that if $\mathbf{p}_1$, $\mathbf{p}_2$, and $\mathbf{p}_3$ are of unit length

$$X^1 = Y^1 \cos \alpha + Y^2 \cos \beta,$$

$$X^2 = Y^1 \sin \alpha + Y^2 \sin \beta,$$

$$X^3 = Y^3,$$

   where α is measured counterclockwise from X^1 to Y^1, whereas β is measured counterclockwise from X^1 to Y^2.
   Hint: Start with $X^j = b_k{}^j Y^k$ and also use $\mathbf{p}_j = B_j{}^k \boldsymbol{\iota}_k$.

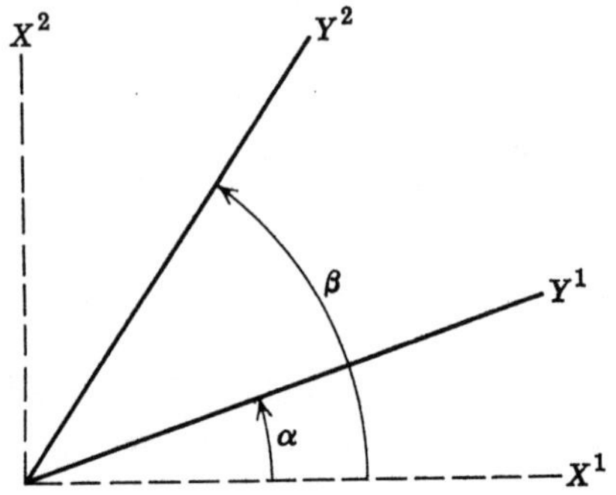

(b) Show that the transformation equations inverse to those of (a) are

$$Y^1 = \frac{X^1 \sin\beta - X^2 \cos\beta}{\sin(\beta - \alpha)},$$

$$Y^2 = \frac{-X^1 \sin\alpha + X^2 \cos\alpha}{\sin(\beta - \alpha)},$$

$$Y^3 = X^3.$$

(c) Write out the equations in (a) and (b) if $\alpha = \pi/6, \quad \beta = \pi/3$.

(d) If α and β are given as in (a), write out the expressions for $\mathbf{p}_1$, $\mathbf{p}_2$ and $\mathbf{p}^1$, $\mathbf{p}^2$ in terms of $\boldsymbol{\iota}_1$, $\boldsymbol{\iota}_2$, $\boldsymbol{\iota}_3$.

6. Let X^j represent rectangular Cartesian coordinates. Suppose that

$$X^1 = -\sqrt{3}\,Y^1 + Y^2,$$
$$X^2 = Y^1 - Y^2,$$
$$X^3 = Y^3.$$

(a) Show that $b = \sqrt{3} - 1$

(b) Show that

$$Y^1 = -\left(\frac{1 + \sqrt{3}}{2}\right)X^1 - \left(\frac{1 + \sqrt{3}}{2}\right)X^2,$$

$$Y^2 = -\left(\frac{1 + \sqrt{3}}{2}\right)X^1 - \left(\frac{3 + \sqrt{3}}{2}\right)X^2,$$

$$Y^3 = X^3.$$

(c) Show that $B = 1/(\sqrt{3} - 1)$, hence $Bb = 1$.

(d) Show that $\mathbf{p}_1 = -\sqrt{3}\,\boldsymbol{\iota}_1 + \boldsymbol{\iota}_2$,

$\mathbf{p}_2 = \boldsymbol{\iota}_1 - \boldsymbol{\iota}_2$,

$\mathbf{p}_3 = \boldsymbol{\iota}_3$.

7. What form does the formula for the distance between two points take in the Cartesian system Y^j of Problem 5?

8. (a) Show that the set of rectangular Cartesian rotations form a group.

(b) Is the set of all translations a group?

9. Show that

(a) $$h_{jk} = \sum_{p=1}^{3} \frac{\partial X^p}{\partial Y^j} \frac{\partial X^p}{\partial Y^k},$$

(b) $$h^{jk} = \sum_{p=1}^{3} \frac{\partial Y^j}{\partial X^p} \frac{\partial Y^k}{\partial X^p},$$

(c) $$h^{jk} = h^{kj}.$$

10. Suppose that the X^j coordinates are rectangular Cartesian and that they are related to general Cartesian coordinates Y^j by means of the transformation equations

$$X^1 = 2Y^1, \qquad X^2 = 3Y^2, \qquad X^3 = 4Y^3.$$

Determine
(a) whether $\mathbf{p}_1$, $\mathbf{p}_2$, $\mathbf{p}_3$ are of unit length,
(b) expressions for $\mathbf{p}^1$, $\mathbf{p}^2$, $\mathbf{p}^3$ in terms of ι_1, ι_2, ι_3,
(c) the metric tensor components (h_{jk}),
(d) the components (h^{jk}).

11. Determine general expressions for the magnitude of a vector $\mathbf{A}$ in the coordinate systems of Problems 5 and 6.

12. Show that A^jA_j is a scalar with respect to the centered affine group.

13. With respect to transformations (1-5*.9a), show that
(a) $\bar{G}^k = B_j{}^k G^j$. *Hint:* Start with (1-5*.15b).
(b) $h^{jk} = B_p{}^j B_q{}^k h^{pq}$.

6. $\mathcal{E}$ *Systems and Determinants*

Besides vector addition and the inner product there is one other binary operation that is a significant part of the algebraic structure of vector analysis—the vector product or cross product. Development of its fundamental properties is facilitated by using certain notational devices and by a firm understanding of determinant theory. Therefore this section is devoted to the introduction of these underlying notions.

The major notational innovation, the so-called $\mathcal{E}$ systems, or indicators, is introduced in the following definition.

Definition 1-6.1.[32]

$$(1\text{-}6.1) \qquad E^{ijk} = \mathcal{E}_{ijk} = \begin{cases} 1 \\ -1 \\ 0 \end{cases} \text{if } i, j, k \text{ is an } \begin{cases} \text{even,} \\ \text{odd permutation of 1, 2, 3,} \\ \text{otherwise.} \end{cases}$$

[32] The use of two symbols E^{ijk} and $\mathcal{E}_{ijk}$ to represent the same set of numbers has no significance at this point other than, perhaps, facilitating the use of the summation notation. In Section 7* it is shown that E^{ijk} and $\mathcal{E}^{ijk}$ are not equal, and therefore use of two symbols avoids an inconsistency.

The term otherwise refers to a repetition of a number. For those not familiar with the terminology of the definition, the following example may be of some help.

Example 1-6.1. Consider the set of numbers 1, 2, 3. These three numbers, used without repetition, may be arranged in six ways:

$$\begin{matrix} 1, 2, 3 & 2, 3, 1 & 3, 1, 2 \\ 2, 1, 3 & 1, 3, 2 & 3, 2, 1 \end{matrix}$$

Any of the first three arrangements can be put in the order of 1, 2, 3 by an even number of interchanges of adjacent elements. Hence these are called the even permutations of 1, 2, 3. The last three combinations, which can be put in the order 1, 2, 3 by an odd number of adjacent interchanges, are called odd permutations of 1, 2, 3. Since the triple of symbols i, j, k can be freely replaced by the numbers 1, 2, 3, it is also the case that a combination involving repetition, such as 1, 1, 3, might appear. The expression otherwise refers to this situation. In particular,

$$\mathcal{E}_{123} = \mathcal{E}_{231} = \mathcal{E}_{312} = 1$$
$$\mathcal{E}_{213} = \mathcal{E}_{132} = \mathcal{E}_{321} = -1,$$

whereas all other symbols have the value zero. An important property of the $\mathcal{E}$ systems, and one that makes them especially valuable for determinant representation, is that of skew symmetry in any adjacent pair of indices. This is the property illustrated by

$$\mathcal{E}_{ijk} = -\mathcal{E}_{jik}.$$

Corresponding remarks can be made with respect to the system E^{ijk}, which is introduced in order to facilitate use of the summation notation. The purpose of a different kernel letter (i.e., E instead of $\mathcal{E}$) is discussed in Section 7*.

If one of the indices in (1-6.1) is suppressed, we define in an analogous way

$$E^{jk} = \mathcal{E}_{jk} = \begin{cases} 1 & \\ -1 & \text{if } j, k \text{ represents} \\ 0 & \end{cases} \begin{cases} 1, 2, \\ 2, 1, \\ \text{otherwise.} \end{cases} \tag{1-6.2}$$

This simplified version is employed in an example illustrating the usage of indicators in determinant representation.

Example 1-6.2. Consider the second-order determinant

$$|a_j^{\ k}| = \begin{vmatrix} a_1^{\ 1} & a_1^{\ 2} \\ a_2^{\ 1} & a_2^{\ 2} \end{vmatrix} = a_1^{\ 1}a_2^{\ 2} - a_1^{\ 2}a_2^{\ 1}.$$

This determinant can be represented by either of the forms

$$(1\text{-}6.3)\qquad \begin{aligned} &\text{(a)}\quad a = |a_j^{\ k}| = \mathcal{E}_{jk} a_1^{\ j} a_2^{\ k}, \\ &\text{(b)}\quad a = E^{jk} a_j^{\ 1} a_k^{\ 2}. \end{aligned}$$

The validity of the statement can be verified by writing out the right-hand members of the equations. Furthermore, we can write

$$(1\text{-}6.4)\qquad \begin{aligned} &\text{(a)}\quad \mathcal{E}_{pq} a = \mathcal{E}_{jk} a_p^{\ j} a_q^{\ k}, \\ &\text{(b)}\quad E^{pq} a = E^{jk} a_j^{\ p} a_k^{\ q}. \end{aligned}$$

If $p = 1$, $q = 2$, then the factor $\mathcal{E}_{12} = 1$ is not detectable and (1-6.4a) reduces to the form (1-6.3a). If $p = 2$, $q = 1$, the left member of (1-6.4a) contains the factor $\mathcal{E}_{21} = -1$. However, the right-hand member of the equation is expressed in a way that corresponds to the interchange of two rows; that is, a change of sign is introduced in the right-hand member also. Similar remarks can be made about the relation (1-6.4b).

It is interesting to note that the determinant[33] concept played an outstanding role in the mathematics of the eighteenth and nineteenth centuries. The names of many famous mathematicians appear in a historical development of the theory. Leibniz (1646–1716, German), who originated the concept, Cramer (1704–1752, Swiss), and Bezout (1730–1783, French) set forth rules for solving simultaneous linear equations which touched on the determinant idea. Improved notations and certain useful identities were introduced by Vandermonde (1735–1796, French) and Lagrange. The structure of determinant theory was completed by the detailed work of Jacobi (1804–1851, German), Cayley,[34] Sylvester, and others. Felix Klein[35] credits Cayley with having said that if he had fifteen lectures to devote to mathematics he would devote one of them to determinants. Klein's own opinion of the place of determinant theory in the field of mathematics was not so high, but he did feel that they were vital in general considerations and as a part of the theory of invariants.

The popularity of tensor algebra, brought about by the advent of relativity theory, put in the foreground a notation that in many ways made trivial the great body of theory that had been developed. This notation, which includes the concepts of summation convention and $\mathcal{E}$ systems, is used in order to put at our disposal the fundamental facts of determinant theory.

---

[33] The name is due to Cauchy.

[34] The symbolizing of a determinant by a square array with bars about it is the handiwork of Cayley.

[35] Klein, *op. cit.*, p. 143.

Note that the $\mathcal{E}$ systems are nicely designed to fit the definition of a determinant. If the order of $|a_j^k|$ is p, then we expect its expansion to be a sum of $p!$ terms, each a product of p factors. Exactly one element from each row and one element from each column should appear as a factor in every term. For example, if the superscripts are kept in the order $1 \cdots p$, then each term involving an even permutation of $1 \cdots p$ with respect to the subscripts is prefixed with a plus sign. If there is an odd permutation, a minus sign is appended.

It is clear from the mode of presentation of these notes that some familiarity with the determinant concept has been assumed. However, if the reader has had no previous experience with determinants, the following formal definition can serve as a starting point. For simplicity we restrict ourselves to third-order determinants. However, the transition to the general theory can pretty much be made by replacing 3 by p.

Definition 1-6.2. A determinant of order 3 is a three-by-three array of elements.

(1-6.5)
$$|a_j^k| = \begin{vmatrix} a_1^{\ 1} & a_1^{\ 2} & a_1^{\ 3} \\ a_2^{\ 1} & a_2^{\ 2} & a_2^{\ 3} \\ a_3^{\ 1} & a_3^{\ 2} & a_3^{\ 3} \end{vmatrix},$$

with which a numerical value a is associated.

The subscripts indicate row and the superscripts indicate column. The numerical value a is obtained as follows:

(1-6.6)
$$\mathcal{E}_{ijk}a = \mathcal{E}_{pqr}a_i^{\ p}a_j^{\ q}a_k^{\ r}.$$

Example 1-6.3. The third-order determinant (1-6.5) has the value

(1-6.7a)
$$\begin{aligned}\mathcal{E}_{123}a = a = \mathcal{E}_{123}a_1^{\ 1}a_2^{\ 2}a_3^{\ 3} &+ \mathcal{E}_{231}a_1^{\ 2}a_2^{\ 3}a_3^{\ 1} \\ &+ \mathcal{E}_{312}a_1^{\ 3}a_2^{\ 1}a_3^{\ 2} + \mathcal{E}_{213}a_1^{\ 2}a_2^{\ 1}a_3^{\ 3} \\ &+ \mathcal{E}_{132}a_1^{\ 1}a_2^{\ 3}a_3^{\ 2} + \mathcal{E}_{321}a_1^{\ 3}a_2^{\ 2}a_3^{\ 1};\end{aligned}$$

that is

(1-6.7b)
$$\begin{aligned}a = a_1^{\ 1}a_2^{\ 2}a_3^{\ 3} &+ a_1^{\ 2}a_2^{\ 3}a_3^{\ 1} + a_1^{\ 3}a_2^{\ 1}a_3^{\ 2} \\ &- (a_1^{\ 2}a_2^{\ 1}a_3^{\ 3} + a_1^{\ 1}a_2^{\ 3}a_3^{\ 2} + a_1^{\ 3}a_2^{\ 2}a_3^{\ 1}).\end{aligned}$$

The student who is unfamiliar with the determinant concept will profit by letting $i, j, k = 2, 1, 3$ and then writing out (1-6.6).

The determinant representation (1-6.6) is by no means unique. A second representation in terms of E^{ijk} is introduced by the following theorem.

Theorem 1-6.1. A third-order determinant $|a_j^{\ k}|$ has the numerical representation

(1-6.8) $$E^{ijk}a = E^{pqr}a_p^{\ i}a_q^{\ j}a_r^{\ k}.$$

PROOF. The order of superscripts i, j, k in (1-6.8) represents a chosen column arrangement. To ascertain the equivalence of (1-6.8) and (1-6.6), we can write out (1-6.6), as in Example 1-6.3, and then permute the factors of each term until the column indices are in the order i, j, k. Corresponding permutations of subscripts occur simultaneously. Examination reveals that the form (1-6.8) has been obtained.

Some of the simpler properties of determinants are implicit in the notational representation. These properties are stated in the following theorem.

Theorem 1-6.2.

(1-6.9a) If all of the elements of any row (or column) of a determinant are zero, the value of the determinant is zero.

(1-6.9b) If all of the elements of a given row (or column) are multiplied by the same factor, then the value of the new determinant is that of the original multiplied by this factor.

(1-6.9c) An even number of interchanges of rows (or columns) of a determinant produces a new determinant whose value is the same as that of the original. If an odd number of inter-changes of rows (or columns) is made, the value of the resulting determinant differs from the original by a factor (-1).

(1-6.9d) If multiples of the elements of any row (column) are added to corresponding elements of another row (column), the new determinant has the same value as the original.

PROOF. (1-6.9a) is an immediate consequence of the fact that every term of the determinant expansion [see (1-6.6)] contains an element from each column and each row.

(1-6.9b) follows from the fact that exactly one element from each row and each column appears in each term of the expansion.

To prove (1-6.9c) for columns, consider (1-6.8). Interchange of columns corresponds to interchange of the superscripts i, j, k. The $\mathcal{E}$ system is such that under an even permutation of i, j, k the sign remains the same, whereas with respect to an odd permutation it changes. Since the value of the determinant is represented by $E^{ijk}a$, the proof is complete. A corresponding proof for rows follows from (1-6.6).

Proofs of (1-6.9b,c) are left to the reader. To prove (1-6.9d) with respect to rows, consider (1-6.6). We have

$$\mathcal{E}_{ijk}a = \mathcal{E}_{pqr}a_i^{\ p}a_j^{\ q}a_k^{\ r}.$$

Suppose that a multiple, $\beta a_j^{\ p}$, of the jth row were added to the ith row; the form corresponding to the right-hand side of (1-6.6) is then

$$\mathcal{E}_{pqr}(a_i^{\ p} + \beta a_j^{\ p})a_j^{\ q}a_k^{\ r},$$
$$\mathcal{E}_{pqr}a_i^{\ p}a_j^{\ q}a_k^{\ r} + \beta\mathcal{E}_{pqr}a_j^{\ p}a_j^{\ q}a_k^{\ r}.$$

Because of the skew symmetry of the $\mathcal{E}$ system in p, q, the term

(1-6.10) $$\mathcal{E}_{pqr}a_j^{\ p}a_j^{\ q}a_k^{\ r} = 0.$$

(See Problem 1, Section 6.) This completes the proof.

The $\mathcal{E}$ systems when subjected to summation and multiplication with one another have some interesting and useful properties. They are presented in the following theorem.

Theorem 1-6.3. $\mathcal{E}$ systems of order 3 satisfy the properties

(1-6.11)
$$\text{(a)}\ E^{ijk}\mathcal{E}_{ipq} = 2\delta_{pq}^{[jk]},$$
$$\text{(b)}\ E^{ijk}\mathcal{E}_{ijq} = 2\delta_q^{\ k};$$
$$\text{(c)}\ E^{ijk}\mathcal{E}_{ijk} = 3!$$

where by definition

$$\delta_{pq}^{[jk]} = \tfrac{1}{2}(\delta_p^{\ j}\,\delta_q^{\ k} - \delta_p^{\ k}\,\delta_q^{\ j}).$$

(See Problem 1, Section 3.)

PROOF. All three proofs can be made by examining the validity of the equalities for the various possible values of the indices. For example, to prove (1-6.11a), we can write out the left- and right-hand sides as follows:

(1-6.12)
$$\text{(a)}\ E^{ijk}\mathcal{E}_{ipq} = E^{1jk}\mathcal{E}_{1pq} + E^{2jk}\mathcal{E}_{2pq} + E^{3jk}\mathcal{E}_{3pq}.$$
$$\text{(b)}\quad 2\delta_{pq}^{[jk]} = \delta_p^{\ j}\,\delta_q^{\ k} - \delta_p^{\ k}\,\delta_q^{\ j}.$$

In both (1-6.12a,b) j, k must be a distinct pair of numbers chosen from 1, 2, 3, otherwise the expressions will have the value zero. The same statement holds for the pair p, q. Furthermore, the pairs j, k and p, q must take on the same numbers, but not necessarily in the same order, or again both (1-6.12a,b) will have the value zero. To illustrate a nonzero evaluation of expressions (1-6.12a,b) let

$$j = 1, \qquad k = 2, \qquad p = 2, \qquad q = 1.$$

Then

$$E^{ijk}\mathcal{E}_{ipq} = -1.$$
$$2\delta_{pq}^{[jk]} = -1.$$

The proof can be completed by considering the other possible choices of number pairs.

The properties of the $\mathcal{E}$ systems stated in the last theorem facilitate consideration of determinant multiplication. A determinant whose elements are constants has a specific numerical value. Of course, there is no problem involved in multiplying two such numerical values together. Therefore the task before us is that of finding a third determinant whose numerical value is the same as the product of numerical values of two given determinants. This third determinant arises from the first two by a process that involves both multiplication and summation. We arbitrarily designate the operation by the name multiplication. There is no claim that this operation, introduced by the next theorem, is unique. In fact, we shall subsequently find that it is not.

Theorem 1-6.4. Let $|a_j^k|$ and $|b_j^k|$ be given determinants of order 3. A determinant $|c_j^k|$ whose elements arise from those of $|a_j^k|$ and $|b_j^k|$ by means of

(1-6.13a) $$c_i^j = a_i^k b_k^j$$

has the numerical value

(1-6.13b) $$c = ab.$$

PROOF. According to the relations in (1-6.8) and (1-6.6), we have

(1-6.14)
$$\text{(a)}\quad E^{ijk} a_i^p a_j^q a_k^r = aE^{pqr},$$
$$\text{(b)}\quad \mathcal{E}_{def} b_p^d b_q^e b_r^f = b\mathcal{E}_{pqr}.$$

Formation of the sum of the products in (1-6.14a,b) results in

(1-6.14c) $$E^{ijk}\mathcal{E}_{def}(a_i^p b_p^d)(a_j^q b_q^e)(a_k^r b_r^f) = ab3!$$

When the identification (1-6.13a) is used, (1-6.14c) takes the form

(1-6.14d) $$E^{ijk}(\mathcal{E}_{def} c_i^d c_j^e c_k^f) = ab3!$$

According to (1-6.6), this relation is equivalent to

$$E^{ijk}\mathcal{E}_{ijk} c = ab3!$$

Therefore employment of (1-6.11c) produces the result (1-6.13b), as was to be shown.

Definition 1-6.3. A determinant $|c_j^k|$ whose elements arise from those of $|a_j^k|$ and $|b_j^k|$ by means of the rules (1-6.13a) is called the product determinant of the determinants $|a_j^k|$ and $|b_j^k|$.

Example 1-6.4. Determinant multiplication is characterized according to (1-6.13a) by a row-by-column multiplication.

$$\begin{vmatrix} 1 & 3 & 5 \\ 2 & 4 & 1 \\ 3 & 1 & 2 \end{vmatrix} \begin{vmatrix} -2 & 5 & 4 \\ 1 & 2 & 3 \\ 1 & 4 & 1 \end{vmatrix} = \begin{vmatrix} 6 & 31 & 18 \\ 1 & 22 & 21 \\ -3 & 25 & 17 \end{vmatrix}. \tag{1-6.15a}$$

The reader can evaluate these determinants and observe that the numerical values satisfy the relation

$$ab = c.$$

The numerical values are just real numbers, hence they commute; that is,

$$ba = c.$$

However, it is of some interest to note that the product $|b_j^{\ k}|\,|a_j^{\ k}|$ does not produce the same square array as $|a_j^{\ k}|\,|b_j^{\ k}|$. In particular,

$$\begin{vmatrix} -2 & 5 & 4 \\ 1 & 2 & 3 \\ 1 & 4 & 1 \end{vmatrix} \begin{vmatrix} 1 & 3 & 5 \\ 2 & 4 & 1 \\ 3 & 1 & 2 \end{vmatrix} = \begin{vmatrix} 20 & 18 & 3 \\ 14 & 14 & 13 \\ 12 & 20 & 11 \end{vmatrix} \tag{1-6.15b}$$

Of course, the numerical value of the right member of (1-6.15b) is the same as that of the right-hand member of (1-6.15a). This noncommutative aspect of determinant multiplication will come to our attention again in Section 8 where matrix theory is discussed.

We have consistently emphasized the transformation concept that underlies many aspects of vector and tensor analysis. We shall consider the behavior of the systems $\mathcal{E}_{ijk}$ with respect to the orthogonal Cartesian group. Similar considerations can be made for the systems E^{ijk}; however, they will be deferred until Section 7*, since there are no distinctions between the contravariant and covariant forms in rectangular Cartesian systems. (See Section 5*.)

We assume the existence of sets of numbers $\mathcal{E}_{ijk}$, $\bar{\mathcal{E}}_{ijk}$, etc., in corresponding rectangular Cartesian systems X^j, $\bar{X}^j$, etc. Each set is defined by (1-6.1). The following theorem establishes the way in which two such sets are related when the coordinate relation is

$$X^j = a_k^{\ j}\bar{X}^k.$$

Theorem 1-6.5. We have

$$\mathcal{E}_{ijk} = aA_i^{\ p}A_j^{\ q}A_k^{\ r}\bar{\mathcal{E}}_{pqr}, \qquad a = 1. \tag{1-6.16}$$

PROOF. On the right-hand side of (1-6.16) we have

$$aA_i{}^pA_j{}^qA_k{}^r\bar{\mathcal{E}}_{pqr} = aA\bar{\mathcal{E}}_{ijk} = \bar{\mathcal{E}}_{ijk};$$

that is, the relation (1-6.16) is simply an identity. This completes the proof.

The determinants A and a both have the value 1, so that the inclusion of the symbol a in relation (1-6.16) might be questioned. The purpose of carrying the symbol is to provide consistency with those considerations involving a transformation group of more generality than the orthogonal Cartesian group and to identify properly the nature of the $\mathcal{E}$ systems.

Definition 1-6.4. The class of $\mathcal{E}$ systems $\{\mathcal{E}_{ijk}\}$, the elements of which satisfy the transformation rule (1-6.16), is said to be a covariant tensor density of order 3 and weight $+1$ with respect to the rectangular Cartesian group of transformations.

In one or two spots before this point in the book the concept of a cofactor of an element in a determinant has been used in numerical calculations. The foundation of determinant theory set forth in this section makes it possible to formulate algebraically the cofactor concept.

Definition 1-6.5. The cofactor $D_i{}^r$ of $d_r{}^i$ in the determinant $|d_r{}^i|$ is given by

(1-6.17) $$\mathcal{E}_{rpq}\, D_i{}^r = \mathcal{E}_{ijk}\, d_p{}^j\, d_q{}^k.$$

Example 1-6.5a. Consider the expression (1-6.17) Suppose we were interested in the cofactor of $d_2{}^3$. The indices p, q, and i are free and therefore may be chosen. In order to obtain $D_3{}^2$, pick $i = 3$, $p = 3$, and $q = 1$. The relation (1-6.17) reduces to

$$\mathcal{E}_{231}\, D_3{}^2 = \mathcal{E}_{312}\, d_3{}^1\, d_1{}^2 + \mathcal{E}_{321}\, d_3{}^2\, d_1{}^1,$$

or, upon evaluating the $\mathcal{E}$ symbols,

$$D_3{}^2 = d_3{}^1\, d_1{}^2 - d_3{}^2\, d_1{}^1.$$

Example 1-6.5b. Consider the set of simultaneous linear homogeneous equations

(1-6.18) $$d_t{}^k X^t = 0.$$

This set of equations has the trivial solution (0, 0, 0). If $|d_t{}^k| \neq 0$, the trivial solution is unique. If $|d_t{}^k| = 0$, then, in general, the equations have nontrivial solutions. In particular, they are satisfied by $D_i{}^1$, $D_i{}^2$, $D_i{}^3$, that is, the cofactors of the elements $d_1{}^i$, $d_2{}^i$, and $d_3{}^i$ of any column. (See Section 2.) The veracity of this remark can be demonstrated as follows.

By multiplying and summing (1-6.17) with E^{tpq}, we have

$$D_i^{\ t} = \tfrac{1}{2}E^{tpq}\mathcal{E}_{ijk}\, d_p^{\ j}\, d_q^{\ k}.$$

The cofactors $D_i^{\ t}$ can be put in place of the X^t in the left side of (1-6.18)

$$\begin{aligned} d_t^{\ k}\, D_i^{\ t} &= \tfrac{1}{2}d_t^{\ k}E^{tpq}\mathcal{E}_{ijs}\, d_p^{\ j}\, d_q^{\ s} \\ &= \tfrac{1}{2}\mathcal{E}_{ijs}(E^{tpq}\, d_t^{\ k}\, d_p^{\ j}\, d_q^{\ s}). \end{aligned}$$

Since the parenthetic expression represents the value of the determinant, and that has been assumed to be zero, it follows that the cofactors satisfy the system of equations.

Example 1-6.6. Under certain linear transformations of coordinates the components P^j of a vector remain fixed or change at most by a common factor of proportionality. The algebraic conditions that specify this behavior are

(1-6.19a) $$\lambda p^j = c_k^{\ j} p^k.$$

The values of λ which satisfy this relation are commonly called eigenvalues of $c_k^{\ j}$ and the corresponding vectors P^k are called eigenvectors.

In order to find the values of λ which satisfy (1-6.19a), we rewrite the set in the form

$$\lambda\delta_k^{\ j}P^k = c_k^{\ j}P^k$$

or

(1-6.19b) $$(c_k^{\ j} - \lambda\delta_k^{\ j})P^k = 0.$$

As indicated in the discussion of Example 1-6.5b, this system of equations has nontrivial solutions (P^1, P^2, P^3) if and only if

(1-6.19c) $$|c_k^{\ j} - \lambda\delta_k^{\ j}| = 0.$$

Let us proceed under the assumption that the determinant is zero. Then it can be written in the form

(1-6.19d) $$\mathcal{E}_{ijk}E^{pqr}(c_p^{\ i} - \lambda\delta_p^{\ i})(c_q^{\ j} - \lambda\delta_q^{\ j})(c_r^{\ k} - \lambda\delta_r^{\ k}) = 0.$$

By expanding this equation by multiplying out the parenthetic expressions and using the multiplicative properties of the $\mathcal{E}$ systems, we obtain the third-order equation in λ.

(1-6.19e) $$\lambda^3 - c_s^{\ s}\lambda^2 + c_s^{\ [s}c_t^{\ t]}\lambda - c = 0,$$

where as previously indicated the brackets imply $c_s^{\ [s}c_t^{\ t]} = \tfrac{1}{2}(c_s^{\ s}c_t^{\ t} - c_s^{\ t}c_t^{\ s})$. This is called the characteristic equation. The eigenvector P^j corresponding to the eigenvalues λ obtained from this equation can be specified as cofactors in the determinant (1-6.19c), since such cofactors satisfy the set (1-6.19b).

Problems

1. Prove (1-6.10). *Hint:* Show that the expression is equal to its negative by using the facts that p, q are dummy indices and skew symmetric on $\mathcal{E}_{pqr}$; that is, $\mathcal{E}_{pqr} = -\mathcal{E}_{qpr}$.

2. Complete the proof of Theorem 1-6.2.

3. Complete the proof of Theorem 1-6.3.

4. (a) Suppose that $\mathcal{E}_{ijkp}$ were defined in analogy to (1-6.1). How many nonzero components does the symbol possess? (i, j, k, p taken from the set 1, 2, 3, 4.)

(b) What can be said if i, j, k, p take values from the set 1, 2, 3?

5. By writing out the summations involved, show that

$$E^{jk} = \delta^{jp}\,\delta^{kq}\mathcal{E}_{pq}.$$

6. Show that the following representation is valid for a third-order determinant $|a_j^{\ k}|$:

$$a = \frac{1}{3!}\,\mathcal{E}_{ijk}E^{pqr}a_p^{\ i}a_q^{\ j}a_r^{\ k}.$$

7. Show that a skew symmetric ($a_j^{\ k} = -a_k^{\ j}$) determinant of order 3 has the value zero.

8. Generalize the representations (1-6.6) and (1-6.8) to determinants of order p.

9. (a) How can the representation in Problem 6 be generalized to higher order determinants?

(b) Can Problem 7 be generalized?

10. (a) Multiply the determinants

$$\begin{vmatrix} 1 & 3 & -2 \\ 5 & 1 & 6 \\ 4 & -2 & 1 \end{vmatrix}\begin{vmatrix} 2 & 1 & 5 \\ 1 & 3 & -4 \\ 2 & 4 & 1 \end{vmatrix}.$$

(b) Give a simple example showing that determinant multiplication is not commutative, that is, that the array arising from $|a_j^{\ k}|\,|b_k^{\ q}|$ is in general different from the one produced by $|b_j^{\ k}|\,|a_k^{\ q}|$. (The resulting numerical values are the same, of course.)

11. Explain why $b_j^{\ k}B_k^{\ p} = \delta_j^{\ p}$ implies $bB = 1$.

12. Start with the cofactor representation

$$\mathcal{E}_{rpq}\,D_i^{\ r} = \mathcal{E}_{ijk}\,d_p^{\ j}\,d_q^{\ k}.$$

Show how multiplication and summation of the relation with $d_s^{\ i}$ and E^{tpq} leads to

$$d_s^{\ i}\,D_i^{\ t} = \delta_s^{\ t}\,d.$$

7. *The Cross Product*

Again consider the introductory relation (1-5.1a), which is

$$\mathbf{PQ} = \sum_{j,k} P^j Q^k \boldsymbol{\iota}_j \boldsymbol{\iota}_k.$$

In Section 5 the binary relation, inner multiplication, was defined for vectors **P** and **Q**. Now we introduce an operation called the cross product of **P** and **Q**. The mode of introduction follows the path indicated by Grassman's work in that we begin with algebraic statements. It can also be said that the delineation of ideas is not too far removed from the order used by Hamilton, for his development of quaternions (from which the theory of the cross product can be extracted) started with a search for meanings to be associated with the "product" of two directed line segments. He then set forth algebraic properties, geometric meanings, and physical interpretations in that order.

A meaning is given to (1-5.1) in the following definition.

Definition 1-7.1. A binary relation between vectors **P** and **Q**, called the cross product of **P** and **Q**, is denoted by

$$\mathbf{P} \times \mathbf{Q} = \sum_{r=1}^{3} \boldsymbol{\iota}_r \mathcal{E}_{rst} P^s Q^t. \tag{1-7.1}$$

By writing out the right-hand side of (1-7.1), we have

(1-7.2a)

$$\mathbf{P} \times \mathbf{Q} = (P^2Q^3 - P^3Q^2)\boldsymbol{\iota}_1 + (P^3Q^1 - P^1Q^3)\boldsymbol{\iota}_2 + (P^1Q^2 - P^2Q^1)\boldsymbol{\iota}_3.$$

The relation (1-7.2a) is easily remembered when expressed in the symbolic determinant form:

$$\mathbf{P} \times \mathbf{Q} = \begin{vmatrix} \boldsymbol{\iota}_1 & \boldsymbol{\iota}_2 & \boldsymbol{\iota}_3 \\ P^1 & P^2 & P^3 \\ Q^1 & Q^2 & Q^3 \end{vmatrix}. \tag{1-7.2b}$$

The determinant is said to be symbolic because one row contains vectors, and therefore a numerical value cannot be associated with it. It is interesting to note that we can expand the symbolic expression (1-7.2b) around any row or column just as it is possible to do to an ordinary determinant. The meaning of various other determinant operations is not clear. For example, although interchange of the second and third rows is intuitively useful (see 1-7.4a), an interchange of the first and second rows produces a state of confusion.[36]

[36] It is not implied that meanings cannot be given to ordinary determinant operations for this symbolic determinant but only that such meanings do not already exist.

From either this form or its expansion (1-7.2a) we can determine the following table of cross products of the unit arrows.

(1-7.3)

	ι_1	ι_2	ι_3
ι_1	0	ι_3	$-\iota_2$
ι_2	$-\iota_3$	0	ι_1
ι_3	ι_2	$-\iota_1$	0

The table should be read from the left-hand column to the top row. For example, $\iota_2 \times \iota_3 = \iota_1$.

Example 1-7.1. Let $\mathbf{P} = 2\iota_1 + 3\iota_2$, $\quad \mathbf{Q} = \iota_1 + 5\iota_2$; then

$$\mathbf{P} \times \mathbf{Q} = \begin{vmatrix} \iota_1 & \iota_2 & \iota_3 \\ 2 & 3 & 0 \\ 1 & 5 & 0 \end{vmatrix} = 7\iota_3.$$

Example 1-7.2. Let $\mathbf{P} = \beta\mathbf{Q}$; then

$$\mathbf{P} \times \mathbf{Q} = \begin{vmatrix} \iota_1 & \iota_2 & \iota_3 \\ \beta Q^1 & \beta Q^2 & \beta Q^3 \\ Q^1 & Q^2 & Q^3 \end{vmatrix} = \mathbf{0}.$$

Hence, whenever **P** and **Q** are parallel, the cross product has the value zero. Assuming that neither **P** nor **Q** is the zero vector, we can show that the converse also holds by writing out the determinant in the form (1-7.2a) and by making use of the linear independence of ι_1, ι_2, and ι_3.

The next objective is to determine the fundamental algebraic properties the cross product satisfies and those it does not.

Theorem 1-7.1a. The cross product is

(1-7.4a) (*Anticommutative*) $\quad \mathbf{P} \times \mathbf{Q} = -\mathbf{Q} \times \mathbf{P}.$

(1-7.4b) (*Distributive*) $\quad \mathbf{P} \times (\mathbf{Q} + \mathbf{R}) = \mathbf{P} \times \mathbf{Q} + \mathbf{P} \times \mathbf{R}.$

(1-7.4c) (*Real number associative*) $\quad \alpha(\mathbf{P} \times \mathbf{Q}) = (\alpha\mathbf{P}) \times \mathbf{Q} = \mathbf{P} \times \alpha\mathbf{Q}.$

PROOF. Relation (1-7.4a) is an immediate consequence of the skew symmetry exhibited by the $\mathcal{E}$ symbol in (1-7.1) or the parenthetic expressions in (1-7.2a). Relations (1-7.4b,c) can be proved by straightforward computation starting with either (1-7.1) or (1-7.2a). The details are left to the reader.

Theorem 1-7.1b. (Closure). If **P** and **Q** are Cartesian vectors, the components of **P** × **Q** transform as vector components with respect to the orthogonal Cartesian group of transformations.

PROOF. Consider the equation of rotational transformation,

$$X^j = a_k{}^j \bar{X}^k,$$

and recall that

$$\mathcal{E}_{rst} = \bar{\mathcal{E}}_{rst}.$$

We have

$$\mathcal{E}_{rst} P^s Q^t = (\bar{\mathcal{E}}_{rst} a_j{}^s a_k{}^t) \bar{P}^j \bar{Q}^k.$$

If we employ the representation for cofactor as expressed in Definition 1-6.5, the right-hand side of the expression will take the form

$$a A_r{}^i \bar{\mathcal{E}}_{ijk} \bar{P}^j \bar{Q}^k.$$

Since $a = 1$ and $A_r{}^i = a_i{}^r$, we have

$$\mathcal{E}_{rst} P^s Q^t = \sum_{i=1}^{3} a_i{}^r \bar{\mathcal{E}}_{ijk} \bar{P}^j \bar{Q}^k. \tag{1-7.5a}$$

The correspondence of the form (1-7.5a) with the appropriate transformation law becomes visually clearer if T_r and $\bar{T}_i$ are used to replace $\mathcal{E}_{rst} P^s Q^t$ and $\bar{\mathcal{E}}_{ijk} \bar{P}^j \bar{Q}^k$, respectively. We have

$$T_r = \sum_{i=1}^{3} a_i{}^r \bar{T}_i. \tag{1-7.5b}$$

Consideration of the proof for the group of translations is left to the reader. Although this completes the proof, the following pair of remarks may be helpful.

1. The coefficients of transformation $A_r{}^i$ are cofactors of the elements $a_i{}^r$ in the determinant of the small a's only because $a = 1$. More generally, the cofactors are designated by $aA_r{}^i$. This remark not only clarifies that step in the proof involving cofactors but also points out the interesting observation that the components of the cross product satisfy the law of transformation for vector components only because of the special transformation group under consideration.

2. The replacements

$$T^r = T_r, \qquad \bar{T}^i = \bar{T}_i$$

enable us to put (1-7.5b) in the nicer notational form

$$T^r = a_i{}^r \bar{T}^i. \tag{1-7.5c}$$

Theorem 1-7.1c. The cross product is not associative in general; that is,

$$\mathbf{P} \times (\mathbf{Q} \times \mathbf{R}) \neq (\mathbf{P} \times \mathbf{Q}) \times \mathbf{R}.$$

PROOF. In order to show that a property does not hold in general, all we need to do is give a contradictory example. Therefore let

$$\mathbf{P} = \iota_1, \qquad \mathbf{Q} = \iota_1, \qquad \text{and} \quad \mathbf{R} = \iota_2.$$

Then

$$\mathbf{P} \times (\mathbf{Q} \times \mathbf{R}) = \iota_1 \times (\iota_1 \times \iota_2) = -\iota_1,$$

whereas

$$(\mathbf{P} \times \mathbf{Q}) \times \mathbf{R} = (\iota_1 \times \iota_1) \times \iota_2 = \mathbf{0} \times \iota_2 = \mathbf{0}.$$

This completes the proof.

The existence of an identity element, $\mathbf{X}$, such that $\mathbf{A} \times \mathbf{X} = \mathbf{A}$ for any $\mathbf{A}$ must be ruled out. (See Problem 1 at the end of the section.) Since there is no identity element, the possibility of an inverse is negated.

Our consideration of the geometric properties of the cross product really begins with Problem 18 of Section 5. The perpendicularity conditions

(1-7.6a)
$$\mathbf{P} \cdot \mathbf{C} = P^1C^1 + P^2C^2 + P^3C^3 = 0,$$
$$\mathbf{Q} \cdot \mathbf{C} = Q^1C^1 + Q^2C^2 + Q^3C^3 = 0,$$

are satisfied by

(1-7.6b)
$$C_1 = \gamma \begin{vmatrix} P^2 & P^3 \\ Q^2 & Q^3 \end{vmatrix}, \qquad C_2 = \gamma \begin{vmatrix} P^3 & P^1 \\ Q^3 & Q^1 \end{vmatrix}, \qquad C_3 = \gamma \begin{vmatrix} P^1 & P^2 \\ Q^1 & Q^2 \end{vmatrix},$$

where γ is any real number. As an aid to clearly relating this problem to the cross product, note that the determinant of the components of ι_1, ι_2, ι_3 has the value

$$\begin{vmatrix} 1 & 0 & 0 \\ 0 & 1 & 0 \\ 0 & 0 & 1 \end{vmatrix} = 1 > 0,$$

whereas that of ι_1, ι_2, $-\iota_3$ is

$$\begin{vmatrix} 1 & 0 & 0 \\ 0 & 1 & 0 \\ 0 & 0 & -1 \end{vmatrix} = -1 < 0.$$

The following definition is consistent with our previous statements concerning right- and left-hand coordinate systems. Furthermore, it includes the foregoing introductory illustration.

Definition 1-7.2. Suppose **P**, **Q**, **R** are linearly independent vectors. If

(1-7.7) $$\varepsilon_{ijk}P^iQ^jR^k = \begin{vmatrix} P^1 & P^2 & P^3 \\ Q^1 & Q^2 & Q^3 \\ R^1 & R^2 & R^3 \end{vmatrix} > 0,$$

the vectors **P**, **Q**, **R**, in that order, are said to form a right-hand system. If the value of the determinant is less than zero, the vectors in the given order determine a left-hand system.

Theorem 1-7.2. The sign of the determinant of the vectors **P**, **Q**, **C**, in that order, of Problem 18, Section 5, is the same as the sign of the factor γ.

PROOF. In order to facilitate the use of the summation notation, let $P^r = P_r$, $Q^s = Q_s$. Then

$$\varepsilon_{ijk}P^iQ^jC^k = \varepsilon_{ijk}P^iQ^j\gamma E^{krs}P_rQ_s$$

$$= \gamma 2\,\delta^{[rs]}_{ij}P^iQ^jP_rQ_s = \gamma[(P^rP_r)(Q^sQ_s) - (P^sQ_s)(P^rQ_r)];$$

that is

$$\varepsilon_{ijk}P^iQ^jC^k = \gamma\,|\mathbf{P}|^2\,|\mathbf{Q}|^2\,(1 - \cos^2\theta) = \gamma\,|\mathbf{P}|^2\,|\mathbf{Q}|^2\sin^2\theta,$$

where θ is the angle determined by **P** and **Q**, $0 \leq \theta \leq \pi$. Since γ clearly determines the sign of the right-hand member of the foregoing expression, the proof is complete.

When $\gamma = 1$, the components (1-7.6b) become precisely those of the cross product. Therefore, as a consequence of the preceding definition and theorem, it follows that **P**, **Q** and $\mathbf{P} \times \mathbf{Q}$, in that order, form a right-hand system. Furthermore, $\mathbf{P} \times \mathbf{Q}$ is perpendicular to both **P** and **Q**. The definition (1-7.1) of the cross product does not demand that **P** and **Q** be linearly independent. If **P** and **Q** are proportional, then $\mathbf{P} \times \mathbf{Q} = \mathbf{0}$, as illustrated in Example 1-7.2. If $\mathbf{P} = \mathbf{0}$ or $\mathbf{Q} = \mathbf{0}$, it is easily seen that again $\mathbf{P} \times \mathbf{Q} = \mathbf{0}$.

Our considerations are restricted to right-handed rectangular Cartesian coordinate systems; therefore $a = 1$. The vector character of the cross product with respect to the transformation group relating these systems carries with it the implication that the geometric interpretation of the cross product does not depend on a particular one of the coordinate frames. This fact is helpful in associating a geometric meaning with the magnitude of the cross product. To this purpose, let a coordinate system X^j be chosen such that the linearly independent vectors **P** and **Q** have components $(P^1, 0, 0)$, $(Q^1, Q^2, 0)$, respectively.

Theorem 1-7.3. The magnitude of the cross product is

(1-7.8) $$|\mathbf{P} \times \mathbf{Q}| = |\mathbf{P}|\,|\mathbf{Q}|\sin\theta,$$

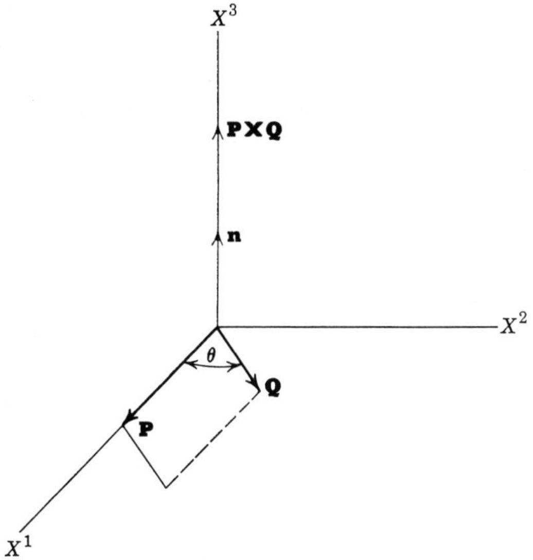

Fig. 1-7.1

where θ is the angle determined by the positive senses of **P** and **Q** such that $0 \leq \theta \leq \pi$. (See Fig. 1-7.1.)

PROOF. According to (1-7.2b),

$$\mathbf{P} \times \mathbf{Q} = \begin{vmatrix} \iota_1 & \iota_2 & \iota_3 \\ P^1 & 0 & 0 \\ Q^1 & Q^2 & 0 \end{vmatrix} = P^1 Q^2 \iota_3. \tag{1-7.9a}$$

From examination of Fig. 1-7.1, we observe that $|Q^2| = |\mathbf{Q}| \sin \theta$. Therefore

$$|\mathbf{P} \times \mathbf{Q}| = [(\mathbf{P} \times \mathbf{Q}) \cdot (\mathbf{P} \times \mathbf{Q})]^{1/2} = |P^1 Q^2| = |\mathbf{P}|\,|\mathbf{Q}| \sin \theta. \tag{1-7.9b}$$

This completes the proof.

Theorem 1-7.3 points toward the natural association of the cross product with a plane area, in particular, the parallelogram area determined by the vectors **P** and **Q**. In order to correlate this result with the preceding facts concerning the cross product, we simply abstract the plane area number $|\mathbf{P}|\,|\mathbf{Q}| \sin \theta$ and reinterpret it as the length of a vector perpendicular to **P** and **Q**. Then

$$\boxed{\mathbf{P} \times \mathbf{Q} = |\mathbf{P}|\,|\mathbf{Q}| \sin \theta \, \mathbf{n},} \tag{1-7.10}$$

where **n** is a unit vector perpendicular to **P** and **Q** and chosen so that **P**, **Q**, **n**, in that order, determine a right-hand system. In most texts (1-7.10) is used as the definition of the cross product.

The geometric interpretation of the magnitude of **P** × **Q** as a plane area points toward meanings not yet expressed. These remaining thoughts have both mathematical and physical significance.

From the mathematical viewpoint, the components

$$2P^{[j}Q^{k]} = P^jQ^k - P^kQ^j \tag{1-7.11a}$$

of the cross product determine a system of the second order. In general, the number of components of such a system in three-space is nine. The skew symmetry of the set (1-7.11a) implies that only six of these components are nonzero and that these six can be paired off according to the arrangement

$$P^{[1}Q^{2]} = -P^{[2}Q^{1]}, \quad P^{[2}Q^{3]} = -P^{[3}Q^{2]}, \quad P^{[1}Q^{3]} = -P^{[3}Q^{1]}.$$

In other words, the system of components is specified completely by an appropriate choice of three components. Thus the system can be associated with a vector by means of the identification

$$C_i = \varepsilon_{ijk}P^jQ^k. \tag{1-7.11b}$$

As we have seen, it is this vector that is called the cross product of **P** and **Q**. It should be noted that the identification (1-7.11b) has no analogue outside three-space. For example, in two-space we might consider identifying $2P^{[j}Q^{k]}$ by the relation

$$\varepsilon_{ij}P^iQ^j, \tag{1-7.11c}$$

but this expression produces only a single component, not the pair of components needed to determine a two-space vector.

From the physical point of view the cross product is used to represent entities associated with rotational motion. A simple illustration is given by the next example.

Example 1-7.3. Suppose that **F** is a force applied to a body at a point *P*. The vector moment of the force related to another point *O* is defined by

$$\mathbf{M} = \mathbf{r} \times \mathbf{F}. \tag{1-7.12}$$

(See Fig. 1-7.2.) The moment of force is then a measure of the rotational effect of **F** in the plane of **r** and **F**. The magnitude of this effect is geometrically expressed by the area of the parallelogram of **r** and **F**. The sense of rotation is determined by the orientation with which **F** endows the parallelogram. Therefore

$$\mathbf{M} = \mathbf{r} \times (-\mathbf{F}) = \mathbf{F} \times \mathbf{r}$$

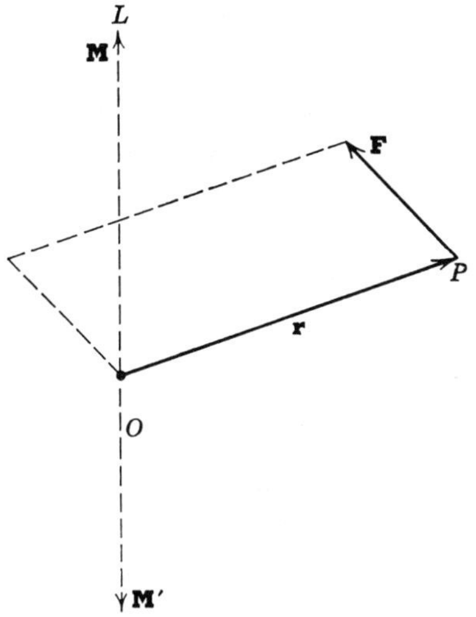

Fig. 1-7.2

represents a moment of the same magnitude as **M** but with the opposite rotation sense. The relationship between **M** and **M**′ represents a *combined geometric and physical* interpretation of the anticommutative property of the cross product. (For further remarks see the problem section.)

Physicists have distinguished vectors (i.e., the cross product) associated with plane areas by calling them axial vectors, while using the term polar vector for our ordinary free vector.

There are two important algebraic expressions that involve interrelationships in some fashion between the dot and cross products. The first of these, the so-called triple scalar product, $\mathbf{P} \cdot \mathbf{Q} \times \mathbf{R}$, is examined in the next theorem.

Theorem 1-7.4. The triple scalar product $\mathbf{P} \cdot \mathbf{Q} \times \mathbf{R}$ (for rectangular Cartesian systems) is represented by

(1-7.13a)
$$\boxed{\mathbf{P} \cdot \mathbf{Q} \times \mathbf{R} = \begin{vmatrix} P^1 & P^2 & P^3 \\ Q^1 & Q^2 & Q^3 \\ R^1 & R^2 & R^3 \end{vmatrix}},$$

or

(1-7.13b)
$$\boxed{\mathbf{P} \cdot \mathbf{Q} \times \mathbf{R} = \mathcal{E}_{ijk} P^i Q^j R^k.}$$

PROOF. To prove (1-7.13a), we simply note that

$$\mathbf{Q} \times \mathbf{R} = \begin{vmatrix} \iota_1 & \iota_2 & \iota_3 \\ Q^1 & Q^2 & Q^3 \\ R^1 & R^2 & R^3 \end{vmatrix} = \iota_1 \begin{vmatrix} Q^2 & Q^3 \\ R^2 & R^3 \end{vmatrix} + \iota_2 \begin{vmatrix} Q^3 & Q^1 \\ R^3 & R^1 \end{vmatrix} + \iota_3 \begin{vmatrix} Q^1 & Q^2 \\ R^1 & R^2 \end{vmatrix}.$$

Therefore

$$\mathbf{P} \cdot \mathbf{Q} \times \mathbf{R} = P^1 \begin{vmatrix} Q^2 & Q^3 \\ R^2 & R^3 \end{vmatrix} + P^2 \begin{vmatrix} Q^3 & Q^1 \\ R^3 & R^1 \end{vmatrix} + P^3 \begin{vmatrix} Q^1 & Q^2 \\ R^1 & R^2 \end{vmatrix}.$$

The right member of this last expression is equivalent to the determinant representation in (1-7.13a).

The form (1-7.13b) likewise follows directly from the definitions of dot and cross products with respect to an orthogonal basis.

It is interesting to note that the triple scalar product represents a determinant. This fact is clearly exhibited by either (1-7.13a) or (1-7.13b).

The triple scalar product is more accurately described as a scalar density, but with respect to the rectangular Cartesian transformation group it has a valid claim to the simpler designation scalar.

Theorem 1-7.5. The quantity

$$\mathbf{P} \cdot \mathbf{Q} \times \mathbf{R} = \mathcal{E}_{ijk} P^i Q^j R^k$$

is a scalar with respect to the rectangular Cartesian transformation group.

PROOF. We have

$$\begin{aligned} \mathcal{E}_{ijk} P^i Q^j R^k &= a A_i{}^r A_j{}^s A_k{}^t a_e{}^i a_f{}^j a_g{}^k \bar{\mathcal{E}}_{rst} \bar{P}^e \bar{Q}^f \bar{R}^g \\ &= a\, \delta_e{}^r \delta_f{}^s \delta_g{}^t \bar{\mathcal{E}}_{rst} \bar{P}^e \bar{Q}^f \bar{R}^g \\ &= a \bar{\mathcal{E}}_{rst} \bar{P}^r \bar{Q}^s \bar{R}^t. \end{aligned}$$

Since $a = 1$, the proof is complete.

The geometric interpretation of the triple scalar product presents the possibility of an interplay between algebraic and geometric modes of thought.

Example 1-7.4. The absolute value of the triple scalar product $\mathbf{P} \cdot \mathbf{Q} \times \mathbf{R}$ can be interpreted geometrically as the volume of a parallelepiped determined by vectors $\mathbf{P}$, $\mathbf{Q}$, and $\mathbf{R}$. To see this interpretation clearly (Fig. 1-7.3), recall that the volume of the parallelepiped is determined by multiplying the area of the base by the height. According to the definition of the dot product [see (1-5.3)],

$$\mathbf{P} \cdot \mathbf{Q} \times \mathbf{R} = |\mathbf{P}|\, |\mathbf{Q} \times \mathbf{R}| \cos \theta = |\mathbf{Q} \times \mathbf{R}|(\,|\mathbf{P}| \cos \theta).$$

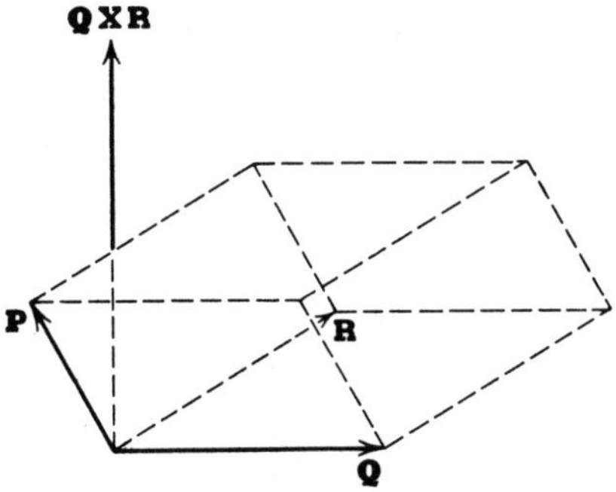

Fig. 1-7.3

Then $|\mathbf{Q} \times \mathbf{R}|$ is numerically equal to the area of the base, whereas $|\mathbf{P}|\,|\cos\theta|$ represents the height of the parallelepiped.

It is an interesting sidelight to observe that the volume of a parallelepiped is algebraically represented by a determinant. This fact is the basis for volume representations when the space dimension is greater than 3.

Example 1-7.5. We might be concerned about the ordering of the vectors **P**, **Q**, **R** in the representation of the triple scalar product. In particular, the question may be asked whether this ordering has any significance with respect to the geometric interpretation of Example 1-7.4. It can be answered in the negative except for sign, for from either (1-7.13a) or (1-7.13b) we can conclude that

$$\text{(1-7.14a)} \quad \mathbf{P}\cdot\mathbf{Q}\times\mathbf{R} = \mathbf{Q}\cdot\mathbf{R}\times\mathbf{P} = \mathbf{R}\cdot\mathbf{P}\times\mathbf{Q}$$
$$= -\mathbf{P}\cdot\mathbf{R}\times\mathbf{Q} = -\mathbf{R}\cdot\mathbf{Q}\times\mathbf{P} = -\mathbf{Q}\cdot\mathbf{P}\times\mathbf{R}.$$

By commuting **R** and $\mathbf{P} \times \mathbf{Q}$ in the last member of the first line and then by equating it to the first member of that line, we get the rather interesting algebraic relation

$$\text{(1-7.14b)} \quad \boxed{\mathbf{P}\cdot\mathbf{Q}\times\mathbf{R} = \mathbf{P}\times\mathbf{Q}\cdot\mathbf{R}.}$$

The implication of this expression is an algebraic property of interchange of the dot and cross.

During the discussion of left- and right-hand coordinate systems, which took place in Section 1, it was indicated that more information would be supplied in this section. The remarks of the following example pertain to that discussion.

Example 1-7.6. Suppose that X^j is a right-hand rectangular Cartesian coordinate system and that $\bar{X}^j$ were left-handed. (See Fig. 1-7.4.) Let the transformation equations relating the systems be

$$\text{(1-7.15)} \quad \bar{X}^1 = X^2, \qquad \bar{X}^2 = X^1, \qquad \bar{X}^3 = X^3.$$

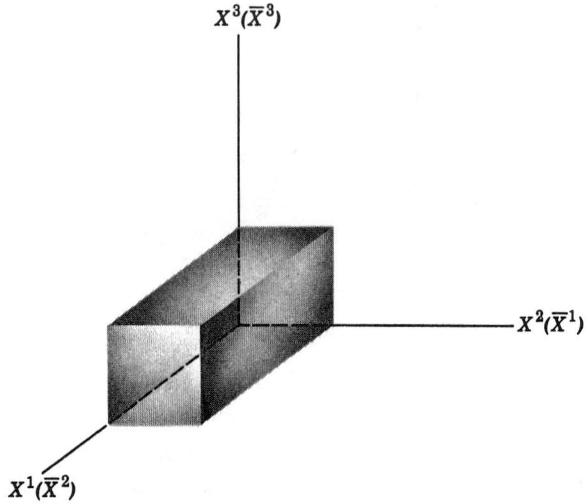

Fig. 1-7.4

The determinant of the transformation is

(1-7.16a)
$$a = \begin{vmatrix} 0 & 1 & 0 \\ 1 & 0 & 0 \\ 0 & 0 & 1 \end{vmatrix} = -1,$$

and

(1-7.16b)
$$\mathbf{P} \cdot \mathbf{Q} \times \mathbf{R} = -\bar{\mathbf{P}} \cdot \bar{\mathbf{Q}} \times \bar{\mathbf{R}}.$$

Therefore the numerical value of the triple scalar product is not invariant with respect to an inversion. Under this circumstance the geometric interpretation of the triple scalar product as a volume is not completely satisfactory. In fact, inversions are classified as transformations that do not preserve volume. The restriction to right-hand systems eliminates the difficulty and the orthogonal Cartesian group with the condition $a = 1$ associated is said to be volume preserving.

It was indicated that there are two algebraic relations of significance to the development of the subject. The second is the so-called "triple vector product." It has been shown that, with respect to the rectangular Cartesian transformation group, $\mathbf{Q} \times \mathbf{R}$ could be considered a vector. Hence $\mathbf{P} \times (\mathbf{Q} \times \mathbf{R})$ is also interpretable as a vector. In a specific instance this vector can be computed as follows:

Example 1-7.7. Let $\mathbf{P} = 2\iota_1 + 3\iota_2 - \iota_3$, $\mathbf{Q} = \iota_1 - 4\iota_2 + \iota_3$, $\mathbf{R} = 3\iota_1 - 2\iota_2 + 2\iota_3$. Then

$$\mathbf{Q} \times \mathbf{R} = \begin{vmatrix} \iota_1 & \iota_2 & \iota_3 \\ 1 & -4 & 1 \\ 3 & -2 & 2 \end{vmatrix} = -6\iota_1 + \iota_2 + 10\iota_3$$

and

$$\mathbf{P} \times (\mathbf{Q} \times \mathbf{R}) = \begin{vmatrix} \iota_1 & \iota_2 & \iota_3 \\ 2 & 3 & -1 \\ -6 & 1 & 10 \end{vmatrix} = 31\iota_1 - 14\iota_2 + 20\iota_3.$$

The computation of Example 1-7.7, although not difficult, is a little tedious and certainly does not lend itself to theoretical considerations. This fact increases the importance of the following discovery.

Theorem 1-7.6. The triple vector cross product has the representation

$$\boxed{\mathbf{P} \times (\mathbf{Q} \times \mathbf{R}) = \mathbf{Q}(\mathbf{P} \cdot \mathbf{R}) - \mathbf{R}(\mathbf{P} \cdot \mathbf{Q}).} \tag{1-7.17a}$$

The component form of this expression is

$$\boxed{E^{rsi}P_s\mathcal{E}_{ijk}Q^jR^k = Q^r(P_tR^t) - R^r(P_tQ^t),} \tag{1-7.17b}$$

where

$$P_s = \delta_{st}P^t = P^s. \tag{1-7.17c}$$

PROOF. The fact that (1-7.17b) is the component form of (1-7.17a) follows from Definition 1-7.1 and appropriate use of the summation notation. The relation (1-7.17c) is introduced to facilitate this usage of the summation process.

The theorem is proved by starting with (1-7.17b). Beginning with the left member,

$$E^{rsi}P_s\mathcal{E}_{ijk}Q^jR^k = E^{rsi}\mathcal{E}_{ijk}P_sQ^jR^k = E^{irs}\mathcal{E}_{ijk}P_sQ^jR^k.$$

Employing (1-6.11a), we can continue the string of equalities

$$\begin{aligned} &= 2\delta_{jk}^{[rs]}P_sQ^jR^k = 2P_sQ^{[r}R^{s]} = P_s(Q^rR^s - Q^sR^r) \\ &= Q^r(P_sR^s) - R^r(P_sQ^s). \end{aligned}$$

Any difficulty on the part of the reader in eliminating the $\delta_{jk}^{[rs]}$ can be resolved by expanding the bracketed expressions and applying the definition of the deltas.

Although the index notation employed in the last proof may seem somewhat involved to the beginner, some idea of its power and usefulness can be obtained by comparing the simplicity of this proof with any of those usually given in texts on vector analysis.[37]

Example 1-7.8. The computation of Example 1-7.6 can be accomplished as follows:

$$\mathbf{P} \cdot \mathbf{R} = 6 - 6 - 2 = -2, \qquad \mathbf{P} \cdot \mathbf{Q} = 2 - 12 - 1 = -11.$$

$$\mathbf{P} \times (\mathbf{Q} \times \mathbf{R}) = \mathbf{Q}(\mathbf{P} \cdot \mathbf{R}) - \mathbf{R}(\mathbf{P} \cdot \mathbf{Q}) = -2\mathbf{Q} - 11\mathbf{R} = -2(\iota_1 - 4\iota_2 + \iota_3)$$
$$+ 11(3\iota_1 - 2\iota_2 + 2\iota_3) = 31\iota_1 - 14\iota_2 + 20\iota_3.$$

In Theorem 1-7.1c it was shown that the cross product is not associative. The rectangular Cartesian representation of the triple vector product makes possible a statement of the relation between $\mathbf{P} \times (\mathbf{Q} \times \mathbf{R})$ and $(\mathbf{P} \times \mathbf{Q}) \times \mathbf{R}$.

Theorem 1-7.7. We have

$$(\mathbf{P} \times \mathbf{Q}) \times \mathbf{R} = \mathbf{Q}(\mathbf{P} \cdot \mathbf{R}) - \mathbf{P}(\mathbf{R} \cdot \mathbf{Q}). \tag{1-7.18a}$$

Therefore

$$\mathbf{P} \times (\mathbf{Q} \times \mathbf{R}) - (\mathbf{P} \times \mathbf{Q}) \times \mathbf{R} = \mathbf{P}(\mathbf{Q} \cdot \mathbf{R}) - \mathbf{R}(\mathbf{P} \cdot \mathbf{Q}). \tag{1-7.18b}$$

PROOF. Making use of the anticommutative property of the cross product and Theorem 1-7.8, we have

$$(\mathbf{P} \times \mathbf{Q}) \times \mathbf{R} = -\mathbf{R} \times (\mathbf{P} \times \mathbf{Q}) = -[\mathbf{P}(\mathbf{R} \cdot \mathbf{Q}) - \mathbf{Q}(\mathbf{R} \cdot \mathbf{P})]$$
$$= \mathbf{Q}(\mathbf{R} \cdot \mathbf{P}) - \mathbf{P}(\mathbf{R} \cdot \mathbf{Q}).$$

This completes the proof of (1-7.18a). The relation (1-7.18b) is obtained by subtracting.

It seems appropriate to conclude this section with a few words about the possibility of vector division.

The operation of division is usually thought of as the inverse of multiplication. The two types of multiplication we have considered are dot and cross products. Hence it is natural to ask whether it is practical to introduce inverse operations with respect to these processes.

A fundamental property of real-number multiplication is its uniqueness characteristic; that is, the solution to the problem is unique. It is this property that makes possible a simple and useful process of division. Now suppose that with respect to the dot product we considered

$$\mathbf{P} \cdot \mathbf{Q} = 0,$$

[37] A variety of proofs can be found in Murray S. Klamkin's article "On the Vector Triple Product," *American Mathematical Monthly*, December 1954.

where $\mathbf{P}$ is given. In particular, suppose $\mathbf{P} = 2\iota_1 - 3\iota_2 + 4\iota_3$. Then $\mathbf{Q} = 3\iota_1 + 2\iota_2$ or $\mathbf{Q} = 4\iota_2 + 3\iota_3$ satisfy the relation. Since Q cannot be uniquely determined, division is *not* defined with respect to the dot product. For the same reason division is *not* defined with respect to the cross product.

There are still other possibilities for the operation of division. Another that is ruled out of the present considerations is the quotient constructed from the components of a pair of vectors; that is, given vector components P^i and Q^j, $Q^j \neq 0$, we can certainly form the quotients P^i/Q^j. This is just a real-number division. The number of possible quotients involved (for $n = 3$) is, however, nine. Therefore the resulting set of components cannot represent a vector. For that reason this sort of quotient does not enter into our present work.

A type of pseudo-division with a place in the algebra of vectors is a special case of the so-called "quotient law" which is usually stated in a development of the algebra of tensors.

Theorem 1-7.8. Suppose that for any vector $\mathbf{P}$

$$\mathbf{P} \cdot \mathbf{Q} = \mathbf{P} \cdot \mathbf{R}; \tag{1-7.19a}$$

then

$$\mathbf{Q} = \mathbf{R}. \tag{1-7.19b}$$

PROOF. From (1-7.19a)

$$\mathbf{P} \cdot (\mathbf{Q} - \mathbf{R}) = P^1(Q^1 - R^1) + P^2(Q^2 - R^2) + P^3(Q^3 - R^3) = 0.$$

$\mathbf{P}$ is arbitrary; therefore we can make the successive component choices

$$(1, 0, 0), \qquad (0, 1, 0), \quad \text{and} \quad (0, 0, 1).$$

With respect to these choices,

$$Q^j = R^j, \qquad j = 1, 2, 3.$$

This completes the proof.

Problems

1. (a) Prove (1-7.4b) and (1-7.4c).
(b) Show that the relation $\mathbf{A} \times \mathbf{X} = \mathbf{A}$ is satisfied if and only if $\mathbf{A} = \mathbf{0}$. *Hint:* Use the form $\varepsilon_{ijk}A^jX^k = A_i$ and show that $\mathbf{A}$ is self-perpendicular.

2. (a) Determine a vector perpendicular to $\mathbf{A} = 2\iota_1 + 3\iota_2 - \iota_3$ and $\mathbf{B} = \iota_1 - \iota_2 + 2\iota_3$.
(b) Compute $\mathbf{B} \times \mathbf{A}$ and then compare with the result of (a).

(c) Search the mathematical literature for other examples of noncommutative products.

3. Compute the area of the parallelogram determined by the vectors $\mathbf{A} = 5\iota_1 - 2\iota_2 + 3\iota_3$, $\mathbf{B} = \iota_1 - 3\iota_2 + \iota_3$

4. Show that $\mathbf{P} \times \mathbf{Q}$ is perpendicular to $\mathbf{P}$ and $\mathbf{Q}$ by starting with the component expression $\varepsilon_{ijk}P^jQ^k$.

5. Use the cross product to determine an equation of the plane passing through the points $P_1(1, 2, 5)$, $P_2(0, 1, 4)$, and $P_3(1, 0, 3)$.

6. How many components has the system $2P^{[j}Q^{k]}$ in a four-dimensional space? How many are nonzero? If appropriately chosen, what is the least number needed to determine them all?

7. Show that the moments of Example 1-7.3 satisfy the relation $\mathbf{M} = -\acute{\mathbf{M}}$. Write a statement interpreting the sense of the moment vector.

8. Transform $\mathbf{r} \times \mathbf{F}$ with respect to an inversion

$$\bar{X}^1 = X^1,$$
$$\bar{X}^2 = -X^2,$$
$$\bar{X}^3 = X^3.$$

9. The representations of two lines are given parametrically by

$$X^1 = 1 + 3t, \qquad X^1 = 7s,$$
$$X^2 = 2 + 5t, \qquad X^2 = 4 + 3s,$$
$$X^3 = 3 - 2t, \qquad X^3 = 2 + 3s.$$

(a) Find the parametric equations of a line through (1, 5, 2) and perpendicular to each of the two given lines.
(b) Show that the given lines are skew (i.e., nonintersecting).
(c) Find the representation of a line perpendicular to each of the given lines and intersecting each of the given lines.

10. Prove that in three-space there is a unique line that is perpendicular to both and intersects each of two given skew lines. (Parallel lines are an exception to this statement.)

11. Given

$$\mathbf{P} = 2\iota_1 - 5\iota_2 + \iota_3, \qquad \mathbf{Q} = \iota_1 + 3\iota_2, \qquad \mathbf{R} = 3\iota_1 + \iota_2 + \iota_3,$$

(a) compute $\mathbf{P} \cdot \mathbf{Q} \times \mathbf{R}$;
(b) compute $\mathbf{P} \times \mathbf{Q} \cdot \mathbf{R}$ and compare with the result of (a). Why are the results the same?

12. Find the volume of the parallelepiped determined by vectors

$$\mathbf{P} = 2\iota_1 + 3\iota_2, \qquad \mathbf{Q} = 3\iota_1 + 4\iota_2, \qquad \mathbf{R} = 4\iota_3.$$

13. Given

$$\mathbf{P} = \iota_1 - 4\iota_2 + 4\iota_3, \qquad \mathbf{Q} = 2\iota_1 - 3\iota_2 + \iota_3, \qquad \mathbf{R} = 4\iota_2 + 2\iota_3,$$

(a) compute $\mathbf{P} \times (\mathbf{Q} \times \mathbf{R})$ directly;
(b) compute $\mathbf{P} \times (\mathbf{Q} \times \mathbf{R})$, using (1-7.17a);
(c) compute $(\mathbf{P} \times \mathbf{Q}) \times \mathbf{R}$.

14. Show that $(\mathbf{P} \times \mathbf{Q}) \times \mathbf{R} = \mathbf{Q}(\mathbf{P} \cdot \mathbf{R}) - \mathbf{P}(\mathbf{Q} \cdot \mathbf{R})$ by showing that

$$E^{ijk}(\mathcal{E}_{jst}P^sQ^t)R_k = Q^i(P^jR_j) - P^i(Q^jR_j).$$

15. Give an example showing that the solution for $\mathbf{Q}$ in $\mathbf{P} \times \mathbf{Q} = \mathbf{R}$ is not unique. (Choose $\mathbf{P}$ and $\mathbf{R}$ arbitrarily.)

16. Given $\mathbf{A} = 5\iota_1 + 3\iota_2 - 2\iota_3$ and $\mathbf{A} \cdot \mathbf{B} = 15$, show that $\mathbf{B}$ is not unique.

17. Given $\mathbf{B} = 3\iota_1 + 5\iota_3$ and $\mathbf{A} \cdot \mathbf{B} = \mathbf{A} \cdot \mathbf{C}$ for all $\mathbf{A}$, find $\mathbf{C}$.

18. Show that (identity of Lagrange)

$$(\mathbf{A} \times \mathbf{B}) \cdot (\mathbf{C} \times \mathbf{D}) = (\mathbf{A} \cdot \mathbf{C})(\mathbf{B} \cdot \mathbf{D}) - (\mathbf{A} \cdot \mathbf{D})(\mathbf{B} \cdot \mathbf{C}).$$

19. Let P_1 be a given point and $\mathbf{r} = \mathbf{r}_0 + \mathbf{B}t$ a given line. Show that the distance h from point to line is given by

$$h = \frac{|(\mathbf{r}_1 - \mathbf{r}_0) \times \mathbf{B}|}{|\mathbf{B}|},$$

where P_1 is the end point of the position vector $\mathbf{r}_1$.

20. Suppose that unit vectors $\mathbf{A}$ and $\mathbf{B}$ determine an angle θ. Show that

$$|\sin\theta| = \left(\begin{vmatrix} A^2 & A^3 \\ B^2 & B^3 \end{vmatrix}^2 + \begin{vmatrix} A^3 & A^1 \\ B^3 & B^1 \end{vmatrix}^2 + \begin{vmatrix} A^1 & A^2 \\ B^1 & B^2 \end{vmatrix}^2 \right)^{1/2}.$$

21. The determinant

$$\begin{vmatrix} \mathbf{A} . \mathbf{A} & \mathbf{A} \cdot \mathbf{B} \\ \mathbf{B} \cdot \mathbf{A} & \mathbf{B} \cdot \mathbf{B} \end{vmatrix}$$

is a second-order example of Gram's determinant, or the Gramian of the system $\mathbf{A}$, $\mathbf{B}$. Show that a necessary and sufficient condition for the linear dependence of $\mathbf{A}$ and $\mathbf{B}$ is that the Gramian has the value zero.
Hint: Method 1. Make use of Lagrange's identity (Problem 18).
Method 2. Start with the algebraic expression $\alpha\mathbf{A} + \beta\mathbf{B} = \mathbf{0}$ and dot with $\mathbf{A}$ and then with $\mathbf{B}$.

7*. $\mathcal{E}$ *Systems and the Cross Product in General Cartesian Systems*

The discussion of $\mathcal{E}$ systems and the cross product in Sections 6 and 7 was general enough that much to be said in this section borders on repetition. It is important, however, that certain concepts be clarified and that other ideas be refined.

The first considerations deal with the seemingly strange procedure of using a different kernel letter (E) for the contravariant components E^{ijk} from that used for the covariant components $\mathcal{E}_{ijk}$. Recall that in Section

5* covariant components G_j were associated with contravariant components by means of the rules

$$\text{(1-7*.1)}\qquad \begin{aligned} &\text{(a)}\quad G_j = h_{jk}G^k, \\ &\text{(b)}\quad G^k = h^{kj}G_j. \end{aligned}$$

Of course, in Section 5* the linear combinations (G_j) of the G^k were introduced in conjunction with a change to a new basis. The relations in (1-7*.1a,b) also serve as a starting point in the development of an algebra of components.

The covariant metric tensor and the associated contravariant metric tensor are used to introduce linear combinations of given components. The processes involved are designated by the terms "lowering indices" and "raising indices." For example, associated with the covariant components $\mathcal{E}_{ijk}$ we have

$$\text{(1-7*.2a)}\qquad \mathcal{E}^{pqr} = h^{pi}h^{qj}h^{rk}\mathcal{E}_{ijk}.$$

Associated with the contravariant components E^{ijk} is the system

$$\text{(1-7*.2b)}\qquad E_{pqr} = h_{pi}h_{qj}h_{rk}E^{ijk}.$$

Theorem 1-7*.1. We have

$$\text{(1-7*.3)}\qquad \begin{aligned} &\text{(a)}\quad \mathcal{E}^{pqr} = |h^{jk}|\, E^{pqr}, \\ &\text{(b)}\quad E_{pqr} = |h_{jk}|\, \mathcal{E}_{pqr}. \end{aligned}$$

PROOF. If we apply the determinant definition to relations (1-7*.2a,b), both results follow.

Theorem (1-7*.1) explains the use of different kernel letters. The values of the determinants of the contravariant and covariant metric tensors are, in general, different from 1; hence the values of $\mathcal{E}^{pqr}$ and E_{pqr} are not defined by (1-6.1).

Suppose that $\mathcal{E}_{ijk}$ and E^{ijk} satisfy Definition 1-6.1 in every Cartesian coordinate system. The laws of transformation are given by the next theorem.

Theorem 1-7*.2. If $Y^j = b_k{}^j\,\overline{Y}^k$, we have

$$\text{(1-7*.4)}\qquad \begin{aligned} &\text{(a)}\quad \mathcal{E}_{ijk} = bB_i{}^pB_j{}^qB_k{}^r\bar{\mathcal{E}}_{pqr}, \\ &\text{(b)}\quad E^{ijk} = Bb_p{}^ib_q{}^jb_r{}^k\bar{E}^{pqr}. \end{aligned}$$

PROOF. As in theorem (1-6.5), these relations are identities.

Definition 1-7*.1a. The class of $\mathcal{E}$ systems $\{\mathcal{E}_{ijk}\}$, the elements of which satisfy the transformation relation (1-7*.4a), is said to be a covariant tensor density of order 3 and weight +1 with respect to the centered affine group.

Definition 1-7*.1b. The class of $\mathcal{E}$ systems $\{E^{ijk}\}$, the elements of which satisfy the transformation rule (1-7*.4b), is said to be a contravariant tensor density of order 3 and weight -1 with respect to the centered affine group.

Now let us turn our attention to the cross product. A rectangular Cartesian system with coordinates X^j can be conveniently used as a connecting link in determining the general rule of transformation. Let a rectangular Cartesian system be identified by coordinates X^j, and suppose that $\overline{Y}^j$, $\overline{\overline{Y}}{}^j$ are coordinates of arbitrary Cartesian systems. Denote the equations of transformation as follows:

$$\text{(1-7*.5)} \qquad \begin{aligned} &\text{(a)} \quad X^j = c_k{}^j \overline{Y}^k, \qquad X^j = d_p{}^j \overline{\overline{Y}}{}^p, \\ &\text{(b)} \quad \overline{Y}^k = C_j{}^k X^j = C_j{}^k d_p{}^j \overline{\overline{Y}}{}^p = b_p{}^k \overline{\overline{Y}}{}^p, \end{aligned}$$

where

$$\text{(1-7*.5c)} \qquad b_p{}^k = C_j{}^k d_p{}^j.$$

Theorem 1-7*.3. The cross product of a pair of contravariant vectors is a covariant vector density of weight $+1$ with respect to the centered affine group of transformations.

PROOF. If **P** and **Q** are given vectors, then, in a manner analogous to that employed in Theorem 1-7.1b, we obtain

$$\text{(1-7*.6)} \qquad \begin{aligned} &\text{(a)} \quad \mathcal{E}_{rst} P^s Q^t = c C_r{}^j \overline{\mathcal{E}}_{jkl} \overline{P}^k \overline{Q}^l, \\ &\text{(b)} \quad \mathcal{E}_{rst} P^s Q^t = d D_r{}^j \overline{\overline{\mathcal{E}}}_{jfg} \overline{\overline{P}}{}^f \overline{\overline{Q}}{}^g. \end{aligned}$$

Therefore

$$c C_r{}^j \overline{\mathcal{E}}_{jkl} \overline{P}^k \overline{Q}^l = d D_r{}^j \overline{\overline{\mathcal{E}}}_{jfg} \overline{\overline{P}}{}^f \overline{\overline{Q}}{}^g.$$

By multiplying this relation by $C c_u{}^r$ and using the properties $cC = 1$ (see Problem 11, Section 6) and $c_u{}^r C_r{}^j = \delta_u{}^j$, we obtain,

$$\overline{\mathcal{E}}_{ukl} \overline{P}^k \overline{Q}^l = C d B_u{}^j \overline{\overline{\mathcal{E}}}_{jfg} \overline{\overline{P}}{}^f \overline{\overline{Q}}{}^g.$$

According to the theorem on determinant multiplication, $Cd = b$. This completes the proof.

The preceding theorem implies that the appropriate representation of the cross product is in terms of covariant components. This conclusion is enforced further by carrying out the transformation of $\mathbf{P} \times \mathbf{Q}$ rather than the components, with respect to the equations,

$$X^j = c_k{}^j \overline{Y}^k.$$

Then

$$\begin{aligned} \mathbf{P} \times \mathbf{Q} &= \sum_{r=1}^{3} \boldsymbol{\iota}_r \mathcal{E}_{rst} P^s Q^t = \sum_{r=1}^{3} C_r{}^q C_r{}^j c \overline{\mathcal{E}}_{jkl} \overline{P}^k \overline{Q}^l \mathbf{p}_q \\ &= h^{qj} \mathbf{p}_q c \overline{\mathcal{E}}_{jkl} \overline{P}^k \overline{Q}^l = \mathbf{p}^j c \overline{\mathcal{E}}_{jkl} \overline{P}^k \overline{Q}^l. \end{aligned}$$

The cross product has its simplest representation in terms of the reciprocal basis $\mathbf{p}^j$ (i.e., in terms of covariant components).

Establishment of the transformation law

$$\mathcal{E}_{ukl}\bar{P}^k\bar{P}^l = bb_u{}^j\bar{\bar{\mathcal{E}}}_{jfg}\bar{\bar{P}}^f\bar{\bar{Q}}^g \tag{1-7*.7}$$

for the covariant components of the cross product sets the ground work for consideration of the triple scalar and triple vector products.

Theorem 1-7*.4. The classes of triple scalar products $\{\mathcal{E}_{ijk}P^iQ^jR^k\}$, $\{E^{ijk}P_iQ_jR_k\}$ are scalars of weights $+1$ and -1, respectively, with respect to the centered affine group of transformations.

PROOF. The result is obtained by employing the transformation laws of the factors involved.

Theorem 1-7*.5. The class $\{E^{ijk}P_j\mathcal{E}_{krs}Q^rR^s\}$ is a contravariant vector of weight zero with respect to the centered affine group.

The proof is obtained by straightforward computation.

Problems

1. If $Y^j = b_k{}^j\bar{Y}^k$, determine the transformation rule for

$$|h_{jk}|, \qquad |h^{jk}|, \qquad \mathcal{E}^{ijk}, \qquad E_{ijk}.$$

2. Construct a tensor of weight zero from $|h_{jk}|$ and E_{ijk}.

3. Prove Theorem 1-7*.4.

4. Prove Theorem 1-7*.5.

5. Prove that the class $E^{ijk}P_jQ_k$ is a vector density of weight -1.

8. *The Algebra of Matrices*

In the preceding development of vector algebra use was made of matrices of numbers such as

$$(a_j{}^k) = \begin{pmatrix} a_1{}^1 & a_1{}^2 & a_1{}^3 \\ a_2{}^1 & a_2{}^2 & a_2{}^3 \\ a_3{}^1 & a_3{}^2 & a_3{}^3 \end{pmatrix}, \qquad (P^1, P^2, P^3).$$

The square matrix $(a_j{}^k)$ was employed in a variety of ways. It represented a set of coefficients of a system of linear equations, a set of transformation equation coefficients, and a set of elements of a determinant.

The utility of such sets of elements had become quite apparent by the middle of the nineteenth century. Determinants and linear transformations were under intensive study at that time. The rather natural, and very important, step of formally developing the algebra of matrices of numbers was taken by the English mathematician Cayley.

The following definition serves the purpose of introducing the matrix concept.

Definition 1-8.1. Any array of elements arranged in m rows and n columns is called a matrix.

In this book the elements are real numbers or real-valued functions.

The concepts of vector algebra could be developed without mentioning the term matrix. However, the matrix and the algebra of matrices have become an important part of modern-day mathematics. Therefore a discussion of the basic language and some of the ideas of matrix algebra should be valuable to the reader in his assimilation of the materials presented here and elsewhere.

The matrices of concern are mostly square, one row, or one column arrays. However, as indicated by Definition 1-8.1, a matrix is not restricted to any one of these forms.

Example 1-8.1. It is quite simple to think of matrices formed from everyday experiences. For example, the weekly array of prices of three "come on" articles, A, B, and C, sold in a supermarket would appear as follows:

(1-8.1)

	S	*M*	*T*	*W*	*TH*	*F*	*S*
A	99	99	99	99	69	69	69
B	87	87	87	87	49	49	49
C	98	98	98	98	79	79	79

Some residents of northern California may even be able to identify the items.

The following convention corresponds the notation of this book with that commonly used in the development of matrix algebra.

Convention 1-8.1. A matrix of m rows and n columns is denoted either by capital letters A, B, $\cdots$, etc., or by $(a_j^{\,k})$, (b_j), (c^k), $\cdots$. When the index notation is used, the subscript represents row and the superscript represents column.

The fundamental laws of matrices are defined as follows:

1. When a matrix is multiplied by a number or function, each individual element is multiplied by that number or function, that is,

$$c(b_j^{\,k}) = (cb_j^{\,k}). \tag{1-8.2a}$$

2. The sum of two matrices is a new matrix, the elements of which are sums of corresponding elements of the original matrices.

$$(b_j^{\,k}) + (c_j^{\,k}) = (b_j^{\,k} + c_j^{\,k}). \tag{1-8.2b}$$

According to (1-8.2b), matrix addition has meaning only when the two matrices involved have the same number of rows and the same number of columns.

The law of multiplication for matrices is analogous to the rule for determinant multiplication. If the matrices are square, there is no distinction.

3. The product of two matrices $(b_j^{\,k})$ and $(c_j^{\,k})$ [note that the number of columns of $(b_k^{\,j})$ must agree with the number of rows in $(c_j^{\,p})$] is a new matrix $(b_k^{\,j}c_j^{\,p})$, summed on j. The product matrix may also be written in the form BC.

Definition 1-8.2. A matrix whose elements are zero is called a zero matrix.

It is important to have the meaning of the equality sign, as used in connection with matrices, clearly in mind.

Definition 1-8.3. Two matrices A and B are equal if and only if

$$a_j^{\,k} = b_j^{\,k}.$$

Clearly two matrices cannot be equal unless they agree with respect to numbers of rows and columns.

Example 1-8.2. As an illustration of matrix multiplication consider

$$(b_j^{\,k}) = \begin{pmatrix} 1 & 2 & 3 \\ 0 & -1 & -2 \\ -3 & 0 & 4 \end{pmatrix}, \qquad (c_j^{\,k}) = \begin{bmatrix} 2 \\ 4 \\ 6 \end{bmatrix}.$$

Then

$$(b_k^{\,j}c_j^{\,p}) = \begin{pmatrix} 2 + 8 + 18 \\ -4 - 12 \\ -6 + 24 \end{pmatrix}, \quad = \begin{bmatrix} 28 \\ -16 \\ 18 \end{bmatrix}.$$

Matrix multiplication is not defined in other situations than that indicated by 3. Therefore consideration of the foregoing example leads to the observation that this type of multiplication is another illustration of a noncommutative process.

In Example 1-8.2 there is no meaning to CB, since the number of columns of $(c_j^{\,k})$ does not agree with the number of rows of $(b_j^{\,k})$. However, the noncommutivity of the binary operation does not depend on this circumstance. Even if both matrices were square, we could not expect the product to commute. This fact is pointed out by the following simple example.

Example 1-8.3. Let

$$(b_j^{\ k}) = \begin{pmatrix} 1 & 1 \\ 0 & 0 \end{pmatrix}, \qquad (c_j^{\ k}) = \begin{pmatrix} 1 & 0 \\ 1 & 0 \end{pmatrix}.$$

Then

$$(b_k^{\ j} c_j^{\ p}) = \begin{pmatrix} 2 & 0 \\ 0 & 0 \end{pmatrix},$$

whereas

$$(c_k^{\ j} b_j^{\ p}) = \begin{pmatrix} 1 & 1 \\ 1 & 1 \end{pmatrix}.$$

The two resultant matrices are clearly not equal; therefore

$$BC \neq CB.$$

Example 1-8.4. If the product of two real numbers is zero, then it follows that one or both of the numbers is zero. The cross product of two vectors is an example of a binary operation which does not share this property. If the cross product $\mathbf{P} \times \mathbf{Q} = \mathbf{0}$, it is possible that neither $\mathbf{P}$ nor $\mathbf{Q}$ is zero but rather that $\mathbf{P}$ is proportional to $\mathbf{Q}$.

The matrix operation of multiplication presents another example of a type of multiplication such that

$$AB = 0,$$

but neither A nor B is necessarily zero. This fact is illustrated by the following:

$$\begin{pmatrix} 1 & 1 \\ 0 & 0 \end{pmatrix}\begin{pmatrix} -1 & -1 \\ 1 & 1 \end{pmatrix} = \begin{pmatrix} 0 & 0 \\ 0 & 0 \end{pmatrix}.$$

The zero matrix is a rather special type. Others are introduced in the next two definitions.

Definition 1-8.4. The square matrix $(\delta_i^{\ j})$ is called the identity matrix. Let the symbol I denote this matrix. Then

$$I = (\delta_i^{\ j}) = \begin{pmatrix} 1 & 0 & \cdots & 0 \\ 0 & 1 & 0 & 0 \\ \cdots & \cdots & \cdots & \cdots \\ 0 & \cdots & 0 & 1 \end{pmatrix}.$$

Definition 1-8.5. The matrix A' obtained from A by interchanging rows and columns is called the transpose of A.

A number of interesting facts are concerned with these special matrices. Some of the ideas relating to them have been touched on previously in terms of the index notation. For example, the Kronecker delta, used to introduce the identity matrix, has been a standard tool.

The following theorems and examples illustrate both properties of the special matrices and their relation to the development of preceding sections.

Theorem 1-8.1. Transpose matrices are subject to the following sum and product rules:

$$\text{(1-8.3)} \qquad \begin{aligned} &\text{(a)} \quad (A + B)' = A' + B'. \\ &\text{(b)} \quad (\alpha A') \quad = \alpha A'; \quad \alpha \text{ is a real number.} \\ &\text{(c)} \quad (AB)' \quad = B'A'. \end{aligned}$$

PROOF. The first two proofs are left to the reader. To prove (1-8.3c), let

$$a_j'^k = a_k{}^j, \qquad b_p'^j = b_j{}^p, \qquad c_k'^p = c_p{}^k,$$

where $c_p{}^k = a_p{}^j b_j{}^k$ and $c_k'^p$ are the elements of the transpose of C. Then, starting with the right-hand side of (1-8.3c), we have

$$b_p'^j a_j'^k = b_j{}^p a_k{}^j = a_k{}^j b_j{}^p.$$

This completes the proof.

Example 1-8.5a. The identity matrix is its own transpose.

Example 1-8.5b. The identity matrix commutes with any other square matrix of the same order, for

$$\delta_k{}^j a_j{}^p = a_k{}^p = a_k{}^j\, \delta_j{}^p.$$

Definition 1-8.6. The inverse A^{-1} of a square matrix A is a matrix that satisfies the conditions

$$\text{(1-8.4)} \qquad AA^{-1} = A^{-1}A = I.$$

Theorem 1-8.2. If the determinant of a square matrix A is different from zero (i.e., $a = |a_j{}^k| \neq 0$), the inverse matrix has elements

$$A_k{}^j = \frac{\text{cofactor } a_j{}^k \text{ in } |a_p{}^q|}{a}.$$

PROOF. This result was obtained in Section 3. [See (1-3.17).]

Example 1-8.6. In rectangular Cartesian rotations the inverse of the matrix of transformation coefficients is the same as the transpose. In Section 3 it was shown that

$$A_k^{\ j} = a_j^{\ k}.$$

Example 1-8.7. Suppose that X^j, $\overline{X}^j$, $\overline{\overline{X}}^j$ are coordinates of rectangular Cartesian systems with a common origin. In Section 5* it was pointed out that the transformations relating such systems determined a transformation group. These fundamental group properties are satisfied by the matrices associated with the transformations. In fact, this statement is nothing more than a shift in emphasis from the transformations to the arrays of transformation coefficients. In particular, if the transformations are denoted by

$$X^j = a_i^{\ j}\overline{X}^i, \qquad \overline{X}^i = b_k^{\ i}\overline{\overline{X}}^k, \qquad X^j = c_k^{\ j}\overline{\overline{X}}^k$$

and A, B, C represent the corresponding matrices, it can be shown that the matrices satisfy the group properties. (See Definition 1-5*.1.) We illustrate with the closure property. Suppose that $(a_i^{\ j})$, $(b_i^{\ j})$ were orthogonal transformation coefficients and $c_i^{\ k} = a_i^{\ j}b_j^{\ k}$. Then

$$\sum_{i=1}^{3} c_i^{\ k}c_i^{\ p} = \sum_{i=1}^{3} a_i^{\ j}b_j^{\ k}a_i^{\ q}b_q^{\ p} = \sum_{i=1}^{3} a_i^{\ j}a_i^{\ q}b_j^{\ k}b_q^{\ p}$$
$$= \delta^{jq}b_j^{\ k}b_q^{\ p} = \delta^{kp}.$$

The corresponding proof in matrix notation would appear as follows. We have (as a translation to matrix form of (1-3.21h)

$$A'A = I, \qquad B'B = I,$$

where the $'$ indicates inverse as well as transpose. Therefore

$$C'C = (AB)'(AB) = B'A'AB = B'IB = B'B = I,$$

as was to be shown.

Definition 1-8.7. The set of coefficients of transformation $(b_j^{\ k})$ associated with a pair of Cartesian coordinate systems is called a rotation matrix. The $(a_j^{\ k})$ associated with rectangular Cartesian systems is called an orthogonal rotation matrix.

In studying Section 3 the reader may have been impressed with the seeming difficulty of even making up a specific example of an orthogonal rotation matrix.

The nine cosines, $a_j^{\ k}$, must meet the rather stringent conditions

$$\sum_{j=1}^{3} a_i^{\ j}a_k^{\ j} = \delta_{ik}, \qquad a = 1.$$

However, construction of examples is made quite easy by the process of successive transformations mentioned in example (1-8.7). A two-dimensional rotation is dependent on only one angle, hence is simple to determine. By carrying out successive plane rotations, three-dimensional transformations can be accomplished.

Example 1-8.8. Plane rotations

$$\begin{pmatrix} 1 & 0 & 0 \\ 0 & \frac{1}{2} & \frac{\sqrt{3}}{2} \\ 0 & \frac{-\sqrt{3}}{2} & \frac{1}{2} \end{pmatrix}, \quad \begin{pmatrix} \frac{1}{2} & \frac{\sqrt{3}}{2} & 0 \\ \frac{-\sqrt{3}}{2} & \frac{1}{2} & 0 \\ 0 & 0 & 1 \end{pmatrix}$$

give rise to the three-space rotation

$$\begin{pmatrix} \frac{1}{2} & \frac{\sqrt{3}}{2} & 0 \\ \frac{-\sqrt{3}}{4} & \frac{1}{4} & \frac{\sqrt{3}}{2} \\ \frac{3}{4} & \frac{-\sqrt{3}}{4} & \frac{1}{2} \end{pmatrix}$$

The question whether any orthogonal rotation matrix can be obtained as a product of plane rotation matrices naturally arises. This question is answered by the following theorem.

Theorem 1-8.3. Any orthogonal rotation matrix can be obtained as a product of not more than three plane orthogonal rotation matrices.

PROOF. It must be shown that any orthogonal rotation matrix D can be obtained according to the following relation:

(1-8.5)

$$\begin{pmatrix} d_1{}^1 & d_1{}^2 & d_1{}^3 \\ d_2{}^1 & d_2{}^2 & d_2{}^3 \\ d_3{}^1 & d_3{}^2 & d_3{}^3 \end{pmatrix} = \begin{pmatrix} 1 & 0 & 0 \\ 0 & a_2{}^2 & a_2{}^3 \\ 0 & a_3{}^2 & a_3{}^3 \end{pmatrix} \left[\begin{pmatrix} b_1{}^1 & 0 & b_1{}^3 \\ 0 & 1 & 0 \\ b_3{}^1 & 0 & b_3{}^3 \end{pmatrix} \begin{pmatrix} c_1{}^1 & c_1{}^2 & 0 \\ c_2{}^1 & c_2{}^2 & 0 \\ 0 & 0 & 1 \end{pmatrix} \right].$$

There is no claim that the representation in (1-8.5) is unique.

The orthogonality conditions (1-3.21a,b) are not independent. Therefore only one set should be considered in determining the number of parameters of $d_j{}^k$ that are independent. There are six conditions imposed on nine unknowns d_j^k; hence choice of an appropriate triple of the d_j^k along with the orthogonality conditions determines the set.

This means that we must show that knowledge of an appropriate three of the d_j^k determines the second-order matrices A, B, and C.

A first step in this procedure is to carry out the multiplication on the right of (1-8.5). We obtain

$$(d_j^k) = \begin{pmatrix} b_1^{\ 1}c_1^{\ 1} & b_1^{\ 1}c_1^{\ 2} & b_1^{\ 3} \\ a_2^{\ 2}c_2^{\ 1} + a_2^{\ 3}a_3^{\ 1}c_1^{\ 1} & a_2^{\ 2}c_2^{\ 2} + a_2^{\ 3}b_3^{\ 1}c_1^{\ 2} & a_2^{\ 3}b_3^{\ 3} \\ a_3^{\ 2}c_2^{\ 1} + a_3^{\ 3}b_3^{\ 1}c_1^{\ 1} & a_3^{\ 2}c_2^{\ 2} + a_3^{\ 3}b_3^{\ 1}c_1^{\ 2} & a_3^{\ 3}b_3^{\ 3} \end{pmatrix}. \tag{1-8.6a}$$

By definition of matrix equality

$$d_1^{\ 3} = b_1^{\ 3}. \tag{1-8.6b}$$

Each of the three matrices A, B, C must satisfy the orthogonality conditions (1-3.21a,b,c). By substituting $d_1^{\ 3}$ for $b_1^{\ 3}$ the elements of the first row of (1-8.6a) are seen to satisfy the relation

$$(b_1^{\ 1})^2[(c_1^{\ 1})^2 + (c_1^{\ 2})^2] + (d_1^{\ 3})^2 = 1. \tag{1-8.6c}$$

But

$$(c_1^{\ 1})^2 + (c_1^{\ 2})^2 = 1.$$

Therefore

$$(b_1^{\ 1})^2 + (d_1^{\ 3})^2 = 1$$

and

$$b_1^{\ 1} = \epsilon_1^{\ 1}[1 - (d_1^{\ 3})^2]^{1/2}, \qquad \epsilon_1^{\ 1} = \pm 1. \tag{1-8.6d}$$

If we employ the orthogonality conditions to which B is subject, we can determine $b_3^{\ 1}$, $b_1^{\ 3}$, and $b_3^{\ 3}$ up to sign. The elements of C can be evaluated by assuming knowledge of a second value d_j^k. For example, from (1-8.6a)

$$b_1^{\ 1}c_1^{\ 1} = d_1^{\ 1}.$$

Therefore

$$c_1^{\ 1} = \frac{d_1^{\ 1}}{b_1^{\ 1}} = \frac{d_1^{\ 1}}{\epsilon_1^{\ 1}[1 - (d_1^{\ 3})^2]^{1/2}}. \tag{1-8.6e}$$

The other elements of C are obtained as a result of the orthogonality conditions. Finally, a third value d_j^k leads to the determination of the element of A. The ϵ's must be chosen so that each of the matrices A, B, C has a determinant with value $+1$. This completes the proof.

Problems

1. Make up an example of a matrix from everyday life.

2. (a) Add the matrices

$$A = \begin{pmatrix} 2 & 3 & 1 \\ 5 & 4 & 3 \\ 2 & 1 & 5 \\ -2 & 0 & 6 \end{pmatrix} \qquad B = \begin{pmatrix} 1 & 3 & 4 \\ 2 & 5 & 1 \\ 6 & 1 & 4 \\ 0 & -1 & 2 \end{pmatrix}.$$

(b) Can the matrices be multiplied?

(c) Determine the product AB if it exists.

3. Find the rotation matrix obtained as a product of

$$\begin{pmatrix} \frac{\sqrt{2}}{2} & \frac{-\sqrt{2}}{2} & 0 \\ \frac{\sqrt{2}}{2} & \frac{\sqrt{2}}{2} & 0 \\ 0 & 0 & 1 \end{pmatrix} \begin{pmatrix} 1 & 0 & 0 \\ 0 & \frac{\sqrt{3}}{2} & -\frac{1}{2} \\ 0 & \frac{1}{2} & \frac{\sqrt{3}}{2} \end{pmatrix}$$

4. Show that the matrices of Problem 3 do not commute; hence rotations are not commutative.

5. Is the determinant of the product of two square matrices equal to the product of the determinants of the matrices?

6. Show that an arbitrary three-space rotation cannot be obtained by means of two plane rotations.
Hint: Set up the product $(a_j{}^k)(b_j{}^k) = (d_j{}^k)$ where the matrices on the left represent plane rotations. Show that three values $(d_j{}^k)$ cannot be arbitrarily determined.

7. Write $X^j = a_k{}^j \bar{X}^k$ in matrix form.

chapter 2 the differentiation of vectors

1. The Differentiation of Vectors

In this chapter many of the ideas of the differential calculus are amalgamated with the algebraic theory of vectors previously presented. From a modern-day viewpoint two concepts, limit and function, assume fundamental importance in its development. The reader who has been subjected to a rigorous course in calculus is aware of the inherent difficulties associated with these ideas. Indeed, the difficulties are not peculiar to beginning students alone. Isaac Newton (1642–1727, English) and Gottfreid Wilhelm Leibniz are given the principal credit for the origination and compilation of the basic concepts of differential calculus. Leibniz originated the term function. Yet, according to the standards of our time, neither had a clear idea of the meaning of either function or limit. The same can be said of a host of other outstanding mathematicians [e.g., Leonard Euler (1707–1783, Swiss)] who lived in the eighteenth century. In fact not until the nineteenth century and the advent of a serious study of the foundations of analysis did the limit concept receive a rigorous treatment. Then Augustin Cauchy (1789–1857, French) and his contemporaries began a study of the foundations that has extended into the twentieth century.

The first section in this chapter deals with functions of one variable and their differentiation. Because the meaning of the term function has gone through a long evolutionary process and is still used with various shades of meaning, a formal statement of the meaning of the term as applied in this book is presented. The concept is employed with this significance throughout the notes.

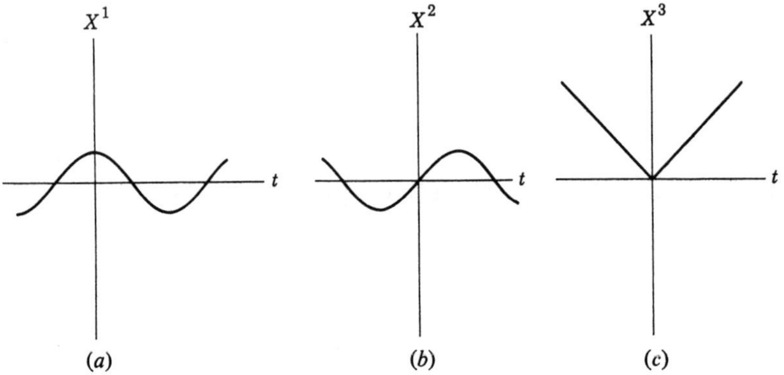

Fig. 2-1.1

Definition 2-1.1. A function f is a set of ordered pairs, no two of which have the same first element. The set of first elements of the pairs is called the domain of the function, whereas the set of second elements of the pairs is called the range. The domain and range elements are related by a given rule.

In this section the domain is either the set of real numbers or some subset of real numbers. The independent variable representing the domain elements is often referred to as a parameter. (See Figs. 2-1a,b,c.)

Example 2-1.1. The following three functions will be useful in the future for illustrative purposes. The symbol t is used to represent the parameter in each instance.

(a) Domain: the set of all real numbers.
Rule: $X^1(t) = \cos t$.
Range: the set of values $X^1(t)$ such that $-1 \leq X^1(t) \leq 1$.

(b) Domain: the set of real numbers.
Rule: $X^2(t) = \sin t$.
Range: the set of values $X^2(t)$ such that $-1 \leq X^2(t) \leq 1$.

(c) Domain: the set of real numbers.
Rule: $X^3(t) = |t|$.
Range: the set of values $X^3(t)$ such that $X^3(t) \geq 0$.

From a geometric standpoint, the main interest of this section centers around curves in Euclidean three-space. It is assumed that the analytic forms for these curves are expressed with respect to a rectangular Cartesian coordinate system.

The kinematical viewpoint of the path traced by a moving particle presents us with an intuitive representation of a curve. The following three-part definition gives a preciseness to the meaning of the concept.

Definition 2-1.2a. Let functions X^1, X^2, and X^3 have continuous derivatives of order 1, at least, on a closed interval $a \leq t \leq b$. Suppose that the first derivatives are not simultaneously zero. Then the representation

$$X^j = X^j(t), \qquad j = 1, 2, 3 \tag{2-1.1a}$$

is said to be an allowable parameter representation.

Definition 2-1.2b. Suppose that a function

$$t = t(t^*) \tag{2-1.1b}$$

is defined on an interval $a^* \leq t^* \leq b^*$ with corresponding range values in $a \leq t \leq b$ and such that $t(a^*) = a$, $t(b^*) = b$ or $t(a^*) = b$, $t(b^*) = a$. If the function in (2-1.1b) has a continuous first derivative, at least, which is different from zero everywhere on $a^* \leq t^* \leq b^*$, then (2-1.1b) is said to be an allowable parameter transformation.

Definition 2-1.2b makes it possible to construct equivalence classes of allowable representations. Two representations are said to be in the same equivalence class if they are related by a transformation satisfying Definition 2-1.2b.

Definition 2-1.2c.[1] The set of all points specified by the allowable parameter representations of an equivalence class is said to be a smooth curve. (Note that a smooth curve is ordered in a certain manner.)

Quite often a smooth curve[2] serves as a mathematical model of the path of motion of a physical particle; the parameter t represents time values and the coordinates X^j represent the position of the particle. Mathematical models of physical motion are developed later in the book and then, as just indicated, t should be interpreted as representative of time. However, it would be unwise of the reader to allow this connotation to become too firmly fixed in his mind. There will be occasions when it will be convenient to interpret the parameter otherwise.

---

[1] See Kreyszig, Erwin, *Differential Geometry*, University of Toronto Press, 1959.

[2] The definition given is not the most general description of a curve as indicated by the adjective "smooth." The fact that we are studying the elementary calculus of vectors evokes the need for the property of differentiability. It will be found that a well-defined tangent at each point of the curve is the geometric manifestation of the property. It is cumbersome to continue to attach the word *smooth*; therefore we shall do so only when the emphasis seems useful. The curves in this text are assumed to be smooth unless it is specifically stated otherwise.

The following examples of smooth space curves should familiarize the reader with the concept.

Example 2-1.2. A circular helix is represented by the parametric equations

$$X^1 = \cos t,$$
$$X^2 = \sin t,$$
$$X^3 = t.$$

This curve can be thought of intuitively as a spiral on a circular cylinder. (See Fig. 2-1.2.) If the domain values t are restricted so that $t > 0$, a curve lying entirely above the X^1, X^2 plane is obtained. Compare with Example 2-1.1. The equations

$$X^1 = \cos (t^*)^3$$
$$X^2 = \sin (t^*)^3$$
$$X^3 = (t^*)^3$$

form another allowable parameter representation of that part of the circular helix corresponding to t and t^* greater than zero. Note that the representations cannot be thought of as in the same equivalence class if the domain of t^* is extended to include zero for $dt/dt^* = 0$ at $t^* = 0$. On the other hand the derivative is different from zero on every closed domain of positive t^* values.

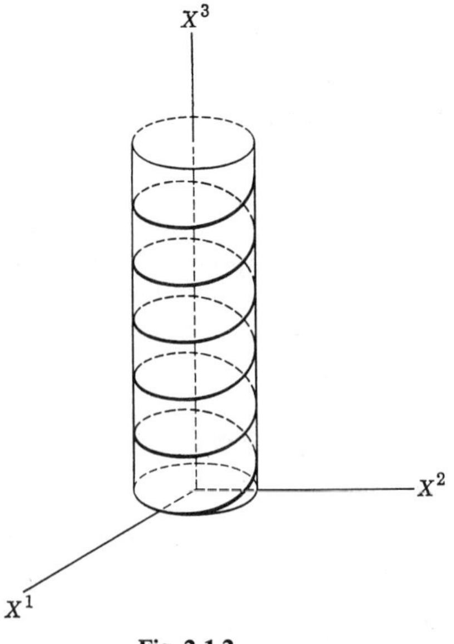

Fig. 2-1.2

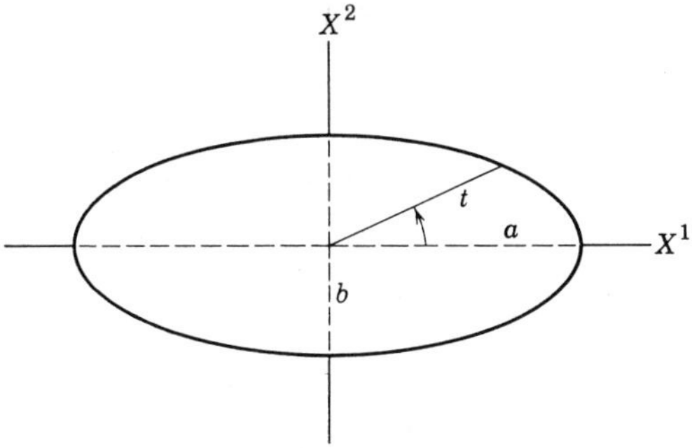

Fig. 2-1.3

Example 2-1.3. Parametric equations of an ellipse, in the plane $X^3 = 0$, and with its center at the origin are

$$X^1 = a \cos t,$$
$$X^2 = b \sin t, \qquad 0 \le t < 2\pi,$$
$$X^3 = 0.$$

If t is interpreted geometrically as a measure of the central angle, then $0 \le t \le 2\pi$ gives rise to the complete ellipse. (See Fig. 2-1.3.)

In the study of differential calculus we investigate the properties of a given curve by examining the behavior of the tangent (i.e., by considering the slope function of the curve). Physically it is found that velocity and acceleration vectors have a relationship to the path of motion of a particle. We prepare the way for a study of these and other considerations by introducing the idea of vector and scalar fields with respect to a smooth space curve.

Definition 2-1.3. Let $X^j = X^j(t)$, $t_1 \le t \le t_2$, represent parametric equations of a smooth space curve C. Suppose that Φ is a function, defined along C, which at any point of C transforms according to the scalar law of transformation. (See Chapter 1, Section 4.) Then $\Phi(X^j)$ is the X^j coordinate system representative of a scalar field $\{\Phi\}$ with respect to the curve C.

The function Φ is assumed to be differentiable, and therefore continuous, unless otherwise stated.

Definition 2-1.4. Let $X^j = X^j(t)$ represent a set of parametric equations of a smooth space curve C. Let U^j be three functions defined on C. Furthermore, suppose that at any point of C the set of three functions transform as components of a vector **U**. Then **U** is said to represent a vector field with respect to C.

Unless otherwise stated, the functions U^j are assumed to be differentiable.

Example 2-1.4. Let

$$\Phi(X^1, X^2, X^3) = \ln(\mathbf{P} \cdot \mathbf{Q}),$$

where

$$\mathbf{P} = \frac{1}{a^2}\boldsymbol{\iota}_1 + \frac{1}{b^2}\boldsymbol{\iota}_2, \qquad \mathbf{Q} = (X^1)^2\boldsymbol{\iota}_1 + (X^2)^2\boldsymbol{\iota}_2.$$

Then

$$\mathbf{P} \cdot \mathbf{Q} = \frac{(X^1)^2}{a^2} + \frac{(X^2)^2}{b^2}$$

is the X^j coordinate system representation of a scalar with respect to the orthogonal Cartesian group. Since $\mathbf{P} \cdot \mathbf{Q}$ is a scalar, it follows that $\ln(\mathbf{P} \cdot \mathbf{Q})$ is a scalar.

If $\mathbf{P} \cdot \mathbf{Q}$ is defined along the curve C of Example 2-1.3, $\Phi = 0$ at every point of C; that is, the scalar field is constant over C.

Example 2-1.5. Let a space curve C be given by parametric equations

$$X^1 = t, \qquad 0 \le t \le 1,$$
$$X^2 = (t)^2,$$
$$X^3 = 0.$$

Let

$$U^1 = 1,$$
$$U^2 = 2X^1,$$
$$U^3 = 0.$$

The curve is the segment of the parabola $X^2 = (X^1)^2$ defined on the interval $0 \le X^1 \le 1$. (See Fig. 2-1.4.) Since $U^2/U^1 = 2X^1$ (i.e., the slope of the tangent line at any point of the parabola), it follows that the vector field $\mathbf{U}$ is a tangent vector field along C.

The tools of elementary calculus suffice for exhibiting the derivative properties when dealing with scalar fields.

If Φ is thought of as a function constructed from $\Phi = \Phi(X^1, X^2, X^3)$ and $X^j = X^j(t)$, the ordinary definition of derivative may be employed; that is, if Φ is differentiable,

$$\boxed{\frac{d\Phi(t)}{dt} = \lim_{\Delta t \to 0} \frac{\Phi(t + \Delta t) - \Phi(t)}{\Delta t}}. \tag{2-1.2a}$$

On the other hand, when Φ is treated as a composite function $\Phi = \Phi[X^j(t)]$,

$$\boxed{\frac{d\Phi}{dt} = \frac{\partial\Phi}{\partial X^1}\frac{dX^1}{dt} + \frac{\partial\Phi}{\partial X^2}\frac{dX^2}{dt} + \frac{\partial\Phi}{\partial X^3}\frac{dX^3}{dt} = \frac{\partial\Phi}{\partial X^j}\frac{dX^j}{dt}}. \tag{2-1.2b}$$

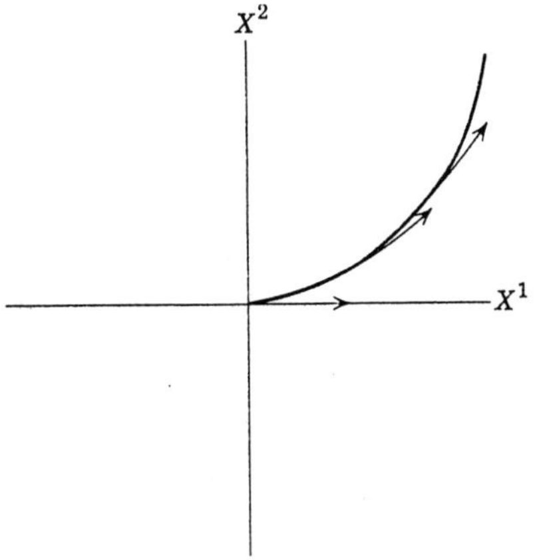

Fig. 2-1.4

The d/dt notation of Leibniz predominates throughout these notes.

Example 2-1.6. Consider the scalar $\mathbf{P} \cdot \mathbf{Q}$ of Example 2-1.4. Suppose that a curve is given by parametric equations

$$X^1 = \cos t, \qquad 0 \le t \le \pi,$$
$$X^2 = \sin t,$$
$$X^3 = 0.$$

Then

$$\Phi(X^j(t)) = \ln\left(\frac{(X^1)^2}{a^2} + \frac{(X^2)^2}{b^2}\right) = \ln\left(\frac{\cos^2 t}{a^2} + \frac{\sin^2 t}{b^2}\right)$$

is a scalar field along the curve. (See Fig. 2-1.5.) The derivative of the scalar field can be calculated from either the middle or the last member of the foregoing relation. If we differentiate the last member, we have

$$\frac{d\Phi}{dt} = \frac{1}{[(\cos^2 t/a^2) + (\sin^2 t/b^2)]}\left(\frac{-2\cos t \sin t}{a^2} + \frac{2 \sin t \cos t}{b^2}\right)$$
$$= \frac{2\cos t \sin t(a^2 - b^2)}{b^2 \cos^2 t + a^2 \sin^2 t}.$$

The process of differentiating a vector field, defined along a curve C, is an immediate abstraction from that of scalar field differentiation.

Definition 2-1.5. Let $\mathbf{U}$ represent a vector field along a curve C. When the limit exists

$$\frac{d\mathbf{U}}{dt} = \lim_{\Delta t \to 0} \frac{\mathbf{U}(t + \Delta t) - \mathbf{U}(t)}{\Delta t} \tag{2-1.3}$$

is called the derivative of the vector field $\mathbf{U}$.

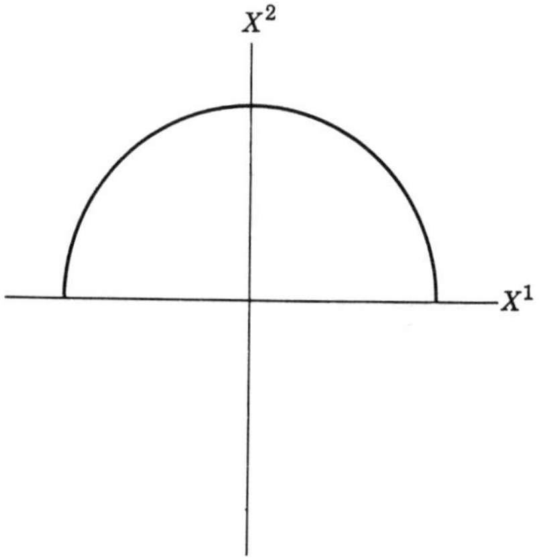

Fig. 2-1.5

If the elements of the n-tuple basis to which $\mathbf{U}$ is referred are constants, as are $\boldsymbol{\iota}_1$, $\boldsymbol{\iota}_2$, and $\boldsymbol{\iota}_3$, the differentiation (2-1.3) can be carried out component by component. We have

$$\lim_{\Delta t \to 0} \frac{\mathbf{U}(t + \Delta t) - \mathbf{U}(t)}{\Delta t}$$

$$= \lim_{\Delta t \to 0} \frac{[U^1(t + \Delta t)\boldsymbol{\iota}_1 + U^2(t + \Delta t)\boldsymbol{\iota}_2 + U^3(t + \Delta t)\boldsymbol{\iota}_3] - [U^1(t)\boldsymbol{\iota}_1 + U^2(t)\boldsymbol{\iota}_2 + U^3(t)\boldsymbol{\iota}_3]}{\Delta t}$$

$$= \lim_{\Delta t \to 0} \left\{ \frac{[U^1(t + \Delta t) - U^1(t)]\boldsymbol{\iota}_1}{\Delta t} + \frac{[U^2(t + \Delta t) - U^2(t)]\boldsymbol{\iota}_2}{\Delta t} + \frac{[U^3(t + \Delta t) - U^3(t)]\boldsymbol{\iota}_3}{\Delta t} \right\}$$

$$= \left[\lim_{\Delta t \to 0} \frac{U^1(t + \Delta t) - U^1(t)}{\Delta t} \right] \boldsymbol{\iota}_1 + \left[\lim_{\Delta t \to 0} \frac{U^2(t + \Delta t) - U^2(t)}{\Delta t} \right] \boldsymbol{\iota}_2 + \left[\lim_{\Delta t \to 0} \frac{U^3(t + \Delta t) - U^3(t)}{\Delta t} \right] \boldsymbol{\iota}_3$$

$$= \frac{dU^1}{dt} \boldsymbol{\iota}_1 + \frac{dU^2}{dt} \boldsymbol{\iota}_2 + \frac{dU^3}{dt} \boldsymbol{\iota}_3.$$

In other words, for a constant basis $\boldsymbol{\iota}_1$, $\boldsymbol{\iota}_2$, $\boldsymbol{\iota}_3$

$$\boxed{\frac{d\mathbf{U}}{dt} = \frac{dU^1}{dt} \boldsymbol{\iota}_1 + \frac{dU^2}{dt} \boldsymbol{\iota}_2 + \frac{dU^3}{dt} \boldsymbol{\iota}_3 = \frac{dU^j}{dt} \boldsymbol{\iota}_j.} \tag{2-1.4}$$

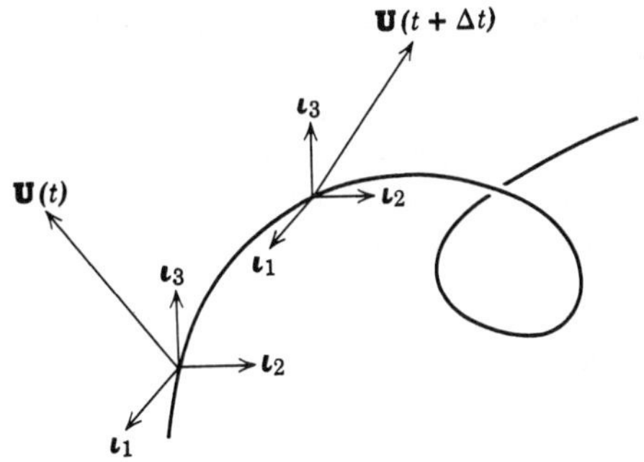

Fig. 2-1.6

The derivation of the relation (2-1.4) depended on two important facts. First of all the vectors $\mathbf{U}(t + \Delta t)$ and $\mathbf{U}(t)$ were compounded in a componentwise manner. This procedure is justified by the fact that the basis is independent of the curve point under consideration. Second, the basis n-tuples were not affected by the limiting process. (See Fig. 2-1.6.)

In a Euclidean space a constant n-tuple basis may always be introduced. Therefore there is meaning associated with the difference $\mathbf{U}(t + \Delta t) - \mathbf{U}(t)$ and with the vector field derivative $d\mathbf{U}/dt$ as defined in (2-1.3). If the space is not Euclidean, a re-examination of the derivative concept is necessary. However, even in a Euclidean space, if systems of representation are used in which the basis is not constant, the basis elements enter into the differentiation process and (2-1.4) is not valid.

According to (2-1.4), whenever a vector field $\mathbf{U}$ is referred to a rectangular Cartesian coordinate system, and more to the point, to the associated orthogonal unit basis $\boldsymbol{\iota}_1, \boldsymbol{\iota}_2, \boldsymbol{\iota}_3$, the vector field differentiation is a component-by-component differentiation. An immediate consequence of this is that the usual laws for differentiation of sums and products carry over from calculus. This information is formally set forth in the following theorem.

Theorem 2-1.1. Let $\mathbf{U}$ and $\mathbf{V}$ be differentiable vector fields defined on a curve C. Then

$$\text{(2-1.5)} \quad \begin{aligned} &\text{(a)} \quad \frac{d(\mathbf{U} + \mathbf{V})}{dt} = \frac{d\mathbf{U}}{dt} + \frac{d\mathbf{V}}{dt}, \\ &\text{(b)} \quad \frac{d\mathbf{U} \cdot \mathbf{V}}{dt} = \left(\frac{d\mathbf{U}}{dt}\right) \cdot \mathbf{V} + \mathbf{U} \cdot \left(\frac{d\mathbf{V}}{dt}\right), \\ &\text{(c)} \quad \frac{d\mathbf{U} \times \mathbf{V}}{dt} = \left(\frac{d\mathbf{U}}{dt}\right) \times \mathbf{V} + \mathbf{U} \times \left(\frac{d\mathbf{V}}{dt}\right), \\ &\text{(d)} \quad \frac{df\mathbf{U}}{dt} = \left(\frac{df}{dt}\right)\mathbf{U} + f\left(\frac{d\mathbf{U}}{dt}\right). \end{aligned}$$

PROOF. Our considerations assume the framework of a rectangular Cartesian coordinate system and a constant unit orthogonal basis $\boldsymbol{\iota}_1$, $\boldsymbol{\iota}_2$, and $\boldsymbol{\iota}_3$. Therefore the properties in (2-1.5) follow immediately from the corresponding elementary properties concerning derivatives of functions. To illustrate this fact, consider (2-1.5c). In order to facilitate use of the summation convention, let $U_j = \delta_{jk}U^k$ and $V_j = \delta_{jk}V^k = V^j$. Then

$$\frac{d(\mathbf{U} \times \mathbf{V})}{dt} = \frac{d}{dt}(\boldsymbol{\iota}_p E^{pjk} U_j V_k) = \boldsymbol{\iota}_p E^{pjk} \frac{d}{dt}(U_j V_k)$$

$$= \boldsymbol{\iota}_p E^{pjk}\left(\frac{dU_j}{dt} V_k + U_j \frac{dV_k}{dt}\right),$$

$$= \frac{d\mathbf{U}}{dt} \times \mathbf{V} + \mathbf{U} \times \frac{d\mathbf{V}}{dt}.$$

The proof of relations (2-1.5a,b,d) is left to the reader.

Note carefully that the order of **U** and **V** must be meticulously observed in (2-1.5c). This is because the cross product is anticommutative. In (2-1.5a,b,d) the ordering is not so important because the sum and products have the commutative property! However, it is worthwhile to form the habit of maintaining order.

The derivative formulas (2-1.5a), (2-1.5c), and (2-1.5d) can be extended in the usual manner; that is, for three vectors **U**, **V**, and **W**, (2-1.5a) is extended to

$$\frac{d(\mathbf{U} + \mathbf{V} + \mathbf{W})}{dt} = \frac{d\mathbf{U}}{dt} + \frac{d\mathbf{V}}{dt} + \frac{d\mathbf{W}}{dt}.$$

for a set of n vectors $\mathbf{U}_1 \cdots \mathbf{U}_n$.

(2-1.6a)
$$\boxed{\frac{d\sum_{j=1}^{n}\mathbf{U}_j}{dt} = \sum_{j=1}^{n}\frac{d\mathbf{U}_j}{dt}.}$$

Extension of (2-1.5c) involves consideration of the triple vector product. Its derivative has the form

(2-1.6b)
$$\frac{d[\mathbf{U} \times (\mathbf{V} \times \mathbf{W})]}{dt} = \frac{d\mathbf{U}}{dt} \times (\mathbf{V} \times \mathbf{W}) + \mathbf{U} \times \left(\frac{d\mathbf{V}}{dt} \times \mathbf{W}\right) + \mathbf{U} \times \left(\mathbf{V} \times \frac{d\mathbf{W}}{dt}\right).$$

Observe that the cross product is not associative. Therefore the association made on the left of (2-1.6b) must be maintained in the right member of that expression. The proof of (2-1.6b) is left to the reader. If, in relation

(2-1.5d), f is a product of functions, (2-1.5d) can be applied and followed by the product formula of elementary calculus with respect to the factor df/dt.

Finally, we observe that there is no meaning to a dot product of more than two vectors; hence the derivative formula (2-1.5b) is not subject to extension.

Two questions which may or may not have occurred to the reader are asked. Assuming that **U** and **V** are vectors and f is a scalar, are the derivatives expressed in (2-1.5a,c,d) of vector character? Is the derivative (2-1.5b) a scalar? These questions are answered for orthogonal Cartesian transformations by the following theorem.

Theorem 2-1.2. Let **U** and **V** be vector fields along a curve C and let f be a scalar field along this curve. Assume that the derivatives (2-1.5a,b,c,d) are well defined on C. Then $(d\mathbf{U} + \mathbf{V})/dt$, $[d(\mathbf{U} \times \mathbf{V})]/dt$, $df\,\mathbf{U}/dt$ are vector fields on C and $(d\mathbf{U} \cdot \mathbf{V})/dt$ is a scalar field on C with respect to the group of orthogonal Cartesian transformations.

PROOF. Each proof follows from the fact that the coefficient of transformation are constants. For example, to prove that $d(\mathbf{U} \times \mathbf{V})]/dt$ has vector character (see Theorem 1-7.1b), we simply note that

$$\frac{d(\mathcal{E}_{jkp}U^kV^p)}{dt} = \frac{d\sum_{r=1}^{3} a_r^{\ j}\,\bar{\mathcal{E}}_{rst}\bar{U}^s\bar{V}^t}{dt} = \sum_{r=1}^{3} a_r^{\ j}\,\frac{d\bar{\mathcal{E}}_{rst}\bar{U}^s\bar{V}^t}{dt}.$$

In other words, the components of the derivative of the cross product satisfy the vector transformation law. It is left to the reader to show that the other derivatives satisfy the appropriate vector or scalar transformation law.

One way of studying a curve is by investigating the behavior of the tangential field along it. This is the viewpoint of differential calculus. The question before us is that of extending our analytic tools so that curves in three dimensions can be included in these investigations.

The vector field

$$\frac{d\mathbf{r}}{dt} = \lim_{\Delta t \to 0} \frac{\mathbf{r}(t + \Delta t) - \mathbf{r}(t)}{\Delta t} \tag{2-1.7}$$

comes to mind for at least two reasons. First of all, in a special case in which the curve lies in the X^1,X^2 plane, the ratio of the components dX^1/dt and dX^2/dt of $d\mathbf{r}/dt$ is dX^2/dX^1. We recognize this derivative as the

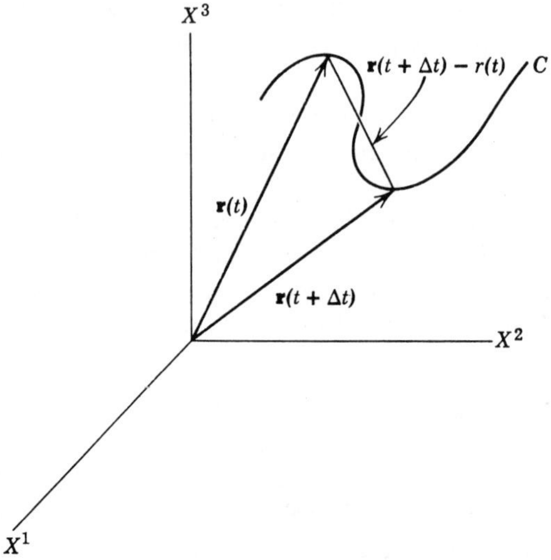

Fig. 2-1.7

representative of a slope function. Second, a geometric argument for the significance of $d\mathbf{r}/dt$ can be made on the basis of Fig. 2-1.7.

Definition 2-1.6. Let a smooth curve C have the vector representation $\mathbf{r} = \mathbf{r}(t)$. Then

(2-1.8a)
$$\frac{d\mathbf{r}}{dt} = \lim_{\Delta t \to 0} \frac{\mathbf{r}(t + \Delta t) - \mathbf{r}(t)}{\Delta t}$$

is called a tangential vector field along C. A line

(2-1.8b)
$$\mathbf{Y} = \mathbf{r}_0 + \left(\frac{d\mathbf{r}}{dt}\right)_0 u$$

is said to be the tangent line to C at a point P_0 of C. The symbols u and $\mathbf{Y}$ are chosen to represent the parameter values and coordinates, respectively, along the line in order to distinguish them from the corresponding entities for the curve.

The direction of the tangent line at a P_0 of C is unique, but the sense and magnitude of a tangential vector depends on the parameterization of the curve. (See Fig. 2-1.8.)

Example 2-1.7. Parametric equations of a parabola in the X^1,X^2 plane are

$$X^1 = 1 + t^2,$$
$$X^2 = t,$$
$$X^3 = 0.$$

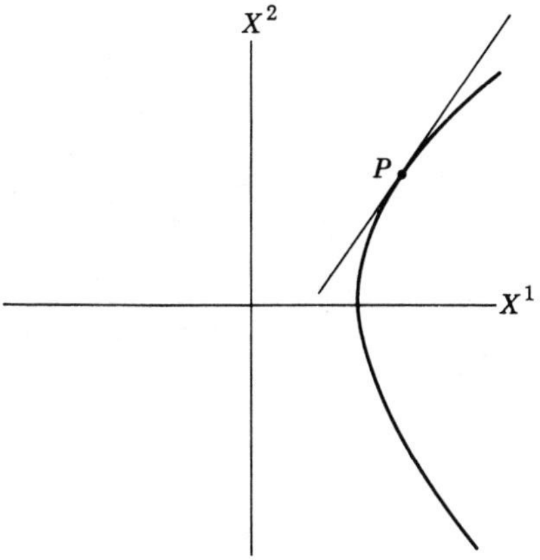

Fig. 2-1.8

By differentiating, we obtain

$$\frac{dX^1}{dt} = 2t,$$

$$\frac{dX^2}{dt} = 1,$$

$$\frac{dX^3}{dt} = 0.$$

Therefore the tangent vector field is expressed by

$$\frac{d\mathbf{r}}{dt} = 2t\boldsymbol{\iota}_1 + \boldsymbol{\iota}_2.$$

The tangent line to the parabola at $t = 3$ is

$$X^1 = 10 + 6u,$$
$$X^2 = 3 + u,$$
$$X^3 = 0.$$

If the parameter t is replaced by $v = -t$,

$$(X^1, X^2, X^3) = (1 + v^2, -v, 0),$$

and

$$\left(\frac{dX^1}{dv}, \frac{dX^2}{dv}, \frac{dX^3}{dv}\right) = (2v, -1, 0).$$

A comparison of these components with those previously determined reveals that the change in sign of the parameter corresponds to a change in sense of the tangent vector field.

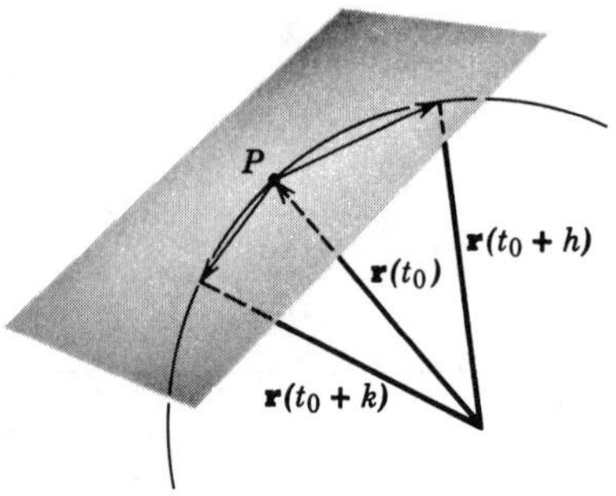

Fig. 2-1.9

With few exceptions, intuition adequately guides our notions concerning tangent lines to plane curves. However, visualization of space-curve tangents is not always so easy. Our intuition is helped by the discovery that, roughly speaking, the tangent line at a curve point P_0 lies in a plane determined by P_0 and two neighboring curve points. In the following paragraphs the equation of the plane is developed, and it is shown that the tangent lies in it. (See Fig. 2-1.9.)

Let $\mathbf{r}_0 = \mathbf{r}(t_0)$ be a position vector with end point P_0 on C. Let $\mathbf{r}(t_0 + h)$, $\mathbf{r}(t_0 + k)$ have end points on C in a neighborhood of P. If $\mathbf{r}(t_0 + h) - \mathbf{r}_0$ and $\mathbf{r}(t_0 + k) - \mathbf{r}_0$ are linearly independent, they determine a plane through P. This plane is also determined by a pair of linear combinations of the two vectors. In particular, choose

$$\frac{\mathbf{r}(t_0 + h) - \mathbf{r}_0}{h}, \frac{\dfrac{1}{h}[\mathbf{r}(t_0 + h) - \mathbf{r}_0] - \dfrac{1}{k}[\mathbf{r}(t_0 + k) - \mathbf{r}_0]}{h - k}.$$

If we use the Taylor expansion[3] for $\mathbf{r}(t_0 + h)$ and $\mathbf{r}(t_0 + k)$, these expressions can be written

$$\frac{\mathbf{r}(t_0 + h) - \mathbf{r}_0}{h} = \mathbf{r}_0' + \mathbf{r}_0'' \frac{h}{2} + \cdots.$$

$$\frac{\dfrac{1}{h}[\mathbf{r}(t_0 + h) - \mathbf{r}_0] - \dfrac{1}{k}[\mathbf{r}(t_0 + k) - \mathbf{r}_0]}{h - k} = \frac{\mathbf{r}_0''}{2} + \frac{h + k}{3!}\mathbf{r}_0''' + \cdots,$$

where the prime denotes differentiation with respect to t. As $h \to 0, k \to 0$, the two vectors approach $\mathbf{r}_0', \frac{1}{2}\mathbf{r}_0''$. If $\mathbf{r}_0'$ and $\mathbf{r}_0''$ are linearly independent, then $\mathbf{r}' \times \mathbf{r}''$ is perpendicular to their plane. These remarks serve as background for the following definition.

[3] This development assumes the existence of higher ordered derivatives of $\mathbf{r}$.

Definition 2-1.7. The plane

$$(\mathbf{R} - \mathbf{r}_0) \cdot \mathbf{r}_0' \times \mathbf{r}_0'' = 0 \tag{2-1.9}$$

is called the osculating plane of a curve C at a point P. $\mathbf{R}$ represents a position vector whose end point assumes values $(\mathrm{X}^1, \mathrm{X}^2, \mathrm{X}^3)$ associated with the plane points.

The word osculating, which means kissing, was introduced by Tinseau around 1780. The term is appropriately chosen, for the plane (2-1.9) is the plane of closest contact with C in a neighborhood of P_0. This fact can be established by introducing the concept of order of contact. Such studies are traditionally a part of differential geometry and are not further pursued in this book.

The relation (2-1.9) can be put into the form

$$\begin{vmatrix} \mathrm{X}^1 - X_0^{\ 1} & \mathrm{X}^2 - X_0^{\ 2} & \mathrm{X}^3 - X_0^{\ 3} \\ \left(\dfrac{dX^1}{dt}\right)_0 & \left(\dfrac{dX^2}{dt}\right)_0 & \left(\dfrac{dX^3}{dt}\right)_0 \\ \left(\dfrac{d^2X^1}{dt^2}\right)_0 & \left(\dfrac{d^2X^2}{dt^2}\right)_0 & \left(\dfrac{d^2X^3}{dt^2}\right)_0 \end{vmatrix} = 0. \tag{2-1.10}$$

It is clear that a curve C has a unique osculating plane at a point P_0 if and only if $\mathbf{r}_0'$ and $\mathbf{r}_0''$ exist and are linearly independent.

Theorem 2-1.3. Let P_0 be a point on a curve C at which there is an osculating plane. The tangent line to C at P lies in the osculating plane.

PROOF. If the vector representation of the curve is $\mathbf{r} = \mathbf{r}(t)$, the tangent line may be represented by

$$\mathbf{Y} = \mathbf{r}_0 + \left(\frac{d\mathbf{r}}{dt}\right)_0 u,$$

where the zero subscript implies evaluation at P_0. It is seen by direct substitution of

$$\mathbf{Y} - \mathbf{r}_0 = \mathbf{r}_0' u$$

into (2-1.9) or (2-1.10) that the tangent line lies in the osculating plane.

Problems

1. Construct a diagram of the curve represented by these parametric equations:
   (a) $X^1 = \cosh t,\ X^2 = \sinh t,\ X^3 = t,\ t \geq 0.$
   (b) $X^1 = a\cos t,\ X^2 = a\sin t,\ X^3 = e^t,\ t \geq 0,\ a > 0.$
   (c) $X^1 = a\cos t,\ X^2 = a\sin t,\ X^3 = e^{\sin t},\ t \geq 0,\ a > 0.$

2. Suppose that a line segment with end points (2, 3, 1) and (5, 7, 9) represented the path of an oscillating particle. Complete the mathematical model by finding a possible set of parametric equations for the motion.

3. Determine the tangent vector field $d\mathbf{r}/dt$ along each of the curves of Problem 1.

4. Find the derivatives of the scalar fields $\mathbf{r} \cdot (d\mathbf{r}/dt)$ associated with each curve of Problem 1.

5. Prove the differentiation rules (2-1.5a,b,c,d) directly from definition; that is, in (2-1.5a,c,d) start with (2-1.3) and use the methods employed in an elementary calculus class rather than the results of function differentiation.

6. If $\mathbf{A} = \cos t\iota_1 + \sin t\iota_2 + t^2\iota_3$, $\mathbf{B} = \sin t\iota_1 + \cos t\iota_2 + t\iota_3$,
(a) find $(d\mathbf{A} \cdot \mathbf{B})/dt$ in two ways;
(b) find $(d\mathbf{A} \times \mathbf{B})/dt$ in two ways.

7. A vector field $\mathbf{V}$ has a constant magnitude, that is, $\mathbf{V} \cdot \mathbf{V} =$ constant for all t. Show that if $d\mathbf{V}/dt \neq 0$ it is perpendicular to $\mathbf{V}$.

8. A third-order determinant

$$\begin{vmatrix} A^1 & A^2 & A^3 \\ B^1 & B^2 & B^3 \\ C^1 & C^2 & C^3 \end{vmatrix}$$

can be represented in either of the forms

$$\mathbf{A} \cdot \mathbf{B} \times \mathbf{C} = \mathcal{E}_{ijk}A^iB^jC^k.$$

Assume that the vectors $\mathbf{A}$, $\mathbf{B}$, and $\mathbf{C}$ are differentiable functions along some curve and show that the derivative of the determinant is given as follows:

$$\frac{d}{dt}\begin{vmatrix} A^1 & A^2 & A^3 \\ B^1 & B^2 & B^3 \\ C^1 & C^2 & C^3 \end{vmatrix} = \begin{vmatrix} \dfrac{dA^1}{dt} & \dfrac{dA^2}{dt} & \dfrac{dA^3}{dt} \\ B^1 & B^2 & B^3 \\ C^1 & C^2 & C^3 \end{vmatrix} + \begin{vmatrix} A^1 & A^2 & A^3 \\ \dfrac{dB^1}{dt} & \dfrac{dB^2}{dt} & \dfrac{dB^3}{dt} \\ C^1 & C^2 & C^3 \end{vmatrix} + \begin{vmatrix} A^1 & A^2 & A^3 \\ B^1 & B^2 & B^3 \\ \dfrac{dC^1}{dt} & \dfrac{dC^2}{dt} & \dfrac{dC^3}{dt} \end{vmatrix}$$

9. Determine parametric equations of the tangent line to the elliptic helix

$$X^1 = 2\cos t, \qquad X^2 = 3\sin t, \qquad X^3 = t$$

at $(\sqrt{2}, 3/\sqrt{2}, \pi/4)$. Also find the equation of the osculating plane at this point.

10. In Example 2-1.6 find components of **P** and **Q** in a system with coordinates $\bar{X}^j$ if $\bar{X}^j = A_k{}^j X^k$ and

$$(A_k{}^j) = \begin{pmatrix} \frac{\sqrt{3}}{2} & \frac{1}{2} & 0 \\ \frac{-1}{2} & \frac{\sqrt{3}}{2} & 0 \\ 0 & 0 & 1 \end{pmatrix}.$$

How is the scalar Φ expressed in the barred system?

11. Find the tangent line to

$$\begin{aligned} X^1 &= t, \\ X^2 &= t^3, \qquad \text{at } (0, 0, 0). \\ X^3 &= 0, \end{aligned}$$

12. Find the osculating plane at a point P_0 for each of the curves of Problem 1.

2. *Geometry of Space Curves*

The purpose of this section is to extend the discussion of geometric concepts concerning space curves, begun in Section 1. These ideas are used, at least in part, to construct mathematical models of physical phenomena. The discussion is facilitated by the introduction of the concept of length of a space curve. Recall that a smooth space curve C is algebraically represented by functions with continuous derivatives that are not simultaneously zero.

Theorem 2-2.1. Suppose that the finite smooth space curve C were represented algebraically by functions $X^j = X^j(t)$ $t_0 \leq t \leq t_p$. Then a number L, given by the following formula, can be associated with C.

$$L = \int_{t_0}^{t_p} \left(\delta_{jk} \frac{dX^j}{dt} \frac{dX^k}{dt} \right)^{1/2} dt, \tag{2-2.1}$$

where $X_0{}^j = X^j(t_0)$ and $X_p{}^j = X^j(t_p)$, respectively, represent the coordinates of the initial and terminal points of C.

PROOF. A partition $(t_0, \cdots, t_n)$ of the common domain of the functions X^j (see Fig. 2-2.1) gives rise to a partition $[X^k(t_0), \cdots, X^k(t_n)]$ of C.

A polygonal path is determined by joining the points of the partition on C by straight-line segments. Denote the length of this path by L_n. Then

$$L_n = \sum_{j=1}^{n} (\delta_{kq}\, \Delta X_j{}^k \Delta X_j{}^q)^{1/2}, \tag{2-2.2a}$$

where

$$\Delta X_j{}^k = X^k(t_j) - X^k(t_{j-1}). \tag{2-2.2b}$$

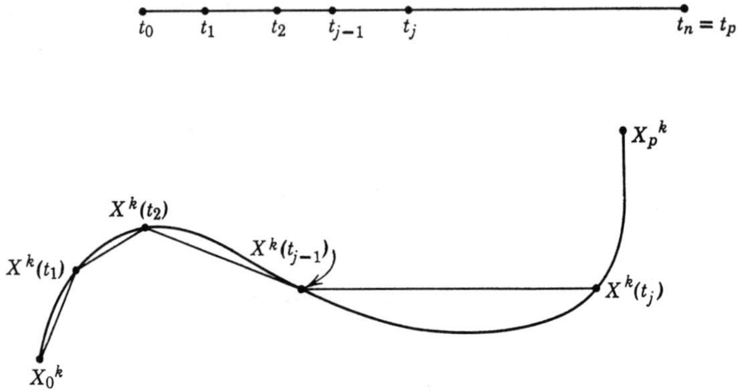

Fig. 2-2.1

According to the mean value theorem for derivatives,

$$\text{(2-2.2c)} \quad \Delta X_j{}^k = X^k(t_j) - X^k(t_{j-1}) = \frac{dX^k}{dt}(\tau_j{}^k)\,\Delta t, \qquad t_{j-1} < \tau_j{}^k < t_j.$$

Since the derivatives are continuous on closed intervals, (2-2.2c) can be replaced by

$$\text{(2-2.2d)} \qquad \Delta X_j{}^k = \left(\frac{dX^k}{dt}(\tau_j) + \epsilon_j{}^k\right)\Delta t;$$

that is, the intermediate domain values $\tau_j{}^k$, where j and k denote dependence on the subinterval of the partition and the particular function, respectively, can be replaced by values τ_j which depend only on subinterval of the partition.[4] When the expression (2-2.2d) is substituted into (2-2.2a), we obtain

$$\text{(2-2.2e)} \quad L_n = \sum_{j=1}^{n}\left[\delta_{kq}\left(\frac{dX^k}{dt}(\tau_j) + \epsilon_j{}^k\right)\left(\frac{dX^q}{dt}(\tau_j) + \epsilon_j{}^q\right)\right]^{1/2} \Delta t.$$

Then, as refinements of the partition $(t_0, \cdots, t_n)$ are taken, hence with corresponding refinements of $(X_0{}^j, \cdots, X_n{}^j)$, the continuity of the functions on closed intervals implies that the summations involving ϵ terms approach zero. In the limit we obtain a number given by (2-2.1). This completes the proof.

Definition 2-2.1. The number

$$L = \int_{t_0}^{t_p}\left(\delta_{jk}\frac{dX^j}{dt}\frac{dX^k}{dt}\right)^{1/2} dt$$

is called the length of the smooth space curve C.

[4] The theory relating to continuity of a function on a closed interval (i.e., a uniformly continuous function) is properly a part of advanced calculus. It is assumed in this text without proof. (See Angus E. Taylor, *Advanced Calculus*, Ginn, 1955.)

The vector field $d\mathbf{r}/dt$ associated with a curve C was investigated in Chapter 2, Section 1. We found that $d\mathbf{r}/dt$ was a tangential vector field along C and that in many circumstances it is advantageous that it have a constant magnitude of 1. This property can be brought about by representing the equation of the curve in terms of a parameter whose values represent arc lengths from a fixed point P_0 of C to P on C. This parameter is denoted by the symbol s.

Theorem 2-2.2. If C is represented by functions whose domain values represent arc lengths s, then $d\mathbf{r}/ds$ is a unit tangent vector field.

PROOF. From (2-2.1) it follows that

(2-2.3a) $$s = \int_{t_0}^{t_p} \left(\frac{d\mathbf{r}}{dt} \cdot \frac{d\mathbf{r}}{dt}\right)^{1/2} dt.$$

Therefore

(2-2.3b) $$\frac{ds}{dt} = \left(\frac{d\mathbf{r}}{dt} \cdot \frac{d\mathbf{r}}{dt}\right)^{1/2}.$$

In other words, ds/dt is the magnitude of the tangent vector field $d\mathbf{r}/dt$. If $t = s$, this magnitude is 1. This completes the proof.

The properties of space curves to be pointed out in this section constitute only a small fragment of those intensively investigated in the nineteenth century. Joseph Liouville (1809–1882, French) and several of his associates made a thorough study of curves and surfaces by using the methods of analysis. The major features of their study of space curves can be developed nicely by the introduction of an orthogonal triad of unit vector fields on a curve C. The primary field of this triad is the unit tangent field $d\mathbf{r}/ds$. The other two members of the set, and some concepts which naturally appear during their introduction, are represented by the following theorems and definitions.

For simplicity of notation let us use the symbol $\mathbf{t}$ in place of the unit tangent field $d\mathbf{r}/ds$ along C.

Theorem 2-2.3. If $d\mathbf{t}/ds \neq 0$,

(2-2.4) $$\mathbf{n} = \frac{1}{\kappa}\frac{d\mathbf{t}}{ds} \quad \text{where} \quad \kappa = \left(\frac{d\mathbf{t}}{ds} \cdot \frac{d\mathbf{t}}{ds}\right)^{1/2}$$

is a unit vector field orthogonal to $\mathbf{t}$ for all s.

PROOF. Because $\mathbf{t}$ is a unit vector field for all s, we can write

(2-2.5) $$\mathbf{t} \cdot \mathbf{t} = 1.$$

By differentiating this expression, we see that $\mathbf{t}$ and $d\mathbf{t}/ds$ are orthogonal, for

(2-2.5b) $$\mathbf{t} \cdot \frac{d\mathbf{t}}{ds} = 0.$$

Since the factor κ represents the magnitude of $d\mathbf{t}/ds$, it follows that $\mathbf{n}$ is also a unit vector field.

Definition 2-2.2. The vector field $\mathbf{n}$ defined on a curve C is called the principal normal field on C.

In a plane there is only one direction perpendicular to a given direction. However, in three dimensions the situation is quite different; there is a plane of directions perpendicular to a given direction. Therefore there is reason to give a special name to a chosen one of the set of vectors orthogonal to $\mathbf{t}$. The following theorem gives still greater meaning to the designation of the vector field $\mathbf{n}$ of Definition 2-2.2 as the principal normal field.

Theorem 2-2.4. At each point of C the principal normal $\mathbf{n}$ lies in the osculating plane.

PROOF. In Chapter 2, Section 1, it was seen that the equation of the osculating plane at a point P_0 of C could be put in the form

$$\text{(2-2.6a)} \qquad \begin{vmatrix} X^1 - X_0^{\ 1} & X^2 - X_0^{\ 2} & X^3 - X_0^{\ 3} \\ \left(\dfrac{dX^1}{ds}\right)_0 & \left(\dfrac{dX^2}{ds}\right)_0 & \left(\dfrac{dX^3}{ds}\right)_0 \\ \left(\dfrac{d^2X^1}{ds^2}\right)_0 & \left(\dfrac{d^2X^2}{ds^2}\right)_0 & \left(\dfrac{d^2X^3}{ds^2}\right)_0 \end{vmatrix} = 0.$$

Parametric equations of a line through P_0 and with the direction of $\mathbf{n}$ are

$$\text{(2-2.6b)} \qquad Y^j = X_0^{\ j} + \left(\frac{d^2X^j}{ds^2}\right)_0 u.$$

We see immediately that the coordinates of any point on this line satisfy (2-2.6a). This completes the proof.

According to this theorem, the osculating plane, or, intuitively speaking, the plane that comes closest to containing the part of the curve in a neighborhood of P_0, is the plane of the tangent vector $\mathbf{t}$ and the principal normal $\mathbf{n}$.

The magnitude κ of the principal normal is significant in itself. Roughly speaking, $d\mathbf{t}/ds$ is the change in the tangent vector $\mathbf{t}$ caused by a change in the arc length value s. Since $\mathbf{t}$ is a unit vector, this change deals with the direction of $\mathbf{t}$.

Definition 2-2.3. $\kappa = |d\mathbf{t}/ds|$ is called the curvature of C at the point P with coordinates $X^j(s)$.

Example 2-2.1. Consider a straight line with the vector representation

$$\mathbf{r} = \mathbf{r}_0 + \mathbf{B}s, \qquad |B| = 1,$$

where **B** is a vector constant. By differentiating, we obtain

$$\mathbf{t} = \frac{d\mathbf{r}}{ds} = \mathbf{B}.$$

Therefore

$$\frac{d\mathbf{t}}{ds} = \mathbf{0}.$$

Hence the curvature, κ, of a straight line is zero at every point of the line.

Example 2-2.2. A circle of radius a can be represented by the vector form

$$\mathbf{r} = a \cos \frac{s}{a}\, \boldsymbol{\iota}_1 + a \sin \frac{s}{a}\, \boldsymbol{\iota}_2,$$

where a is the radius of the circle and $\boldsymbol{\iota}_1$, $\boldsymbol{\iota}_2$ is a pair of orthogonal unit n-tuples. Then

$$\mathbf{t} = \frac{d\mathbf{r}}{ds} = -\sin \frac{s}{a}\, \boldsymbol{\iota}_1 + \cos \frac{s}{a}\, \boldsymbol{\iota}_2,$$

$$\frac{d\mathbf{t}}{ds} = -\frac{1}{a}\left(\cos \frac{s}{a}\, \boldsymbol{\iota}_1 + \sin \frac{s}{a}\, \boldsymbol{\iota}_2\right).$$

Hence

(2-2.7) $$\kappa^2 = \frac{1}{a^2}, \qquad \kappa = \frac{1}{a}.$$

Note that, by defining $\kappa \geq 0$, **n** has the same sense as $d\mathbf{t}/ds$. In the circle **n** points inward. (See Fig. 2-2.2.)

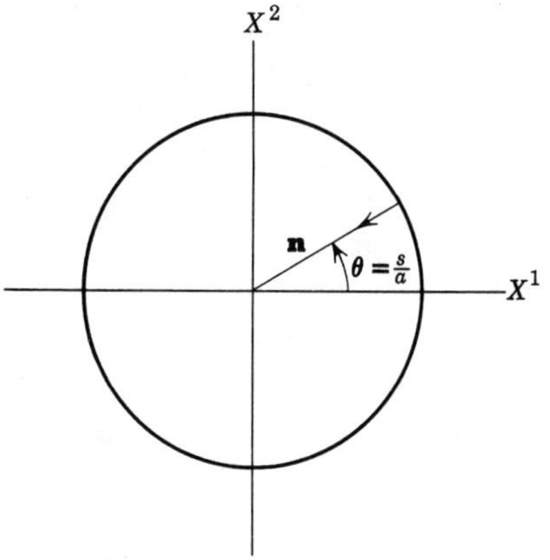

Fig. 2-2.2

Examples 2-2.1 and 2-2.2 facilitate a good intuitive grasp of the meaning of the curvature κ. A straight line has curvature zero at any point. The curvature of a circle increases without bound as the radius of the circle decreases. In general $\kappa(s)$ is a scalar field (with respect to rectangular Cartesian transformations) along C. The numerical value $\kappa(s)$ at a point of C is small if the curve does not appreciably deviate from the tangent line; the number representing κ increases as the deviation from the tangent line becomes more pronounced. Another way of expressing this idea is to say that the curvature of C at a point is the same as that of the circle which best fits the curve at that point. Indeed, the study of curves by examination of the best-fitting circle at each point is a well-developed part of differential geometry.

The concept of curvature, with its associated vector ideas, suffices for a study of plane curves. However, to describe space curves completely, a third unit vector must be introduced. Since **t** and **n** are unit orthogonal vector fields, the third vector completes the formation of a unit orthogonal triad.

Definition 2-2.4. The vector field

$$\mathbf{b} = \mathbf{t} \times \mathbf{n} \tag{2-2.8}$$

is called the binormal vector field along C.

Theorem 2-2.5. **t**, **n**, **b**, in that order, form a right-hand orthogonal unit triad of vectors at each point of a curve C.

PROOF. **b** is orthogonal to **t** and **n** by definition. Its magnitude of 1 can be computed directly from (2-2.8). The fact that

$$1 = \mathbf{b} \cdot \mathbf{b} = \mathbf{b} \cdot \mathbf{t} \times \mathbf{n} = \mathbf{t} \times \mathbf{n} \cdot \mathbf{b} = \mathbf{t} \cdot \mathbf{n} \times \mathbf{b}$$

is also a consequence of (2-2.8). Since the triple scalar product of the last member of this expression is equivalent to the determinant of **t**, **n**, and **b**, in that order, and since the value of the determinant is greater than zero, it follows that the system is right-handed.

The magnitude of the vector field $d\mathbf{b}/ds$ also has a geometric significance; **b** is a unit field, so that the change, $d\mathbf{b}/ds$, of **b** with respect to change in the value of s has to do with direction. In this case the deviation in direction can be tied in with the variation of the curve from the plane of **t** and **n**. In particular, we measure the projection of $d\mathbf{b}/ds$ onto the principal normal.

Definition 2-2.5. $\tau(s) = -\mathbf{n} \cdot (d\mathbf{b}/ds)$ is called the torsion[5] field with respect to a curve C.

[5] According to Erwin Kreyszig, *op. cit.*, p. 38, the name "torsion" was introduced by L. I. de la Vallée in 1825.

It is not the purpose of this book to delve deeply into the field of differential geometry. Therefore, having introduced the fundamental tools used in differential geometry to study smooth space curves, this section is brought to a close with a statement of a basic set of relationships. These are the so-called Frenet-Serret formulas given by Joseph A. Serret (1819–1885, French) and F. Frenet.[6]

Theorem 2-2.6. (Frenet-Serret formulas)

(2-2.9)

$$\text{(a)} \quad \frac{d\mathbf{t}}{ds} = \kappa\mathbf{n},$$

$$\text{(b)} \quad \frac{d\mathbf{n}}{ds} = -\kappa\mathbf{t} + \tau\mathbf{b},$$

$$\text{(c)} \quad \frac{d\mathbf{b}}{ds} = -\tau\mathbf{n}.$$

PROOF. Relation (2-2.9a) has already been discussed. [See (2-2.4b).] To obtain (2-2.9c), first of all note that $\mathbf{b} \cdot \mathbf{b} = 1$ implies $(d\mathbf{b}/ds) \cdot \mathbf{b} = 0$. Hence $d\mathbf{b}/ds$ lies in the plane of $\mathbf{t}$ and $\mathbf{n}$ and can be expressed in the form

(2-2.10a) $$\frac{d\mathbf{b}}{ds} = \alpha(s)\mathbf{t} + \beta(s)\mathbf{n},$$

where $\alpha(s)$ and $\beta(s)$ are unknown. Next note that $\mathbf{b} \cdot \mathbf{t} = 0$ implies

(2-2.10b) $$\mathbf{t} \cdot \frac{d\mathbf{b}}{ds} = -\mathbf{b} \cdot \frac{d\mathbf{t}}{ds} = -\kappa\mathbf{b} \cdot \mathbf{n} = 0.$$

On the other hand, from (2-2.10a) we see that

(2-2.10c) $$\mathbf{t} \cdot \frac{d\mathbf{b}}{ds} = \alpha(s).$$

Comparison of (2-2.10b) and (2-2.10c) leads to the evaluation

$$\alpha(s) = 0.$$

With $\alpha(s) = 0$, it follows as a consequence of Definition 2-2.5 that $\beta = -\tau$.

To prove (2-2.10b), we note that $d\mathbf{n}/ds$ is perpendicular to $\mathbf{n}$, hence can be written

(2-2.10d) $$\frac{d\mathbf{n}}{ds} = \gamma(s)\mathbf{t} + \delta(s)\mathbf{b}.$$

---

[6] Dirk J. Struik (*Concise History of Mathematics*, Vol. II, Dover, 1948) attributes the French School of mathematicians headed by Liouville with having developed the formulas around 1847. On the other hand, Coxeter (*Introduction to Geometry*, Wiley, 1961) states that the formulas were presented by Serret in 1851 and Frenet in 1852.

To evaluate γ and δ, we must differentiate both

$$\mathbf{t}\cdot\mathbf{n} = 0 \quad \text{and} \quad \mathbf{b}\cdot\mathbf{n} = 0.$$

These differentiations lead to

$$\mathbf{t}\cdot\frac{d\mathbf{n}}{ds} = -\mathbf{n}\cdot\frac{d\mathbf{t}}{ds} = -\kappa\mathbf{n}\cdot\mathbf{n} = -\kappa, \tag{2-2.10e}$$

$$\mathbf{b}\cdot\frac{d\mathbf{n}}{ds} = -\mathbf{n}\cdot\frac{d\mathbf{b}}{ds} = +\tau\mathbf{n}\cdot\mathbf{n} = +\tau. \tag{2-2.10f}$$

By making use of (2-2.10e,f), we find that

$$\gamma(s) = -\kappa(s), \qquad \delta(s) = \tau(s).$$

This completes the proof of (2-2.9b).

In Section 3 we apply some of these ideas in a study of the kinematics of motion.

Problems

1. Find the length of each of the following space curves:

(a) $X^1 = t, \quad X^2 = t^2, \quad 0 \le t \le \frac{1}{2}$. Ans. $\sqrt{\frac{2}{4}} + \frac{1}{4}\ln(\sqrt{2} + 1)$.

(b) $X^1 = \cos t, \quad X^2 = \sin t, \quad X^3 = t, \quad 0 \le t \le t_p$. Ans. $\sqrt{2}t_p$.

2. Express the parametric equations of the circular helix (Problem 1b) in terms of arc length.

3. Find the magnitude of $d\mathbf{r}/dt$ in each of the following cases:

(a) $X^1 = 1 + t, \quad X^2 = t, \quad X^3 = 0$.
(b) $X^1 = t, \quad X^2 = t^2, \quad X^3 = t^3$.
(c) $X^1 = a \sin t/a, \quad X^2 = a \cos t/a, \quad X^3 = 0$.
(d) $X^1 = t, \quad X^2 = t, \quad X^3 = \sin t$.

4. Does the parameter t represent arc length in any of the parts of Problem 3?

5. Assume that s and an arbitrary parameter t are functionally related. Furthermore, assume the existence of all necessary derivatives. Show that

(a) $$\frac{d^2\mathbf{r}}{ds^2} = \left[\left(\frac{d\mathbf{r}}{dt}\cdot\frac{d\mathbf{r}}{dt}\right)\frac{d^2\mathbf{r}}{dt^2} - \left(\frac{d\mathbf{r}}{dt}\cdot\frac{d^2\mathbf{r}}{dt^2}\right)\frac{d\mathbf{r}}{dt}\right]\left(\frac{d\mathbf{r}}{dt}\cdot\frac{d\mathbf{r}}{dt}\right)^{-2}.$$

(b) $$\frac{d^3\mathbf{r}}{ds^3} = \frac{d^3\mathbf{r}}{dt^3}\left(\frac{dt}{ds}\right)^3 + 3\frac{d^2\mathbf{r}}{dt^2}\frac{dt}{ds}\frac{d^2t}{ds^2} + \frac{d\mathbf{r}}{dt}\frac{d^3t}{ds^3}.$$

6. Using the results of Problem 5 show, that

(a) $$\kappa^2 = \frac{(d\mathbf{r}/dt \times d^2\mathbf{r}/dt^2)\cdot(d\mathbf{r}/dt \times d^2\mathbf{r}/dt^2)}{(d\mathbf{r}/dt \cdot d\mathbf{r}/dt)^3},$$

(b) $$\tau = \frac{d\mathbf{r}/dt \cdot d^2\mathbf{r}/dt^2 \times d^3\mathbf{r}/dt^3}{(d\mathbf{r}/dt \times d^2\mathbf{r}/dt^2)\cdot(d\mathbf{r}/dt \times d^2\mathbf{r}/dt^2)}.$$

Hint: In (a) employ the triple vector product and the Lagrange identity, Section 7, problem set.

7. Compute (or comment) κ^2 and τ in each part of Problem 3.

8. Compute (or comment) κ^2 and τ if

(a) $X^1 = 2\cos t, \quad X^2 = 3\sin t, \quad X^3 = 0.$
(b) $X^1 = \cosh t, \quad X^2 = \sinh t, \quad X^3 = 0.$
(c) $X^1 = \cos 5t, \quad X^2 = \sin 5t, \quad X^3 = e^t.$

9. Find the equation of the osculating plane, the parametric equations of the tangent line, and the parametric equations of the normal line to each of the curves in Problem 3 at the point corresponding to $t = \pi/4$. Let $a = \frac{1}{2}$ in (c).

3. *Kinematics*

Kinematics is sometimes called the geometry of motion. In this approach to mechanics the motion of a physical particle or group of particles is represented by means of a mathematical model. The pictorial representation of the model plays something of an intermediate role; that is, both the physical happening and the corresponding mathematical analysis have a common geometric interpretation.

In this section the motion of a single particle will be studied. In a kinematical investigation, the particle is idealized as a point. No considerations of mass enter into the discussion. The main objective is to describe the motion of a particle by means of a smooth space curve (intuitively thought of as traced out by the particle) and then to examine the concepts of velocity and acceleration in terms of this representation. Various frames of reference prove to be appropriate for such developments. In this section we consider three examples: a rectangular Cartesian coordinate system and an orthogonal triad of unit vectors which vary along the path of motion, a rectangular Cartesian system and the constant bases ι_1, ι_2, and ι_3, and polar coordinates in a plane along with an appropriate set of basis vectors.

Definition 2-3.1. Let C be a smooth space curve with the vector representation $\mathbf{r} = \mathbf{r}(t)$, where t represents a measure of time. Assume that at least second derivatives of $\mathbf{r}$ exist on C. Then

$$\mathbf{V} = \frac{d\mathbf{r}}{dt}, \qquad \mathbf{a} = \frac{d^2\mathbf{r}}{dt^2}, \tag{2-3.1}$$

respectively, are called the velocity and acceleration vector fields on C.

Velocity can be thought of as the rate of change of position with respect to time. In the case of one-dimensional motion (say along the X^1 axis) $d\mathbf{r}/dt$ reduces to $(dX^1/dt$, on ι_1, an expression that might be recognized from elementary calculus. Of course, we must realize that direction and sense, as well as magnitude, play a part in the present usage of the term velocity. Similar remarks hold for the acceleration concept.

Theorem 2-3.1. Let $s = s(t)$ be an increasing differentiable function. Then we have

$$\text{(2-3.2)} \qquad \text{(a)} \quad \boxed{\mathbf{V} = \frac{ds}{dt}\mathbf{t} = |\mathbf{V}|\,\mathbf{t},} \qquad \text{(b)} \quad \boxed{\mathbf{a} = \frac{d^2s}{dt^2}\mathbf{t} + \kappa\,|\mathbf{V}|^2\mathbf{n}.}$$

PROOF. According to Definition 2-3.1,

$$\text{(2-3.3)} \qquad \mathbf{V} = \frac{d\mathbf{r}}{dt} = \frac{d\mathbf{r}}{ds}\frac{ds}{dt} = \mathbf{t}\frac{ds}{dt}.$$

Since $\mathbf{t}$ is a unit vector, $|ds/dt|$ must represent the magnitude of $\mathbf{V}$. Because $s = s(t)$ is an increasing function, $ds/dt > 0$, and therefore (2-3.2a) is proved.

Relation (2-3.2b) is obtained by differentiating (2-3.2a):

$$\mathbf{a} = \frac{d\mathbf{V}}{dt} = \frac{d^2\mathbf{r}}{ds^2}\left(\frac{ds}{dt}\right)^2 + \frac{d\mathbf{r}}{ds}\frac{d^2s}{dt^2} = \kappa\mathbf{n}\left(\frac{ds}{dt}\right)^2 + \mathbf{t}\frac{d^2s}{dt^2}.$$

This completes the proof.

From (2-3.2a) we observe that the velocity vector is a tangential vector. The magnitude $|\mathbf{V}| = |ds/dt|$ is usually called the speed. It is interesting to note the fact that $\mathbf{V}$ is a tangential vector correlates nicely with Newton's first law, which indicates that a particle tends to maintain a straight-line motion unless outside influence is brought to bear.

From the relation in (2-3.2b) we learn that the acceleration vector always lies in the osculating plane, that is, the plane of the tangent vector and the principal normal.

The reader should also note that the basis $\mathbf{t}$, $\mathbf{n}$, $\mathbf{b}$ differs from any we have previously considered because the elements $\mathbf{t}$, $\mathbf{n}$, and $\mathbf{b}$ are not constant. At each point of C the triple of vector fields $\mathbf{t}$, $\mathbf{n}$, and $\mathbf{b}$ is a unit orthogonal triad, but each element, in general, varies in direction along C.

Most of the development of these notes has centered around rectangular Cartesian coordinates and a basis $\boldsymbol{\iota}_1$, $\boldsymbol{\iota}_2$, and $\boldsymbol{\iota}_3$. In many instances it is quite convenient to describe the kinematical aspects of particle motion in terms of such a system. This description is rather straightforward.

The position of the particle is given by the vector representation

(2-3.4a) $$\mathbf{r} = X^1\boldsymbol{\iota}_1 + X^2\boldsymbol{\iota}_2 + X^3\boldsymbol{\iota}_3,$$

where

(2-3.4b) $$X^j = X^j(t).$$

Since the basis n-tuples are constants, the rectangular Cartesian representations of the velocity and acceleration fields are obtained by straightforward differentiations.

$$\mathbf{V} = \frac{d\mathbf{r}}{dt} = \frac{dX^1}{dt}\boldsymbol{\iota}_1 + \frac{dX^2}{dt}\boldsymbol{\iota}_2 + \frac{dX^3}{dt}\boldsymbol{\iota}_3. \tag{2-3.4c}$$

$$\mathbf{a} = \frac{d^2\mathbf{r}}{dt^2} = \frac{d^2X^1}{dt^2}\boldsymbol{\iota}_1 + \frac{d^2X^2}{dt^{\ 2}}\boldsymbol{\iota}_2 + \frac{d^2X^3}{dt^2}\boldsymbol{\iota}_3. \tag{2-3.4d}$$

Note that the expressions for velocity and acceleration can also be written in the concise forms

(2-3.4e) $$\mathbf{V} = \frac{dX^j}{dt}\boldsymbol{\iota}_j, \qquad \mathbf{a} = \frac{d^2X^j}{dt^2}\boldsymbol{\iota}_j.$$

The following examples illustrate particle motions, as described in rectangular Cartesian representations.

Example 2-3.1. Let the motion of a particle be represented by the parametric equations

$$X^j = X_0^{\ j} + B^j e^t, \qquad 0 \le t \le 1,$$

where

$$(X_0^{\ 1}, X_0^{\ 2}, X_0^{\ 3}) = (2, 1, 3), \qquad (B^1, B^2, B^3) = (-1, 3, 1),$$

and e^t is the exponential function. Then

$$\begin{aligned} \mathbf{r} &= (2 - e^t)\boldsymbol{\iota}_1 + (1 + 3e^t)\boldsymbol{\iota}_2 + (3 + e^t)\boldsymbol{\iota}_3, \\ \mathbf{V} &= -e^t\boldsymbol{\iota}_1 + 3e^t\boldsymbol{\iota}_2 + e^t\boldsymbol{\iota}_3, \\ \mathbf{a} &= -e^t\boldsymbol{\iota}_1 + 3e^t\boldsymbol{\iota}_2 + e^t\boldsymbol{\iota}_3. \end{aligned}$$

Furthermore,

$$\begin{aligned} |\mathbf{V}| &= \sqrt{\mathbf{V}\cdot\mathbf{V}} = \sqrt{11}\, e^t, \\ |\mathbf{a}| &= \sqrt{\mathbf{a}\cdot\mathbf{a}} = \sqrt{11}\, e^t. \end{aligned}$$

Therefore the particle is at (1, 4, 4) at $t = 0$ and has a speed numerically represented by $\sqrt{11}$. (See Fig. 2-3.1.) The acceleration is also indicated by $\sqrt{11}$ at $t = 0$. The motion is along a straight-line path, with both speed and acceleration increasing exponentially. Note that both the velocity and acceleration vector fields vary in magnitude along the curve but are constant in direction.

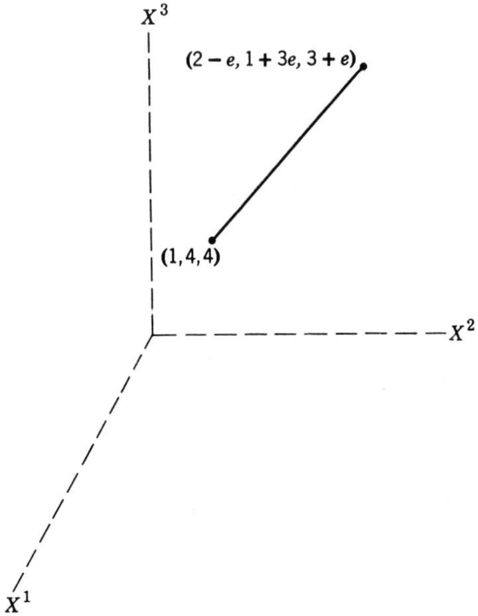

Fig. 2-3.1

Example 2-3.2. Suppose that a particle moves along the circular helix described in Example 2-1.2. Then

$$\begin{aligned}\mathbf{r} &= \cos t\boldsymbol{\iota}_1 + \sin t\boldsymbol{\iota}_2 + t\boldsymbol{\iota}_3,\\ \mathbf{V} &= -\sin t\boldsymbol{\iota}_1 + \cos t\boldsymbol{\iota}_2 + \boldsymbol{\iota}_3,\\ \mathbf{a} &= -\cos t\boldsymbol{\iota}_1 - \sin t\boldsymbol{\iota}_2.\end{aligned}$$

Furthermore,

$$|\mathbf{V}| = \sqrt{\mathbf{V}\cdot\mathbf{V}} = \sqrt{2}, \qquad |\mathbf{a}| = \sqrt{1},$$

and

$$\mathbf{V}\cdot\mathbf{a} = 0.$$

We observe that the magnitude of the velocity, as well as the magnitude of the acceleration field, is constant. Both fields, however, vary in direction along the curve.

Another point of interest is that the velocity and acceleration vectors are perpendicular for all t.

A third way of illustrating kinematical ideas is in terms of a polar coordinate system. This model represents particle motion which takes place in a plane and presents another example of the usage of a non-constant vector basis.

A polar coordinate system is constructed by choosing a pole O and a polar OP. (See Fig. 2-3.2.) A point is designated by a number pair (ρ, θ).

The parametric equations of a curve are expressed in the form

(2-3.5) $$\rho = \rho(t), \qquad \theta = \theta(t).$$

(See Fig. 2-3.3.) As in the preceding discussions, the parametric equations are dually interpreted as equations of motion of a particle and as equations of a curve.

The basis associated with the polar coordinate system consists of the pair **R**, a unit vector with the sense and direction of the position vector **r**, and **P**, a unit vector orthogonal to **R**.

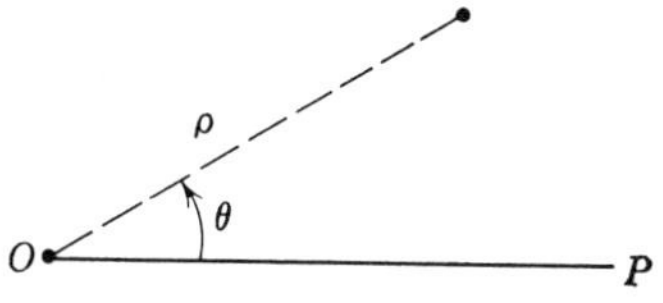

Fig. 2-3.2

Theorem 2-3.2. $d\mathbf{R}/d\theta$ is a unit vector perpendicular to **R**.

PROOF. Since **R** is a unit vector for all θ,

(2-3.6a) $$\mathbf{R} \cdot \mathbf{R} = 1.$$

By differentiating with respect to θ, we obtain

(2-3.6b) $$2\frac{d\mathbf{R}}{d\theta} \cdot \mathbf{R} = 0.$$

Therefore $d\mathbf{R}/d\theta$ is perpendicular to **R**.

R can be represented in terms of a constant orthogonal basis (with $\boldsymbol{\iota}_1$ along OP), as follows:

(2-3.6c) $$\mathbf{R} = \cos\theta\boldsymbol{\iota}_1 + \sin\theta\boldsymbol{\iota}_2.$$

Therefore

(2-3.6d) $$\frac{d\mathbf{R}}{d\theta} = -\sin\theta\boldsymbol{\iota}_1 + \cos\theta\boldsymbol{\iota}_2.$$ [7]

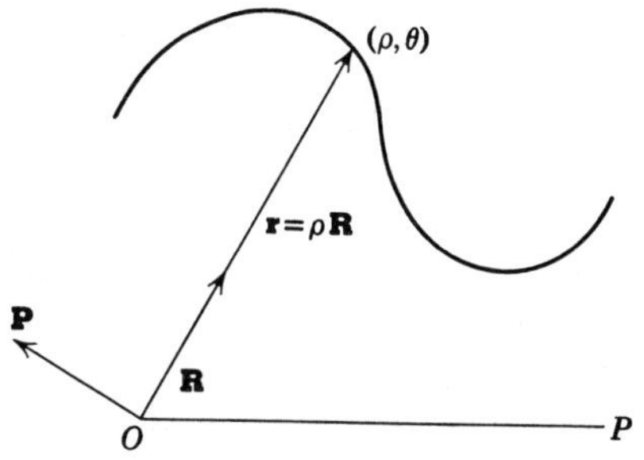

Fig. 2-3.3

[7] The representations of **R** and $d\mathbf{R}/d\theta$ in terms of a basis $\boldsymbol{\iota}_1$ and $\boldsymbol{\iota}_2$ identifies them as Cartesian vectors and therefore justifies the use of the term vector in this discussion of polar coordinates. Considerations of a vector as a collection of n-tuples associated with nonlinear systems, such as polar coordinate systems, are treated in Chapter 3.

From (2-3.6d) it can be determined that the magnitude of $d\mathbf{R}/d\theta$ is 1. As a matter of fact, the orthogonality of $\mathbf{R}$ and $d\mathbf{R}/d\theta$ also follows from (2-3.6c) and (2-3.6d). This completes the proof.

Definition 2-3.2. $\mathbf{R}$ and $\mathbf{P} = d\mathbf{R}/d\theta$ are the basis vectors associated with a polar coordinate system.

As a matter of future convenience, the following relations are introduced at this time.

Theorem 2-3.3. We have

$$\frac{d^2\mathbf{R}}{d\theta^2} = -\mathbf{R}. \tag{2-3.7}$$

PROOF. When (2-3.6d) is differentiated, we obtain the result

$$\frac{d^2\mathbf{R}}{d\theta^2} = -\cos\theta\boldsymbol{\iota}_1 - \sin\theta\boldsymbol{\iota}_2 = -\mathbf{R}.$$

This completes the proof.

The expressions for velocity and acceleration of a moving particle can be obtained by straightforward differentiation. In the following theorem it is assumed that the derivatives exist.

Theorem 2-3.4. The velocity and acceleration vector fields of a particle, with parametric equations $\rho = \rho(t)$, $\theta = \theta(t)$, are

(2-3.8)

$$\text{(a)}\quad \mathbf{V} = \frac{d\rho}{dt}\mathbf{R} + \rho\frac{d\theta}{dt}\mathbf{P}.$$

$$\text{(b)}\quad \mathbf{a} = \left(\frac{d^2\rho}{dt^2} - \rho\left(\frac{d\theta}{dt}\right)^2\right)\mathbf{R} + \left(\rho\frac{d^2\theta}{dt^2} + 2\frac{d\rho}{dt}\frac{d\theta}{dt}\right)\mathbf{P}.$$

PROOF. The position of a particle is given by

$$\mathbf{r} = \rho\mathbf{R},$$

where ρ is a function with domain t and $\mathbf{r}$ is a vector function of a function; that is, $\mathbf{R}$ depends on θ and θ in turn depends on the parameter t. Therefore

$$\mathbf{V} = \frac{d\mathbf{r}}{dt} = \frac{d\rho}{dt}\mathbf{R} + \rho\frac{d\mathbf{R}}{d\theta}\frac{d\theta}{dt} = \frac{d\rho}{dt}\mathbf{R} + \rho\frac{d\theta}{dt}\mathbf{P}.$$

When $\mathbf{V}$ is differentiated with respect to t, we obtain $\mathbf{a}$; that is,

$$\mathbf{a} = \frac{d^2\rho}{dt^2}\mathbf{R} + \frac{d\rho}{dt}\frac{d\mathbf{R}}{d\theta}\frac{d\theta}{dt} + \frac{d\rho}{dt}\frac{d\theta}{dt}\mathbf{P} + \rho\frac{d^2\theta}{dt^2}\mathbf{P} + \rho\left(\frac{d\theta}{dt}\right)^2\frac{d^2\mathbf{R}}{d\theta^2}.$$

Making use of (2-3.7) in the last member of this expression and collecting the components of **R** and **P**, we obtain the result (2-3.8b).

Example 2-3.3. If the acceleration is purely radial, then

$$\rho \frac{d^2\theta}{dt^2} + 2\frac{d\rho}{dt}\frac{d\theta}{dt} = 0;$$

that is (multiplying by an integrating factor),

$$\frac{d}{dt}\left(\rho^2 \frac{d\theta}{dt}\right) = 0,$$

and

$$\rho^2 \frac{d\theta}{dt} = \text{constant}.$$

Now the formula for area, as expressed in polar coordinates, is

$$\text{area} = \tfrac{1}{2}\int \rho^2 \, d\theta.$$

Therefore

$$\frac{d \text{ area}}{dt} = \tfrac{1}{2}\rho^2 \frac{d\theta}{dt} = \text{constant}.$$

This relation is known as the "law of areas": If the acceleration of a particle is always directed toward a fixed point O, the position vector will sweep out area at a constant rate.

Problems

1. Suppose that the parametric equations of motion of a particle were given by
$X^1 = \cos t, \quad X^2 = \sin t, \quad X^3 = t, \quad t \geq 0.$
(a) Find **V**, **a**.
(b) Show that **V** and **a** are perpendicular for all t.
(c) Compute τ.
(d) Express s in terms of t.

2. Find **V** and **a** if the parametric equations of motion are given as
(a) $X^1 = a \cos s/a, \quad X^2 = a \sin s/a, \quad X^3 = 0.$
(b) $X^1 = t, \quad X^2 = t^2, \quad X^3 = 0, \quad t \geq 0.$
(c) Find ds/dt in part (b).

3. Write out the expressions for **r**, **V**, **a** if the parametric equations of the motion of a particle are
(a) $X^1 = 1 + t^2, \quad X^2 = t, \quad X^3 = 0.$
(b) $X^1 = t, \quad X^2 = t^2, \quad X^3 = t^3.$
(c) $X^1 = a \sin t/a, \quad X^2 = a \cos t/a, \quad X^3 = 0.$
(d) $X^1 = t, \quad X^2 = t, \quad X^3 = \sin t.$

(e) $X^1 = 2 \cos t, \qquad X^2 = 3 \sin t, \qquad X^3 = 0.$
(f) $X^1 = \cosh t, \qquad X^2 = \sinh t, \qquad X^3 = 0, \qquad t \geq 0.$
(g) $X^1 = \cos 5t, \qquad X^2 = \sin 5t, \qquad X^3 = e^t.$

4. If **A**, **B**, and ω are constants, **A** and **B** are linearly independent and

$$\mathbf{r} = \mathbf{A} \cos \omega t + \mathbf{B} \sin \omega t$$

show that the acceleration vector is directed toward the origin and has a magnitude proportional to the magnitude of **r**.

5. If the motion of a particle is uniform and rectilinear along a line through (X_0^1, X_0^2, X_0^3) and with direction (B^1, B^2, B^3), write down a representation for the position vector. Then show that the velocity vector is constant and thc acceleration vector is zero.

6. Construct the representation of a motion that traces out the same path as that of Problem 5, but for which the velocity and acceleration vectors are not constants.

7. The parametric equations of motion of a particle are

$$X^1 = 3 \cos t, \qquad X^2 = 3 \sin t, \qquad X^3 = t.$$

(a) Write down expressions for **r**, **V**, and **a**.
(b) Construct a unit vector field perpendicular to the plane of **r** and **a** (for all t).
(c) Briefly describe the motion.
(d) What angle does **V** make with **a**?

8. Answer (a), (b), (c) of Problem 7 for a particle with parametric representation

$$X^1 = e^t, \qquad X^2 = e^t, \qquad X^3 = \sin t.$$

9. Make a proof of

$$\frac{d^2\mathbf{R}}{d\theta^2} = \left(\mathbf{R} \cdot \frac{d^2\mathbf{R}}{d\theta^2}\right)\mathbf{R},$$

assuming that

$$\frac{d^2\mathbf{R}}{d\theta^2} = \alpha\mathbf{R} + \beta\mathbf{P}$$

and computing α and β.

10. Show that the magnitude of the velocity field can be expressed in the form

$$|\mathbf{V}| = (\mathbf{V} \cdot \mathbf{V})^{1/2} = \left[\left(\frac{d\rho}{d\theta}\right)^2 + \rho^2\right]^{1/2} \left|\frac{d\theta}{dt}\right|.$$

11. Find **r**, **V**, and **a** for the motions with the following parametric representations:

(a) $\rho = e^{bt}, \qquad \theta = t, \quad \text{if} \quad \begin{matrix} b > 0. \\ b < 0. \end{matrix}$
(b) $\rho = \sin t, \qquad \theta = \cos t.$
(c) $\rho = \cosh t, \qquad \theta = e^t.$

12. Discuss briefly the nature of the motion in each part of 11.

4. Moving Frames of Reference

In preceding sections it was assumed that the coordinate systems described fixed frames of reference. In this section transformation relationships between rectangular Cartesian coordinate systems describing reference frames in relative motion are investigated.

A principle of relativity of motion, that is, a statement of the fact that the motion of a body has meaning only with respect to surrounding bodies, has played a fundamental role in the theoretical development of mechanics at least since the time of Newton.[8] In particular, Newton, in his development of gravitational theory, postulated a principle of relativity for bodies in uniform rectilinear motion with respect to one another.

Today, with man at the threshold of a period of solar and universe exploration, mathematical models of bodies in relative motion are of great importance.

Suppose that a frame of reference $\bar{O}$ is assumed fixed and that a frame O is in rotational and translational motion with respect to $\bar{O}$. (See Fig. 2-4.1.)

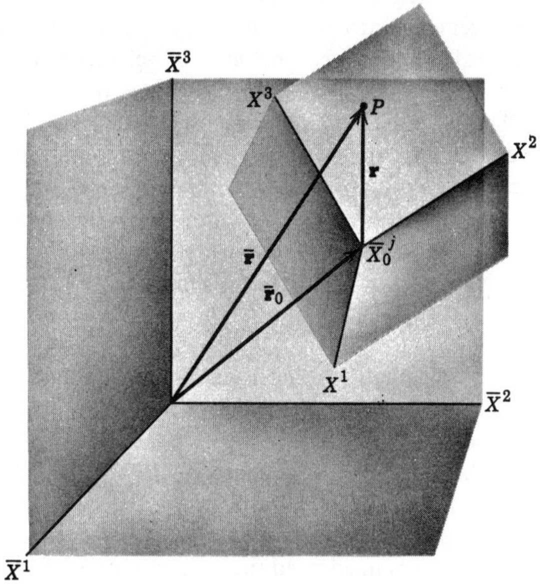

Fig. 2-4.1

[8] Theoretical knowledge of the nonobjective nature of motion dates back to the Greeks. Aristotle defines "place" as the relation of a body to the bodies in its neighborhood.

Furthermore, let an object P be in motion with respect to O. A kinematical study of the relationships between these bodies can be made by setting up the following mathematical model.

Rectangular Cartesian coordinates $\bar{X}^j$ and X^j, respectively, describe the systems $\bar{O}$ and O. It is assumed that there is a universal time. (This is the Newtonian assumption. Its denial leads to Einstein's special theory of relativity.)

$$t = \bar{t}. \tag{2-4.1}$$

Therefore the parametric equations of a particle in motion, that is

$$\text{(a) } \bar{X}^j = \bar{X}^j(t), \qquad \text{(b) } X^k = X^k(t), \tag{2-4.2}$$

can be expressed in terms of a common domain of values t.

Our assumption is that the $\bar{X}^j$ coordinate system is fixed. From the physical point of view this means that the laws and concepts at the base of classical mechanics take on their natural forms when referred to these coordinates. In particular, the components of velocity and acceleration vectors can be represented as projections on the coordinate axes. Observations and measurements intrinsic to the X^j system cannot be interpreted so simply in general, for the measurements of velocity and acceleration exhibit deviations. The outside observer attributes these to the motion; the intrinsic observer thinks of them as pseudo-effects. To determine these effects, we start with the laws of transformation relating the systems $\bar{O}$ and O. Then appropriate expressions are obtained for velocity and acceleration in the moving system by computing the standard forms, that is, the first and second derivatives, in the $\bar{O}$ system.

According to the results of Chapter 1, Section 3,

$$\bar{X}^j(t) = A_k^{\ j}(t)\, X^k(t) + \bar{X}_0^{\ j}(t), \tag{2-4.3a}$$

where $\bar{X}_0^{\ j}$ are the $\bar{O}$ coordinates of the origin of the coordinate system of O. The relation (2-4.3a) can also be described as the component form of the vector equation $\bar{\mathbf{r}} = \bar{\mathbf{r}}_0 + \mathbf{r}$. (See Fig. 2-4.1.) Because of the motion of O with respect to $\bar{O}$, the translation components $\bar{X}_0^{\ j}$ and the rotation coefficients $A_k^{\ j}$ are functions with domain t. The variability of these quantities distinguishes this situation from preceding developments.

Since the $\bar{O}$ system is assumed fixed (i.e., the motion of other systems is judged with the $\bar{O}$ system as reference), the components of velocity are obtained by differentiation of (2-4.3a).

$$\frac{d\bar{X}^j}{dt} = \frac{dA_k^{\ j}}{dt} X^k + A_k^{\ j} \frac{dX^k}{dt} + \frac{d\bar{X}_0^{\ j}}{dt}. \tag{2-4.3b}$$

Since it is assumed that the components $d\bar{X}^j/dt$ determine the true velocity, it is apparent from (2-4.3b) that this velocity cannot be obtained from the functions dX^j/dt alone. We must take into account functions $d\bar{X}_0{}^j/dt$ and $dA_k{}^j/dt$, respectively, because of translational and rotational effects. The terms $A_k{}^j(dX^k/dt)$ are usually called components of the "apparent velocity," whereas the $(dA_k{}^j/dt)X^k$ are designated as components of an "angular velocity." For the moment, we stop the discussion of velocity with the remark that it is not, when defined in the usual sense, a vector concept. This fact is pointed out by comparing the transformation rule (2-4.3b) with (1-4.1). A way in which velocity can be expressed mathematically by means of components that satisfy the transformation law (1-4.1), even when the coefficients of the orthogonal Cartesian transformations are dependent on a parameter t, is discussed in Section 4.* Our present objective is to investigate the acceleration concept. The components of acceleration arise by differentiation of (2-4.3b).

$$\text{(2-4.3c)} \quad \frac{d^2\bar{X}^j}{dt^2} = \frac{d^2A_k{}^j}{dt^2}X^k + 2\frac{dA_k{}^j}{dt}\frac{dX^k}{dt} + A_k{}^j\frac{d^2X^k}{dt^2} + \frac{d^2\bar{X}_0{}^j}{dt^2}.$$

The components $d^2\bar{X}^j/dt^2$ of the fixed $\bar{O}$ system (it is assumed that they denote the "true" acceleration) do not arise from the components d^2X^k/dt^2 alone. Therefore acceleration, as previously defined, does not fit into the vector concept with respect to the group of variable rectangular Cartesian transformations now under discussion if a requirement of that concept is the transformation law (1-4.1).

The remaining theoretical development of this section is devoted to expressing the velocity and acceleration in a classical form found in most texts on vector analysis. We call it the arrow form. In Section 4* it will be seen how the concepts of velocity and acceleration can be redefined so that they once again fit into the fold of vector concepts.

In most texts that deal with moving frames of reference we find the relations

$$\bar{\mathbf{V}} = \bar{\mathbf{V}}_0 + \mathbf{V}_A + \boldsymbol{\omega} \times \mathbf{r},$$

$$\mathbf{a} = \mathbf{a}_0 + \mathbf{a}_A + 2\boldsymbol{\omega} \times \mathbf{V}_A + \boldsymbol{\omega} \times (\boldsymbol{\omega} \times \mathbf{r}) + \frac{d\boldsymbol{\omega}}{dt} \times \mathbf{r},$$

where $\mathbf{V}_A$ and $\mathbf{a}_A$ represent apparent velocity and apparent acceleration, respectively, whereas the components of $\boldsymbol{\omega}$ measure angular velocity. These relations are algebraically equivalent to the component forms (2-4.3b) and (2-4.3c); however, they are expressed in terms of angular velocity. The following definitions and theorems, which introduce and develop the algebraic properties associated with the angular velocity concept, establish the connecting link between the foregoing representations and the component forms.

Definition 2-4.1. Functions ω_{ki} are defined by

(2-4.4a)
$$\omega_{ki} = \sum_{j=1}^{3} A_k{}^j \frac{dA_i{}^j}{dt}.$$

Theorem 2-4.1a. The matrix (ω_{ki}) is skew symmetric, that is,

(2-4.4b)
$$\omega_{ki} = -\omega_{ik}, \quad \text{or} \quad \begin{pmatrix} 0 & \omega_{12} & \omega_{13} \\ -\omega_{12} & 0 & \omega_{23} \\ -\omega_{13} & -\omega_{23} & 0 \end{pmatrix}.$$

PROOF. For any domain value t the rectangular Cartesian systems $\bar{X}^j$ and X^j are related (with respect to the rotation) by the orthogonal transformation equations of Chapter 1, Section 3. Therefore the orthogonality conditions hold; that is,

(2-4.5a)
$$\sum_{j=1}^{3} A_k{}^j A_i{}^j = \delta_{ki}.$$

By differentiating (2-4.5a) with respect to t, we obtain

(2-4.5b)
$$\sum_{j=1}^{3} \left(\frac{dA_k{}^j}{dt} A_i{}^j + A_k{}^j \frac{dA_i{}^j}{dt} \right) = 0.$$

The relation (2-4.5b) can then be put in the form

$$\omega_{ik} = \sum_{j=1}^{3} A_i{}^j \frac{dA_k{}^j}{dt} = -\sum_{j=1}^{3} A_k{}^j \frac{dA_i{}^j}{dt} = -\omega_{ki}.$$

This completes the proof.

In general, the matrix (ω_{ki}) has six nonzero components. Because of the skew symmetry property, these six components can be specified by appropriately choosing three of them. A set of three components emanates by means of the following definition.

Definition 2-4.2. The components ω^p are defined by

(2-4.6a)
$$\omega^p = \tfrac{1}{2} E^{pki} \omega_{ki}.$$

The way in which the matrix elements ω_{ki} are represented in terms of the ω^p is stated in the next theorem.

Theorem 2-4.1b. We have

(2-4.6b)
$$\omega_{rs} = \mathcal{E}_{prs} \omega^p.$$

PROOF. When (2-4.6) is multiplied and summed with $\mathcal{E}_{prs}$, the result (2-4.6b) follows as a consequence of the algebraic properties of the $\mathcal{E}$ systems developed in Chapter 1, Section 6.

We are now in a position to bring about a connection between the two forms of expression for velocity and acceleration. The next theorem introduces the necessary computations.

Theorem 2-4.1c. We have

(2-4.7)

$$\text{(a)}\quad \frac{dA_k^{\ j}}{dt} = a_j^{\ i}\mathcal{E}_{pik}\omega^p = \sum_{i=1}^{3} A_i^{\ j}\,\mathcal{E}_{pik}\omega^p,$$

$$\text{(b)}\quad \frac{d^2A_k^{\ j}}{dt^2} = -2\delta_{pk}^{[rq]}A_q^{\ j}\omega_r\omega^p + a_j^{\ i}\mathcal{E}_{pik}\frac{d\omega^p}{dt}.$$

Proof. According to (2-4.4a) and (2-4.6b),

(2-4.8a)
$$\sum_{q=1}^{3} A_i^{\ q}\frac{dA_k^{\ q}}{dt} = \omega_{ik} = \mathcal{E}_{pik}\omega^p.$$

Multiplying and summing (2-4.8a) with $a_j^{\ i}$ produces the result

(2-4.8b)
$$\sum_{q=1}^{3} \delta_j^{\ q}\frac{dA_k^{\ q}}{dt} = a_j^{\ i}\mathcal{E}_{pik}\omega^p.$$

When the definition of the Kronecker delta is employed in the left member of (2-4.8b), the result in (2-4.7a) is obtained.

The first step toward realization of the form in (2-4.7b) consists in differentiating (2-4.7a). We obtain

(2-4.8c)
$$\frac{d^2A_k^{\ j}}{dt^2} = \mathcal{E}_{pik}\left(\frac{da_j^{\ i}}{dt}\omega^p + a_j^{\ i}\frac{d\omega^p}{dt}\right).$$

Since $a_j^{\ i} = A_i^{\ j}$, the first term of the parenthetic expression on the right of (2-4.8c) can be replaced according to (2-4.7a). Then

(2-4.8d)
$$\frac{d^2A_k^{\ j}}{dt^2} = \sum_{i=1}^{3}\mathcal{E}_{pik}\left(a_j^{\ q}\mathcal{E}_{rqi}\omega^r\omega^p + a_j^{\ i}\frac{d\omega^p}{dt}\right).$$

When, for convenience of notation, we let $\omega_r = \omega^r$ and replace $\mathcal{E}_{rqi}$ by E^{rqi} and $a_j^{\ q}$ by $A_q^{\ j}$, (2-4.8d) has the form

(2-4.8e)
$$\frac{d^2A_k^{\ j}}{dt^2} = A_q^{\ j}\mathcal{E}_{pik}E^{rqi}\omega_r\omega^p + a_j^{\ i}\,\mathcal{E}_{pik}\frac{d\omega^p}{dt}$$

$$= -2A_q^{\ j}\,\delta_{pk}^{[rq]}\omega_r\omega^p + a_j^{\ i}\mathcal{E}_{pik}\frac{d\omega^p}{dt}.$$

This completes the proof.

The arrow form of the velocity and acceleration expressions (2-4.3b) and (2-4.3c) now can be obtained.

Theorem 2-4.2. The components of velocity $d\bar{X}^j/dt$ and of acceleration $d^2\bar{X}^j/dt^2$ in an $\bar{O}$ frame of reference (which is assumed fixed) are obtained from measurements in an O system (which is in rotational and translational motion with respect to $\bar{O}$) according to the following relations:

$$\text{(2-4.9)}\quad \text{(a)}\quad \frac{d\bar{X}^j}{dt} = \frac{d\bar{X}_0{}^j}{dt} + A_k{}^j\frac{dX^k}{dt} + \sum_{i=1}^{3} A_i{}^j \mathcal{E}_{pik}\omega^p X^k.$$

$$\text{(b)}\quad \frac{d^2\bar{X}^j}{dt^2} = \frac{d^2\bar{X}_0{}^j}{dt^2} + A_k{}^j\frac{d^2X^k}{dt^2} + 2\sum_{i=1}^{3} A_i{}^j \mathcal{E}_{pik}\omega^p \frac{dX^k}{dt}$$
$$+ A_q{}^j[(\omega_r X^r)\omega^q - (\omega_r\omega^r)X^q] + \sum_{i=1}^{3} A_i{}^j \mathcal{E}_{pik}\frac{d\omega^p}{dt}X^k.$$

PROOF. Relation (2-4.9a) is an immediate consequence of substituting for the first term of the right-hand member of (2-4.3b), according to the result (2-4.7a).

In order to obtain (2-4.9b), we must appropriately substitute both (2-4.7a) and (2-4.7b) into (2-4.3c). This action results in the form

$$\text{(2-4.10a)}\quad \frac{d^2\bar{X}^j}{dt^2} = \frac{d^2\bar{X}_0{}^j}{dt^2} + A_k{}^j\frac{d^2X^k}{dt^2} + 2\sum_{i=1}^{3} A_i{}^j \mathcal{E}_{pik}\omega^p \frac{dX^k}{dt}$$
$$+ \left(-2\delta_{pk}^{[rq]}A_q{}^j\omega_r\omega^p + \sum_{i=1}^{3} A_i{}^j \mathcal{E}_{pik}\frac{d\omega^p}{dt}\right)X^k.$$

The relation (2-4.10a) also can be given the form

$$\text{(2-4.10b)}\quad \frac{d^2\bar{X}^j}{dt^2} = \frac{d^2\bar{X}_0{}^j}{dt^2} + A_k{}^j\frac{d^2X^k}{dt^2} + 2\sum_{i=1}^{3} A_i{}^j \mathcal{E}_{pik}\omega^p \frac{dX^k}{dt}$$
$$+ A_q{}^j[(\omega_r X^r)\omega^q - (\omega_r\omega^r)X^q] + \sum_{i=1}^{3} A_i{}^j \mathcal{E}_{pik}\frac{d\omega^p}{dt}X^k.$$

This completes the proof.

As previously indicated, the results in (2-4.9a) and (2-4.9b) are commonly found in the arrow form,

$$\text{(2-4.11)}\quad \text{(a)}\quad \bar{\mathbf{V}} = \bar{\mathbf{V}}_0 + \mathbf{V}_A + \boldsymbol{\omega} \times \mathbf{r},$$

$$\text{(b)}\quad \bar{\mathbf{a}} = \bar{\mathbf{a}}_0 + \mathbf{a}_A + 2\boldsymbol{\omega} \times \mathbf{V}_A + \boldsymbol{\omega} \times (\boldsymbol{\omega} \times \mathbf{r}) + \frac{d\boldsymbol{\omega}}{dt} \times \mathbf{r},$$

where $\mathbf{V}_A$ signifies an apparent velocity and $\mathbf{a}_A$ indicates an apparent acceleration. $\boldsymbol{\omega}$, with components $-\omega^j$, is referred to as an angular velocity "vector." We have already noted that the components of velocity and acceleration do not transform according to the law (1-4.1). Later it will be seen that the same statement can be made concerning angular velocity.

In comparing (2-4.9a,b) with (2-4.11a,b), we must make the following correspondences:

Arrow	*Components*	*Common Name*
$\bar{\mathbf{v}}_0$	$\dfrac{d\bar{X}_0{}^j}{dt}$	Translation velocity of O
$\mathbf{v}_A$	$A_k{}^j \dfrac{dX^k}{dt}$	Apparent velocity
$\boldsymbol{\omega} \times \mathbf{r}$	$\sum_{i=1}^{3} A_i{}^j \varepsilon_{pik} \omega^p X^k$	Apparent velocity of rotation
$\bar{\mathbf{a}}_0$	$\dfrac{d^2\bar{X}_0{}^j}{dt^2}$	Translational acceleration of O
$\mathbf{a}_A$	$A_k{}^j \dfrac{d^2X^k}{dt^2}$	Apparent acceleration
$2\boldsymbol{\omega} \times \mathbf{V}_A$	$2\sum_{i=1}^{3} A_i{}^j \varepsilon_{pik} \omega^p \dfrac{dX^k}{dt}$	Coriolis acceleration
$\boldsymbol{\omega} \times (\boldsymbol{\omega} \times \mathbf{r})$	$A_q{}^j[(\omega_r X^r)\omega^q - \omega_r \omega^r X^q]$	Centripetal acceleration
$\dfrac{d\boldsymbol{\omega}}{dt} \times \mathbf{r}$	$\sum_{i=1}^{3} A_i{}^j \varepsilon_{pik} \dfrac{d\omega^p}{dt} X^k$	

The component forms make possible a clear geometric interpretation of each arrow. The translational velocity and acceleration terms are straightforward projections on the axes of the $\bar{O}$ system. In all other cases a set of values referred to the O system are operated on by $A_k{}^j$. This corresponds to projecting arrow representatives in the O system onto the axes of the $\bar{O}$ system. Of course, these projections are dependent on t.

Now attention will be turned to examples in which relations (2-4.3b,c) or (2-4.9a,b) are used to illustrate given kinematical situations.

Example 2-4.1. Let a rigid body be rotated about a fixed axis. Suppose the fixed axis is parallel to the $\bar{X}^3$ axis of the coordinates of a fixed frame of reference. Furthermore, let an X^j coordinate system be associated with the rigid body in such a way that the origin is on the axis and the X^3 coordinate line corresponds to the axis. (See Fig. 2-4.2.) Now what is the motion from the viewpoint of the fixed system of a surface point P of the body that lies on the X^1 axis a distance r from the origin O? The parametric equations of such a point with respect to the X^j system are simply

(2-4.12a) $$X^1 = r, \qquad X^2 = 0, \qquad X^3 = 0.$$

The parametric equations in the $\bar{X}^j$ system are

(2-4.12b) $$\bar{X}^j = A_k{}^j X^k + \bar{X}_0{}^j,$$

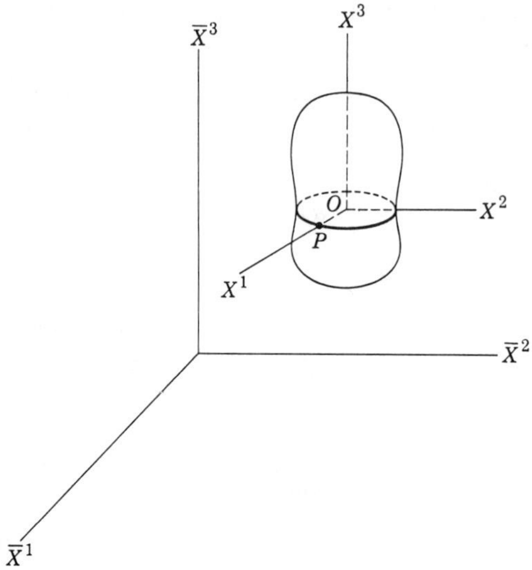

Fig. 2-4.2

where $\bar{X}_0^j$ are the coordinates of O. If

$$(A_k^{\ j}) = \begin{pmatrix} \cos\theta & -\sin\theta & 0 \\ \sin\theta & \cos\theta & 0 \\ 0 & 0 & 1 \end{pmatrix}, \tag{2-4.12c}$$

then

$$\begin{aligned} \bar{X}^1 &= r\cos\theta + \bar{X}_0^{\ 1}, \\ \bar{X}^2 &= -r\sin\theta + \bar{X}_0^{\ 2}, \\ \bar{X}^3 &= 0 + \bar{X}_0^{\ 3}. \end{aligned} \tag{2-4.12d}$$

Note that it is quite convenient to think in terms of matrix[9] multiplication in order to obtain the relations in (2-4.12d) as well as some of those that follow.

In order to obtain the velocity components, we must first compute $dA_k^{\ j}/dt$.

$$\frac{dA_k^{\ j}}{dt} = \begin{pmatrix} -\sin\theta & -\cos\theta & 0 \\ \cos\theta & -\sin\theta & 0 \\ 0 & 0 & 0 \end{pmatrix} \frac{d\theta}{dt}. \tag{2-4.12e}$$

[9] We would put the first term on the right of (2-4.12b) in the form $X^k A_k^{\ j}$. This corresponds to the matrix form

$$(X^1, X^2, X^3)\begin{pmatrix} A_1^{\ 1} & A_1^{\ 2} & A_1^{\ 3} \\ A_2^{\ 1} & A_2^{\ 2} & A_2^{\ 3} \\ A_3^{\ 1} & A_3^{\ 2} & A_3^{\ 3} \end{pmatrix},$$

By substituting into (2-4.3b) we have

$$\left(\frac{d\bar{X}^1}{dt}, \frac{d\bar{X}^2}{dt}, \frac{d\bar{X}^3}{dt}\right) = \frac{d\theta}{dt}(r, 0, 0)\begin{pmatrix} -\sin\theta & -\cos\theta & 0 \\ \cos\theta & -\sin\theta & 0 \\ 0 & 0 & 0 \end{pmatrix}$$

$$+ (0, 0, 0)\begin{pmatrix} \cos\theta & -\sin\theta & 0 \\ \sin\theta & \cos\theta & 0 \\ 0 & 0 & 1 \end{pmatrix} + (0, 0, 0),$$

that is,

$$\frac{d\bar{X}^1}{dt} = -r\sin\theta\frac{d\theta}{dt},$$

(2-4.12f)
$$\frac{d\bar{X}^2}{dt} = -r\cos\theta\frac{d\theta}{dt},$$

$$\frac{d\bar{X}^3}{dt} = 0.$$

In order to obtain the components of acceleration, we must compute $d^2A_k^{\ j}/dt^2$.

(2-4.12g)
$$\left(\frac{d^2A_k^{\ j}}{dt^2}\right) = \begin{pmatrix} -\cos\theta & \sin\theta & 0 \\ -\sin\theta & -\cos\theta & 0 \\ 0 & 0 & 0 \end{pmatrix}\left(\frac{d\theta}{dt}\right)^2$$

$$+ \begin{pmatrix} -\sin\theta & -\cos\theta & 0 \\ \cos\theta & -\sin\theta & 0 \\ 0 & 0 & 0 \end{pmatrix}\frac{d^2\theta}{dt^2}.$$

The only term of the right-hand member of (2-4.3c) that contributes to the result is the first. This is because the particle is fixed in the X^j system. Hence

$$\frac{d^2\bar{X}^1}{dt^2} = -r\cos\theta\left(\frac{d\theta}{dt}\right)^2 - r\sin\theta\frac{d^2\theta}{dt^2},$$

(2-4.12h)
$$\frac{d^2\bar{X}^2}{dt^2} = +r\sin\theta\left(\frac{d\theta}{dt}\right)^2 - r\cos\theta\frac{d^2\theta}{dt^2},$$

$$\frac{d^2\bar{X}^3}{dt^2} = 0.$$

When considering problems involving rotating frames of reference, it is standard practice to start with knowledge of the angular velocity components ω^j and a set of given initial conditions. For example, in considering the motion of a particle in relation to the earth's surface, we might assume that a frame of reference, fixed with respect to the earth, rotated uniformly in a system associated with the sun. The results from which the computations of the last example were made are expressed in terms of rotation coefficients $A_j^{\ k}$. The way in which the $A_j^{\ k}$ are produced by means of an initial state of rotation and a prescribed constant angular velocity is indicated in the next theorem.

Theorem 2-4.3. If the angular velocity components ω^j are constant and $(A_j^{\ k})_0$ is a given set of values $A_j^{\ k}$ at $t = 0$, then

$$A_j^{\ k} = \left[(A_j^{\ k})_0 + \frac{1}{K^2}\left(\frac{d^2A_j^{\ k}}{dt^2}\right)_0\right] - \frac{1}{K^2}\left(\frac{d^2A_j^{\ k}}{dt^2}\right)_0 \cos Kt + \frac{1}{K}\left(\frac{dA_j^{\ k}}{dt}\right)_0 \sin Kt, \tag{2-4.13a}$$

where

$$K = [(\omega^1)^2 + (\omega^2)^2 + (\omega^3)^2]^{1/2}. \tag{2-4.13b}$$

PROOF. From (2-4.6b) and (2-4.7a) we obtain

$$\frac{dA_k^{\ q}}{dt} - \omega^j_{\ k} A_j^{\ q} = 0, \tag{2-4.14a}$$

where

$$\omega^j_{\ k} = \delta^{jp}\omega_{pk} = \omega_{jk} \tag{2-4.14b}$$

is introduced as a notational convenience. The relation (2-4.14a) can be put in the form

$$\left(\delta_k^{\ j}\frac{d}{dt} - \omega^j_{\ k}\right)A_j^{\ q} = 0. \tag{2-4.14c}$$

For each value of q ($q = 1, 2, 3$), relation (2-4.14c) consists of a set of three homogeneous first-order differential equations with constant coefficients. The solutions of this system are annihilated by $\left(\delta_k^{\ j}\frac{d}{dt} - \omega^j_{\ k}\right)$. Since

$$\det.\left(\delta_k^{\ j}\frac{d}{dt} - \omega^j_{\ k}\right) = \frac{d^3}{dt^3} + K^2\frac{d}{dt} = \frac{d}{dt}\left(\frac{d^2}{dt^2} + K^2\right), \tag{2-4.14d}$$

it follows that

$$A_j^{\ k} = c_j^{\ k} + d_j^{\ k}\cos Kt + e_j^{\ k}\sin Kt, \tag{2-4.14e}$$

where $c_j^{\ k}$, $d_j^{\ k}$, $e_j^{\ k}$ are constants of integration which must be determined from the initial conditions. (See the problems at the end of the section for details. Also see the discussion of eigenvalues at the end of Chapter 1, Section 6.) Since the $A_k^{\ q}$ are known at $t = 0$ and the $\omega^j{}_k$ are given, from (2-4.14e) and the result obtained by differentiating this expression we can compute $dA_k^{\ q}/dt$ and $d^2A_k^{\ q}/dt^2$ at $t = 0$. This enables us to calculate the unknown coefficients. At $t = 0$

$$(A_j^{\ k})_0 = c_j^{\ k} + d_j^{\ k},$$

$$\left(\frac{dA_j^{\ k}}{dt}\right)_0 = Ke_j^{\ k}, \tag{2-4.14f}$$

$$\left(\frac{d^2A_j^{\ k}}{dt^2}\right)_0 = -K^2d_j^{\ k}.$$

When these relations are substituted into (2-4.14e), the result (2-4.13a) follows.

Problems

1. Determine the components of $\bar{\mathbf{r}}$, $\bar{\mathbf{V}}$, and $\bar{\mathbf{a}}$ as well as the components ω_{ik} and ω^p if the parametric equations of a particle in the X^j system are

(a)

$$X^1 = 2 + 3t,$$
$$X^2 = 1 + 4t,$$
$$X^3 = 3 + 5t,$$

(i.e., straight-line motion) with

$$(A_j^{\ k}) = \begin{pmatrix} 1 & 0 & 0 \\ 0 & \cos\theta & -\sin\theta \\ 0 & \sin\theta & \cos\theta \end{pmatrix}, \quad \theta = \theta(t).$$

(b)

$$X^1 = t,$$
$$X^2 = t^2,$$
$$X^3 = 0,$$

(i.e., parabolic motion), with

$$(A_j^{\ k}) = \begin{pmatrix} \cos\theta & -\sin\theta & 0 \\ \sin\theta & \cos\theta & 0 \\ 0 & 0 & 1 \end{pmatrix}, \quad \theta = \theta(t).$$

In each case let

$$\bar{X}_0^{\ 1} = t,$$
$$\bar{X}_0^{\ 2} = 1 + 2t,$$
$$\bar{X}_0^{\ 3} = 2 + 3t.$$

2. Repeat Problem 1 if

$$X^1 = r\sin\phi, \phi = t$$
$$X^2 = 0,$$
$$X^3 = r\cos\phi,$$

where r is constant, ϕ is measured from X^3,

$$(A_j{}^k) = \begin{pmatrix} \cos\theta & -\sin\theta & 0 \\ \sin\theta & \cos\theta & 0 \\ 0 & 0 & 1 \end{pmatrix}, \theta = \theta(t)$$

and $(\bar{X}_0{}^j) = (0, 0, 0)$.

3. Show that $-\frac{1}{2}\omega^r{}_q\omega^q{}_r = K^2$ where K^2 is defined by (2-4.13b).

4. Show that det. $[\delta_k{}^j(d/dt) - \omega^j{}_k] = (d^3/dt) + K^2(d/dt)$.

Hint: Start with det. $(\delta_k{}^j(d/dt) - \omega^j{}_k)$

$$= \frac{1}{3!}\,\varepsilon_{ijk}E^{pqr}\left(\delta_p{}^i\frac{d}{dt} - \omega^i{}_p\right)\left(\delta_q{}^j\frac{d}{dt} - \omega^j{}_q\right)\left(\delta_r{}^k\frac{d}{dt} - \omega^k{}_r\right).$$

5. If $\omega^1 = \omega^2 = 0$, $\omega^3 = \omega$, and $(A_j{}^k)_0 = \delta_j{}^k$, show that

$$(A_j{}^k) = \begin{pmatrix} \cos\omega t & -\sin\omega t & 0 \\ \sin\omega t & \cos\omega t & 0 \\ 0 & 0 & 1 \end{pmatrix}.$$

6. If $\omega^1 = \omega^2 = 0$, $\omega^3 = \omega$, show that the Coriolis components are

$$2\omega\left(-\cos\omega t\,\varepsilon_{31k}\frac{dX^k}{dt} - \sin\omega t\,\varepsilon_{32k}\frac{dX^k}{dt}\right),$$

$$2\omega\left(\sin\omega t\,\varepsilon_{31k}\frac{dX^k}{dt} - \cos\omega t\,\varepsilon_{32k}\frac{dX^k}{dt}\right),\quad 0,$$

and, when $dX^1/dt = f(t)$, $dX^2/dt = dX^3/dt = 0$, these components reduce to $2\omega(df/dt\sin\omega t, df/dt\cos\omega t, 0)$

4*. *A Tensor Formulation of the Theory of Rotating Frames of Reference*

In this section the allowable transformation group is the set of orthogonal Cartesian rotations with time-dependent coefficients

$$\bar{X}^j = A_k{}^j(t)X^k. \tag{2-4*.1}$$

The goal of this section is to determine mathematical formulations of the concepts of velocity and acceleration such that the respective components transform according to the law (1-4.1). The attainment of this objective makes possible the generalization of the vector concept to systems of orthogonal Cartesian coordinates related by transformations with variable coefficients. The importance of representing physical concepts, such as velocity and acceleration, in vector form lies in the adaptability of these forms to the expression of physical laws in a universal manner. This mode of thought is illustrated in Section 6 of this chapter.

In a sense, the set of components ω_{ki} comprises the key to the mathematical formulations to be developed in this chapter. Therefore, as a first step in the development, the relations (2-4.9a,b) are expressed in terms of the components ω_{ki}.

Theorem 2-4*.1. The relations in (2-4.9a,b), for rotations alone, can be written

(2-4*.2)

$$\text{(a)}\quad \frac{d\bar{X}^j}{dt} = A_k{}^j\left(\frac{dX^k}{dt} + \omega^k{}_i X^i\right),$$

$$\text{(b)}\quad \frac{d^2\bar{X}^j}{dt^2} = A_k{}^j\left(\frac{d^2X^k}{dt^2} + 2\omega^k{}_i\frac{dX^i}{dt} + \omega^k{}_i\omega^i{}_q X^q + \frac{d\omega^k{}_q}{dt}X^q\right),$$

where $\omega^k{}_i = \omega_{ki}$.

PROOF. The relations (2-4*.2a,b) can be obtained most easily by computing $dA_k{}^j/dt^2$ from (2-4.4a) and then appropriately plugging the result in (2-4.3b) and (2-4.3c); that is, according to (2-4.4a) and also using the fact that $A_p{}^j = a_j{}^p$,

$$a_j{}^p\frac{dA_k{}^j}{dt} = \omega^p{}_k, \tag{2-4*.3a}$$

where $\omega^p{}_k$ is written rather than ω_{pk} simply for notational convenience. Summing the products of (2-4*.3a) with $A_p{}^q$, we obtain

$$\frac{dA_k{}^q}{dt} = A_p{}^q\omega^p{}_k. \tag{2-4*.3b}$$

Another differentiation yields

$$\frac{d^2A_k{}^q}{dt^2} = \frac{dA_p{}^q}{dt}\omega^p{}_k + A_p{}^q\frac{d\omega^p{}_k}{dt} \tag{2-4*.3c}$$

$$= A_r{}^q\omega^r{}_p\omega^p{}_k + A_p{}^q\frac{d\omega^p{}_k}{dt}.$$

As previously indicated, when we substitute the results (2-4*.3b) and (2-4*.3c) into (2-4.3b,c) the relations in (2-4*.2a,b) follow. This completes the proof.

The next theorem investigates the mode of transformation of the components $\omega^j{}_k$. As in the considerations of Section 3, it is assumed that $\bar{O}$ is a fixed frame of reference; O and $\bar{\bar{O}}$ are frames in rotational motion with respect to $\bar{O}$. X^j, $\bar{X}^j$, and $\bar{\bar{X}}^j$ are rectangular Cartesian coordinates associated with O, $\bar{O}$, and $\bar{\bar{O}}$, respectively. Furthermore, $A_k^{\ j}$ and $B_k^{\ j}$ are coefficients of orthogonal transformations such that

(2-4*.4) $$\bar{X}^j = A_k^{\ j} X^k, \qquad \bar{X}^j = B_k^{\ j} \bar{\bar{X}}^k.$$

As usual, $a_k^{\ j}$ and $b_k^{\ j}$ represent the inverse transformation coefficients.

Theorem 2-4*.2. The transformation rule relating the components $\omega^i_{\ k}$ and $\bar{\bar{\omega}}^i_{\ k}$ is

(2-4*.5) $$\boxed{\omega^i_{\ k} = \frac{\partial X^i}{\partial \bar{\bar{X}}^q} \frac{\partial \bar{\bar{X}}^p}{\partial X^k} \bar{\bar{\omega}}^q_{\ p} + \frac{\partial X^i}{\partial \bar{\bar{X}}^q} \frac{d(\partial \bar{\bar{X}}^q / \partial X^k)}{dt}}.$$

PROOF. From (2-4*.4) we observe that

(2-4*.6a) $$\frac{\partial \bar{X}^j}{\partial X^k} = A_k^{\ j}, \qquad \frac{\partial \bar{X}^j}{\partial \bar{\bar{X}}^k} = B_k^{\ j}.$$

Therefore

(2-4*.6b) $$\begin{aligned} \omega^i_{\ k} &= a_j^{\ i} \frac{dA_k^{\ j}}{dt} = \frac{\partial X^i}{\partial \bar{X}^j} \frac{d}{dt}\left(\frac{\partial \bar{X}^j}{\partial X^k}\right) \\ &= \frac{\partial X^i}{\partial \bar{\bar{X}}^p} \frac{\partial \bar{\bar{X}}^p}{\partial \bar{X}^j} \frac{d}{dt}\left(\frac{\partial \bar{X}^j}{\partial \bar{\bar{X}}^q} \frac{\partial \bar{\bar{X}}^q}{\partial X^k}\right) \\ &= \frac{\partial X^i}{\partial \bar{\bar{X}}^p} \frac{\partial \bar{\bar{X}}^p}{\partial \bar{X}^j}\left[\frac{d(\partial \bar{X}^j / \partial \bar{\bar{X}}^q)}{dt} \frac{\partial \bar{\bar{X}}^q}{\partial X^k} + \frac{\partial \bar{X}^j}{\partial \bar{\bar{X}}^q} \frac{d(\partial \bar{\bar{X}}^q / \partial X^k)}{dt}\right] \\ &= \frac{\partial X^i}{\partial \bar{\bar{X}}^p} \frac{\partial \bar{\bar{X}}^q}{\partial X^k} b_j^{\ p} \frac{dB_q^{\ j}}{dt} + \delta_q^{\ p} \frac{\partial X^i}{\partial \bar{\bar{X}}^p} \frac{d(\partial \bar{\bar{X}}^q / \partial X^k)}{dt}. \end{aligned}$$

The desired result is a consequence of making the replacement

$$b_j^{\ p} \frac{dB_q^{\ j}}{dt} = \bar{\bar{\omega}}^p_{\ q},$$

in (2-4*.6b).

It is left as an exercise for the reader to show that the transformation rule (2-4*.5) is symmetric in nature. In other words,

(2-4*.7) $$\boxed{\bar{\bar{\omega}}^i_{\ k} = \frac{\partial \bar{\bar{X}}^i}{\partial X^q} \frac{\partial X^p}{\partial \bar{\bar{X}}^k} \omega^q_{\ p} + \frac{\partial \bar{\bar{X}}^i}{\partial X^q} \frac{d(\partial X^q / \partial \bar{\bar{X}}^k)}{dt}}.$$

The reader should note that the $\bar{O}$ system is assumed fixed. Physically, this can be interpreted to mean that $\bar{O}$ has no rotational motion and therefore the set of components $\bar{\omega}^i_{\ k}$ (which measure the rotation) is a zero set, that is,

(2-4*.8) $$\bar{\omega}^i_{\ k} = 0.$$

From the standpoint of the transformation concept we could say that the components $\bar{\omega}^i_{\ k}$ are obtained by consideration of the identity transformation; that is,

$$\omega^i{}_k = \delta_j{}^i \frac{d\delta_k{}^j}{dt} = 0.$$

Now that the transformation rule for the components $\omega^i{}_k$ is determined, we can turn our attention to the problem of expressing velocity and acceleration in a form in which the respective components will transform consistently with the law (1-4.1). If this law is satisfied, the entities are said to be of vector character.

Definition 2-4*.1. Let $\{U^k\}$ be a collection of n-tuples U^k, $\bar{U}^k$, etc., one n-tuple associated with each rectangular Cartesian coordinate system. The collection $\{U^k\}$ is said to be a contravariant[10] vector with respect to the orthogonal Cartesian group with time-dependent coefficients of transformation. The elements (U^1, U^2, U^3) are components of the vector.

The solution to the problem of formulating expressions for velocity and acceleration that transform appropriately lies in relation (2-4*.2a) and is indicated by the following definition.

Definition 2-4*.2. Let U^k be the components of a contravariant vector. Then

$$\frac{\mathfrak{D}U^k}{dt} = \frac{dU^k}{dt} + \omega^k{}_i U^i \tag{2-4*.9}$$

are called the components of the rotational derivative of the vector **U**.

Theorem 2-4*.3. The rotational derivative $\mathfrak{D}U^k/dt$ is a contravariant vector.

PROOF. Consider the coordinate systems related by

$$X^j = c_k{}^j \bar{\bar{X}}^k = \frac{\partial X^j}{\partial \bar{\bar{X}}^k} \bar{\bar{X}}^k.$$

Of course, they are both rectangular Cartesian and the transformation is orthogonal. By substituting for U^k and $\omega^k{}_i$ according to their transformation rules, we obtain

(2-4*.10a)

$$\begin{aligned}
\frac{\mathfrak{D}U^k}{dt} &= \frac{dU^k}{dt} + \omega^k{}_i U^i \\
&= \frac{d[(\partial X^k/\partial \bar{\bar{X}}^j)\bar{\bar{U}}^j]}{dt} + \left[\frac{\partial X^k}{\partial \bar{\bar{X}}^p}\frac{\partial \bar{\bar{X}}^q}{\partial X^i}\bar{\bar{\omega}}^p{}_q + \frac{\partial X^k}{\partial \bar{\bar{X}}^p}\frac{d(\partial \bar{\bar{X}}^p/\partial X^i)}{dt}\right]\frac{\partial X^i}{\partial \bar{\bar{X}}^j}\bar{\bar{U}}^j \\
&= \frac{\partial X^k}{\partial \bar{\bar{X}}^p}\left(\frac{d\bar{\bar{U}}^p}{dt} + \bar{\bar{\omega}}^p{}_j \bar{\bar{U}}^j\right) + \left(\frac{d(\partial X^k/\partial \bar{\bar{X}}^j)}{dt} + \frac{\partial X^k}{\partial \bar{\bar{X}}^p}\frac{\partial X^i}{\partial \bar{\bar{X}}^j}\frac{d(\partial \bar{\bar{X}}^p/\partial X^i)}{dt}\right)\bar{\bar{U}}^j.
\end{aligned}$$

[10] For a discussion of the term contravariant see Chapter 1, Section 5*.

To show that the theorem is valid, we must verify that the second parenthetic expression on the right of (2-4*.10a) has the value zero. This fact follows by the differentiation with respect to t of

$$\frac{\partial X^k}{\partial \overline{X}^j}\frac{\partial \overline{X}^j}{\partial X^p} = \delta_p{}^k.$$

When the parenthetic expression is set equal to zero, (2-4*.10a) reduces to

$$\frac{\mathfrak{D}U^k}{dt} = \frac{\partial X^k}{\partial \overline{X}^j}\frac{\mathfrak{D}\overline{U}^j}{dt}. \tag{2-4*.10b}$$

This completes the proof.

Since the rotational derivative of a vector is a vector, it follows that rotational derivatives of any order are vectors. This fact puts us in a position to consider the concepts of velocity and acceleration. In particular, recall that the transformations being dealt with are rotations, hence X^j are components of a position vector **r**. Because the X^j are vector components, we can take a rotational derivative.

Definition 2-4*.3. Let the velocity and acceleration concepts be defined by means of the respective sets of components

$$\frac{\mathfrak{D}X^j}{dt}, \qquad \frac{\mathfrak{D}^2X^j}{dt^2} = \frac{\mathfrak{D}(\mathfrak{D}X^j/dt)}{dt}.$$

Theorem 2-4*.4. Velocity and acceleration as defined in Definition 2-4*.3 are vector concepts. Furthermore, in a given coordinate system they reduce to the usual forms (2-4.9a,b).

PROOF. The rotational derivative was modeled after the parenthetic expression in (2-4*.2a). Furthermore, the components $\bar{\omega}^k{}_i$ equal zero. Therefore the components of velocity satisfy Definition 2-4*.2. The computation showing that the components of acceleration satisfy (2-4*.2b) is straightforward and is left to the reader.

The problem of determining "universal laws of nature" is one of the major concerns of theoretical physics. Vector and tensor analysis have gained prominence as valuable scientific tools because they, more than any other mathematical discipline, lend themselves to this cause.

Problems

1. Derive the transformation law (2-4*.7),
 (a) repeating the procedure used to obtain (2-4*.5), and
 (b) starting with (2-4*.5).

2. Is the transformation rule for the components $\omega^i{}_k$ consistent with the tensor transformation rule stated in Chapter 1, Section 5*?

3. Show that the components of acceleration $\mathfrak{D}^2X^j/dt^2$ are in agreement with (2-4*.2b).

5. *Newtonian Orbits*

A problem in dynamics is investigated in this section. Studies made under the heading of dynamics differ from kinematical investigations in that the concept of mass is considered. Particles are still idealized as points, and their paths are represented by space curves, but because mass (m) is taken into account the study of velocity and acceleration fields along the curves is replaced by a study of momentum ($m\mathbf{v}$) and force fields ($m\mathbf{a}$).

To the ancients the earth was a plane in a vast sea. Observations concerning the nature of the earth and its relationship to the other heavenly bodies were made over a long period. By the time of the Greek civilization important facts concerning these relations had been established. Hipparchus, 160 B.C., and Ptolemy, A.D. 100, developed a theory of epicycles which described the motions of the known planets from an earth-centered viewpoint. More than a thousand years later Copernicus[11] brought about a philosophic revolution with the observation that paths of planets could be represented more simply by taking the sun as the center of reference.

By using observed data, obtained by himself and his predecessor Tycho Brahe (1546–1601, Danish), Johann Kepler (1571–1630, German) elaborated on this view and determined empirically the motion of the planets with respect to the sun. Half a century later Isaac Newton published his *Principia* in which he set forth a theory of gravitation that formed a theoretical foundation for Kepler's empirical results. No doubt Newton fashioned his theory to include Kepler's laws. But, as every student of physics knows, the Newtonian theory was not sterile. It has been the working model of the physical sciences for more than two hundred years and has formed a foundation stone for much of modern-day physics, including Einstein's general theory of relativity. Discrepancies in observation and Newtonian gravitational theory led to the discovery of the planet Neptune. Adams (English) and Leverrier (French) obtained sufficiently accurate mathematical results to enable astronomers to locate the planet. (Neptune was first recognized by Galle (German) in 1846.) In fact, still further astronomical investigations, spurred by the discovery of Neptune, led to the discovery of Pluto by C. W. Tombaugh in 1930.

The objective of this section is to illustrate the way in which the Kepler laws of planetary motion can be obtained from the Newtonian theory.

---

[11] His proper name was Nicolas Copernik and he was born in Polish Russia. Copernicus is the Latinized form of Copernik.

These laws are as follows:

(2-5.1)
- (a) The planets describe ellipses with the sun at one focus.
- (b) The radius vector from the sun to a planet sweeps out equal areas in equal times.
- (c) The squares of the periods of the planets are proportional to the cubes of their mean distances[12] from the sun.

Our starting point is Newton's law of gravitation.

Newton's Law of Gravitation. Any two bodies whose dimensions are negligible in comparison with their distance apart attract each other with forces directed along their joining line. The force on either body is directly proportional to the product of the masses of the bodies and inversely proportional to the square of the distances between them.

The algebraic structure of vectors was not developed until the nineteenth century, whereas Newton lived from 1642–1727. Therefore it is clear that the vector methods used in this development were not included in the original work. To see the degree of simplification brought about by the vector concepts, we need only begin writing out the equations in complete detail.

Regard the sun, of mass M, as fixed in space and let the point representing it be the initial point of the position vector $\mathbf{r} = r\mathbf{R}$. ($\mathbf{R}$ is a unit radial vector.) Let the mass of a planetary body be represented by m. On the one hand, the force vector field associated with the path of planetary motion is defined to be $\mathbf{F} = m\mathbf{a}$. On the other hand, according to Newton's law of gravitation,

$$\mathbf{F} = \frac{-mMG\mathbf{R}}{r^2}, \tag{2-5.2}$$

where G is a constant of proportionality, the so-called gravitational constant. Therefore we have

$$m\frac{d\mathbf{v}}{dt} = \frac{-mMG\mathbf{R}}{r^2} \tag{2-5.3a}$$

or

$$\mathbf{a} = \frac{-MG\mathbf{R}}{r^2}. \tag{2-5.3b}$$

Theroem 2-5.1. The planetary motion is planar.

PROOF. By cross multiplication of both members of (2-5.3b) with $\mathbf{r}$, we obtain

$$\mathbf{r} \times \mathbf{a} = \mathbf{0}. \tag{2-5.4a}$$

---

[12] The major semiaxis $a = \frac{1}{2}(r_1 + r_2)$ is called the planet's mean distance from the sun. For earth $a \approx 92{,}900{,}000$ miles.

The relation (2-5.4a) is equivalent to

(2-5.4b)
$$\frac{d(\mathbf{r} \times \mathbf{v})}{dt} = \mathbf{0},$$

since

$$\frac{d(\mathbf{r} \times \mathbf{v})}{dt} = \left(\frac{d\mathbf{r}}{dt} \times \mathbf{v}\right) + \left(\mathbf{r} \times \frac{d\mathbf{v}}{dt}\right) = (\mathbf{v} \times \mathbf{v}) + (\mathbf{r} \times \mathbf{a}) = \mathbf{r} \times \mathbf{a}.$$

Therefore by integration of (2-5.4b) we obtain

(2-5.4c)
$$\mathbf{r} \times \mathbf{v} = \mathbf{h},$$

where $\mathbf{h}$ is a vector constant. Since $\mathbf{r}$ and $\mathbf{v}$ are perpendicular to the same direction $\mathbf{h}$ for all t, the motion is planar.

Theorem 2-5.2. The equation of planetary motion is

(2-5.5)
$$r = \frac{h^2/MG}{1 + e\cos\theta}$$

where $h = \sqrt{\mathbf{h} \cdot \mathbf{h}}$ and e, the magnitude of a vector constant of integration, represents the eccentricity of a conic section.

PROOF. As a first step in obtaining the desired equation, cross multiply (2-5.4c) with the acceleration vector $\mathbf{a}$.

(2-5.6a)
$$\mathbf{a} \times (\mathbf{r} \times \mathbf{v}) = \mathbf{a} \times \mathbf{h}.$$

On the one hand, the left member of (2-5.6a) can be written

(2-5.6b)
$$\begin{aligned}\mathbf{a} \times (\mathbf{r} \times \mathbf{v}) &= \frac{-MG}{r^2}\mathbf{R} \times \left(r\mathbf{R} \times \frac{dr\mathbf{R}}{dt}\right) = \frac{-MG}{r^2}\mathbf{R} \times \left(r^2\mathbf{R} \times \frac{d\mathbf{R}}{dt}\right) \\ &= -MG\mathbf{R} \times \left(\mathbf{R} \times \frac{d\mathbf{R}}{dt}\right) = -MG\left[\mathbf{R}\left(\mathbf{R} \cdot \frac{d\mathbf{R}}{dt}\right) - \frac{d\mathbf{R}}{dt}(\mathbf{R} \cdot \mathbf{R})\right] \\ &= MG\frac{d\mathbf{R}}{dt}.\end{aligned}$$

In developing the result (2-5.6b), we must use the triple vector product as well as the facts

$$\mathbf{R} \times \mathbf{R} = \mathbf{0}, \qquad \mathbf{R} \cdot \mathbf{R} = 1, \qquad \mathbf{R} \cdot \frac{d\mathbf{R}}{dt} = 0.$$

On the other hand, the right-hand member of (2-5.6a) can be written

(2-5.6c)
$$\mathbf{a} \times \mathbf{h} = \frac{d\mathbf{v}}{dt} \times \mathbf{h} = \frac{d(\mathbf{v} \times \mathbf{h})}{dt}.$$

Therefore, by replacing the members of (2-5.6a) according to (2-5.6b,c) and putting the terms on the same side of the resulting vector equation, we have

(2-5.6d) $$MG\frac{d\mathbf{R}}{dt} - \frac{d(\mathbf{v}\times\mathbf{h})}{dt} = \frac{d}{dt}(MG\mathbf{R} - \mathbf{v}\times\mathbf{h}) = \mathbf{0}.$$

By integrating,

(2-5.6e) $$MG\mathbf{R} - \mathbf{v}\times\mathbf{h} = -MG\mathbf{e},$$

where the arbitrary constant of integration is written with the factor $-MG$ simply for convenience of the algebraic manipulation that follows. From (2-5.6e),

(2-5.6f) $$\mathbf{v}\times\mathbf{h} = MG(\mathbf{R} + \mathbf{e}).$$

When this expression is dotted with $\mathbf{r}$, we obtain

(2-5.6g) $$\mathbf{r}\cdot\mathbf{v}\times\mathbf{h} = MG(r + re\cos\theta),$$

where $\mathbf{r}\cdot\mathbf{e} = re\cos\theta$. By using the algebraic property of interchange of dot and cross on the left of (2-5.6g), as well as relation (2-5.4c), the expression (2-5.6g) can be put in the form

$$h^2 = MGr(1 + e\cos\theta).$$

The result of the theorem follows by simple algebraic manipulation.

The relation (2-5.5) is the polar form of a conic section with eccentricity e and with one focus at the origin. The orbit is an ellipse, parabola, or hyperbola for $e < 1$, $e = 1$, and $e > 1$, respectively.

Since the orbit of a planet is closed, it must be an ellipse. The closest point of the planet's orbit to the focus is called the perihelion, whereas the farthest point is called the aphelion. (See Fig. 2-5.1.)

The second Kepler law is satisfied, since, by assumption [see (2-5.3b) and Example 2-3.3], the acceleration is always pointed toward a fixed position. This fact may also be determined in the context of the present problem, as indicated in the following example.

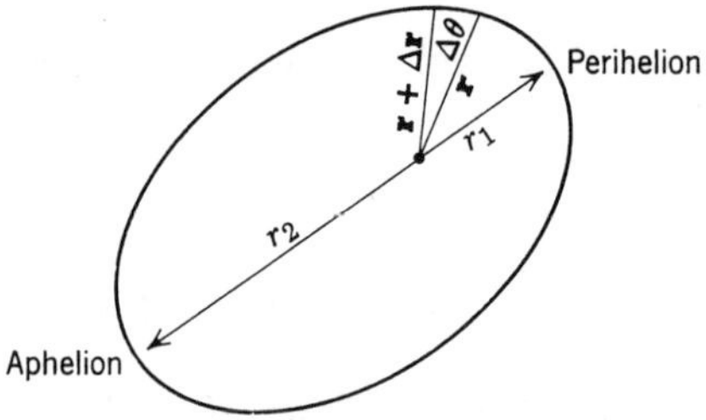

Fig. 2-5.1

Example 2-5.1. The law of areas can be intuitively determined by considering the element of area swept out in Fig. 2-5.1.

(2-5.7)
$$\Delta A = \tfrac{1}{2}\,|\mathbf{r}|\,|\mathbf{r} + \Delta\mathbf{r}|\,|\sin\Delta\theta| = \tfrac{1}{2}\,|\mathbf{r} \times (\mathbf{r} + \Delta\mathbf{r})|$$
$$= \tfrac{1}{2}\,|\mathbf{r} \times \Delta\mathbf{r}|.$$

Therefore

$$\frac{\Delta A}{\Delta t} = \frac{1}{2}\left|\mathbf{r} \times \frac{\Delta\mathbf{r}}{\Delta t}\right|, \qquad \Delta t > 0.$$

Then, assuming the existence of the limit,

$$\frac{dA}{dt} = \lim_{t\to 0}\frac{\Delta A}{\Delta t} = \tfrac{1}{2}\lim\left|\mathbf{r} \times \frac{\Delta\mathbf{r}}{\Delta t}\right| = \tfrac{1}{2}\,|\mathbf{r} \times \mathbf{v}|$$
$$= \tfrac{1}{2}\,|\mathbf{h}|.$$

Therefore dA/dt is constant.

The third Kepler law is derived as follows. Because the sectorial speed $\tfrac{1}{2}\,|\mathbf{h}|$ is constant, the period of revolution P in an elliptic orbit is obtained by dividing the area of the ellipse by the sectorial speed. As indicated in Fig. 2-5.1, the length of the major axis (i.e., $2a$) is $r_1 + r_2$, where r_1 and r_2 are the perihelion ($\theta = 0$) and aphelion ($\theta = \pi$) distances, respectively. From (2-5.5),

(2-5.8a)
$$r_1 = \frac{h^2/MG}{1 + e}, \qquad r_2 = \frac{h^2/MG}{1 - e}, \qquad h = |\mathbf{h}|.$$

Therefore

$$2a = r_1 + r_2 = \frac{2h^2}{MG(1 - e^2)}$$

and

(2-5.8b)
$$1 - e^2 = \frac{h^2}{aMG}.$$

The area of an ellipse of semiaxis a, b is

(2-5.8c)
$$\pi ab = \pi a^2(1 - e^2)^{1/2}.$$

Hence the period of revolution is

$$P = \frac{\pi a^2(1 - e^2)^{1/2}}{\tfrac{1}{2}h} = \frac{2\pi a^2}{h}\frac{h}{(aMG)^{1/2}} = \frac{2\pi a^{3/2}}{(MG)^{1/2}}.$$

Therefore

(2-5.8d)
$$\frac{P^2}{a^3} = \frac{4\pi^2}{MG}.$$

The right-hand member of (2-5.8d) is constant. Furthermore, the value is not dependent on the particular planet under consideration. Hence Kepler's third law is valid for all the planets.

Problems

1. If $\mathbf{h} = \mathbf{0}$ in (2-5.4c), what, then, is the nature of the motion?

2. The angular momentum of a particle about a point, denoted by L, is defined as

$$\mathbf{L} = \mathbf{r} \times m\mathbf{v}.$$

The moment of force or torque of a particle about a point is

$$\mathbf{H} = \mathbf{r} \times \mathbf{F},$$

where the force

$$\mathbf{F} = \frac{dm\mathbf{v}}{dt}.$$

Show that

$$\mathbf{H} = \frac{d\mathbf{L}}{dt}.$$

3. If the total torque is zero, show that the angular momentum is conserved.

6. *An Introduction to Einstein's Special Theory of Relativity*

Einstein's initial paper concerning the special theory of relativity, "On the Electrodynamics of Moving Bodies," was published in 1905. As with any other theory of major importance, it was not the product of one man; rather, Einstein's work culminated the thoughts of many individuals. To examine its origins, we must trace the histories of classical mechanics, electromagnetic theory, and the related mathematical and philosophical developments. It is not the purpose of this book to go into great detail concerning these histories, but a few relevant facts may be useful in setting the scene for those aspects of special relativity that are to be presented.

Alongside the knowledge of the solar system, discussed in Chapter 2, Section 5, Galileo placed the mechanics of freely falling bodies. In his turn, Newton combined the physical knowledge of the seventeenth century with his own physical intuition and mathematical learning to formulate the foundations of what is known today as classical mechanics. At the base of Newton's theory of mechanics were two assumptions of specific importance to the development of the ideas presented in this section. According to the theory:

1. Measurement of time does not depend on a physical frame of reference; that is, time has a universal character.
2. Uniform rectilinear motion cannot be intrinsically determined by mechanical experiments.

Relativity of motion was not a new concept with Newton. The fact that an individual's observation of the motion of an object depended on his own circumstances[13] was known to the Greeks. On the other hand, the idea that uniform rectilinear motion could not be determined mechanically without reference to another physical object was new.

From the mathematical point of view, the facts of immediate concern may be put as follows: It is assumed that a frame of reference (called by the physicist an inertial frame) exists in which Newton's laws of classical mechanics hold.[14] Any frame in uniform rectilinear motion in relation to the original system can be shown to be an inertial frame. We associate a rectangular Cartesian coordinate system with each reference frame. The transformations relating coordinates, as specified by these various systems, form a group, and it is required that the fundamental laws of classical mechanics be invariant with respect to this group.[15] These laws are invariant under the Galilean transformations

$$\bar{X}^j = A_k^{\ j} X^k - v^j t, \qquad |A_k^{\ j}| = 1,$$

where the coefficients of rotation $A_k^{\ j}$ and the velocity components v^j are constants. It is assumed that at $t = 0$ the origins of the orthogonal systems coincide. The path of motion of one frame with respect to the other is illustrated in Fig. 2-6.1.

Let us turn to the history of electromagnetic theory and, in particular, to the developments of the nineteenth century. Late in the eighteenth century Charles Augustin Coulomb (1736–1806, French) experimentally confirmed an inverse square law for the electrostatic force field. Around 1820 Hans Christian Oersted (1777–1851, Danish) discovered that an electric current produces a magnetic field, and a little later in the century Michael Faraday (1791–1867, English) and Joseph Henry (1799–1878, American) independently determined that a magnetic flux produces an electric field, thereby establishing the principle of electromagnetic induction. The body of knowledge of electromagnetic theory was climaxed by Clerk

[13] Are you moving at this instant?

[14] I. Every body tends to remain in a state of rest or of uniform rectilinear motion unless compelled to change its state by action of an impressed force.

II. The "rate of change of motion," that is, the rate of change of momentum, is proportional to the impressed force and occurs in the direction of the applied force.

III. To every action there is an equal and opposite reaction, that is, the mutual actions of two bodies are equal and opposite.

[15] Klein's *Erlanger Programm* of 1872 classified geometries in terms of invariants of transformation groups. The close relationship of this principle for classifying geometries to the demand that physical laws have invariant mathematical forms with respect to a given transformation group was pointed out by Klein. See *Entwicklung der Mathematik II*, Chelsea, New York, 1950.

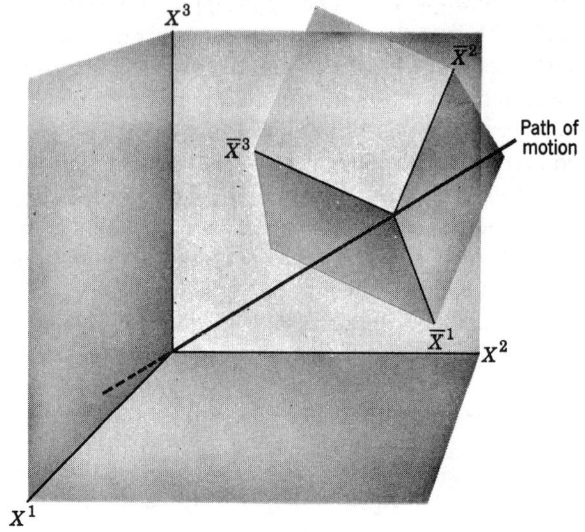

Fig. 2-6.1

Maxwell (1831–1879, English). His famous equations put the theory on a postulational basis and opened the door to mathematical speculation.

The phenomena of electricity and magnetism were destined to have a profound influence on the philosophical development of man, for electromagnetic theory set forth the physical properties of nonmaterial substances. By application of his mathematical talents Lorentz determined that although the Maxwell equations were not invariant to the Galilean transformations they did preserve their form under a special set of transformations, which subsequently were named after him. For the motion of one rectangular Cartesian frame along a z axis common with the $\bar{z}$ axis of another these Lorentz transformations have the form

$$\bar{x} = x \qquad \bar{y} = y$$

$$\bar{z} = \frac{z - vt}{[1 - (v/c)^2]^{1/2}} \qquad \bar{t} = \frac{t - (v/c^2)z}{[1 - (v/c)^2]^{1/2}},$$

where v is representative of a constant velocity and c is the velocity of light in vacuo. Although these transformation equations preserve the form of Maxwell's equations, they also suggest from the classical point of view a strange relation between space and time. It was not Lorentz's destiny to explain this relationship in a satisfactory manner. If he had done so, much of the scientific glory which later went to Einstein might have been his.

To these brief accounts of classical mechanics and electromagnetic theory let us add an experimental fact. It was hoped, because of the nonmaterial aspect of electromagnetic phenomena, that through it the uniform rectilinear motion of a physical frame of reference might be determined intrinsically (i.e., without reference to any outside body). In particular optical experiments were attempted, the most famous of which was the series carried out by Albert A. Michelson (1852–1931, American) and Edward Williams Morley (1838–1923, American) in the last part of the nineteenth century. In these experiments an attempt was made to determine the velocity of the earth (making the assumption that the earth moved along the tangent line to its path of motion at a given instant, hence could be considered in uniform rectilinear motion) by compounding its velocity with the velocity of light, first in the direction of the earth's motion and then orthogonal to this direction. No positive result was achieved; that is, the compounding of velocities, in the classical sense, failed. Furthermore, these experiments and others pointed to the possibility that the measurement of the velocity of light did not depend on uniform motion of either the source or the receiving agent.

We have before us two theories, classical mechanics and electromagnetics, and the experimental facts of the failure of the composition of velocities when light is involved and the constancy of the velocity of light. Let us add to these the theoretical physicists' desire for a universal physical theory and a suggestion by the French mathematician Henri Poincaré (1854–1912) that perhaps neither mechanical nor electromagnetic experiments could intrinsically detect uniform rectilinear motion. The stage is set for a new theory; in this instance, Einstein's special theory of relativity.

Let us turn our attention to the mathematical development of the previously stated ideas.

If the axes of two rectangular Cartesian coordinate systems, O and $\bar{O}$, are parallel and the O system is in uniform translatory motion with respect to $\bar{O}$, the associated Galilean transformations take the form (Fig. 2-6.2)

$$\bar{X}^j = X^j - v^j t, \qquad \bar{t} = t \tag{2-6.1a}$$

in which throughout the discussion we make the identifications $X^1 = x$, $X^2 = y$, $X^3 = z$ and correspondingly in the barred system. To interpret these equations physically, we associate the coordinates X^j and $\bar{X}^j$ with a particle P moving uniformly and rectilinearly with respect to either system. It is assumed that time is measured in a common way for all systems related by the transformation equations. Under this assumption a differentiation of the Galilean transformation equations leads to the classical

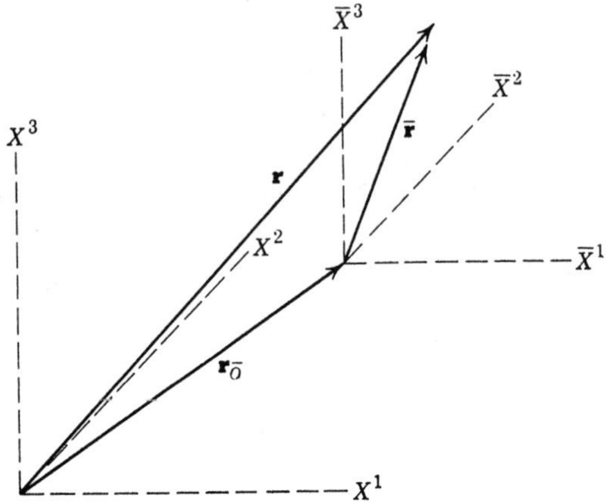

Fig. 2-6.2

law of composition of velocities

$$\frac{d\bar{X}^j}{dt} = \frac{dX^j}{dt} - v^j.$$

This law is a central concept in classical mechanics, but, as indicated in the introductory remarks, the phenomenon of light does not conform to it. Before beginning the search for transformation equations that produce a composition of velocities consistent with the physical facts concerning light, it is constructive to digress on a brief investigation of the geometric interpretation of the Galilean transformations.

The equations in (2-6.1a) were introduced as equations of motion of a particle in a three-dimensional Euclidean space. This interpretation involves motion and therefore is primarily physical, rather than geometric, in nature. A static model can be developed by extending the dimension of the space to four. Assume the space to be Euclidean and refer it to a rectangular Cartesian coordinate system (x, y, z, t). Then the transformation equations in (2-6.1a), which can also be written in the form

(2-6.1b) $$\bar{X}^\alpha = X^\alpha - v^j \delta_j{}^\alpha X^4$$

or

(2-6.1c) $$\bar{X}^\alpha = B_\beta{}^\alpha X^\beta,$$

are centered affine transformations. In (2-6.1b) the Greek index α ranges over values 1, 2, 3, 4, and $X^4 = t$, whereas $\bar{X}^4 = \bar{t}$. Throughout the remainder of this section indices indicated by Greek letters take on values

1, 2, 3, 4. The matrix of transformation coefficients is

(2-6.1d) $$(B_\beta{}^\alpha) = \begin{pmatrix} 1 & 0 & 0 & 0 \\ 0 & 1 & 0 & 0 \\ 0 & 0 & 1 & 0 \\ -v^1 & -v^2 & -v^3 & 1 \end{pmatrix}.$$

As long as some one of the velocity components is different from zero, this matrix will not satisfy the orthogonality conditions in (1-3.21e). Therefore the coordinate system founded on it will not be rectangular. Specifically, the x, y, and z axes are not changed under the transformation; however the $\bar{t}$ axis differs from the t axis. It is impossible to construct a four-dimensional diagram to illustrate the foregoing remarks, but our intuition can be stimulated by suppressing one of the space dimensions and then constructing the diagram. This is done in Fig. 2-6.3a. In Fig. 2-6.3b two-space dimensions are suppressed.

The transformation relation $\bar{t} = t$ points out the universal nature of time in classical space time. This fact is illustrated geometrically by a plane of simultaneity in Fig. 2-6.3a. This plane, which is parallel to the coincident y, z, and $\bar{y}$, $\bar{z}$ planes, consists of the set of all points with a common time coordinate. The existence of planes of simultaneity, illustrated in Fig. 2-6.3a, and of lines of simultaneity, shown in Fig. 2-6.3b, are specifically pointed out so that comparisons can be made with corresponding representations in the relativistic situation. A word of warning when interpreting the diagrams: the reader must realize that the unit of distance is not the same along the t and $\bar{t}$ axes. (See Problem 2b.)

In both Fig. 2-6.3a and Fig. 2-6.3b a line, called a world line, has been constructed parallel to the $\bar{t}$ axis. This line consists of the set of all points with the same space coordinates in the barred coordinate system. A corresponding world line for the unbarred system must be drawn parallel to the t axis. The world lines illustrate the fact that the concept of rest has no objective meaning in classical space-time.

The cone drawn in Fig. 2-6.3a degenerates to a pair of intersecting lines in Fig. 2-6.3b. In both illustrations the lines of the cone represent emanations of light rays from a source fixed at the origin of coordinates. The cone is symmetric to the t axis but it is not symmetric to the $\bar{t}$ axis. This fact is expressed algebraically by the variance of the equation of the cone under a centered affine transformation. (See Problem 3.) These observations concerning the light cone conclude the digression and, at the same time, initiate the discussion of the ideas underlying the special theory of relativity.

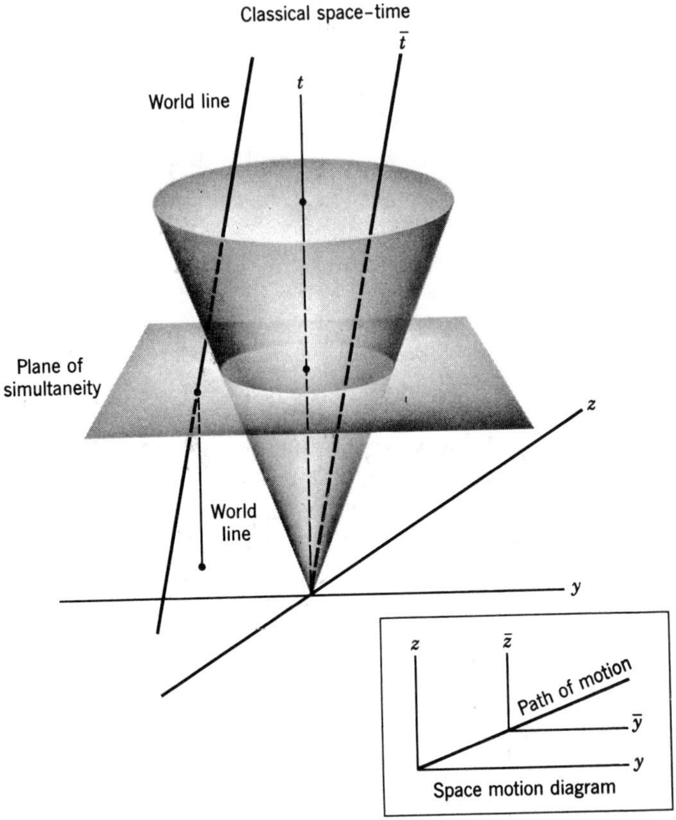

Fig. 2-6.3a. ***The concept of rest has no objective meaning in classical space time but simultaneity has.***

Experiments such as those of Michelson and Morley indicated that the measure of the velocity of light is independent of uniform rectilinear motions. From the mathematical viewpoint the question is one of obtaining the transformation equations relating coordinate systems O and $\bar{O}$ in uniform rectilinear motion, with respect to one another, under the condition[16]

(2-6.2a)
$$-(X^1)^2 - (X^2)^2 - (X^3)^2 + (X^4)^2 = -(\bar{X}^1)^2 - (\bar{X}^2)^2 - (\bar{X}^3)^2 + (\bar{X}^4)^2$$

or

(2-6.2b)
$$h_{\alpha\beta}X^\alpha X^\beta = h_{\lambda\mu}\bar{X}^\lambda \bar{X}^\mu,$$

[16] More precisely, the condition is $h_{\alpha\beta}X^\alpha X^\beta = p^2\bar{h}_{\lambda\mu}\bar{X}^\lambda\bar{X}^\mu$, but it can be shown that $p = 1$. See W. Rindler, *Special Relativity*, Interscience, 1960, p. 17.

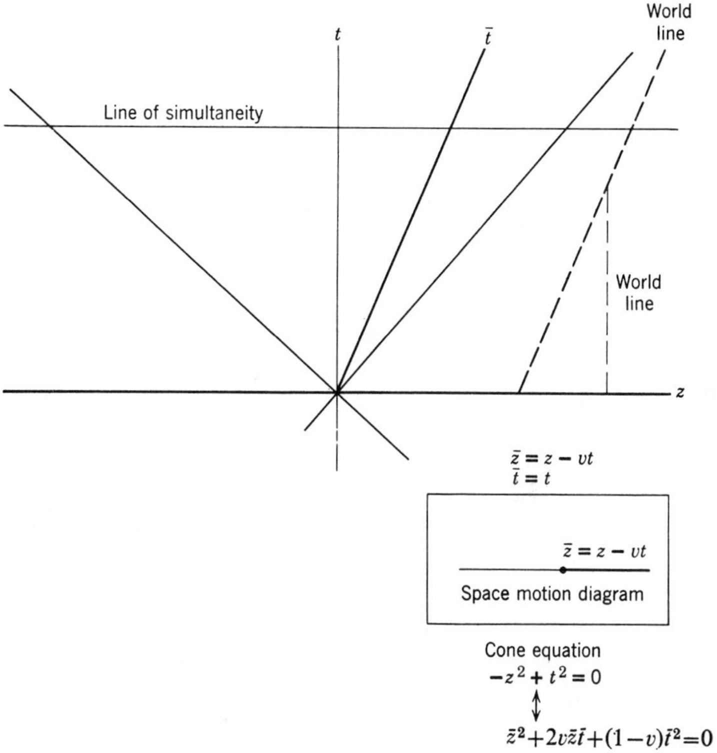

Fig. 2-6.3b

where

$$(h_{\alpha\beta}) = (\bar{h}_{\alpha\beta}) = \begin{pmatrix} -1 & 0 & 0 & 0 \\ 0 & -1 & 0 & 0 \\ 0 & 0 & -1 & 0 \\ 0 & 0 & 0 & 1 \end{pmatrix}.$$

From the physical point of view, we can think of light radiating from a point source (the origin of coordinates) in a spherical wave. (See Fig. 2-6.4.)

If the velocity of light is denoted by c, and X^4 represents the time elapsed while a point of the wave reaches the position P, then

$$(X^1)^2 + (X^2)^2 + (X^3)^2 = c^2(X^4)^2.$$

By taking our unit of length as the distance traversed by light in one second we impose the normalization $c = 1$; the relation then has the form

$$-(X^1)^2 - (X^2)^2 - (X^3)^2 + (X^4)^2 = 0.$$

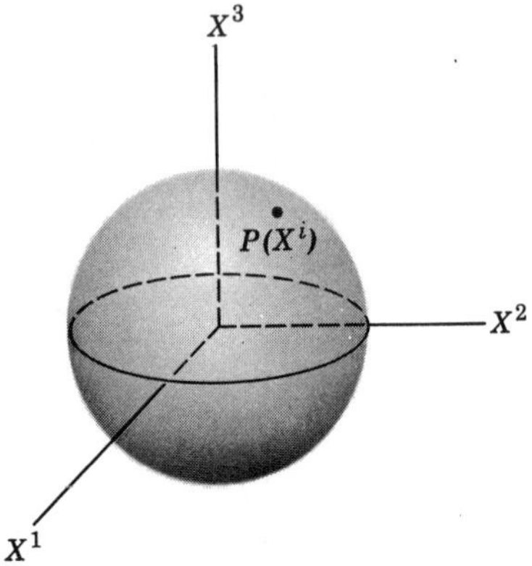

Fig. 2-6.4

The requirement that precisely the same observations must be made from the point of view of an $\bar{O}$ coordinate system (the origin of which corresponds to that of the O system at the emission of the light wave) leads to the relation

$$h_{\alpha\beta}X^\alpha X^\beta = \bar{h}_{\lambda\mu}\bar{X}^\lambda \bar{X}^\mu.$$

The fact that the systems O and $\bar{O}$ are in uniform rectilinear motion with respect to one another implies that the coordinates are linearly related; that is

$$X^\alpha = c_\beta{}^\alpha \bar{X}^\beta, \tag{2-6.3}$$

where the constant coefficients of transformation $c_\beta{}^\alpha$ are not known. (See Problems 1, 5, 6.) In order that the form $h_{\alpha\beta}X^\alpha X^\beta$ may be invariant under the transformation, it must be that

$$h_{\alpha\beta} = C_\alpha{}^\lambda C_\beta{}^\mu \bar{h}_{\lambda\mu}.$$

Since the matrices $(h_{\alpha\beta})$ and $(\bar{h}_{\alpha\beta})$ are identical, the foregoing conditions remind us of the orthogonality conditions associated with the transformations relating rectangular Cartesian systems. Indeed, these conditions are taken as the orthogonality conditions of the four-dimensional space which is now before us. It is usually called a Minkowski space because the geometrical aspects of the concepts presently under investigation were first pointed out by H. Minkowski in 1908. Before pursuing this trend of thought, it is convenient to make the following simplifying assumptions.

Let the axes of the three-space systems O and $\bar{O}$ be parallel. Furthermore, assume that the uniform rectilinear motion of the $\bar{O}$ system is along the positive sense of the X^3 axis of the O frame and that initially the origins coincide. For simplicity of notation we also use X^3 and X^4 interchangeably

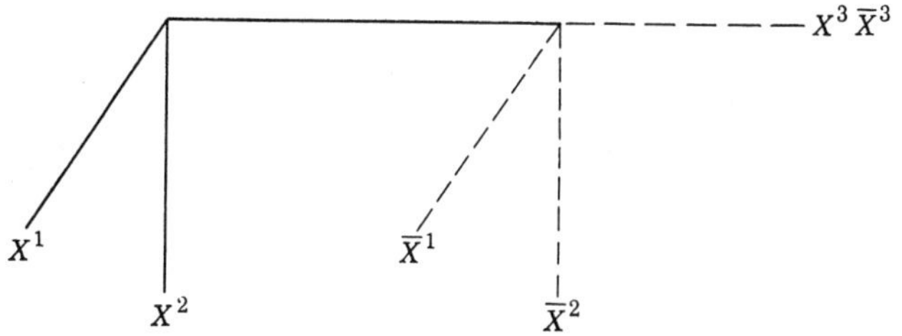

Fig. 2-6.5

with z and t. The corresponding relations hold in the $\bar{O}$ system. (See Fig. 2-6.5.)

Then

$$\bar{X}^1 = X^1, \qquad \bar{X}^2 = X^2,$$
$$\bar{X}^3 = C_3{}^3X^3 + C_4{}^3X^4, \qquad \bar{X}^4 = C_3{}^4X^3 + C_4{}^4X^4,$$

and the orthogonality conditions $h_{\alpha\beta} = C_\alpha{}^\lambda C_\beta{}^\mu \bar{h}_{\lambda\mu}$ reduce to

$$-1 = -(C_1{}^1)^2$$
$$-1 = -(C_2{}^2)^2$$
$$-1 = -(C_3{}^3)^2 + (C_3{}^4)^2$$
$$0 = -C_3{}^3C_4{}^3 + C_3{}^4C_4{}^4$$
$$1 = -(C_4{}^3)^2 + (C_4{}^4)^2.$$

If we concentrate attention on the relations

$$(C_3{}^3)^2 - (C_3{}^4)^2 = 1,$$
$$(C_4{}^4)^2 - (C_4{}^3)^2 = 1,$$

a marked resemblance to the orthogonality conditions on plane rotations is easily noted. In that case the condition to be satisfied was of the form

$$(X^4)^2 + (X^3)^2 = (\bar{X}^4)^2 + (\bar{X}^3)^2,$$

whereas in the present case

$$(X^4)^2 - (X^3)^2 = (\bar{X}^4)^2 - (\bar{X}^3)^2.$$

In the case of the rectangular Cartesian rotations, the coefficients of transformation were parameterized by circular functions $\sin\theta$ and $\cos\theta$. The foregoing considerations suggest that the coefficients of transformation associated with the Minkowski metric be parameterized by hyperbolic functions sinh and cosh. In particular, choose

$$C_3{}^3 = \cosh\chi, \qquad C_3{}^4 = -\sinh\chi,$$
$$C_4{}^4 = \cosh\chi, \qquad C_4{}^3 = -\sinh\chi,$$

and the conditions on the coefficients will be satisfied. We have

$$\begin{aligned} \bar{z} &= z \cosh \chi - t \sinh \chi. \\ \bar{t} &= -z \sinh \chi + t \cosh \chi. \end{aligned} \tag{2-6.4a}$$

If we consider a particle fixed at the origin of $\bar{O}$, then $\bar{z} = 0$ and

$$v = \frac{z}{t} = \frac{\sinh \chi}{\cosh \chi} = \tanh \chi;$$

that is, the uniform velocity of the $\bar{O}$ system with respect to the O system is represented by $\tanh \chi$. Furthermore, we can write the transformation equations in the form

$$\bar{z} = \cosh \chi(z - t \tanh \chi),$$

$$\bar{t} = \cosh \chi(-z \tanh \chi + t).$$

Use of the identity

$$1 - \tanh^2 \chi = \operatorname{sech}^2 \chi$$

immediately leads to a representation of the transformation equations in terms of the velocity v; that is,

$$\begin{aligned} \bar{z} &= \frac{z - vt}{(1 - v^2)^{1/2}}, \\ \bar{t} &= \frac{-zv + t}{(1 - v^2)^{1/2}}. \end{aligned} \tag{2-6.4b}$$

These are the special Lorentz transformations mentioned in the introductory remarks. The mathematical development just presented implies that the measure of light velocity is an invariant of the Lorentz group of transformations. In his paper, "Electromagnetic Phenomena in a System Moving with Any Velocity Less than That of Light" (English version in Proceedings of the Academy of Sciences of Amsterdam, **6**, 1904), Lorentz showed that the Maxwell equations were invariant with respect to these transformations; that is, the Lorentz transformations had the same characterizing relation to electromagnetic theory as the Galilean transformations had to Newtonian mechanics.

The invariance of the form $F = (X^4)^2 - (X^3)^2$ led to the special Lorentz transformation group just as the invariance of $G = (X^4)^2 + (X^3)^2$ led to the orthogonal Cartesian group. G imposes the ordinary Euclidean metric geometry in the plane, but F imposes a hyperbolic geometry.[17] The development of special relativity as expressed in this hyperbolic geometry was first expounded by H. Minkowski in 1908. The plane in which the metric form F holds sway is accordingly called the Minkowski

[17] The forms F and G are commonly called fundamental metric forms.

plane. Minkowski's method of presentation is advantageous in representing the kinematical aspects of special relativity: in particular, the concepts of "contraction of length" and "dilation of time."

Figure 2-6.6 illustrates the fundamental notions of the Minkowski plane. Suppose that from the ordinary Euclidean point of view a particle is in uniform rectilinear motion along the z axis of a coordinate system O. Let an $\bar{O}$ system be associated with the particle with axes parallel to those of the O system. (See Fig. 2-6.5. As represented in Fig. 2-6.6, the space-time reference system of the frame $\bar{O}$ is denoted by the $\bar{z}$ and $\bar{t}$ axes. Each coordinate pair (z, t) is said to represent an event in space time. From the viewpoint of the O system:

1. All events whose coordinates satisfy $|z^2 - t^2| = L^2$ are at an equal interval from the origin. (This is the set of events represented in the diagram along the branches of the conjugate hyperbolas.)
2. A light particle moving along the z axis in three-space is pictured in space time by means of the events satisfying the equation $z^2 - t^2 = 0$. One branch of this so-called light cone is represented in Fig. 2-6.6.
3. All events represented on a line parallel to the $\bar{z}$ axis take place at the same time in relation to the $\bar{O}$ system. Therefore such a line is called a line of simultaneity with respect to the $\bar{O}$ system.
4. All events represented on a line parallel to the $\bar{t}$ axis take place at the same spot in relation to the $\bar{O}$ system. Such a line is called a world line.

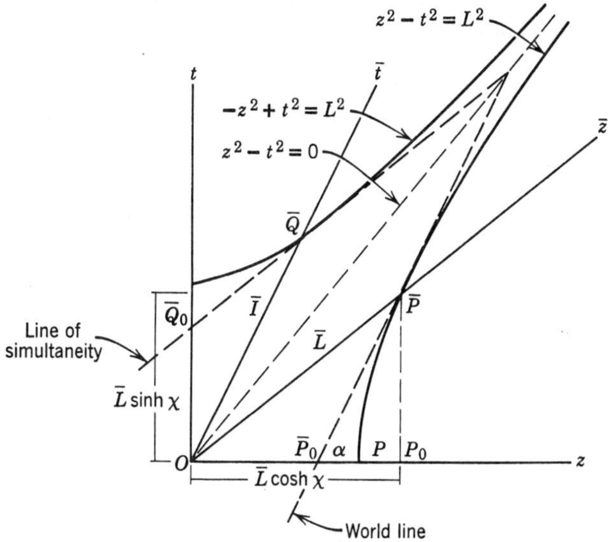

Fig. 2-6.6

5. A particle moves from $\bar{P}_0$ to P_0 in the time interval $P_0\bar{P}$. Therefore

$$ctn\alpha = \frac{z}{t} = v.$$

6. The path of a particle in uniform rectilinear motion with velocity v is represented by a line $\bar{t}$ intermediate to the t axis and the light-cone line $z - t = 0$. Note that the t axis denotes the history of a particle stationary at the origin of coordinates, whereas $z - t = 0$ represents the history of a light particle. The intermediacy of the line corresponding to the motion of a material particle correlates with the assumption that all velocities are less than that of light (i.e., $v < 1$).

7. The $\bar{z}$ and $\bar{t}$ axes must be symmetric with respect to the light-cone line $z - t = 0$, since the equation of the cone is preserved under the Lorentz transformations; that is, $c = \bar{z}/\bar{t} = 1$, or $\bar{z} - \bar{t} = 0$ must be invariant. (See Problem 7.)

A line of simultaneity and a world line associated with the $\bar{O}$ system are represented in Fig. 2-6.6. A striking feature of the special theory of relativity is the breakdown of any absolute meaning to the concept of simultaneity. The time relationship of a pair of events takes on precisely the same relative aspect as the position relationship of a pair of events and is dramatically expressed by the following theorems.

Theorem 2-6.1a. Let $\bar{L}$ be a rod fixed in an $\bar{O}$ coordinate system and in uniform rectilinear motion in relation to an O system. The length measurement of $\bar{L}$ is shorter when determined from the viewpoint of the O system than it is when determined by $\bar{O}$ measurements. In particular,

$$\bar{L}_O = (1 - v^2)^{1/2}\,\bar{L}_{\bar{O}}, \tag{2-6.5}$$

where $\bar{L}_O$, $\bar{L}_{\bar{O}}$, respectively represent the measures of $\bar{L}$ from the viewpoint of O and $\bar{O}$.

PROOF. The end points of the rod must be considered simultaneously to be measured. Therefore the length of the rod $\bar{L}_O$ in the O system is determined by the events O and $\bar{P}_O$. (See Fig. 2-6.6.) Since P_O is the projection of a point of the hyperbola on the z axis, the length OP_O can be conveniently represented by $\bar{L}\cosh\chi$. Furthermore,

$$\tanh\chi = v = ctn\alpha.$$

Hence

$$\begin{aligned}\bar{L}_0 &= O\bar{P}_0 = OP_0 - \bar{P}_0P_0 = \bar{L}\cosh\chi - P_0\bar{P}ctn\alpha,\\ &= \bar{L}\cosh\chi - \bar{L}\sinh\chi\tanh\chi = \bar{L}\frac{(\cosh^2\chi - \sinh^2\chi)}{\cosh\chi},\\ &= \frac{1}{\cosh\chi}\bar{L} = \sqrt{1 - v^2}\,\bar{L}.\end{aligned}$$

Since the rod is fixed in the barred system, $\bar{L}$ and $\bar{L}_{\bar{O}}$ are identical. This completes the proof.

Theorem 2-6.1b. Let $\bar{I}$ be a time interval associated with an $\bar{O}$ coordinate system, which is in uniform rectilinear motion with respect to an O system. The O determination of this time interval is then less than the $\bar{O}$ measurement. We have

$$\bar{I}_0 = (1 - v^2)^{1/2}\, \bar{I}_{\bar{O}}. \tag{2-6.6}$$

PROOF. Let $O\bar{Q}$ represent the $\bar{O}$ time interval. (See Fig. 2-6.6.) The branch of the hyperbola passing through $\bar{Q}$ has the representation $t^2 - z^2 = L^2$. Therefore the coordinates of $\bar{Q}$ can be expressed parametrically by

$$t = \bar{I} \cosh \chi, \qquad z = \bar{I} \sinh \chi.$$

The measurement of the time interval $\bar{I}$ with respect to the O system must occur at a given place; hence we consider the interval $O\bar{Q}_0$. In other words, these considerations with respect to time interval are symmetric to those concerning space interval. To complete the proof, we follow the pattern of the preceding theorem.

The reader should be careful in his interpretation of the result of the last theorem. The fact that the observed time interval is shorter in the O system implies that the passage of time in $\bar{O}$ is judged greater. We can summarize the result of the theorem with the statement that clocks appear to run fastest in the system in which they are fixed.

Let us once more turn our attention to the consideration of composition of velocities. Since the special Lorentz transformations were obtained by denying the classical method of composition, it is clear that they will lead to a new law.

Theorem 2-6.2. Let a particle P have a uniform velocity $\bar{u}$ along the $\bar{z}$ axis of $\bar{O}$. If the $\bar{O}$ system is in uniform rectilinear motion along the z axis of O with velocity v and u represents the velocity of P with respect to O, then

$$u = \frac{\bar{u} + v}{1 + \bar{u}v}. \tag{2-6.7}$$

PROOF. The coordinates of P are (z, t), $(\bar{z}, \bar{t})$ in O and $\bar{O}$, respectively. According to the Lorentz transformations,

$$\bar{z} = \frac{z - vt}{(1 - v^2)^{1/2}}, \qquad \bar{t} = \frac{-vz + t}{(1 - v^2)^{1/2}}.$$

Upon dividing,

$$\bar{u} = \frac{\bar{z}}{\bar{t}} = \frac{z - vt}{-vz + t} = \frac{u - v}{-vu + 1}.$$

The final result is obtained by solving the last equation for u. It is interesting to note that $|\bar{u}| < 1$, $|v| < 1$ implies

$$\left| \frac{\bar{u} + v}{1 + \bar{u}v} \right| < 1.$$

(See the problem section for more information on this statement.)

The preceding theorem deals with a particle in uniform rectilinear motion and therefore is restricted in scope. In order to be in a position to consider velocities and accelerations of particles in general motion, it is necessary to introduce a certain differential invariant of the Lorentz group. This invariant, which is introduced by the next theorem, is analogous to the differential form of the expression for arc length in Euclidean three-space.

Theorem 2-6.3. The form

$$ds^2 = h_{\alpha\beta}\, dX^\alpha\, dX^\beta \tag{2-6.8}$$

is an invariant of the Lorentz transformation group.

PROOF. The linear transformations under consideration are of the form

$$\bar{X}^\gamma = C_\lambda{}^\gamma X^\lambda, \qquad X^\lambda = c_\gamma{}^\lambda \bar{X}^\gamma$$

and the components $h_{\lambda\mu}$ have already been seen to satisfy the transformation rule

$$h_{\lambda\mu} = C_\lambda{}^\alpha C_\mu{}^\beta h_{\alpha\beta}.$$

Therefore we have

$$h_{\lambda\mu}\, dX^\lambda\, dX^\mu = C_\lambda{}^\alpha C_\mu{}^\beta \bar{h}_{\alpha\beta} c_\gamma{}^\lambda\, d\bar{X}^\nu c_\gamma{}^\mu\, d\bar{X}^\gamma$$
$$= \delta_\nu{}^\alpha\, \delta_\gamma{}^\beta \bar{h}_{\alpha\beta}\, d\bar{X}^\nu\, d\bar{X}^\gamma = \bar{h}_{\alpha\beta}\, d\bar{X}^\alpha\, d\bar{X}^\beta.$$

This completes the proof.

The class $\{dX^\lambda/ds\}$ is a vector,[18] since

$$\frac{dX^\lambda}{ds} = \frac{\partial X^\lambda}{\partial \bar{X}^\gamma}\frac{d\bar{X}^\gamma}{ds} = c_\gamma{}^\lambda \frac{d\bar{X}^\gamma}{ds}.$$

Furthermore,

$$\frac{dX^\lambda}{ds} = \frac{dX^\lambda}{dt}\frac{dt}{ds} = \frac{dX^\lambda}{dt}\frac{1}{(1 - v^2)^{1/2}}.$$

For small values of v, dX^λ/ds is approximately the same as dX^λ/dt. For these reasons we make the following definition.

Definition 2-6.1. The vector $\{dX^\lambda/ds\}$ is called the four-space relativistic velocity vector of a particle that is in general motion in Euclidean three-space.

---

[18] See Chapter 1, Section 5.*

With the introduction of a relativistic velocity vector, let us consider the relativistic form of Newton's second law. In particular, we approach the problem by stating criteria a relativistic form of Newton's law must satisfy. We then determine the form of the law by analogy to the classical law. Criteria commonly taken are the following:

(a) The law must have the same form as the classical law.
(b) For low velocities the relativistic form must reduce to the classical form.
(c) Newton's relativistic second law must be Lorentz-invariant.

Theorem 2-6.4. Newton's relativistic second law is

$$\overset{4}{F}{}^{\lambda} = m_0 \frac{d^2X^\lambda}{ds^2}, \tag{2-6.9a}$$

where

$$\text{(b)}\quad \overset{4}{F}{}^{a} = \frac{F^a}{(1 - v^2)^{1/2}},$$

(2-6.9)

$$\text{(c)}\quad \overset{4}{F}{}^{4} = \frac{\sum_{a=1}^{3} F^a v^a}{(1 - v^2)^{1/2}}.$$ [19]

PROOF. The proof consists of showing that the criteria (a), (b), and (c) are satisfied. The first three components of relation (2-6.9a) can be written in the form

$$\frac{F^a}{(1 - v^2)^{1/2}} = m_0 \frac{d\{dX^a/dt[(1 - v^2)^{-1/2}]\}}{dt}(1 - v^2)^{-1/2} \tag{2-6.10a}$$

or

$$F^a = \frac{d\{[m_0/(1 - v^2)^{1/2}]\, dX^a/dt\}}{dt}. \tag{2-6.10b}$$

This expression corresponds to Newton's classical second law if

$$m_0/(1 - v^2)^{1/2}$$

is interpreted as mass. This is done in the development of special relativity. Some of the consequences of this identification are discussed after the completion of this proof.

For very small values of $|v|$ the relation (2-6.10a) approximates

$$F^a = m_0 \frac{d^2X^a}{dt^2}.$$

---

[19] See Problem 9.

This is a classical form of Newton's second law. The fourth component can be expressed as

$$\frac{\sum_{a=1}^{3} F^a v^a}{(1 - v^2)^{1/2}} = m_0 \frac{d[(1 - v^2)^{-1/2}]}{dt} (1 - v^2)^{-1/2}$$

or

$$\sum_{a=1}^{3} F^a v^a = m_0 v \frac{dv}{dt} (1 - v^2)^{-3/2}.$$

When $|v|$ is approximately zero, and therefore the components v^a are also nearly zero, this expression takes the form $0 = 0$.

The Lorentz invariance of (2-6.9a) depends on the assumption that $\overset{4}{F}{}^\lambda$ are components of a four-space vector and the linearity of the special Lorentz transformations. The linearity accounts for the fact that d^2X^λ/ds^2 are vector components. On the one hand,

$$F^\lambda = c_\gamma{}^\lambda \bar{F}^\gamma.$$

On the other,

$$\frac{d^2X^\lambda}{ds^2} = c_\gamma{}^\lambda \frac{d^2\bar{X}^\gamma}{ds^2}.$$

When these relations are applied to (2-6.9a), we have

$$0 = \overset{4}{F}{}^\lambda - m_0 \frac{d^2X^\lambda}{ds^2} = c_\gamma{}^\lambda \left(\overset{4}{\bar{F}}{}^\gamma - m_0 \frac{d^2\bar{X}^\lambda}{ds^2} \right).$$

Note that m_0 is a scalar. Since $|c_\gamma{}^\lambda| \neq 0$, we can sum the products obtained by multiplying the preceding relation with $C_\lambda{}^\mu$. When this is done, we have

$$\overset{4}{\bar{F}}{}^\mu - m_0 \frac{d^2\bar{X}^\mu}{ds^2} = 0.$$

Therefore the law in (2-6.9a) is invariant under special Lorentz transformations. This completes the proof.

As indicated in the foregoing, the first three components of Newton's relativistic second law can be expressed in the form [see (2-6.10b)]

$$F^a = \frac{d\{[m_0/(1 - v^2)^{1/2}]\, dX^a/dt\}}{dt}.$$

The relation (2-6.10) suggests the identification of $m_0/(1 - v^2)^{1/2}$ as an entity in its own right, for when this is done

$$F^a = \frac{d}{dt}(mv^a).$$

This is the classical form of Newton's second law.

Definition 2-6.2. The quantity

$$m = \frac{m_0}{(1 - v^2)^{1/2}} \tag{2-6.11}$$

is called inertial mass; m_0 is the rest mass.

The introduction of inertial mass is responsible for a philosophic revolution in physics. Mass can no longer be thought of as a constant. The inertial mass corresponds to the rest mass if $v = 0$ and increases as $|v|$ increases. Moreover, from (2-69.a) and (2-6.9c) it follows that

$$\frac{dm}{dt} = \sum_{a=1}^{3} F^a v^a. \tag{2-6.12}$$

The right-hand member of (2-6.12) represents the rate at which the force does work on the particle. In classical mechanics a definite integral of the expression denotes the difference between the initial and final kinetic energies of a particle. We make the assumption that relativistic kinetic energy T is defined by

$$\frac{dT}{dt} = \sum_{a=1}^{3} F^a v^a, \tag{2-6.13}$$

with the initial condition that the kinetic energy is zero when the velocity is zero. Then

$$\frac{dT}{dt} = \frac{dm}{dt}, \tag{2-6.14a}$$

and

$$T = m + \text{constant}. \tag{2-6.14b}$$

Since $m = m_0$ when $T = 0$, it follows that

$$T = m - m_0. \tag{2-6.14c}$$

From this relation we see that the kinetic energy increases or decreases as the mass increases or decreases. Furthermore, a search for an analogue to the classical principle of conservation of energy leads us to write (2-6.14c) in the form

$$m = T + m_0. \tag{2-6.14d}$$

Einstein made one of the most important discoveries related to the special theory of relativity when he hypothesized the equivalence of mass and energy. Under this hypothesis we identify not only the excess of mass $m - m_0$ with kinetic energy but also the rest mass m_0 with energy. In particular, m_0 may be thought of as a manifestation of the work done in creating the particle or as potential energy. Then the total energy E is

(2-6.15a) $$E = m = T + m_0.$$

The merit of this hypothesis has been amply demonstrated by the atomic energy experiments of more than a decade. The relation (2-6.15a) is usually expressed in the form

(2-6.15b) $$E = mc^2,$$

where c represents the speed of light. We have chosen the light second as our unit of distance; therefore $c = 1$.

There are many other aspects of special relativity that would be interesting to investigate. It is hoped that the reader will be stimulated to pursue them. However, it would not be consistent with the purpose of this book to carry their development any further.

Problems

1. If the motion of a particle is determined by the parametric equations

$$X^j = X_0^{\ j} + B^j t,$$

that is, it is linear and uniform in the X^j coordinate system, show that it is uniform and linear in a system based on the Galilean transformations (2-6.1).

2. (a) Compute the components

$$\bar{h}_{\alpha\beta} = \sum_{\gamma=1}^{4} \frac{\partial X^\gamma}{\partial \bar{X}^\alpha} \frac{\partial X^\gamma}{\partial \bar{X}^\beta}$$

of the fundamental metric tensor if the transformation equations are denoted by (2-6.1a,b,c). (See Chapter 1, Section 5*.)
Hint: First write down the inverse transformation equations.

(b) Determine the formula from which interval should be computed in the barred system. Find a formula for the special case of an interval along the $\bar{t}$ axis.

3. Show that $-z^2 + t^2 = 0$ transforms to

$$-\bar{z}^2 - 2v\bar{z}\bar{t} + (1 - v^2)\bar{t}^2 = 0$$

under the transformation

$$z = \bar{z} + v\bar{t},$$
$$t = \bar{t}.$$

4. (a) Show that the z axis and $\bar{z}$ axis coincide under the transformation of Problem 3. If $v = 1$, construct the $\bar{z},\bar{t}$ axes under the assumption that the z and t axes are perpendicular to each other.

(b) If $\begin{cases} y = \bar{y} + v^1\bar{t} \\ z = \bar{z} + v^2\bar{t}, \\ t = \bar{t} \end{cases}$ shows that the y axis corresponds to the $\bar{y}$ axis and that the z and $\bar{z}$ axes correspond.

Hint: See Chapter 1, Section 5*.

5. Suppose that $h_{\alpha\beta}\, dX^\alpha\, dX^\beta = \bar{h}_{\lambda\gamma}\, d\bar{X}^\lambda\, d\bar{X}^\gamma$.

(a) Show that

$$h_{\alpha\beta} = \frac{\partial \bar{X}^\lambda}{\partial X^\alpha} \frac{\partial \bar{X}^\gamma}{\partial X^\beta} \bar{h}_{\lambda\gamma}.$$

(b) Show that if the elements $h_{\alpha\beta}$ as well as the values $\bar{h}_{\alpha\beta}$ are constants the transformation is linear.

Hint: Start by differentiating the expression of (a) and arrive at the conclusion $\partial^2\bar{X}^\lambda/\partial X^\alpha\, \partial X^\beta = 0$.

6. (a) Show that the equations (2-6.1c) satisfy the conditions $\partial^2\bar{X}^\lambda/\partial X^\alpha\, \partial X^\beta = 0$.

(b) Show that $\partial^2\bar{X}^\lambda/\partial X^\alpha\, \partial X^\beta = 0$ implies that the $\partial\bar{X}^\lambda/\partial X^\beta$ are constants.

7. Show that $z/t = 1$ is an invariant under the special Lorentz transformations.

8. (a) If a coordinate system $\bar{O}$ is in uniform rectilinear motion with respect to a system O with velocity $v = \frac{3}{4}$ and a particle P has velocity $\bar{u} = \frac{1}{2}$ in relation to $\bar{O}$, find the velocity of the particle with respect to O. (All motions are along a common straight line.)

(b) Consider the formula in (2-6.7) for composition of velocities. If $|\bar{u}| < 1$ and $|v| < 1$, show that $|u| < 1$.

9. Show that

$$\overset{4}{F^4} = \frac{\sum_{a=1}^{3} F^a v^a}{(1 - v^2)^{1/2}}$$

results from the relation $h_{\alpha\beta}\, (dX^\alpha/ds)\, (dX^\beta/ds) = 1$. [See (2-6.8).]

chapter 3 partial differentiation and associated concepts

1. *Surface Representations*

Just as the concept of curve serves as a geometrical bridge between certain physical phenomena and mathematical analysis so does that of surface. The purpose of this section is to introduce the reader to the idea of surface as it is used in this text.

Elementary representations of surfaces usually take one of three forms. That most often met in an analytic geometry or calculus course is the explicit form

$$X^3 = f(X^1, X^2), \tag{3-1.1a}$$

in which X^3 is expressed as a function of two independent variables. The domain of definition is a region of the X^1, X^2 plane, and it can be expressed in terms of inequalities such as

$$a \leq X^1 \leq b, \qquad c \leq X^2 \leq d, \quad \text{or} \quad (X^1)^2 + (X^2)^2 \leq r^2, \quad \text{etc.}$$

A second representation is the implicit form

$$\Psi(X^1, X^2, X^3) = 0, \tag{3-1.1b}$$

due to the mathematician Gaspard Monge (1746–1810, French), who was one of the founders of the differential geometry of curves and surfaces. The form (3-1.1a) can always be expressed implicitly by the simple algebraic expedient of putting all the terms on one side of the equation. The converse is true when the criteria of the implicit function theorem are met. The third form of surface representation, and the one used most often in this book, is the so-called parametric:

$$X^1 = X^1(v^1, v^2),$$
$$X^2 = X^2(v^1, v^2), \qquad (3\text{-}1.1c)$$
$$X^3 = X^3(v^1, v^2).$$

This form played a fundamental part in the development of surface theory by the "father of differential geometry," Carl Fredrich Gauss. It has many advantages for theoretic development, some of which are utilized in the remainder of this book.

To make a detailed study of the concept of surface, its representations, and their relations would be too great a digression. Therefore only the definitions of the previously mentioned representations and some indication of the way in which they are connected are discussed.

Definition 3-1.1. Let $\Psi(X^1, X^2, X^3)$ be a continuous function on a region R. Furthermore, let $\Psi(X^1, X^2, X^3)$ have continuous first partial derivatives on R such that for each triple (X^1, X^2, X^3) at least one of the derivatives does not vanish. Then the geometric locus of points whose coordinates satisfy

$$\Psi(X^1, X^2, X^3) = 0$$

is said to be an implicitly defined surface.[1]

On occasion it will be expedient to follow the lead of Monge and use the implicit surface representation (3-1.1b). However, most theoretical considerations are based on the parametric form indicated in the next definition. This definition is stated in three parts, like the definition of a curve.

Definition 3-1.2a. Let $\bar{D}$ be a domain of real-number pairs (v^1, v^2) such that

(a) functions X^j, expressed as in (3-1.1c) are continuous on $\bar{D}$;
(b) there is a domain D in $\bar{D}$ on which the functions X^j have continuous partial derivatives of at least the first order;
(c) the matrix $(\partial X^j/\partial v^\beta)$, $j = 1, 2, 3$, $\beta = 1, 2$[2] has rank 2 on D.[3] Then

[1] See V. Hlavatý, Differentialgeometrie der Kurven und Flächen und Tensorrechnung, (translated to German by M. Pinl), P. Noordhoff, Groningen, The Netherlands, 1939.

[2] Greek letters used as indices indicate a range 1, 2 in this section.

[3] The rank of the matrix $(\partial X^j/\partial v^\beta)$ is 2 if and only if at least one of the second-order determinants

$$\begin{vmatrix} \dfrac{\partial X^1}{\partial v^1} & \dfrac{\partial X^2}{\partial v^1} \\[2ex] \dfrac{\partial X^1}{\partial v^2} & \dfrac{\partial X^2}{\partial v^2} \end{vmatrix}, \qquad \begin{vmatrix} \dfrac{\partial X^1}{\partial v^1} & \dfrac{\partial X^3}{\partial v^1} \\[2ex] \dfrac{\partial X^1}{\partial v^2} & \dfrac{\partial X^3}{\partial v^2} \end{vmatrix}, \qquad \begin{vmatrix} \dfrac{\partial X^2}{\partial v^1} & \dfrac{\partial X^3}{\partial v^1} \\[2ex] \dfrac{\partial X^2}{\partial v^2} & \dfrac{\partial X^3}{\partial v^2} \end{vmatrix}$$

is different from zero.

(3-1.1c) is said to be an allowable parameter representation. The symbols v^1 and v^2 are called surface parameters. The pairs (v^1, v^2) in D are called regular, whereas those in $\bar{D}$, but not in D, are said to be singular.

The parametric representation of the preceding definition determines a set of points in a three-dimensional Euclidean space; however, it is not the only parameterization that specifies this particular set of points. New parameterizations can be obtained by imposing a transformation

$$v^\alpha = v^\alpha(u^\beta).$$

In order to satisfy assumptions (a), (b), and (c) in Definition (3-1.2a), we stipulate the following.

Definition 3-1.2b. Suppose that the transformation $v^\alpha = v^\alpha(u^\beta)$ satisfied the following conditions:

(a) It is a one-to-one bicontinuous transformation between domains $\bar{E}$ of the u^β and $\bar{D}$ of the v^α.
(b) The functions $v^\alpha(u^\beta)$ have continuous first partial derivatives, at least on a domain E, which corresponds under the transformation to D.
(c) The Jacobian $|\partial v/\partial u|$ is different from zero everywhere on E.

The parameter transformation is then said to be allowable.

The preceding definition makes it possible to classify parameter representations. We specify that two-parameter representations are in the same equivalence class if they are related by a transformation satisfying the conditions of Definition 3-1.2b. With this stipulation in mind, we are in a position to give a meaning to the term surface.

Definition 3-1.2c. The set of all points represented by any one of the allowable representations of an equivalence class is said to be a surface.

If a number pair (v^1, v^2) is singular, one of two possibilities hold. Either the surface is not smooth at the point of consideration (for example, at the vertex of a cone) or the parametric equations give rise to the singularity even though the surface element could be geometrically classified as smooth. (See Example 3-1.2.)

If the implicit form (3-1.1b) is given, then, in a neighborhood of a point at which $\partial\Psi/\partial X^j \neq 0$ for some value of j, (3-1.1c) can be obtained. The theoretical development of this fact makes use of an implicit function theorem.[4]

---

[4] See Angus E. Taylor, *Advanced Calculus*, Ginn, 1955, Chapter VIII.

Our definition points out that a parameterization of the implicit form (3-1.1b) is not unique. Furthermore, a second parameterization satisfying (3-1.1b) may not represent all of the original surface.

The following examples illustrate the preceding statement.

Example 3-1.1. A form of the equation of a plane is

$$\mathbf{A} \cdot (\mathbf{r} - \mathbf{r}_0) = 0.$$

When this implicit surface representation is written out, we have

$$A^1(X^1 - X_0^{\;1}) + A^2(X^2 - X_0^{\;2}) + A^3(X^3 - X_0^{\;3}) = 0.$$

A possible parameterization is

$$\begin{aligned} X^1 &= v^1, \\ X^2 &= v^2, \\ X^3 &= \frac{1}{A^3}[A^3X_0^{\;3} + A^1(X_0^{\;1} - v^1) + A^2(X_0^{\;2} - v^2)], \end{aligned}$$

where $-\infty < v^1 < \infty$, $-\infty < v^2 < \infty$. The reader can verify that the parameteric representation satisfies properties (a), (b), (c) of Definition 3-1.2.

If the foregoing parameterization is replaced by

$$\begin{aligned} X^1 &= \sin {}^*v^1 \\ X^2 &= \sin {}^*v^2 \\ X &= \frac{1}{A^3}[A^3X_0^{\;3} + A^1(X_0^{\;1} - \sin {}^*v^1) + A^2(X_0^{\;2} - \sin {}^*v^2)], \end{aligned}$$

the whole plane is no longer represented. In particular, X^1 and X^2 are restricted to the interval $[-1, 1]$. The problem of finding and verifying the validity of the domain of $({}^*v^1, {}^*v^2)$, for which these two representations belong to the same equivalence class, is left to the student.

Example 3-1.2. The implicit representation of a sphere of radius a and center at the origin is

$$(X^1)^2 + (X^2)^2 + (X^3)^2 - a^2 = 0. \tag{3-1.2a}$$

A standard parameterization is

$$\begin{aligned} X^1 &= a \sin\theta \cos\phi, \\ X^2 &= a \sin\theta \sin\phi, \\ X^3 &= a \cos\theta, \end{aligned} \tag{3-1.2b}$$

where we have let $v^1 = \theta$, $v^2 = \phi$, $0 \le \theta \le \pi$, $0 \le \phi < 2\pi$. (See Fig. 3-1.1.)

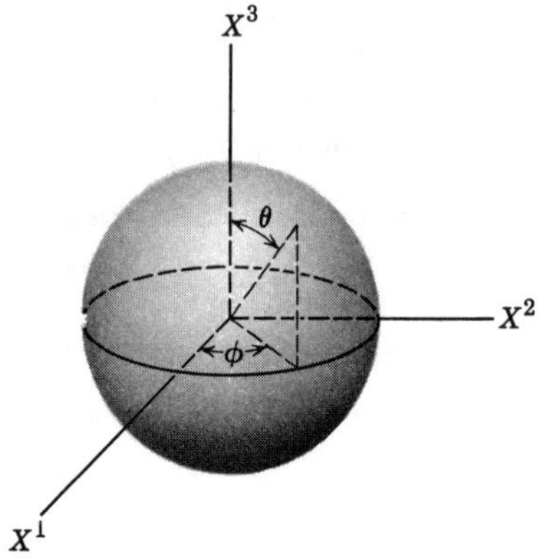

Fig. 3-1.1

We have

$$\left(\frac{\partial X^j}{\partial v^\beta}\right) = a\begin{pmatrix} \cos\theta\cos\phi & \cos\theta\sin\phi & -\sin\theta \\ -\sin\theta\sin\phi & \sin\theta\cos\phi & 0 \end{pmatrix}. \tag{3-1.2c}$$

The second-order determinants associated with (3-1.2c) have the values

$$a^2\cos\theta\sin\theta, \qquad -a^2\sin^2\theta\sin\phi, \qquad -a^2\sin^2\theta\cos\phi.$$

By careful examination of these values it is seen that $(0, \phi)$ and (π, ϕ) are the only singular coordinate pairs. Since the surface is smooth at each corresponding point, it is the parametric representation that produces the singularities.

A second parameterization associated with the sphere is indicated below. In order to satisfy the single-valued property associated with the term function, separate representations must be given for that part of the sphere corresponding to $X^3 > 0$ and that part corresponding to $X^3 < 0$. It will be seen that this parameterization cannot account for $X^3 = 0$ and still meet the requirements of Definition 3-1.2a. Parametric equations representing the upper hemisphere are

$$\begin{aligned} X^1 &= v^1, \\ X^2 &= v^2, \\ X^3 &= [a^2 - (v^1)^2 - (v^2)^2]^{1/2}. \end{aligned} \tag{3-1.3a}$$

Since only real surfaces are to be considered,

$$(v^1)^2 + (v^2)^2 < a^2.$$

Parametric equations of the lower hemisphere are

$$\text{(3-1.3b)}\qquad \begin{aligned} X^1 &= v^1, \\ X^2 &= v^2, \\ X^3 &= -[a^2 - (v^1)^2 - (v^2)^2]^{1/2}, \end{aligned}$$

where again

$$(v^1)^2 + (v^2)^2 < a^2.$$

The matrix $(\partial X^j/\partial v^\beta)$ has the following form with respect to the parametric representation (3-1.3a) of the upper hemisphere:

$$\text{(3-1.3c)}\qquad \left(\frac{\partial X^j}{\partial v^\beta}\right) = \begin{pmatrix} 1 & 0 & \dfrac{-v^1}{[a^2 - (v^1)^2 - (v^2)^2]^{1/2}} \\ 0 & 1 & \dfrac{-v^2}{[a^2 - (v^1)^2 - (v^2)^2]^{1/2}} \end{pmatrix}$$

The matrix has rank 2 on the domain of definition. Note that the domain of definition cannot be extended to

$$(v^1)^2 + (v^2)^2 = a^2.$$

The few facts concerning surfaces so far stated will be valuable in the geometrical interpretation of vector concepts. A knowledge of what is meant by a curve on a surface is also important. The way in which this term is to be used is indicated by the following definition.

Definition 3-1.3.[5] Let v^1 and v^2 be functions, defined on a real number interval $a \le t \le b$, which possess the following properties:

(a) At least the first derivatives are continuous and do not vanish simultaneously.
(b) The coordinate pair $(v^1(t), v^2(t))$ is in D (see Definition 3-1.2) for all t.
(c) Not all elements $\partial X^j/\partial v^\beta$ vanish simultaneously at surface points with coordinates $(v^1(t), v^2(t))$.

The set of points with coordinates $(v^1(t), v^2(t))$ is said to be a surface curve.[6] For our purposes, the most important of these curves are the parameter curves.

Theorem 3-1.1. If a surface element is given by the parametric representation $X^j = X^j(v^\beta)$, the sets of points in D, such that $v^1 =$ constant for all v^2 and the set $v^2 =$ constant for all v^1, represent surface curves.

PROOF. Consider the points with coordinates $v^1 =$ constant for all v^2. We can take as a parameterization

$$v^1 = c, \qquad v^2 = t.$$

The properties of Definition 3-1.4 are easily checked.

---

[5] See V. Hlavatý, *op. cit.*, p. 99.

[6] The curve has other parametric representations. See Definitions 2-1.2a, b, c.

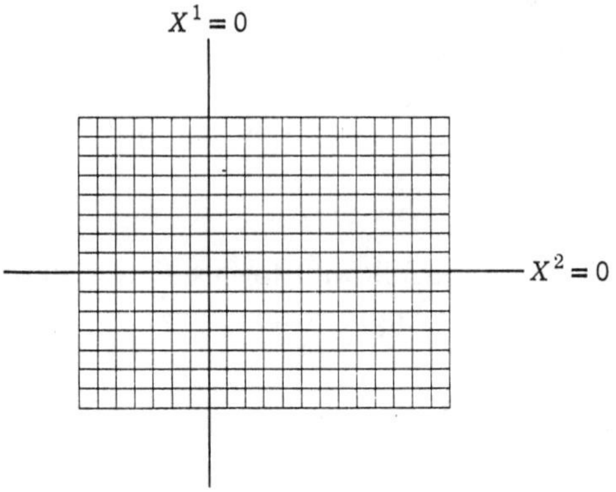

Fig. 3-1.2

Definition 3-1.4. The curves of the families v^1 = constant for all v^2 and v^2 = constant for all v^1 are called parametric surface curves.

Such curves play the same role on a surface as the curves X^1 = constant and X^2 = constant play in the plane, that is, the two families determine a coordinate net for the surface. (See Fig. 3-1.2.)

Example 3-1.3. Consider the surface (3-1.3a); v^1 = constant and v^2 = constant correspond to X^1 = constant and X^2 = constant. Therefore in this situation the parametric curves can be thought of as curves of intersection of the upper hemisphere and the coordinate planes. (See Fig. 3-1.3.)

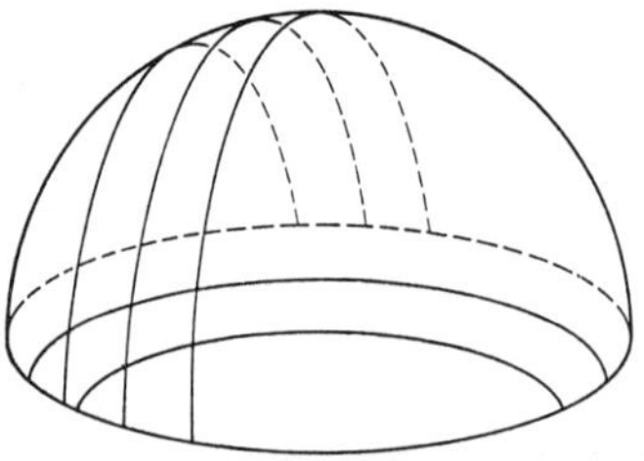

Fig. 3-1.3

Example 3-1.4. Consider the parametric curves associated with the surface representation (3-1.2b). The curves θ = constant are latitudinal circles and can be interpreted geometrically as intersection curves of cones θ = constant with the sphere. The curves ϕ = constant are longitudinal circles which can be thought of as intersections of planes ϕ = constant with the sphere. (See Fig. 3-1.4.)

The next example presents a general surface curve on a sphere.

Example 3-1.5. Consider the parametric equations (3-1.2b) of a sphere. Let

$$\begin{aligned} \theta &= t \\ \phi &= 2t \end{aligned} \qquad 0 < t \leq \frac{\pi}{2}. \tag{3-1.4a}$$

Then

$$X^1 = a \sin t \cos 2t,$$
(3-1.4b) $$X^2 = a \sin t \sin 2t,$$
$$X^3 = a \cos t.$$

The reader can easily verify that Definition 3-1.3 is satisfied. Note that the exclusion of the value $t = 0$ is quite necessary. A rough sketch of the curve is indicated by Fig. 3-1.5.

Definitions 2-1.3 and 2-1.4 introduce the concepts of scalar and vector fields over a space curve C. The following definitions extend these ideas to surfaces and space regions.

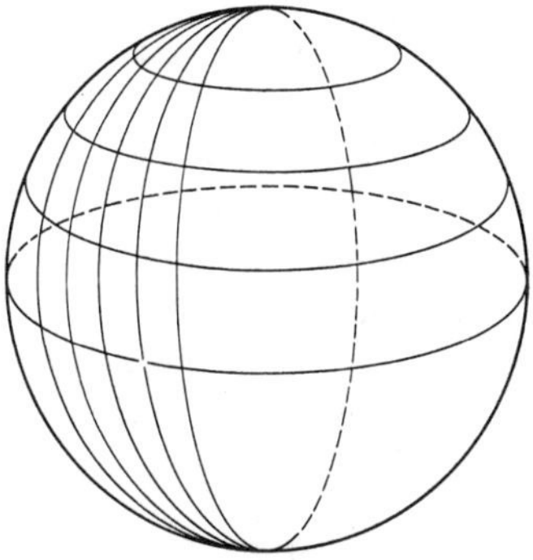

Fig. 3-1.4

Definition 3-1.5a. Suppose that the class of functions $\{\Phi\}$ is defined on a region R and that at each point of R any two elements of $\{\Phi\}$ satisfy Definition 1-4.2. Then $\{\Phi\}$ is said to be a scalar field on R with respect to the allowable transformation group.

If R is a space region, then Φ, a representative of the class $\{\Phi\}$, is a function of the coordinates X^j. On the other hand, if R is a surface or section of a surface, then Φ has a domain consisting of parametric values (v^1, v^2).

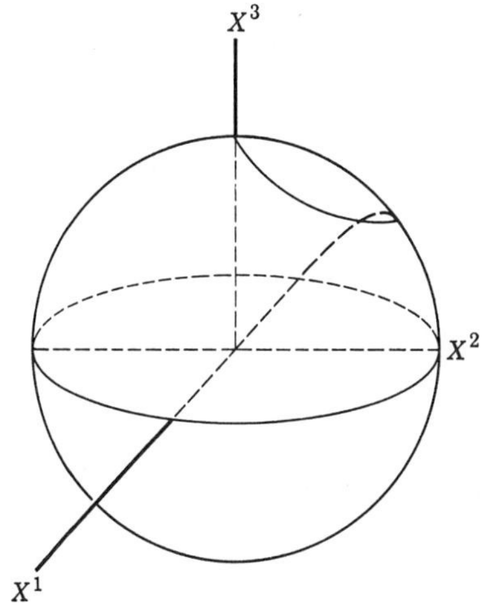

Fig. 3-1.5

Definition 3-1.5b. Suppose that the class of function triples $\{U^1, U^2, U^3\}$ is defined on a region R and at each point of R any two triples of $\{U^1, U^2, U^3\}$ satisfy Definition 1-4.1. Then $\{U^1, U^2, U^3\}$ is said to be a vector field on R in relation to the allowable group of transformations.

The vector field is designated by the symbol $\mathbf{U}$, to be consistent with previous usage.

Example 3-1.6. The Newtonian force field, $\mathbf{F} = (-mMG/r^2)\mathbf{R}$, played a part in Chapter 2, Section 5. It was assumed that there existed a curve C (the geometric representation of the path of a planet) along which $\mathbf{F}$ was defined. We then proceeded to find the polar equation of C. Upon reflection, the viewpoint taken toward the vector field $\mathbf{F}$ must be considered as rather artificial. Actually, the sun sets up a physical condition throughout space such that, if a body of mass $m = 1$ were placed at any point, a force $\mathbf{F} = -MGR/r^2$ would be observed; that is, the physical phenomenon is described by a vector function whose domain is the set of triples (X^1, X^2, X^3) corresponding to every point of space (except $r = 0$).

Example 3-1.7. Consider a windowpane. At a given instant of time there is a temperature associated with each point of the pane. We can introduce a temperature function that is certainly independent of coordinate representation. The domain of the function is a set of number pairs (v^1, v^2) corresponding to the points of a plane region.

Problems

1. Set up parametric equations for a plane surface other than those given in Example 3-1.1. Check to see whether your representation has any singularities.
2. Find parametric representations for each of the following surfaces. Check to see that Definition 3-1.2 is satisfied.

(a) $\dfrac{(X^1)^2}{a^2} + \dfrac{(X^2)^2}{b^2} + \dfrac{(X^3)^2}{c^2} = 1$, ellipsoid.

(b) $\dfrac{(X^1)^2}{a^2} + \dfrac{(X^2)^2}{b^2} - \dfrac{(X^3)^2}{c^2} = 1$, hyperboloid of 1 sheet.

(c) $\dfrac{(X^1)^2}{a^2} + \dfrac{(X^2)^2}{b^2} - \dfrac{(X^3)^2}{c^2} = 0$, quadric cone.

(d) $\dfrac{(X^1)^2}{a^2} - \dfrac{(X^2)^2}{b^2} + 2X^3 = 0$, hyperbolic paraboloid.

(c) $$\frac{(X^1)^2}{a^2} + \frac{(X^2)^2}{b^2} = 1, \text{ elliptic cylinder.}$$

3. Let a sphere be represented parametrically by

$$X^1 = 3 \sin \theta \cos \phi,$$
$$X^2 = 3 \sin \theta \sin \phi,$$
$$X^3 = 3 \cos \theta.$$

Construct a sketch of each of the following surface curves.

(a) $\theta = t, \quad \phi = 2t, \quad 0 \le t \le \pi.$

(b) $\theta = t, \quad \phi = t^2. \quad 0 \le t \le \pi/2.$

4. Suppose the parametric equations of a hemisphere were

$$X^1 = v^1$$
$$X^2 = v^2, \qquad (v^1)^2 + (v^2)^2 \le 4.$$
$$X^3 = [4 - (v^1)^2 - (v^2)^2]^{1/2}.$$

Construct a sketch of each of the surface curves

(a) $v^1 = t, \; v^2 = t^2. \qquad -\sqrt{2} \le t \le \sqrt{2}.$

(b) $v^1 = \cos t, \; v^2 = \sin t. \qquad 0 \le t \le 2\pi.$

(c) $v^1 = \cosh t, \; v^2 = \sinh t. \qquad 0 \le t \le 1.$

How might these curves be thought of as intersections of surfaces?

5. Show that the curve $\phi = \ln (\text{ctn } \theta - \csc \theta)$ makes the same angle with each of the meridians it crosses. Such a curve is called a loxodrome. *Hint:* Show that

$$\frac{d\mathbf{r}_1/d\theta \cdot d\mathbf{r}_2/d\theta}{|d\mathbf{r}_1/d\theta|\,|d\mathbf{r}_2/d\theta|} \text{ is constant.}$$

$d\mathbf{r}_1/d\theta$ and $d\mathbf{r}_2/d\theta$ represent tangential vector fields along a meridian and along the given curve, respectively.

6. With respect to Example 3-1.1, find the domain $(^*v^1, {}^*v^2)$ such that

$$v^1 = \sin {}^*v^1$$
$$v^2 = \sin {}^*v^2$$

is an allowable parameter transformation.

7. Find the appropriate parameter transformation for Example 3-1.2 and then repeat Problem 6.

2. *Vector Concepts Associated with Partial Differentiation*

Let us suppose that a function Φ represents a scalar field along a given curve. The curve may be a space curve, in which case

(3-2.1a) $$\Phi = \Phi[X^j(t)],$$

and the associated derivative field is

$$\frac{d\Phi}{dt} = \frac{\partial\Phi}{\partial X^j}\frac{dX^j}{dt}. \tag{3-2.1b}$$

On the other hand, the curve might lie on a given surface. Then

$$\Phi = \Phi\{X^j[v^\beta(t)]\}, \tag{3-2.2a}$$

and the associated derivative field has the representations

$$\frac{d\Phi}{dt} = \frac{\partial\Phi}{\partial X^j}\frac{\partial X^j}{\partial v^\beta}\frac{dv^\beta}{dt} = \frac{\partial\Phi}{\partial X^j}\frac{dX^j}{dt}. \tag{3-2.2b}$$

Whether the curve in question is a general space curve or one which is restricted to a surface, the factors dX^j/dt appearing in the representations (3-2.1b) and (3-2.2b) are components of the tangent vector field $d\mathbf{r}/dt$ to the curve.

For the moment, suppose that Φ were defined along a space curve $X^j = X^j(t)$. The representation (3-2.1b) has the form of a dot product. Furthermore, the factors dX^j/dt have geometric significance. Therefore it is quite natural to introduce an expression with components $\partial\Phi/\partial X^j$ with the objective of determining its geometric properties. This is the procedure used in the following development.

Definition 3-2.1. The symbol ∇ (read "del") is an operator whose components satisfy the condition

$$\frac{d\Phi}{dt} = \nabla\Phi \cdot \frac{d\mathbf{r}}{dt}. \tag{3-2.3}$$

The operator ∇ was first introduced by Hamilton in the development of the theory of quaternions. The preceding mode of introduction does not coincide with that of Hamilton. The representation introduced by him was the rectangular Cartesian form in Theorem 3-2.1.[7]

Theorem 3-2.1. The rectangular Cartesian form of ∇ is

$$\boxed{\nabla = \iota_1\frac{\partial}{\partial X^1} + \iota_2\frac{\partial}{\partial X^2} + \iota_3\frac{\partial}{\partial X^3} = \sum_{j=1}^{3}\iota_j\frac{\partial}{\partial X^j}.} \tag{3-2.4}$$

Proof. Represent the components of $\nabla\Phi$ by $(\nabla\Phi)_1$, $(\nabla\Phi)_2$, $(\nabla\Phi)_3$. Express the left-hand side of (3-2.3) as indicated by (3-2.1b). Then

$$\frac{\partial\Phi}{\partial X^j}\frac{dX^j}{dt} = (\nabla\Phi)_j\frac{dX^j}{dt}. \tag{3-2.5a}$$

---

[7] See Philip Kelland, Peter G. Tait and C. G. Knott, Introduction to Quaternions, Macmillan, New York, 1904. In this text the symbol ∇ is called nabla.

Relation (3-2.5a) can be written just as well in the form

$$\left[\frac{\partial \Phi}{\partial X^j} - (\nabla\Phi)_j\right]\frac{dX^j}{dt} = 0. \tag{3-2.5b}$$

The representation of ∇ at a point P is not dependent on any one-space curve through P. Therefore the components dX^j/dt can be chosen arbitrarily. Because of this fact the parenthetic expression in (3-2.5b) is equal to zero. This completes the proof.

As far as the considerations of this section are concerned, the form (3-2.4) of ∇ could be introduced as the definition. However, this would lead to future difficulties. In the section on curvilinear coordinates ∇ is used in a form consistent with Definition 3-2.1 but distinct from (3-2.4).

Various uses of the operator ∇ are possible from a purely algebraic point of view. We can consider the application of ∇ to a scalar. On the other hand, ∇ might be employed in conjunction with the processes of dot and cross multiplication. Let us explore these possibilities as well as the geometric and physical concepts they initiate.

Definition 3-2.2. Let Φ be a differentiable scalar function (the domain can be coordinate n-tuples associated with a space curve, a surface element, or a region of space). The quantity $\nabla\Phi$ is called the gradient[8] field derived from Φ. For rectangular Cartesian coordinates

$$\boxed{\nabla\Phi = \iota_1\frac{\partial\Phi}{\partial X^1} + \iota_2\frac{\partial\Phi}{\partial X^2} + \iota_3\frac{\partial\Phi}{\partial X^3}.} \tag{3-2.6}$$

Theorem 3-2.2. The gradient field $\nabla\Phi$ is a vector field with respect to the orthogonal Cartesian group of transformations.

PROOF. Recall the coordinate transformation relations in (1-3.9a), that is, $X^j = a_k^{\ j}\bar{X}^k$. We have noted that

$$a_k^{\ j} = \frac{\partial X^j}{\partial \bar{X}^k}.$$

Because we are dealing with the orthogonal Cartesian transformations (see Section 3),

$$\frac{\partial \bar{X}^k}{\partial X^j} = A_j^{\ k} = a_k^{\ j} = \frac{\partial X^j}{\partial \bar{X}^k}. \tag{3-2.7a}$$

Now the components of $\nabla\Phi$ satisfy the relation

$$\frac{\partial\Phi}{\partial X^j} = \frac{\partial \bar{X}^k}{\partial X^j}\frac{\partial\Phi}{\partial \bar{X}^k} = \sum_{k=1}^{3}\frac{\partial X^j}{\partial \bar{X}^k}\frac{\partial\Phi}{\partial \bar{X}^k}. \tag{3-2.7b}$$

Therefore the proof is complete.

[8] The term "gradient" is taken from meteorology.

It is worth noting that this proof (that the components $\partial\Phi/\partial X^j$ transform in a manner contravariant to the way in which the basis n-tuples ι_j transform) depends specifically on the property (3-2.7a) of the orthogonal Cartesian group. More generally, as indicated by the middle member of (3-2.7b), the components $\partial\Phi/\partial X^j$ transform in a manner covariant[9] to the basis n-tuples.

It is convenient to think of ∇ as a vector operator; in order that it may be so the components of ∇ must satisfy the transformation law

(3-2.7c) $$\frac{\partial}{\partial X^j} = \frac{\partial \bar{X}^k}{\partial X^j}\frac{\partial}{\partial \bar{X}^k}.$$

Because no inconsistency can result from this procedure, it is used whenever convenient in the remainder of the book.

The next theorem serves to determine the major geometric properties of the gradient function.

Theorem 3-2.3. If a scalar function Φ is constant on a space curve C, the gradient $\nabla\Phi$ is perpendicular to the tangent vector $d\mathbf{r}/dt$ on C. (Assume that $\nabla\Phi \neq \mathbf{0}$.)

PROOF. Since Φ is constant on C, $d\Phi/dt = 0$. Therefore

(3-2.8) $$0 = \frac{d\Phi}{dt} = \nabla\Phi \cdot \frac{d\mathbf{r}}{dt},$$

and since (3-2.8) is a necessary and sufficient condition for orthogonality the proof is complete.

In the next theorem it is assumed that a surface is given according to the representation of Monge, that is, in the form

(3-2.9a) $$\Psi(X^1, X^2, X^3) = 0,$$

and that the surface can also be expressed parametrically by

(3-2.9b) $$X^j = X^j(v^\beta).$$

Theorem 3-2.4. At each point of a surface (3-2.9a) where $\nabla\Psi \neq \mathbf{0}$ the gradient vector field $\nabla\Psi$ is normal to the surface.

PROOF. Let P_0 be a point on Ψ with surface coordinates

$$(v_0^1, v_0^2).$$

Then

$$X^j = X^j(v_0^1, v^2)$$

and

$$X^j = X^j(v^1, v_0^2)$$

[9] See Chapter 1, Section 5.*

are parametric curves passing through P_0. The surface representation $\Psi(X^1, X^2, X^3) = 0$ can also be interpreted as a constant scalar field $\Psi[X^j(v_0{}^1, v^2)]$ or $\Psi[X^j(v^1, v_0{}^2)]$ defined on either of the parameter curves. According to Theorem 3-2.3, $\nabla\Psi$ is orthogonal to the tangent vector associated with each parameter curve. Therefore $\nabla\Psi$ is orthogonal or normal to the tangent plane determined by these two tangent vectors. Since orthogonality to the tangent plane is equivalent to orthogonality to the surface, the theorem is proved.

It is possible to have a scalar function Φ defined on a surface $\Psi = 0$ so that

(3-2.10) $$\Phi = \Phi(\Psi).$$

Since Ψ has the value zero for every point on the surface, it follows that the scalar function Φ is constant on the surface. The relation between the respective gradient functions is stated in the next theorem.

Theorem 3-2.5. If Φ is a scalar field defined on a surface $\Psi = 0$ in such a way that (3-2.10) holds:

(3-2.11) $$\nabla\Phi = \frac{\partial\Phi}{\partial\Psi}\nabla\Psi;$$

that is, $\nabla\Phi$ has the same direction as $\nabla\Psi$.

PROOF. We have

(3-2.12) $$\nabla\Phi = \sum_{j=1}^{3}\frac{\partial\Phi}{\partial X^j}\boldsymbol{\iota}_j = \sum_{j=1}^{3}\frac{\partial\Phi}{\partial\Psi}\frac{\partial\Psi}{\partial X^j}\boldsymbol{\iota}_j = \frac{\partial\Phi}{\partial\Psi}\nabla\Psi.$$

This completes the proof.

Example 3-2.1. The gravitational force vector field has the representation (see Chapter 2, Section 5)

(3-2.13a) $$\mathbf{F} = \frac{-mMG\mathbf{R}}{r^2},$$

where $\mathbf{R}$ is a unit vector. For simplicity, assume that the center of the force field is at the origin of a rectangular Cartesian coordinate system; then

$$r = [(X^1)^2 + (X^2)^2 + (X^3)^2]^{1/2}$$

The immediate interest is in the fact that $\mathbf{F}$ is proportional to $1/r^2$; (3-2.13a) can also be written in the form

(3-2.13b) $$\mathbf{F} = \frac{k\mathbf{R}}{r^2} = \frac{k\mathbf{r}}{r^3}, \qquad k = -mMG.$$

The introduction of a gravitational potential function

$$\Phi = \frac{k}{r} \tag{3-2.13c}$$

is standard procedure in a theoretical discussion of gravitational concepts. Then, as we can show by a straightforward computation,

$$\mathbf{F} = -\boldsymbol{\nabla}\Phi. \tag{3-2.13d}$$

The function Φ is defined everywhere except at the origin. In particular, let

$$\Psi(X^1, X^2, X^3) \equiv (X^1)^2 + (X^2)^2 + (X^3)^2 - a^2 = 0. \tag{3-2.13e}$$

The function Φ is constant on the spherical surface (3-2.13e). We have

$$\boldsymbol{\nabla}\Psi = 2(X^1\boldsymbol{\iota}_1 + X^2\boldsymbol{\iota}_2 + X^3\boldsymbol{\iota}_3) = 2\mathbf{r}; \tag{3-2.13f}$$

therefore

$$\boldsymbol{\nabla}\Phi = -\frac{k\mathbf{r}}{r^3} = -\frac{k}{2r^3}\boldsymbol{\nabla}\Psi. \tag{3-2.13g}$$

The last relation may be compared to (3-2.11) in Theorem 3-2.5.

Surfaces such as (3-2.13e), on which a given potential function is constant, are called equipotential. It is clear from Theorem 3-2.5 that the associated vector force field is normal to the equipotential surfaces. This particular fact gives rise to a fruitful geometric interpretation of force fields in terms of lines of force and equipotential surfaces. Such representations were first used by Faraday in his study of electric and magnetic phenomena.

Although potential functions can be associated with many fields of force (static electric, static magnetic, gravitational, etc.), it would be improper to leave the impression that this can always be done. It is more accurate to say that such a representation is possible only in exceptional cases. Because these exceptions are important, a special name has been established. When a force field $\mathbf{F}$ has a representation $\boldsymbol{\nabla}\Phi$, it is said to be a conservative field of force.

Another important property of the gradient of a scalar field Φ is stated in the following theorem.

Theorem 3-2.6. The maximum change of the scalar field Φ takes place in the direction and sense of $\boldsymbol{\nabla}\Phi$.

PROOF. Let P be a point in the region of space on which Φ is defined. Let C be an arbitrary curve through P_0, whose parametric equations are represented in terms of arc length s. We have

$$\frac{d\Phi}{ds} = \boldsymbol{\nabla}\Phi \cdot \frac{d\mathbf{r}}{ds} = |\boldsymbol{\nabla}\Phi| \left|\frac{d\mathbf{r}}{ds}\right| \cos\theta. \tag{3-2.14}$$

The change of Φ with respect to s is represented by the left-hand member of this expression. Examining the right-hand member, we observe that $\nabla\Phi$ depends only on the coordinates X_0^j of P_0 (in particular, not on the choice of C), whereas $|d\mathbf{r}/ds| = 1$ for all C. Therefore $d\Phi/ds$ depends on $\cos\theta$. Since $\cos\theta$ is maximum when $\theta = 0$, it follows that $d\Phi/ds$ is maximum when C is chosen such that the tangential vector $d\mathbf{r}/ds$ has the same sense and direction as $\nabla\Phi$.

If $\theta = \pi$, then

$$\frac{d\Phi}{ds} = -|\nabla\Phi|; \tag{3-2.15}$$

that is $d\Phi/ds$ is minimum in the sense opposite that of $\nabla\Phi$.

Definition 3-2.3. Let a scalar field Φ along a curve C be expressed in terms of arc length s. The change of Φ with respect to s, that is,

$$\frac{d\Phi}{ds} = \nabla\Phi\cdot\frac{d\mathbf{r}}{ds},$$

is called the directional derivative of the scalar field Φ in the direction and sense of $d\mathbf{r}/ds$.

Note that the directional derivative of Φ is just the component (or the projection with the appropriate sign) of $\nabla\Phi$ in the given direction.

Example 3-2.2. Let $\Phi = \Phi(X^1, X^2)$ be a scalar function defined on a region of the X^1, X^2 plane. A line,

$$\begin{aligned} X^1 &= a + s\cos\theta, \\ X^2 &= b + s\sin\theta, \end{aligned} \tag{3-2.16a}$$

expressed in terms of the arc-length parameter s, specifies a direction at a point $P(a, b)$ of this region. The unit tangent vector

$$\left(\frac{dX^1}{ds}, \frac{dX^2}{ds}\right) = (\cos\theta, \sin\theta), \tag{3-2.16b}$$

determines a particular sense. The directional derivative $d\Phi/ds$ is then

$$\begin{aligned} \frac{d\Phi}{ds} &= \frac{\partial\Phi}{\partial X^1}\frac{dX^1}{ds} + \frac{\partial\Phi}{\partial X^2}\frac{dX^2}{ds} \\ &= \frac{\partial\Phi}{\partial X^1}\cos\theta + \frac{\partial\Phi}{\partial X^2}\sin\theta. \end{aligned} \tag{3-2.16c}$$

The values $\partial\Phi/\partial X^1$, $\partial\Phi/\partial X^2$ depend only on the point P; therefore the directional derivative is just dependent on the direction and sense of the tangential vector or, in other words, on the angle θ. (See Figs. 3-2.1 and 3-2.2.)

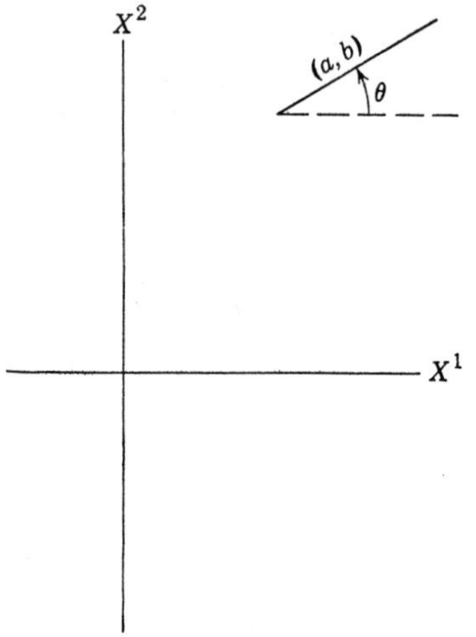

Fig. 3-2.1

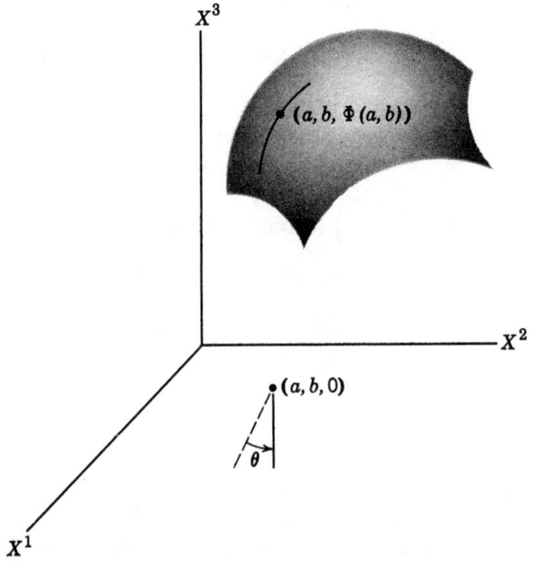

Fig. 3-2.2

In the first part of this example Φ was not interpreted geometrically: neither were $\nabla\Phi$, nor the directional derivative $d\Phi/ds$. Geometric meanings can be supplied by setting $\Phi(X^1, X^2)$ equal to a dependent variable X^3. Then

$$X^3 = \Phi(X^1, X^2) \tag{3-2.17a}$$

is the explicit representation of a surface (under the assumption that the appropriate derivatives exist). The equations in (3-2.16a) represent a plane, and the intersection of that plane with the surface in (3-2.17a) is a surface curve:

$$\begin{aligned} X^1 &= a + s\cos\theta, \\ X^2 &= b + s\sin\theta, \\ X^3 &= \Phi(a + s\cos\theta, b + s\sin\theta). \end{aligned} \tag{3-2.17b}$$

Observe that s does not denote arc length for this curve. (See Problem 15.) If (3-2.17a) is expressed in the form

$$\Psi(X^1, X^2, X^3) = \Phi(X^1, X^2) - X^3, \tag{3-2.17c}$$

then

$$\left(\frac{\partial\Psi}{\partial X^1}, \frac{\partial\Psi}{\partial X^2}, \frac{\partial\Psi}{\partial X^3}\right) = \left(\frac{\partial\Phi}{\partial X^1}, \frac{\partial\Phi}{\partial X^2}, -1\right). \tag{3-2.17d}$$

Therefore $\partial\Phi/\partial X^1$, $\partial\Phi/\partial X^2$ are the first two components of the normal to the surface. A geometric significance for the directional derivative itself can be obtained from the expression for a tangential vector field to the surface curve (3-2.17b), that is,

$$\frac{d\mathbf{r}}{ds} = \cos\theta\,\boldsymbol{\iota}_1 + \sin\theta\,\boldsymbol{\iota}_2 + \left(\frac{\partial\Phi}{\partial X^1}\cos\theta + \frac{\partial\Phi}{\partial X^2}\sin\theta\right)\boldsymbol{\iota}_3. \tag{3-2.17e}$$

When this expression is dotted with $\boldsymbol{\iota}_3$, we have

$$\boldsymbol{\iota}_3 \cdot \frac{d\mathbf{r}}{ds} = \frac{dX^3}{ds} = \frac{\partial\Phi}{\partial X^1}\cos\theta + \frac{\partial\Phi}{\partial X^2}\sin\theta = \frac{d\Phi}{ds}. \tag{3-2.17f}$$

Therefore the directional derivative of Φ in the direction of θ can be interpreted as the projection, with appropriate sign, of the tangent vector onto the X^3 axis.

We conclude this discussion of the application of the vector operator ∇ to scalar functions by considering the differentiation of sums and products. The reader will find these rules easy to assimilate, for they agree with those learned in the study of calculus.

Theorem 3-2.7. Let Φ_1 and Φ_2 be scalar functions defined on a common region of space. If the partial derivatives exist, then

$$\nabla(\Phi_1 + \Phi_2) = (\nabla\Phi_1) + (\nabla\Phi_2). \tag{3-2.18a}$$

$$\nabla(\Phi_1\Phi_2) = (\nabla\Phi_1)\Phi_2 + \Phi_1(\nabla\Phi_2). \tag{3-2.18b}$$

PROOF. As a consequence of the corresponding properties of partial derivatives, we have

$$\nabla(\Phi_1 + \Phi_2) = \sum_{j=1}^{3} \boldsymbol{\iota}_j \frac{\partial}{\partial X^j}(\Phi_1 + \Phi_2) = \sum_{j=1}^{3} \boldsymbol{\iota}_j \frac{\partial \Phi_1}{\partial X^j} + \sum_{j=1}^{3} \boldsymbol{\iota}_j \frac{\partial \Phi_2}{\partial X^j}.$$

The relation (3-2.18b) can be verified in an analogous manner.

If Φ_1 is constant (i.e., $\Phi = c$), then the relation (3-2.18b) reduces to

(3-2.18c) $$\nabla c\Phi_2 = c\,\nabla\Phi_2.$$

A rather obvious next step (at least from the algebraic point of view) is that of examining the properties of the vector operator del in connection with dot and cross multiplication. Actually, the introduction of a new concept is involved. This new entity is presented in analogy with and in generalization of an old one.

Definition 3-2.4. If **G** is a differentiable vector field expressed in terms of rectangular Cartesian components, then

(3-2.19a) $$\nabla \cdot \mathbf{G} = \frac{\partial G^1}{\partial X^1} + \frac{\partial G^2}{\partial X^2} + \frac{\partial G^3}{\partial X^3}.$$

$\nabla \cdot \mathbf{G}$ is read "del dot **G**" and is called the divergence of the vector field **G**.

The name divergence comes about because of the physical uses of the concept. Roughly speaking, $\nabla \cdot \mathbf{G}$ gives a measure of the variation of the field **G** at each point of definition. The name convergence was given to $-\nabla \cdot \mathbf{G}$ by Clerk Maxwell and is still found in some writings on physical science.

On occasion it will be convenient to express $\nabla \cdot \mathbf{G}$ in notational form $\partial_j G^j$, where

(3-2.19b) $$\partial_j = \frac{\partial}{\partial X^j}.$$

This convention lends itself to the application of the summation convention.

The expression (3-2.19a) looks like the ordinary dot product; however, the reader should be wary. In particular, note that $\nabla \cdot \mathbf{G}$ is not commutative; $\nabla \cdot \mathbf{G}$ is a function, whereas $\mathbf{G} \cdot \nabla$ is an operator.

The rules of differentiation associated with the divergence concept agree with those of ordinary differentiation.

Theorem 3-2.8. If Φ is a differentiable scalar field and **G** and **H** are differentiable vector fields, then

(3-2.20)

(a) $$\nabla \cdot (\mathbf{G} + \mathbf{H}) = \nabla \cdot \mathbf{G} + \nabla \cdot \mathbf{H}.$$

(b) $$\nabla \cdot (\Phi\mathbf{G}) = \nabla\Phi \cdot \mathbf{G} + \Phi\nabla \cdot \mathbf{G}.$$

PROOF. The proofs follow from the corresponding properties of partial derivatives. The details are left to the reader.

Ample demonstration of the application of the divergence concept to physics are to be found in the chapter on integration. The following example may serve the immediate purpose of indicating its usefulness.

Example 3-2.3. In Example 3-2.1, as well as in the section on Newtonian orbits, it was pointed out that the gravitational force field of a body is proportional to $\mathbf{r}/r^3$. The same statement is valid for both the electrostatic and magnetostatic force fields (under appropriate conditions). For such fields we have

$$\nabla \cdot \left(\frac{\mathbf{r}}{r^3}\right) = 0, \qquad r \neq 0. \tag{3-2.21a}$$

This fact is demonstrated by the following computations:

$$\nabla \cdot \mathbf{r} = \partial_j X^j = \frac{\partial X^1}{\partial X^1} + \frac{\partial X^2}{\partial X^2} + \frac{\partial X^3}{\partial X^3} = 3 \tag{3-2.21b}$$

and

$$\frac{\partial r^{-3}}{\partial X^j} = \frac{\partial}{\partial X^j}\left[\sum_{k=1}^{3} (X^k)^2\right]^{-3/2} = -\frac{3}{2}\left[\sum_{k=1}^{3} (X^k)^2\right]^{-5/2} \sum_{k=1}^{3} 2X^k \frac{\partial X^k}{\partial X^j}. \tag{3-2.21c}$$

The results in (3-2.21b) and (3-2.21c) can be used to verify (3-2.21a).

The case in which a force field proportional to $\mathbf{r}/r^3$ has a zero divergence everywhere except at $r = 0$ corresponds to an idealized physical situation in which the field is assumed to originate at a single point.

Theorem 3-2.9. The divergence of a vector field $\mathbf{G}$ is a scalar field with respect to the orthogonal Cartesian group of transformations.

PROOF. With the coordinate transformation equations

$$X^j = a_k{}^j \bar{X}^k$$

in mind, we have

$$\nabla \cdot \mathbf{G} = \partial_j G^j = A_j{}^k \frac{\partial}{\partial \bar{X}^k}(a_p{}^j \bar{G}^p)$$

$$= \delta_p{}^k \frac{\partial \bar{G}^p}{\partial \bar{X}^k} = \frac{\partial \bar{G}^p}{\partial \bar{X}^p} = \nabla \cdot \bar{\mathbf{G}}.$$

This completes the proof.

In particular, if $\mathbf{G}$ is a gradient vector, that is,

$$\mathbf{G} = \nabla\Phi,$$

then $\nabla \cdot \nabla\Phi$ is a scalar field. This expression is called the Laplacian and is often written in the symbolic form $\nabla^2\Phi$. Laplace's partial differential equation

$$\nabla^2\Phi = 0$$

and Poisson's equation

$$\nabla^2\Phi = 4\pi\rho,$$

play significant parts in various aspects of theoretical physics.

Let us turn our attention to the concept of cross multiplication as it is related to the operator ∇. The following definition is suggested by the cross product of two vectors.

Definition 3-2.5. If $\mathbf{G}$ is a differentiable vector field referred to rectangular Cartesian coordinates, then

$$\nabla \times \mathbf{G} = \boldsymbol{\iota}_1\left(\frac{\partial G^3}{\partial X^2} - \frac{\partial G^2}{\partial X^3}\right) + \boldsymbol{\iota}_2\left(\frac{\partial G^1}{\partial X^3} - \frac{\partial G^3}{\partial X^1}\right) + \boldsymbol{\iota}_3\left(\frac{\partial G^2}{\partial X^1} - \frac{\partial G^1}{\partial X^2}\right), \tag{3-2.22a}$$

or

$$\nabla \times \mathbf{G} = \begin{vmatrix} \boldsymbol{\iota}_1 & \boldsymbol{\iota}_2 & \boldsymbol{\iota}_3 \\ \dfrac{\partial}{\partial X^1} & \dfrac{\partial}{\partial X^2} & \dfrac{\partial}{\partial X^3} \\ G^1 & G^2 & G^3 \end{vmatrix}. \tag{3-2.22b}$$

The expression $\nabla \times \mathbf{G}$ (read del cross $\mathbf{G}$) is called the curl of $\mathbf{G}$.

The determinant (3-2.22b) is merely symbolic. It provides an easy way to remember the expression in (3-2.22a). Another form of the curl of a vector, which is computationally advantageous, is

$$\nabla \times \mathbf{G} = \sum_{p=1}^{3} \boldsymbol{\iota}_p \mathcal{E}_{pjk}\, \partial^j G^k, \tag{3-2.22c}$$

where

$$\partial^j = \frac{\partial}{\partial X^j}. \tag{3-2.22d}$$

The name "curl" was coined by Clerk Maxwell. Maxwell, in his *Electricity and Magnetism*, published in 1873, made use of some of the notation and terminology of Hamilton's quaternion theory but pretty much avoided the operational methods, probably because he considered them too complicated.

We often find, especially in the German literature, the terms rotor or rotation used for $\nabla \times \mathbf{G}$.

The next theorem points out the vector character of the curl of a vector in relation to the orthogonal Cartesian group of transformations.

Theorem 3-2.10. Let $\mathbf{G}$ be a differentiable vector field. Then $\nabla \times \mathbf{G}$ is a vector field with respect to the orthogonal Cartesian group of transformations.

PROOF. The components of $\nabla \times \mathbf{G}$ are

$$\sum_{j=1}^{3} \mathcal{E}_{pjk} \frac{\partial}{\partial X^j} G^k.$$

We have

$$\begin{aligned}\sum_{j=1}^{3} \mathcal{E}_{pjk} \frac{\partial G^k}{\partial X^j} &= \sum_{j=1}^{3} (A_p{}^q A_j{}^r A_k{}^s \bar{\mathcal{E}}_{qrs}) A_j{}^u \frac{\partial}{\bar{X}^u} (a_v{}^k \bar{G}^v) \\ &= \delta^{ru}\, \delta_v{}^s A_p{}^q \bar{\mathcal{E}}_{qrs} \frac{\partial}{\partial \bar{X}^u} \bar{G}^v \\ &= \sum_{u=1}^{3} A_p{}^q \bar{\mathcal{E}}_{quv} \frac{\partial}{\partial \bar{X}^u} \bar{G}^v.\end{aligned}$$

This completes the proof.

Problems

1. Put the surface representations of Problem 2, Section 1, in the form $\Psi(X^1, X^2, X^3) = 0$ and find $\nabla\Psi$ in each case.

2. (a) Find $\nabla\Psi$ if $\Psi(X^1, X^2, X^3) = (X^1)^2 + (X^2)^2 + (X^3)^2$.

(b) Let $X^1 = 3 \sin\theta \cos\phi$, $X^2 = 3 \sin\theta \sin\phi$, $X^3 = 3 \cos\theta$. Show that

$$\nabla\Psi = 6\,(\sin\theta \cos\phi\, \boldsymbol{\iota}_1 + \sin\theta \sin\phi\, \boldsymbol{\iota}_2 + \cos\theta\, \boldsymbol{\iota}_3).$$

(c) If $\theta = \theta(t)$, $\phi = \phi(t)$ show that

$$\begin{aligned}\frac{d\mathbf{r}}{dt} &= 3\left(\cos\theta \cos\phi \frac{d\theta}{dt} - \sin\theta \sin\phi \frac{d\phi}{dt}\right)\boldsymbol{\iota}_1 \\ &\quad + 3\left(\cos\theta \sin\phi \frac{d\theta}{dt} + \sin\theta \cos\phi \frac{d\phi}{dt}\right)\boldsymbol{\iota}_2 - 3 \sin\theta \frac{d\theta}{dt}\, \boldsymbol{\iota}_3.\end{aligned}$$

[Use the parametric representation of (b).]

(d) Compute $d\mathbf{r}/dt$ in (c) if $\theta = t$, $\phi = t$.

(e) Show by direct computation that $\nabla\Psi \cdot (d\mathbf{r}/dt) = 0$ [by using (b) and (d)].

3. The scalar function $\Phi = \ln[(X^1)^2 + (X^2)^2 + (X^3)^2]^{1/2}$ is constant on the surface

$$(X^1)^2 + (X^2)^2 + (X^3)^2 = a^2,$$

where a is a constant. Show by direct computation that the gradient of the surface and the gradient of Φ are proportional at each point of the surface. In fact,

$$\nabla\Phi = \frac{1}{2a^2} \nabla\Psi.$$

4. Compute $\nabla \cdot \mathbf{G}$ and $\nabla \times \mathbf{G}$ for each of the following vector fields:
   (a) $\mathbf{G} = X^j \iota_j$.
   (b) $\mathbf{G} = X^1X^2\iota_1 + (X^1)^2\iota_2 + X^1X^2X^3\iota_3$.
   (c) $\mathbf{G} = \sin X^1 \cos X^2 \iota_1 + \cos X^2 \iota_2 + \sin X^1 \cos X^3 \iota_3$.
5. Find the equation of the tangent plane to the ellipsoid

$$\frac{(X^1)^2}{4} + \frac{(X^2)^2}{2} + \frac{(X^3)^2}{1} = 1 \quad \text{at} \quad (1, 1, \tfrac{1}{2}).$$

6. Find a vector normal to the surface $X^1X^2 - X^3 = 1$ at the point (2, 1, 1).
7. If $\Phi(X^1, X^2, X^3) = (X^1)^2 + (X^2)^2 + (X^3)^2$, show that $\nabla\Phi = 2\mathbf{r}$.
8. Find $\mathbf{H}$ if $\mathbf{H} = -\nabla\Phi$ and $\Phi = e^{aX^3} \sin bX^1 \cosh cX^2$; a, b, c are constants.
9. Given $f = f(u, v)$, $u = u(X^j)$, $v = v(X^j)$, show that

$$\nabla f = \frac{\partial f}{\partial u}\nabla u + \frac{\partial f}{\partial v}\nabla v.$$

10. If $d\Phi/dt = \mathbf{F} \cdot (d\mathbf{r}/dt)$ for arbitrary $d\mathbf{r}/dt$, prove that $F_j = \partial\Phi/\partial X^j$.
11. Show that the parameter s, used to express the curve equations (3-2.17b), does not denote arc length by showing that $d\mathbf{r}/ds$ is not a unit vector.

3. *Identities Involving* ∇

This section is devoted to a compilation of the algebraic properties of gradient, divergence, and curl. All proofs are referred to rectangular Cartesian coordinates. Later it will be seen that the results obtained in this section can be employed in relation to general coordinate systems. It is assumed throughout the section that the differentiations are defined.

Theorem 3-3.1. We have

(3-3.1)
(a) $\nabla \cdot \mathbf{r} = 3$ (for a three-dimensional space),
(b) $\nabla \times \mathbf{r} = \mathbf{0}$,
(c) $\nabla \cdot r^{-3}\mathbf{r} = 0$.

PROOF. The relations in (3-3.1a) and (3-3.1c) were developed as part of Example 3-2.3. The proof of (3-3.1b) is constructed as follows:

$$\nabla \times \mathbf{r} = \sum_{p=1}^{3} \iota_p \mathcal{E}_{pjk}\, \partial^j X^k = \sum_{p=1}^{3} \iota_p \mathcal{E}_{pjk}\, \delta^{jk} = \mathbf{0},$$

where

$$\partial^j = \frac{\partial}{\partial X^j}.$$

The reader may verify the last equality by writing out the summations. A simpler way of reaching the result is to note that, although δ^{jk} is symmetric in j and k, the symbol $\mathcal{E}_{pjk}$ is skew symmetric (i.e., $\mathcal{E}_{pjk} = -\mathcal{E}_{pkj}$).

Theorem 3-3.2. We have

(a) $\nabla \cdot \nabla \times \mathbf{U} = 0$ (the divergence of the curl is zero).

(3-3.2) (b) $\boxed{\nabla \times (\nabla \times \mathbf{U}) = \nabla(\nabla \cdot \mathbf{U}) - \nabla^2\mathbf{U},}$ $(\nabla^2 = \nabla \cdot \nabla)$.

(c) $(\nabla \times \nabla) \times \mathbf{U} = \mathbf{0}$.

Proof. The proofs of the relations in (3-3.2a) and (3-3.2c) are left to the reader. The expression in (3-3.2b) can be verified as follows: The components of $\nabla \times (\nabla \times \mathbf{U})$ are

$$\begin{aligned} \mathrm{E}^{rsp}\,\partial_s \mathcal{E}_{pjk}\,\partial^j U^k &= \mathrm{E}^{rsp}\mathcal{E}_{pjk}\,\partial_s\,\partial^j U^k \\ &= 2\delta^{[rs]}_{jk}\,\partial_s\,\partial^j U^k = 2\partial_s\,\partial^{[r}U^{s]} \\ &= \partial^r(\partial_s U^s) - (\partial_s\,\partial^s U^r). \end{aligned}$$

The relation in (1-6.11a), as well as the definition of the brackets [as indicated in (1-6.12b)], was used to obtain the equalities. The last member of the string of equalities represents the components of the right-hand side of (3-3.2b). This completes the proof.

Theorem 3-3.3. We have

$$\text{(a)}\quad \mathbf{U} \times (\nabla \times \mathbf{W}) = \nabla_W(\mathbf{W} \cdot \mathbf{U}) - (\mathbf{U} \cdot \nabla)\mathbf{W}$$

where ∇_W symbolizes the fact that ∇ operates on $\mathbf{W}$ alone. The corresponding statement holds for ∇_U.

(3-3.3)

(b) $$\begin{aligned} \nabla(\mathbf{U} \cdot \mathbf{W}) &= \nabla_U(\mathbf{U} \cdot \mathbf{W}) + \nabla_W(\mathbf{U} \cdot \mathbf{W}) \\ &= \mathbf{U} \times (\nabla \times \mathbf{W}) + \mathbf{W} \times (\nabla \times \mathbf{U}) + (\mathbf{U} \cdot \nabla)\mathbf{W} \\ &\quad + (\mathbf{W} \cdot \nabla)\mathbf{U}. \end{aligned}$$

(c) $\nabla \cdot (\mathbf{U} \times \mathbf{W}) = (\nabla \times \mathbf{U}) \cdot \mathbf{W} - (\nabla \times \mathbf{W}) \cdot \mathbf{U}$.

(d) $\nabla \times (\mathbf{U} \times \mathbf{W}) = (\mathbf{W} \cdot \nabla)\mathbf{U} - \mathbf{W}(\nabla \cdot \mathbf{U}) + \mathbf{U}(\nabla \cdot \mathbf{W}) - (\mathbf{U} \cdot \nabla)\mathbf{W}$.

Proof. The validity of (3-3.3a) is demonstrated as follows (for notational convenience let $U_q = U^q$ and $\partial^s = \partial_s$):

$$\begin{aligned} \mathbf{U} \times (\nabla \times \mathbf{W}) &= \iota_p \mathrm{E}^{pqr} U_q \mathcal{E}_{rst}\,\partial^s W^t \\ &= \iota_p 2\delta^{[pq]}_{st} U_q\,\partial^s W^t = \iota_p 2 U_q\,\partial^{[p}W^{q]} \\ &= \iota_p U_q\,\partial^p W^q - \iota_p U_q\,\partial^q W^p \\ &= \nabla_W(\mathbf{W} \cdot \mathbf{U}) - (\mathbf{U} \cdot \nabla)W. \end{aligned}$$

The other proofs are left to the reader. In particular, (3-3.3b) is easily obtained with the help of (3-3.3a).

Theorem 3-3.4. We have

$$\text{(3-3.4)} \qquad \begin{aligned} &\text{(a)} \quad (\mathbf{W}\cdot\nabla)\mathbf{r} = \mathbf{W}, \\ &\text{(b)} \quad \frac{d\mathbf{G}}{dt} = \left(\frac{d\mathbf{r}}{dt}\cdot\nabla\right)\mathbf{G}, \qquad \mathbf{G} = \mathbf{G}[X^j(t)]. \end{aligned}$$

PROOF. The proof of (3-3.4a) is left to the reader. Starting with the right-hand side of (3-3.4b), we have

$$\frac{d\mathbf{r}}{dt}\cdot\nabla\,\mathbf{G} = \frac{dX^j}{dt}\frac{\partial}{\partial X^j}(G^k\boldsymbol{\iota}_k) = \frac{dG^k}{dt}\,\boldsymbol{\iota}_k = \frac{d\mathbf{G}}{dt}.$$

The relation in (3-3.4b) is nothing more than the vector statement of a total derivative formula.

To memorize the identities of this section would be a tedious task. Fortunately, this can be avoided. By close examination the observant reader can determine that the derivative laws already known from calculus actually hold. Of course, a consistency property must be kept in mind; that is, ∇ must operate on the same quantities in both members of an identity.

Problems

1. Prove all identities left to the reader for verification.

2. Given $\nabla \times \mathbf{H} = \mathbf{0}$ and $\nabla\cdot\mathbf{H} = 0$, show that $\nabla^2\mathbf{H} = 0$.

3. Derive an expansion for $\nabla\cdot(f\mathbf{G})$ where $f = f(X^i)$ is a scalar field and $\mathbf{G}$ is a vector field.

4. Derive an expansion for $\nabla \times (f\mathbf{G})$.

5. Show that in a two-dimensional Euclidean space $\nabla\cdot\mathbf{r} = 2$.

6. If $\Phi\mathbf{U}$ satisfies La place's equation ($\nabla^2\Phi\mathbf{U} = 0$) and $\nabla\times(\nabla\times\Phi\mathbf{U}) = \mathbf{0}$ show that the gradient of $(\nabla\Phi)\cdot\mathbf{U} + \Phi\,\nabla\cdot\mathbf{U}$ is zero.

7. Show that $\nabla(\mathbf{a}\cdot\mathbf{r}) = \mathbf{a}$ where $\mathbf{a}$ is a constant vector.

8. Show that $\nabla\times(\mathbf{a}\times\mathbf{r}) = 2\mathbf{a}$, $\mathbf{a}$ constant.

9. Prove that with respect to the rectangular Cartesian group of transformations,
   (a) $\nabla\times(\nabla\times\mathbf{U})$ is a vector,
   (b) $\mathbf{U}\times(\nabla\times\mathbf{W})$ is a vector,
   (c) $\nabla(\mathbf{U}\cdot\mathbf{W})$ is a vector.

4. *Bases in General Coordinate Systems*

In this section the groundwork is laid for the task of expressing previously introduced ideas in terms of coordinate systems that are not Cartesian. Probably the most familiar examples of non-Cartesian coordinates are the cylindrical and spherical coordinate systems. These are introduced by way of example.

Two immediate values of developing the vector concepts in terms of non-Cartesian systems come to mind. First of all, geometric models of many physical problems lend themselves quite naturally to non-Cartesian interpretation. For example, the gravitational field of a particle is such that spheres with the particle at their common center are equipotential surfaces. Therefore a spherical coordinate system affords a simple description of this phenomenon.

Second, the general coordinate systems form a natural framework for the introduction of concepts fundamental to the development of tensor algebra and analysis.

In order to have specific coordinate systems at hand for purposes of illustration, the cylindrical and spherical systems are introduced in the following examples. The equations of transformation relating each of them to rectangular Cartesian coodinates are also presented.

Example 3-4.1. The transformation equations relating cylindrical coordinates to rectangular Cartesian coordinates are

$$\begin{aligned} X^1 &= \rho \cos \theta, \\ X^2 &= \rho \sin \theta, \\ X^3 &= z. \end{aligned} \tag{3-4.1}$$

The cylindrical coordinate system consists of a polar coordinate system in a plane along with a height. The transformation relations can be determined from Fig. 3-4.1.

The coordinate surfaces $\rho =$ constant, $\theta =$ constant, and $z =$ constant are circular cylinders, planes, and planes, respectively.

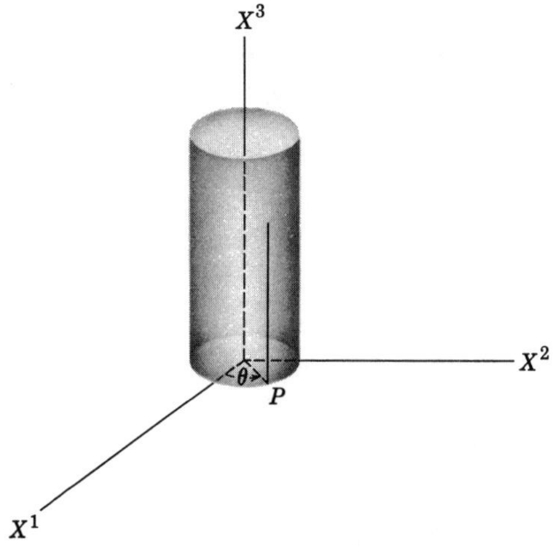

Fig. 3-4.1

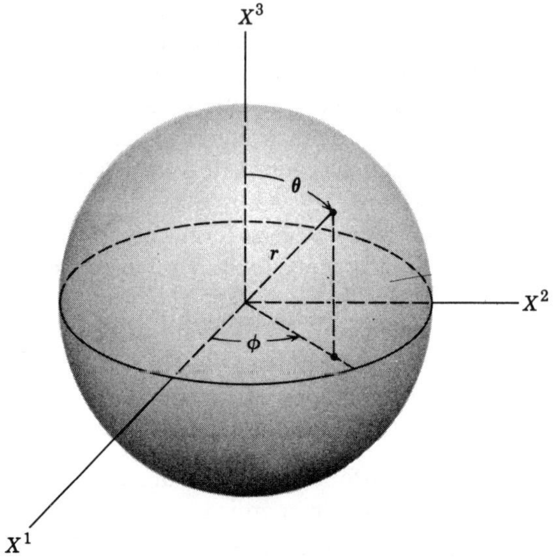

Fig. 3-4.2

Example 3-4.2. The transformation equations relating spherical coordinates and rectangular Cartesian coordinates are

(3-4.2)
$$\begin{aligned} X^1 &= r \sin \theta \cos \phi, \\ X^2 &= r \sin \theta \sin \phi, \\ X^3 &= r \cos \theta. \end{aligned}$$

These relations can be read directly from Fig. 3-4.2.

In this example the coordinate surfaces $r =$ constant, $\theta =$ constant, and $\phi =$ constant are, respectively, spheres, cones, and planes.

The general coordinate transformations discussed in this section are symbolized by the form

(3-4.3)
$$\bar{X}^j = \bar{X}^j(\bar{\bar{X}}^1, \bar{\bar{X}}^2, \bar{\bar{X}}^3), \qquad j = 1, 2, 3.$$

It is assumed that

(a) each of the functions $\bar{X}^j$ has continuous partial derivatives of at least the first order;

(3-4.4) (b) the determinant $|\partial \bar{X}^j / \partial \bar{\bar{X}}^k| \neq 0$ for every coordinate triple of the domain determined by the function $\bar{X}^j$.

It is the province of an advanced calculus course to show that these properties ensure that each function has an inverse with continuous first partial derivatives in a neighborhood of a given point. We shall not enter into the matter here, but certain consequences of the stated assumptions will be derived at the appropriate time.

As has been the case throughout the book, it is assumed that the space is Euclidean. Therefore a rectangular Cartesian coordinate system is always available. As usual the symbols X^j are reserved for representation of a rectangular Cartesian system.

Example 3-4.3. The transformation functions relating cylindrical coordinates and rectangular Cartesian coordinates satisfy the conditions in (3-4.4a,b), except at $\rho = 0$. The determinant $|\partial X^j/\partial \bar{X}^k|$, in which $\bar{X}^1 = \rho$, $\bar{X}^2 = \theta$, and $\bar{X}^3 = z$, is exhibited here.

$$\begin{vmatrix} \dfrac{\partial X^1}{\partial \rho} & \dfrac{\partial X^2}{\partial \rho} & \dfrac{\partial X^3}{\partial \rho} \\ \dfrac{\partial X^1}{\partial \theta} & \dfrac{\partial X^2}{\partial \theta} & \dfrac{\partial X^3}{\partial \theta} \\ \dfrac{X^1}{\partial z} & \dfrac{X^2}{\partial z} & \dfrac{X^3}{\partial z} \end{vmatrix} = \begin{vmatrix} \cos\theta & \sin\theta & 0 \\ -\rho\sin\theta & \rho\cos\theta & 0 \\ 0 & 0 & 1 \end{vmatrix} = \rho.$$

The determinant $|\partial \bar{X}^j/\partial \bar{\bar{X}}^k|$ is called the Jacobian of transformation of the $\bar{X}^j$ coordinate system with respect to the $\bar{\bar{X}}^j$ system. It is named after C. G. J. Jacobi (1804–1851, German), an honor bestowed on him by Sylvester in recognition of his outstanding contributions to mathematics in general and the determinant theory in particular. It is notationally convenient to suppress the indices and write $|\partial \bar{X}/\partial \bar{\bar{X}}|$.

Definition 3-4.1. Allowable coordinate transformations are those of the form

$$\bar{X}^j = \bar{X}^j(\bar{\bar{X}}^k)$$

that satisfy assumptions (3-4.4a,b).

Important properties of the allowable coordinate transformations are indicated by the following three theorems.

Theorem 3-4.1a. The Jacobians $|\partial \bar{X}/\partial \bar{\bar{X}}|$ and $|\partial \bar{\bar{X}}/\partial \bar{X}|$ of an allowable coordinate transformation and its inverse, respectively, satisfy the relation

$$\left|\frac{\partial \bar{X}}{\partial \bar{\bar{X}}}\right| \left|\frac{\partial \bar{\bar{X}}}{\partial \bar{X}}\right| = 1. \tag{3-4.5}$$

PROOF. Since the transformation

$$\bar{X}^j = \bar{X}^j(\bar{\bar{X}}^k) \tag{3-4.6a}$$

has an inverse (at least in a neighborhood of a given point),

(3-4.6b) $$\overline{\overline{X}}^k = \overline{\overline{X}}^k(\overline{X}^j).$$

We can write

(3-4.6)
$$\text{(c)}\quad d\overline{X}^j = \frac{\partial \overline{X}^j}{\partial \overline{\overline{X}}^k}\, d\overline{\overline{X}}^k,$$
$$\text{(d)}\quad d\overline{\overline{X}}^k = \frac{\partial \overline{\overline{X}}^k}{\partial \overline{X}^j}\, d\overline{X}^j.$$

Because the relations (3-4.6c,d) are linear in the differentials, it is easily shown that the sets of partial derivatives satisfy the property. (See Problem 3.)

(3-4.6e) $$\frac{\partial \overline{X}^j}{\partial \overline{\overline{X}}^k}\frac{\partial \overline{\overline{X}}^p}{\partial \overline{X}^j} = \overline{\overline{\delta}}_k{}^p.$$

The definition of determinant multiplication applied to (3-4.6) implies the result (3-4.5), as was to be shown.

Theorem 3-4.1b. If allowable transformations

$$\overline{X}^j = \overline{X}^j(\overline{\overline{X}}^k), \qquad \overline{\overline{X}}^k = \overline{\overline{X}}^k(\overline{\overline{\overline{X}}}^p)$$

are defined on a common region and the composition transformation is indicated by

$$\overline{X}^j = \overline{X}^j(\overline{\overline{\overline{X}}}^p),$$

the Jacobians satisfy the relation

(3-4.7a) $$\left|\frac{\partial \overline{X}}{\partial \overline{\overline{\overline{X}}}}\right| = \left|\frac{\partial \overline{X}}{\partial \overline{\overline{X}}}\right|\left|\frac{\partial \overline{\overline{X}}}{\partial \overline{\overline{\overline{X}}}}\right|.$$

PROOF. As a consequence of the properties of partial derivatives, we have

(3-4.7b) $$\frac{\partial \overline{X}^j}{\partial \overline{\overline{\overline{X}}}^p} = \frac{\partial \overline{X}^j}{\partial \overline{\overline{X}}^k}\frac{\partial \overline{\overline{X}}^k}{\partial \overline{\overline{\overline{X}}}^p}.$$

The desired result is achieved by employing the definition of determinant multiplication.

Theorem 3-4.1c. The set of allowable coordinate transformations forms a group.

PROOF. The properties of Definition 1-5*.1 must be demonstrated. Observe the following:

(a) A composite of continuous functions is continuous; therefore

$$\overline{X}^j = \overline{X}^j[\overline{\overline{X}}^k(\overline{\overline{\overline{X}}}^p)] = \overline{X}^j(\overline{\overline{\overline{X}}}^p)$$

is continuous. Furthermore, from (3-4.7b) we observe that the partial derivatives $\partial \bar{X}^j/\partial \bar{\bar{\bar{X}}}^p$ are continuous (products and sums of continuous functions are continuous), and therefore closure is established.

(b) If $R : \bar{\bar{X}} \to \bar{X}$,

$S : \bar{\bar{\bar{X}}} \to \bar{\bar{X}}$,

$T : \bar{\bar{\bar{\bar{X}}}} \to \bar{\bar{\bar{X}}}$,

then

$$R(ST) = (RS)T;$$

that is, the associative law holds.

(c) The identity transformation $\bar{X}^j = \bar{\bar{X}}^j$ belongs to the set.

(d) The assumptions (3-4.4a,b) imply the existence of an inverse to each transformation of the set. This completes the proof.

For the reader who has investigated the ideas introductory to tensor analysis discussed in the starred sections the remainder of this section will involve a certain amount of repetition. However, the generality of the underlying transformation group makes it a little more than just review. If the text has been used primarily as an introduction to vector analysis, it would be worthwhile to refer to Chapter 1, Section 5*, for comparison and analogy.

As in Section 5*, three bases are to be associated with each coordinate system. The primary purpose of introducing this complication is to obtain a framework that will make possible a careful discussion of gradient, divergence, and curl in terms of general coordinate systems.

The rectangular Cartesian coordinate systems are allowable systems and therefore are at our disposal. The procedure "going out from" these systems, that is, the process of using such a system as a fundamental frame of reference and then transforming away, is of much value.

Consider the position vector $\mathbf{r}$ expressed in terms of rectangular Cartesian coordinates and the basis $\boldsymbol{\iota}_1$, $\boldsymbol{\iota}_2$, $\boldsymbol{\iota}_3$ associated with the axes of the system.

(3-4.8a)
$$\mathbf{r} = X^j \boldsymbol{\iota}_j.$$

Let $X^j = X^j(t)$ and $\bar{X}^j = \bar{X}^j(t)$, respectively, be parametric equations of a curve C in the rectangular Cartesian coordinates X^j and the general coordinates $\bar{X}^j$. Then

(3-4.8b)
$$\frac{d\mathbf{r}}{dt} = \frac{\partial \mathbf{r}}{\partial \bar{X}^j}\frac{d\bar{X}^j}{dt}.$$

The curve C is auxiliary; in fact, (3-4.8b) could just as well be stated in the differential form

$$d\mathbf{r} = \frac{\partial \mathbf{r}}{\partial \bar{X}^j}\, d\bar{X}^j.$$

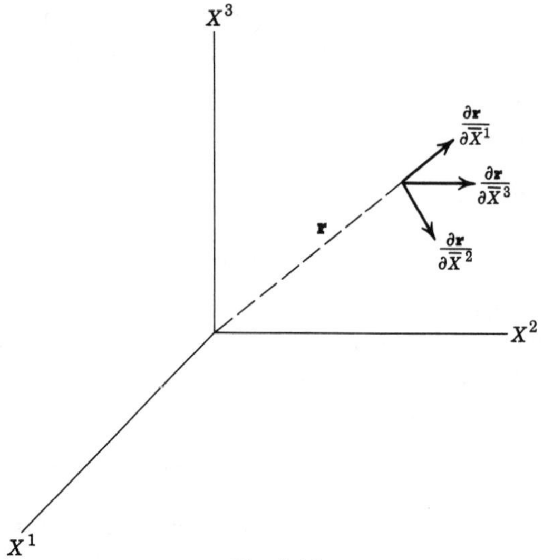

Fig. 3-4.3

The set of partial derivatives $\partial \mathbf{r}/\partial \bar{X}^1$, $\partial \mathbf{r}/\partial \bar{X}^2$, $\partial \mathbf{r}/\partial \bar{X}^3$ is the key to the introduction of a basis for generalized coordinate systems. (See Fig. 3-4.3). Such a set can be associated with each general coordinate system $\bar{X}^j$, $\bar{\bar{X}}^j$, etc. The significance of the collection $\{\partial \mathbf{r}/\partial \bar{X}^j\}$ is illustrated by the following theorem.

Theorem 3-4.2a. The elements of the collection $\{\partial \mathbf{r}/\partial X^j\}$ are related by the rule of transformation

(3-4.9a)
$$\frac{\partial \mathbf{r}}{\partial \bar{X}^j} = \frac{\partial \bar{\bar{X}}^k}{\partial \bar{X}^j} \frac{\partial \mathbf{r}}{\partial \bar{\bar{X}}^k}.$$

If the initial system is rectangular Cartesian, then

(3-4.9b)
$$\boldsymbol{\iota}_j = \frac{\partial \bar{\bar{X}}^k}{\partial X^j} \frac{\partial \mathbf{r}}{\partial \bar{\bar{X}}^k}.$$

PROOF. The equation in (3-4.9a) simply illustrates the usual rule of partial differentiation; (3-4.9b) follows from the fact that in terms of a basis $\boldsymbol{\iota}_1, \boldsymbol{\iota}_2, \boldsymbol{\iota}_3$, $\mathbf{r} = X^1\boldsymbol{\iota}_1 + X^2\boldsymbol{\iota}_2 + X^3\boldsymbol{\iota}_3$, and therefore $\partial \mathbf{r}/\partial X^1 = \boldsymbol{\iota}_1$, etc.

The significance of the theorem lies in the fact that the sets $\partial \mathbf{r}/\partial \bar{X}^j$ and $\partial \mathbf{r}/\partial \bar{\bar{X}}^j$ are algebraically related in a manner analogous to that holding for rectangular Cartesian bases.

Theorem 3-4.2b. The set of n-tuples $\partial \mathbf{r}/\partial \bar{X}^j$ is a linearly independent set.

PROOF. Assume that the set is dependent. Then one of the arrows is dependent on the other two; that is,

$$\frac{\partial \mathbf{r}}{\partial \bar{X}^{j_1}} = \alpha \frac{\partial \mathbf{r}}{\partial \bar{X}^{j_2}} + \beta \frac{\partial \mathbf{r}}{\partial \bar{X}^{j_3}}, \tag{3-4.10a}$$

where α and β are not simultaneously equal to zero and j_1, j_2, j_3 take on distinct values from the set 1, 2, 3. From (3-4.10a) we are able to get a contradiction to the fact that the Jacobian of transformation is different from zero. If (3-4.10a) is cross multiplied with $\partial \mathbf{r}/\partial \bar{X}^{j_2}$ and dot is multiplied with $\partial \mathbf{r}/\partial \bar{X}^{j_3}$, we obtain

$$\pm \left| \frac{\partial X^k}{\partial \bar{X}^j} \right| = \frac{\partial \mathbf{r}}{\partial \bar{X}^{j_3}} \cdot \frac{\partial \mathbf{r}}{\partial \bar{X}^{j_2}} \times \frac{\partial \mathbf{r}}{\partial \bar{X}^{j_1}} = \alpha(0) + \beta(0) = 0. \tag{3-4.10b}$$

Since the assumption of linear dependence leads to a contradiction, the theorem is valid.

In the sequel the partial derivatives $\partial \mathbf{r}/\partial \bar{X}^1$, $\partial \mathbf{r}/\partial \bar{X}^2$, and $\partial \mathbf{r}/\partial \bar{X}^3$ are replaced by $\bar{\mathbf{r}}_1$, $\bar{\mathbf{r}}_2$, and $\bar{\mathbf{r}}_3$, respectively. The same notational convention is followed with respect to $\bar{\bar{X}}^j$, $\bar{\bar{\bar{X}}}^j$, etc., coordinate systems.

$\bar{\mathbf{r}}_1$, $\bar{\mathbf{r}}_2$, and $\bar{\mathbf{r}}_3$ are Cartesian vectors. The preceding theorems imply that such sets will serve as bases in generalized coordinate systems, thereby giving us the opportunity to extend the vector concept to them. Because of their geometric interpretation, it is convenient to refer to $\bar{\mathbf{r}}_1$, $\bar{\mathbf{r}}_2$, and $\bar{\mathbf{r}}_3$ as arrows.

Definition 3-4.2. The arrows $\bar{\mathbf{r}}_1$, $\bar{\mathbf{r}}_2$, and $\bar{\mathbf{r}}_3$ are said to be covariant basis arrows associated with an $\bar{X}^j$ coordinate system. Corresponding statements hold for sets $\bar{\bar{\mathbf{r}}}_j$, $\bar{\bar{\bar{\mathbf{r}}}}_j$, etc.

The basis arrows are of a much more complicated nature than any we have dealt with in that they are functions of position. However, at a given point they have the direction of the corresponding coordinate curves just as the covariant basis arrows associated with the general Cartesian systems had. This statement is easily verified by holding two of the variables, say $\bar{X}^2$ and $\bar{X}^3$, fixed in the expression

$$\mathbf{r} = \mathbf{r}(\bar{X}^1, \bar{X}^2, \bar{X}^3).$$

The resulting representation is that of a coordinate curve along which $\bar{X}^1$ varies; $\partial \mathbf{r}/\partial \bar{X}^1$ corresponds to the ordinary derivative along this curve and therefore is a tangential vector. The extension of the vector concept to general coordinate systems is brought about in a manner similar to the extension to general Cartesian systems. (See Chapter 1, Section 5*.) It should be realized that the Cartesian systems are included in the set of general coordinate systems even though our attention is primarily focused elsewhere.

We assume that triples of functions $\bar{U}^1, \bar{U}^2, \bar{U}^3$; $\bar{\bar{U}}^1, \bar{\bar{U}}^2, \bar{\bar{U}}^3$, etc., are defined in the respective coordinate systems.

Definition 3-4.3. A collection $\{U^1, U^2, U^3\}$ of ordered triples, the elements of which satisfy the transformation law

$$\bar{U}^j = \frac{\partial \bar{X}^j}{\partial \bar{\bar{X}}^k} \bar{\bar{U}}^k, \tag{3-4.11}$$

is said to be a contravariant vector. The triples $\bar{U}^1, \bar{U}^2, \bar{U}^3$, etc., are called contravariant components of the vector in the respective coordinate systems.

The prototype for this definition is the tangential vector field to a curve C, that is, $d\mathbf{r}/dt$. If we supposed that the field were defined in terms of a rectangular Cartesian system X^j, which is related to a general system by means of transformation equations $X^j = X^j(\bar{X}^k)$, then

$$\frac{dX^j}{dt} = \frac{\partial X^j}{\partial \bar{X}^k} \frac{d\bar{X}^k}{dt}. \tag{3-4.12a}$$

First of all it should be noted that this relation is the component form of (3-4.8b). Second, if we write

$$\frac{dX^j}{dt} = \frac{\partial X^j}{\partial \bar{\bar{X}}^k} \frac{d\bar{\bar{X}}^k}{dt}, \tag{3-4.12b}$$

where $X^j = X^j(\bar{\bar{X}}^k)$, then dX^j/dt can be eliminated from (3-4.12a) and (3-4.12b), thereby producing the relation

$$\frac{d\bar{X}^j}{dt} = \frac{\partial \bar{X}^j}{\partial \bar{\bar{X}}^k} \frac{d\bar{\bar{X}}^k}{dt}. \tag{3-4.12c}$$

On the one hand, this expression is independent of a rectangular Cartesian system and therefore quite general. On the other, it is precisely of the form (3-4.11).

Another property implicit in Definition 3-4.11 is the algebraic invariance of the form $U^j\mathbf{r}_j$. (See Problem 13.) In a Cartesian system this form represents a Cartesian vector and is geometrically represented by an arrow (more precisely a family of arrows). Therefore it is consistent to write

$$\mathbf{U} = \bar{U}^j\bar{\mathbf{r}}_j \tag{3-4.13}$$

and to think in the same geometric terms relevant to the discussion of Cartesian systems. Clearly, the components of a contravariant vector are naturally expressed in terms of the covariant basis of Definition 3-4.2.

Two other types of components are useful in this development of vector concepts. The first is associated with the basis introduced in the next theorem.

Theorem 3-4.3. The set of arrows $\bar{\mathbf{r}}^j$, which satisfy the relation

$$\bar{\mathbf{r}}^j \cdot \bar{\mathbf{r}}_k = \bar{\delta}_k{}^j, \tag{3-4.14a}$$

determines a contravariant arrow basis associated with the $\bar{X}^j$ system.

PROOF. We must show that arrows satisfying (3-4.14a) exist and that they determine a linearly independent set. Since (3-4.14a) can be construed as a set of nine nonhomogeneous equations in nine unknowns (the components of $\bar{\mathbf{r}}^1$, $\bar{\mathbf{r}}^2$, $\bar{\mathbf{r}}^3$ expressed in terms of a basis $\boldsymbol{\iota}_j$) with the determinant of coefficients different from zero, it follows that the $\bar{\mathbf{r}}^j$ exist.

In order to prove that the set $\bar{\mathbf{r}}^1$, $\bar{\mathbf{r}}^2$, $\bar{\mathbf{r}}^3$ is linearly independent, assume the contrary. Then

$$\bar{\mathbf{r}}^{j_1} = \alpha\bar{\mathbf{r}}^{j_2} + \beta\bar{\mathbf{r}}^{j_3}, \tag{3-4.14b}$$

where j_1, j_2, and j_3 are distinct. By dot multiplying (3-4.3a) with $\bar{\mathbf{r}}_{j_1}$, we obtain the result

$$1 = \bar{\mathbf{r}}_{j_1} \cdot \bar{\mathbf{r}}^{j_1} = \alpha(0) + \beta(0) = 0 \text{ (no sum on } j_1). \tag{3-4.14c}$$

This is an obvious inconsistency; therefore the set $\bar{\mathbf{r}}^j$ is linearly independent.

From the geometric standpoint, we see that

$$\begin{aligned} &\bar{\mathbf{r}}^1 \text{ is perpendicular to } \bar{\mathbf{r}}_2 \text{ and } \bar{\mathbf{r}}_3,\\ &\bar{\mathbf{r}}^2 \text{ is perpendicular to } \bar{\mathbf{r}}_3 \text{ and } \bar{\mathbf{r}}_1,\\ &\bar{\mathbf{r}}^3 \text{ is perpendicular to } \bar{\mathbf{r}}_1 \text{ and } \bar{\mathbf{r}}_2. \end{aligned} \tag{3-4.14d}$$

For this reason the basis $\bar{\mathbf{r}}^j$ is said to be reciprocal to $\bar{\mathbf{r}}_j$ and conversely.

In case the arrows $\bar{\mathbf{r}}_j$ consist of a mutually orthogonal set of unit elements with a right-hand rotation, the ordered sets $\bar{\mathbf{r}}_j$ and $\bar{\mathbf{r}}^j$ coincide. As mentioned in Chapter 1, Section 5*, when the basis $\bar{\mathbf{r}}_j$ is the set $\boldsymbol{\iota}_1$, $\boldsymbol{\iota}_2$, $\boldsymbol{\iota}_3$ associated with a rectangular Cartesian coordinate system, then, according to (3-4.14d), the reciprocal basis introduces nothing new. This is precisely why the considerations presently being made played no part during the portion of the development of the sections that deal with the orthogonal Cartesian group of transformations. The immediate purpose in introducing a second basis is to obtain a measure of freedom in algebraic manipulation which would not otherwise be possible. This algebraic flexibility depends on the relations indicated in subsequent definitions and theorems.

The connecting links are matrices $(\bar{g}_{ij})$ and $(\bar{g}^{ij})$. These arrays not only relate the bases $\bar{\mathbf{r}}_j$ and $\bar{\mathbf{r}}^j$ but also determine the fundamental metric properties of the space.

Definition 3-4.4a. Suppose that X^j were the coordinates of a rectangular Cartesian system and $X^j = X^j(\bar{X}^k)$ an allowable transformation. The elements of $(\bar{g}_{jk})$ are given by

$$\bar{g}_{jk} = \frac{\partial \mathbf{r}}{\partial \bar{X}^j} \cdot \frac{\partial \mathbf{r}}{\partial \bar{X}^k} = \sum_{i=1}^{3} \frac{\partial X^i}{\partial \bar{X}^j} \frac{\partial X^i}{\partial \bar{X}^k}. \tag{3-4.15a}$$

Definition 3-4.4b. The matrix $(\bar{g}^{jk})$ is inverse to $(\bar{g}_{jk})$, that is,

$$\bar{g}_{jk}\bar{g}^{jp} = \bar{\delta}_k{}^p. \tag{3-4.15b}$$

Example 3-4.4. Rectangular Cartesian and spherical coordinates are related by the transformation equations $X^1 = r \sin\theta \cos\phi$, $X^2 = r \sin\theta \sin\phi$, $X^3 = r \cos\theta$. (See Example 3-4.2.) By direct computation based on (3-4.15a,b), we can show that

$$(\bar{g}_{jk}) = \begin{pmatrix} 1 & 0 & 0 \\ 0 & r^2 & 0 \\ 0 & 0 & r^2 \sin^2\theta \end{pmatrix}, \qquad (\bar{g}^{jk}) = \begin{pmatrix} 1 & 0 & 0 \\ 0 & \dfrac{1}{r^2} & 0 \\ 0 & 0 & \dfrac{1}{r^2 \sin^2\theta} \end{pmatrix} \tag{3-4.16}$$

The next theorem indicates a linear relation between the bases $\bar{\mathbf{r}}_j$ and $\bar{\mathbf{r}}^j$.

Theorem 3-4.4. We have

$$\text{(a)} \quad \bar{\mathbf{r}}_j = \bar{g}_{jk}\bar{\mathbf{r}}^k. \qquad \text{(b)} \quad \bar{\mathbf{r}}^j = \bar{g}^{jk}\bar{\mathbf{r}}_k. \tag{3-4.17}$$

PROOF. According to (3-4.14a),

$$\bar{\mathbf{r}}_j \cdot \bar{\mathbf{r}}^k = \bar{\delta}_j{}^k.$$

Multiplying and summing the members of this relation with $\bar{g}_{pk}$ produces the relation

$$\bar{\mathbf{r}}_j \cdot \bar{g}_{pk}\bar{\mathbf{r}}^k = \bar{g}_{pk}\,\bar{\delta}_j{}^k = \bar{g}_{pj} = \bar{\mathbf{r}}_p \cdot \bar{\mathbf{r}}_j.$$

This expression can be written in the form

$$\bar{\mathbf{r}}_j \cdot (\bar{g}_{pk}\bar{\mathbf{r}}^k - \bar{\mathbf{r}}_p) = 0.$$

Since the elements $\mathbf{r}_j$ are linearly independent, it is not possible that all three of them are perpendicular to a given nonzero vector. Therefore the expression in parentheses must be a zero vector. This completes the proof of (a).

The proof of (b) follows from (a) by making use of (3-4.15b). The details are left to the reader.

To every triple of contravariant components $\bar{U}^k$ there corresponds a second set of elements

$$\bar{U}_j = \bar{g}_{jk}\bar{U}^k. \tag{3-4.18a}$$

By use of the elements $\bar{g}^{pj}$ we are able to solve (3-4.18a) for the contravariant components $\bar{U}^k$. In particular, employment of (3-4.15b) produces the result

(3-4.18b) $$\bar{U}^p = \bar{g}^{pj}\bar{U}_j.$$

The notational convention introduced by (3-4.18a) endows the set of elements $\bar{g}_{jk}$ with an algebraic property usually referred to as "lowering of an index." The remark that an index is lowered by $\bar{g}_{jk}$ is simply an abbreviation for the statement: for purposes of convenience the components $\bar{U}^k$ have been replaced by another set $\bar{U}_j$ obtained through the linear relation (3-4.18a). The process involved in (3-4.18b) is denoted by the terminology "raising of an index."

The elements $\bar{U}_j$ are naturally associated with the contravariant basis $\bar{\mathbf{r}}^j$, for we have

$$\bar{\mathbf{U}} = \bar{U}^j\bar{\mathbf{r}}_j = \bar{g}^{jk}\bar{U}_k\bar{g}_{jp}\bar{\mathbf{r}}^p = \bar{\delta}_p{}^k\bar{U}_k\bar{\mathbf{r}}^p = \bar{U}_p\bar{\mathbf{r}}^p.$$

This fact motivates the following definition.

Definition 3-4.5. A collection $\{U_1, U_2, U_3\}$, one element $(\bar{U}_1, \bar{U}_2, \bar{U}_3)$ from each allowable coordinate system such that

(3-4.19a) $$\bar{U}_j = \frac{\partial \bar{\bar{X}}^k}{\partial \bar{X}^j}\bar{\bar{U}}_k,$$

is said to be a covariant vector. The elements $\bar{U}_j$, etc., are called the covariant components of the vector in the respective coordinate systems.

The rule of transformation (3-4.19a) must be consistent with (3-4.18a) if the previously stated expressions are to fit together. This consistency is verified by carrying out the transformation. We have

$$\bar{U}_j = \bar{g}_{jk}\bar{U}^k = \frac{\partial \mathbf{r}}{\partial \bar{X}^j}\cdot\frac{\partial \mathbf{r}}{\partial \bar{X}^k}\bar{U}^k = \frac{\partial \mathbf{r}}{\partial \bar{\bar{X}}^p}\frac{\partial \bar{\bar{X}}^p}{\partial \bar{X}^j}\cdot\frac{\partial \mathbf{r}}{\partial \bar{\bar{X}}^q}\frac{\partial \bar{\bar{X}}^q}{\partial \bar{X}^k}\frac{\partial \bar{X}^k}{\partial \bar{\bar{X}}^r}\bar{\bar{U}}^r$$
$$= \frac{\partial \bar{\bar{X}}^p}{\partial \bar{X}^j}\bar{\bar{g}}_{pq}\bar{\bar{\delta}}_r{}^q\bar{\bar{U}}^r = \frac{\partial \bar{\bar{X}}^p}{\partial \bar{X}^j}\bar{\bar{g}}_{pr}\bar{\bar{U}}^r = \frac{\partial \bar{\bar{X}}^p}{\partial \bar{X}^j}\bar{\bar{U}}_p.$$

The reader should note that in a Euclidean space a Cartesian vector can be represented in general coordinate systems by components of either a contravariant or a covariant vector. Therefore it is consistent to speak of a vector with contravariant and covariant components. We use this convenient language often.

The prototype of a covariant vector is the gradient $\partial\Phi/\partial X^j$. This statement is initiated by the transformation rule

(3-4.19b) $$\frac{\partial \Phi}{\partial X^j} = \frac{\partial \bar{X}^k}{\partial X^j}\frac{\partial \phi}{\partial \bar{X}^k},$$

which is of the same form as (3-4.19a).

The preceding development made use of the matrix elements $(\bar{g}_{jk})$ and $(\bar{g}^{jk})$ to relate covariant and contravariant sets. Let us investigate these matrices in more detail. The next theorem introduces an explicit form for the elements $\bar{g}^{jk}$.

Theorem 3-4.5. We have

(3-4.20a)
$$\bar{g}^{jk} = \bar{\mathbf{r}}^j \cdot \bar{\mathbf{r}}^k = \sum_{i=1}^{3} \frac{\partial \bar{X}^j}{\partial X^i} \frac{\partial \bar{X}^k}{\partial X^i}.$$

PROOF. By dot multiplying the relation (3-4.17b) with $\bar{\mathbf{r}}^k$, we obtain

$$\bar{\mathbf{r}}^j \cdot \bar{\mathbf{r}}^k = \bar{g}^{jp}\bar{\mathbf{r}}_p \cdot \bar{\mathbf{r}}^k = \bar{g}^{jp}\, \bar{\delta}_p^{\ k} = \bar{g}^{jk}.$$

This completes the first part of the proof. It remains to be shown that the rectangular Cartesian components of $\bar{\mathbf{r}}^j$ are $\partial \bar{X}^j/\partial X^i$. To obtain this result, start with

$$\bar{\mathbf{r}}_j \cdot \bar{\mathbf{r}}^k = \bar{\delta}_j^{\ k}.$$

Since the components of $\bar{\mathbf{r}}_j$ are $\partial X^i/\partial \bar{X}^j$, the equations can be written in the form

$$\frac{\partial X^i}{\partial \bar{X}^j} \bar{r}_i^{\ k} = \bar{\delta}_j^{\ k},$$

where $\bar{r}_i^{\ 1}$, $\bar{r}_i^{\ 2}$, $\bar{r}_i^{\ 3}$, respectively, are the components of $\bar{\mathbf{r}}^1$, $\bar{\mathbf{r}}^2$, $\bar{\mathbf{r}}^3$. By multiplying and summing with $\partial \bar{X}^j/\partial X^p$, we obtain

$$\delta_p^{\ i}\bar{r}_i^{\ k} = \frac{\partial \bar{X}^j}{\partial X^p} \bar{\delta}_j^{\ k}$$

or

(3-4.20b)
$$\bar{r}_p^{\ k} = \frac{\partial \bar{X}^k}{\partial X^p}.$$

The last member of (3-4.20a) is obtained by substituting the result (3-4.20b) in the middle member.

The next theorem introduces the transformation laws for the elements of $(\bar{g}_{jk})$ and $(\bar{g}^{jk})$. It also illustrates a point emphasized throughout the book, that scalar and vector character depend on the allowable group of transformations. In particular, the dot product, which is a scalar with respect to the group of orthogonal Cartesian transformations, is used to introduce a nonscalar entity in relation to the more general group of transformations now at our disposal.

Theorem 3-4.6. Let $\bar{X}^j = \bar{X}^j(\bar{\bar{X}}^k)$ be an allowable coordinate transformation in the sense of Definition 3-4.1. Let matrices $(\bar{g}_{jk})$, $(\bar{g}^{jk})$;

$(\bar{\bar{g}}_{jk})$, $(\bar{\bar{g}}^{jk})$ be associated with the respective systems. The matrix elements then satisfy the transformation rules

(3-4.21)

$$\text{(a)} \quad \bar{g}_{jk} = \frac{\partial \bar{\bar{X}}^p}{\partial \bar{X}^j} \frac{\partial \bar{\bar{X}}^q}{\partial \bar{X}^k} \bar{\bar{g}}_{pq}$$

$$\text{(b)} \quad \bar{g}^{jk} = \frac{\partial \bar{X}^j}{\partial \bar{\bar{X}}^p} \frac{\partial \bar{X}^k}{\partial \bar{\bar{X}}^q} \bar{\bar{g}}^{pq}$$

PROOF. We have

$$\bar{g}_{jk} = \frac{\partial \mathbf{r}}{\partial \bar{X}^j} \cdot \frac{\partial \mathbf{r}}{\partial \bar{X}^k} = \frac{\partial \bar{\bar{X}}^p}{\partial \bar{X}^j} \frac{\partial \bar{\bar{X}}^q}{\partial \bar{X}^k} \frac{\partial \mathbf{r}}{\partial \bar{\bar{X}}^p} \cdot \frac{\partial \mathbf{r}}{\partial \bar{\bar{X}}^q} = \frac{\partial \bar{\bar{X}}^p}{\partial \bar{X}^j} \frac{\partial \bar{\bar{X}}^q}{\partial \bar{X}^k} \bar{\bar{g}}_{pq}.$$

This completes the proof of (a). The proof of (b) follows in the same way from (3-4.20a).

Definition 3-4.6a. Let the set of matrices $(\bar{g}_{jk})$, $(\bar{\bar{g}}_{jk})$, etc., associated with the respective systems $\bar{X}^j$, $\bar{\bar{X}}^j$, etc., be denoted by $\{g_{jk}\}$. Then $\{g_{jk}\}$ is said to be the fundamental metric tensor associated with the allowable transformation group. The tensor is of covariant order 2 and the elements g_{jk} are called covariant components.

Definition 3-4.6b. The set of matrices $\{g^{jk}\}$ is said to be the associated metric tensor with respect to the allowable transformation group. It is, according to its law of transformation (3-4.21b), a contravariant tensor of order 2. The elements g^{jk} are called contravariant components of order 2.

The components of the metric tensor take on a special form in rectangular Cartesian systems. In particular,

(3-4.21c)
$$g_{jk} = \delta_{jk}.$$

Furthermore, the transformation law (3-4.21a) reduces to

(3-4.21d)
$$\delta_{jk} = a_j{}^p a_k{}^q \, \bar{\delta}_{pq} = \bar{\delta}_{jk}.$$

Therefore the components can be interpreted as scalars. This rationale establishes consistency in calling $\bar{\mathbf{r}}_j \cdot \bar{\mathbf{r}}_k$ a set of scalars with respect to rectangular Cartesian transformations and tensor components when referred to a more general transformation group.

Before considering the third of the bases mentioned earlier in the section, we digress by illustrating the previously introduced concepts in the following example.

Example 3-4.5. Let $\bar{X}^j$ be the system of cylindrical coordinates described in Example 3-4.1. Recall that

$$\bar{X}^1 = \rho, \qquad \bar{X}^2 = \theta, \qquad \bar{X}^3 = z.$$

The set of basis arrows $\partial\mathbf{r}/\partial\rho$, $\partial\mathbf{r}/\partial\theta$, $\partial\mathbf{r}/\partial z$ is represented in terms of $\boldsymbol{\iota}_1$, $\boldsymbol{\iota}_2$, $\boldsymbol{\iota}_3$ as follows:

$$\begin{aligned} \bar{\mathbf{r}}_1 &= \cos\theta\boldsymbol{\iota}_1 + \sin\theta\boldsymbol{\iota}_2, \\ \bar{\mathbf{r}}_2 &= -\rho\sin\theta\boldsymbol{\iota}_1 + \rho\cos\theta\boldsymbol{\iota}_2, \\ \bar{\mathbf{r}}_3 &= \boldsymbol{\iota}_3. \end{aligned} \tag{3-4.22a}$$

It can easily be shown that the elements of the set are mutually orthogonal. Furthermore, $\bar{\mathbf{r}}_1$ and $\bar{\mathbf{r}}_3$ are unit arrows. However, $\bar{\mathbf{r}}_2$ has a variable magnitude ρ and therefore has a magnitude of only 1 at those points on the unit sphere, $\rho = 1$.

Computing from (3-4.15a), we have

$$(\bar{g}_{jk}) = \begin{pmatrix} 1 & 0 & 0 \\ 0 & \rho^2 & 0 \\ 0 & 0 & 1 \end{pmatrix}. \tag{3-4.22b}$$

The contravariant components $\bar{g}^{jk}$ are obtained easily from (3-4.15b). The ease of computation is a consequence of the orthogonality of $(\bar{g}_{jk})$ (i.e., the fact that the off-diagonal elements are zero).

Since $\bar{\mathbf{r}}^j = \bar{g}^{jk}\bar{\mathbf{r}}_k$, we have

$$\begin{aligned} \bar{\mathbf{r}}^1 &= \cos\theta\,\boldsymbol{\iota}_1 + \sin\theta\,\boldsymbol{\iota}_2, \\ \bar{\mathbf{r}}^2 &= -\frac{1}{\rho}\sin\theta\,\boldsymbol{\iota}_1 + \frac{1}{\rho}\cos\theta\,\boldsymbol{\iota}_2, \\ \bar{\mathbf{r}}^3 &= \boldsymbol{\iota}_3. \end{aligned} \tag{3-4.22c}$$

The fact that $\bar{\mathbf{r}}^1$, $\bar{\mathbf{r}}^2$, $\bar{\mathbf{r}}^3$ correspond to $\bar{\mathbf{r}}_1$, $\bar{\mathbf{r}}_2$, $\bar{\mathbf{r}}_3$, except with respect to the magnitudes of $\bar{\mathbf{r}}^2$ and $\bar{\mathbf{r}}_2$, agrees with the orthogonality of the bases. Note that the magnitudes of $\bar{\mathbf{r}}^2$ and $\bar{\mathbf{r}}_2$ are reciprocal, as might be expected.

The preceding example brings to mind the observation that

$$(\bar{g}_{jk}) = \begin{pmatrix} \bar{g}_{11} & 0 & 0 \\ 0 & \bar{g}_{22} & 0 \\ 0 & 0 & \bar{g}_{33} \end{pmatrix} \tag{3-4.23a}$$

is equivalent to

$$\bar{\mathbf{r}}_j \cdot \bar{\mathbf{r}}_k = 0. \tag{3-4.23b}$$

In other words, the basis elements are mutually perpendicular or orthogonal if and only if (3-4.23a) holds.

The bases $\bar{\mathbf{r}}_j$ and $\bar{\mathbf{r}}^j$ contribute algebraic flexibility. In particular, they introduce an aesthetic quality to that part of the theory involving transformations. The third type of basis to be introduced in this section lends itself to geometric and physical interpretation.

Definition 3-4.7. Let

(3-4.24a) $$\bar{\mathbf{e}}_1 = \frac{1}{\bar{g}_1}\bar{\mathbf{r}}_1, \qquad \bar{\mathbf{e}}_2 = \frac{1}{\bar{g}_2}\bar{\mathbf{r}}_2, \qquad \bar{\mathbf{e}}_3 = \frac{1}{\bar{g}_3}\bar{\mathbf{r}}_3,$$

where

(3-4.24b) $$\bar{g}_j = \sqrt{\bar{g}_{jj}}.$$

Theorem 3-4.7a. The set $\bar{\mathbf{e}}_1$, $\bar{\mathbf{e}}_2$, $\bar{\mathbf{e}}_3$, associated with a coordinate system $\bar{X}^j$, is a unit arrow basis.

PROOF. The arrows $\bar{\mathbf{e}}_j$ are proportional to the $\bar{\mathbf{r}}_j$. Since each $\bar{\mathbf{e}}_j$ is formed by dividing $\bar{\mathbf{r}}_j$ by its magnitude, they are unit arrows.

Theorem 3-4.7b. A vector $\bar{\mathbf{U}}$ expressed in terms of the basis $\bar{\mathbf{e}}_j$ has the forms

(3-4.25a) $$\bar{\mathbf{W}} = (\bar{W}^1\bar{g}_1)\bar{\mathbf{e}}_1 + (\bar{W}^2\bar{g}_2)\bar{\mathbf{e}}_2 + (\bar{W}^3\bar{g}_3)\bar{\mathbf{e}}_3.$$

(3-4.25b) $$\bar{\mathbf{W}} = (\bar{g}^{1k}\bar{W}_k\bar{g}_1)\bar{\mathbf{e}}_1 + (\bar{g}^{2k}\bar{W}_k\bar{g}_2)\bar{\mathbf{e}}_2 + (\bar{g}^{3k}\bar{W}_k\bar{g}_3)\bar{\mathbf{e}}_3.$$

PROOF. As a consequence of (3-4.24a,b),

$$\bar{\mathbf{W}} = \bar{\mathbf{W}}^j\bar{\mathbf{r}}_j = \bar{W}^1\bar{g}_1\bar{\mathbf{e}}_1 + \bar{W}^2\bar{g}_2\bar{\mathbf{e}}_2 + \bar{W}^3\bar{g}_3\bar{\mathbf{e}}_3.$$

Since $\bar{W}^j = \bar{g}^{jk}\bar{W}_k$, the foregoing expression can also be written as in (3-4.25b).

Definition 3-4.8. The components $\bar{w}_1 = \bar{W}^1\bar{g}_1$, $\bar{w}_2 = \bar{W}^2\bar{g}_2$, $\bar{w}_3 = \bar{W}^3\bar{g}_3$ associated with a basis $\bar{\mathbf{e}}_1$, $\bar{\mathbf{e}}_2$, $\bar{\mathbf{e}}_3$ are called physical components of the vector $\mathbf{W}$.

The physical components of a vector represent displacements in the directions of the unit arrows $\bar{\mathbf{e}}_j$. Furthermore, they are dimensionally correct. These properties make them useful in many geometrical and physical considerations. However, their usage is primarily restricted to a single coordinate system, for transformation laws involving physical components can become cumbersome.

In some texts the physical components of Definition 3-4.8 are said to be of the first kind. Corresponding components of the second kind are introduced as coefficients of a basis

$$\frac{\bar{\mathbf{r}}^1}{|\bar{\mathbf{r}}^1|}, \frac{\bar{\mathbf{r}}^2}{|\bar{\mathbf{r}}^2|}, \frac{\bar{\mathbf{r}}^3}{|\bar{\mathbf{r}}^3|}.$$

If the basis $\bar{\mathbf{r}}_j$ is orthogonal, then so is the basis $\bar{\mathbf{r}}^j$. Under this circumstance, the covariant and contravariant components of a vector $\bar{\mathbf{W}}$ satisfy the relation

(3-4.26) $$W_x = g_{xx}W^x \qquad \text{(no sum on } x\text{)}.$$

The following theorem concerning the physical components is a consequence of this relation.

Theorem 3-4.8. Let $\bar{\mathbf{e}}_j$ be an orthogonal unit basis. The physical components of a vector $\bar{\mathbf{W}}$ can be expressed in terms of the covariant components:

$$\bar{w}_1 = \frac{1}{\bar{g}_1}\,\overline{W}_1, \qquad \bar{w}_2 = \frac{1}{\bar{g}_2}\,\overline{W}_2, \qquad \bar{w}_3 = \frac{1}{\bar{g}_3}\,\overline{W}_3. \tag{3-4.27}$$

PROOF. When (3-4.26) is applied to (3-4.25b), the result follows.

In Section 5 the physical component representations of gradient, divergence, and curl are among the results obtained.

Problems

1. The transformation equations relating spherical and rectangular Cartesian coordinates are given by (3-4.2). With respect to them, complete the following:
   (a) Compute the matrix $(\partial X^j/\partial \bar{X}^k)$.
   (b) Evaluate the Jacobian of transformation $|\partial X/\partial \bar{X}|$.
   (c) In light of the conditions in (3-4.4), which coordinate triples (r, θ, ϕ) must be excluded
   (d) Compute the matrix $\partial \bar{X}^k/\partial X^j$.

   Hint: $$\frac{\partial \bar{X}^k}{\partial X^j} = \frac{\text{cofactor } (\partial X^j/\partial \bar{X}^k) \text{ ln } |\partial X/\partial \bar{X}|}{|\partial X/\partial \bar{X}|}.$$

   (e) Check the property $|\partial X/\partial \bar{X}|\,|\partial \bar{X}/\partial X| = 1$ by direct computation.
   (f) Express $\bar{\mathbf{r}}_j$, $\bar{\mathbf{r}}^j$, and $\bar{\mathbf{e}}_j$ in terms of $\boldsymbol{\iota}_j$.
   (g) Express $\bar{\mathbf{r}}_j$ in terms of $\bar{\mathbf{r}}^k$ and conversely.
2. Repeat Problem 1 with respect to the transformation equations
   $X^1 = \rho \cosh \theta \cos \phi, \qquad X^2 = \rho \cosh \theta \sin \phi, \qquad X^3 = \rho \sinh \theta.$
   Also compute $\bar{g}_{jk}$ and $\bar{g}^{jk}$.
3. Show that relations (3-4.6c,d) imply (3-4.6e).
   Hint: Multiply and sum (3-4.6c) with a set $A_j{}^p$ satisfying the property $A_j{}^p(\partial \bar{X}^j/\partial \bar{\bar{X}}^k) = \delta_k{}^p$. Use (3-4.6d) to show that $A_j{}^k = \partial \bar{\bar{X}}^p/\partial \bar{X}^j$.
4. Generally speaking, little computation is done with the allowable transformations $\bar{X}^j = \bar{X}^j(\bar{\bar{X}}^k)$. Instead, the vector and tensor concepts are built around linear transformations of the form $\bar{U}^j = \partial \bar{X}^j/\partial \bar{\bar{X}}^k$, $\bar{\bar{U}}^k$ arising from the general group. Show that these linear transformations form a group.
5. Show that the orthogonal Cartesian group is a subgroup of the allowable group of Definition 3-4.1.
6. Prove (3-4.17b) (i.e., $\mathbf{r}^j = \bar{g}^{jk}\mathbf{r}_k$).

7. Prove (3-4.18b) (i.e., $\bar{U}^k = \bar{g}^{kj}\bar{U}_j$).

8. Prove (3-4.21b) (i.e., $\bar{g}^{jk} = (\partial\bar{X}^j/\partial\bar{\bar{X}}^p)(\partial\bar{X}^k/\partial X^q)\bar{\bar{g}}^{pq}$).

9. (a) Suppose that $\bar{\mathbf{r}}_1, \bar{\mathbf{r}}_2, \bar{\mathbf{r}}_3$ were an orthogonal basis. Show that the reciprocal basis is given by

$$\bar{\mathbf{r}}^1 = \frac{\bar{\mathbf{r}}_2 \times \bar{\mathbf{r}}_3}{\bar{\mathbf{r}}_1 \cdot \bar{\mathbf{r}}_2 \times \bar{\mathbf{r}}_3}, \quad \bar{\mathbf{r}}^2 = \frac{\bar{\mathbf{r}}_3 \times \bar{\mathbf{r}}_1}{\bar{\mathbf{r}}_1 \cdot \bar{\mathbf{r}}_2 \times \bar{\mathbf{r}}_3}, \quad \bar{\mathbf{r}}^3 = \frac{\bar{\mathbf{r}}_1 \times \bar{\mathbf{r}}_2}{\bar{\mathbf{r}}_1 \times \bar{\mathbf{r}}_2 \cdot \bar{\mathbf{r}}_3}.$$

(b) If the basis $\bar{\mathbf{e}}_1, \bar{\mathbf{e}}_2, \bar{\mathbf{e}}_3$ is a unit orthogonal basis, what is the value of $\bar{\mathbf{e}}_1 \cdot \bar{\mathbf{e}}_2 \times \bar{\mathbf{e}}_3$?

(c) Identify the quantity $\bar{\mathbf{r}}_1 \cdot \bar{\mathbf{r}}_2 \times \bar{\mathbf{r}}_3$.

10. Determine the transformation rule for the elements of the set $\{\delta_j{}^k\}$ where $\bar{\delta}_j{}^k = \bar{g}_{jp}\bar{g}^{pk}$.

11. Show that both $\bar{g}_{jk}$ and $\bar{g}^{jk}$ are symmetric in j, k.

12. If the basis $\bar{\mathbf{r}}_j$ is orthogonal, show that $\bar{g}^{xx} = 1/\bar{g}_{xx}$.

13. (a) Show that the form $\bar{U}^j\bar{\mathbf{r}}_j$ is an algebraic invariant.

(b) Show that the form $\bar{U}_j\bar{\mathbf{r}}^j$ is an algebraic invariant.

5. *Vector Concepts in Curvilinear Orthogonal Coordinate Systems*

Most geometrical and physical problems involve orthogonal coordinate systems. To employ other systems would lead us into involved computations, even on the theoretic level. For this reason the present section is devoted to expressing the concepts of gradient, divergence, and curl in terms of the orthogonal systems. The algebraic foundation established in Section 4 enables us to do the job in a relatively simple way. It also makes possible an approach that later can be extended easily to nonorthogonal systems.

If $\boldsymbol{\iota}_1, \boldsymbol{\iota}_2, \boldsymbol{\iota}_3$ is a unit orthogonal basis associated with a rectangular Cartesian coordinate system X^j, then

$$g_{jk} = \frac{\partial \mathbf{r}}{\partial X^j} \cdot \frac{\partial \mathbf{r}}{\partial X^k} = \boldsymbol{\iota}_j \cdot \boldsymbol{\iota}_k = \delta_{jk}. \tag{3-5.1a}$$

Since the reciprocal basis coincides with $\boldsymbol{\iota}_1, \boldsymbol{\iota}_2, \boldsymbol{\iota}_3$, it is also the case that

$$g^{jk} = \delta^{jk}. \tag{3-5.1b}$$

This fact is important in the next theorem. The transformation rules (3-4.21b) and (3-4.9b) also play a part.

Theorem 3-5.1. The physical components of the vector operator $\boldsymbol{\nabla}$ (in a curvilinear orthogonal coordinate system) are

$$\left(\frac{1}{\bar{g}_1}\frac{\partial}{\partial \bar{X}^1}, \frac{1}{\bar{g}_2}\frac{\partial}{\partial \bar{X}^2}, \frac{1}{\bar{g}_3}\frac{\partial}{\partial \bar{X}^3}\right).$$

Therefore

$$\nabla = \frac{1}{\bar{g}_1}\bar{\mathbf{e}}_1\frac{\partial}{\partial \bar{X}^1} + \frac{1}{\bar{g}_2}\bar{\mathbf{e}}_2\frac{\partial}{\partial \bar{X}^2} + \frac{1}{\bar{g}_3}\bar{\mathbf{e}}_3\frac{\partial}{\partial \bar{X}^3}. \tag{3-5.2}$$

PROOF. Starting with the rectangular Cartesian representation for ∇, we can simply transform to a curvilinear system; that is,

$$\begin{aligned}\nabla &= \sum_{j=1}^{3}\boldsymbol{\iota}_j\frac{\partial}{\partial X^j} = \delta^{jk}\boldsymbol{\iota}_j\frac{\partial}{\partial X^k} = g^{jk}\boldsymbol{\iota}_j\frac{\partial}{\partial X^k}\\ &= \left(\frac{\partial X^j}{\partial \bar{X}^p}\frac{\partial X^k}{\partial \bar{X}^q}\bar{g}^{pq}\right)\left(\frac{\partial \bar{X}^s}{\partial X^j}\frac{\partial \mathbf{r}}{\partial \bar{X}^s}\right)\left(\frac{\partial \bar{X}^t}{\partial X^k}\frac{\partial}{\partial \bar{X}^t}\right)\\ &= \delta_p^{\ s}\,\delta_q^{\ t}\bar{g}^{pq}\frac{\partial \mathbf{r}}{\partial \bar{X}^s}\frac{\partial}{\partial \bar{X}^t} = \bar{g}^{st}\frac{\partial \mathbf{r}}{\partial \bar{X}^s}\frac{\partial}{\partial \bar{X}^t};\end{aligned}$$

$(\bar{g}^{st})$ is diagonal, therefore, according to (3-4.24a) and the fact that $\bar{g}^{xx} = 1/\bar{g}_{xx}$,

$$\boxed{\nabla = \bar{g}^{st}\frac{\partial \mathbf{r}}{\partial \bar{X}^s}\frac{\partial}{\partial \bar{X}^t} = \sum_{j=1}^{3}\frac{1}{\bar{g}_j}\bar{\mathbf{e}}_j\frac{\partial}{\partial \bar{X}^j}.} \tag{3-5.3}$$

This completes the proof.

The reader should note that the second member of (3-5.3) has all the characteristics necessary for a definition of ∇ valid in any of the allowable coordinate systems. (See Chapter 3, Section 4, for a definition of an allowable coordinate system.)

Example 3-5.1. With respect to the cylindrical coordinate system of Examples 3-4.1 and 3-4.5, the covariant components of the fundamental metric tensor are

$$(\bar{g}_{jk}) = \begin{pmatrix}1 & 0 & 0\\ 0 & \rho^2 & 0\\ 0 & 0 & 1\end{pmatrix}. \tag{3-5.4a}$$

Therefore

$$\bar{g}_1 = 1, \qquad \bar{g}_2 = \rho, \qquad \bar{g}_3 = 1,$$

and

$$\nabla = \boldsymbol{\rho}\frac{\partial}{\partial \rho} + \frac{1}{\rho}\boldsymbol{\theta}\frac{\partial}{\partial \theta} + \mathbf{k}\frac{\partial}{\partial z} \tag{3-5.4b}$$

where

$$\bar{\mathbf{e}}_1 = \boldsymbol{\rho}, \qquad \bar{\mathbf{e}}_2 = \boldsymbol{\theta}, \qquad \bar{\mathbf{e}}_3 = \mathbf{k}.$$

The derivations of curvilinear coordinate representations for divergence and curl require knowledge of certain relationships between the Jacobian of transformation and the determinant of the covariant components of the fundamental metric tensor. These facts as well as an extension of the $\mathcal{E}$-system concept are now discussed.

$\mathcal{E}$ systems were introduced in Chapter 1, Section 6. The components of the respective systems transform according to covariant and contravariant laws with appropriate density factors. The major characteristic of these systems is that their components have values 1, -1, 0 in every allowable coordinate system. The next definition extends these systems to the generalized coordinates presently under discussion.

Definition 3-5.1. Let

$$(3\text{-}5.5)\qquad \bar{\mathcal{E}}_{jkq} = \bar{E}^{jkq} = \begin{cases} 1 & \\ -1 & \text{if } jkq \text{ is} \begin{cases}\text{an even permutation,}\\ \text{an odd permutation of } 1, 2, 3,\\ \text{otherwise.}\end{cases} \\ 0 & \end{cases}$$

Furthermore, let the components $\begin{cases}\bar{\bar{\mathcal{E}}}_{pst}\\ \bar{\bar{E}}^{pst}\end{cases}$ in any other general coordinate system be related to the components $\begin{cases}\bar{\mathcal{E}}_{jkq}\\ \bar{E}^{jkq}\end{cases}$ by means of the transformation rule

$$(3\text{-}5.6)\qquad \begin{aligned} &\text{(a)}\quad \bar{\mathcal{E}}_{jkq} = \left|\frac{\partial \bar{X}}{\partial \bar{\bar{X}}}\right| \frac{\partial \bar{\bar{X}}^p}{\partial \bar{X}^j}\frac{\partial \bar{\bar{X}}^s}{\partial \bar{X}^k}\frac{\partial \bar{\bar{X}}^t}{\partial \bar{X}^q}\bar{\bar{\mathcal{E}}}_{pst},\\ &\text{(b)}\quad \bar{E}^{jkq} = \left|\frac{\partial \bar{\bar{X}}}{\partial \bar{X}}\right| \frac{\partial \bar{X}^j}{\partial \bar{\bar{X}}^p}\frac{\partial \bar{X}^k}{\partial \bar{\bar{X}}^s}\frac{\partial \bar{X}^q}{\partial \bar{\bar{X}}^t}\bar{\bar{E}}^{pst}. \end{aligned}$$

The set $\{\bar{\mathcal{E}}_{jkq}\}$ is then said to be a tensor density of covariant order 3 and weight $+1$; $\{\bar{E}^{jkq}\}$ is said to be a tensor density of contravariant order 3 and weight -1.

The Jacobians $|\partial\bar{X}/\partial\bar{\bar{X}}|^{+1}$, $|\partial\bar{X}/\partial\bar{\bar{X}}|^{-1}$ represent the density factors. A comparison with the introduction of $\mathcal{E}$ systems in Chapter 1, Section 6, reveals that Definition 3-5.1 preserves the numerical values under transformation of coordinates; that is,

$$(3\text{-}5.7)\qquad \bar{\bar{\mathcal{E}}}_{pqr} = \bar{\mathcal{E}}_{pqr}, \qquad \bar{\bar{E}}^{pqr} = \bar{E}^{pqr}.$$

A proof of this fact based on Definition 3-5.1 is left to the reader. (See Problem 9.)

Now consider the relationship of the Jacobian and the determinant of the covariant components of the fundamental metric tensor.

Theorem 3-5.2. Let X^j represent a rectangular Cartesian coordinate system and $\bar{X}^j$ represent an allowable general coordinate system. Then

$$(3\text{-}5.8)\qquad \boxed{\left|\frac{\partial X}{\partial \bar{X}}\right| = \sqrt{\bar{g}}.}$$

PROOF. We have

$$\bar{g}_{jk} = \delta_{pq} \frac{\partial X^p}{\partial \bar{X}^j} \frac{\partial X^q}{\partial \bar{X}^k}. \tag{3-5.9a}$$

The right-hand side of (3-5.9a) can be thought of as two successive determinant multiplications[10]; that is, the multiplication $\delta_{pq}(\partial X^p/\partial \bar{X}^j)$ followed by a multiplication of the resultant determinant with $|\partial X^q/\partial \bar{X}^k|$. Hence

$$|\bar{g}_{jk}| = |\delta_{pq}| \left|\frac{\partial X^p}{\partial \bar{X}^j}\right| \left|\frac{\partial X^q}{\partial \bar{X}^k}\right| = \left|\frac{\partial X}{\partial \bar{X}}\right|^2. \tag{3-5.9b}$$

This completes the proof.

The next theorem represents a slight digression. Its usefulness is demonstrated in the sequel.

Theorem 3-5.3. Let $|d_j^{\ i}|$ be a determinant whose elements are independent functions defined in some region of space. Suppose that at least the first partial derivatives were continuous on the domain of definition. Then

$$\frac{\partial\, |d_j^{\ i}|}{\partial d_p^{\ q}} = D_q^{\ p}, \tag{3-5.10a}$$

where

$$D_q^{\ p} = \text{cofactor of } d_p^{\ q} \text{ in } |d|. \tag{3-5.10b}$$

PROOF. We have

$$\mathcal{E}_{rst}\, |d| = \mathcal{E}_{ijk}\, d_r^{\ i}\, d_s^{\ j}\, d_t^{\ k}. \tag{3-5.11a}$$

By partial differentiation of (3-5.11a)

$$\begin{aligned} \mathcal{E}_{rst} \frac{\partial\, |d|}{\partial d_p^{\ q}} &= \mathcal{E}_{ijk}(\delta_r^{\ p}\, \delta_q^{\ i}\, d_s^{\ j}\, d_t^{\ k} + d_r^{\ i}\, \delta_s^{\ p}\, \delta_q^{\ j}\, d_t^{\ k} + d_r^{\ i}\, d_s^{\ j}\, \delta_t^{\ p}\, \delta_q^{\ k}) \\ &= \delta_r^{\ p} \mathcal{E}_{qjk}\, d_s^{\ j}\, d_t^{\ k} + \delta_s^{\ p} \mathcal{E}_{iqk}\, d_r^{\ i}\, d_t^{\ k} + \delta_t^{\ p} \mathcal{E}_{ijq}\, d_r^{\ i}\, d_s^{\ j}. \end{aligned} \tag{3-5.11b}$$

According to Definition 1-6.5, a cofactor has the representation

$$\mathcal{E}_{vst}\, D_q^{\ v} = \mathcal{E}_{qjk}\, d_s^{\ j}\, d_t^{\ k}. \tag{3-5.11c}$$

Therefore by substituting into the last member of (3-5.11b) we have

$$\mathcal{E}_{rst} \frac{\partial\, |d|}{\partial d_p^{\ q}} = \delta_r^{\ p} \mathcal{E}_{vst}\, D_q^{\ v} - \delta_s^{\ p} \mathcal{E}_{vrt}\, D_q^{\ v} + \delta_t^{\ p} \mathcal{E}_{vrs}\, D_q^{\ v}. \tag{3-5.11d}$$

---

[10] In order to bring about row-by-column multiplication as indicated in the definition of determinant multiplication, it is sometimes necessary to consider the transpose of the matrix of determinant elements. This does not effect the numerical value of the product.

If we rewrite the right-hand member of (3-5.11d) with $D_q^{\ v}$ as a factor and multiply and sum with E^{rst}, then

$$(3\text{-}5.11\text{e}) \quad 3!\,\frac{\partial\,|d|}{\partial d_p^{\ q}} = (2\delta_r^{\ p}\,\delta_v^{\ r} + 2\delta_s^{\ p}\,\delta_v^{\ s} + 2\delta_t^{\ p}\,\delta_v^{\ t})\,D_q^{\ v} = 6D_q^{\ p}.$$

The proof is completed by dividing each side of (3-5.11e) by six.

The result (3-5.10a) is immediately employed in the next theorem. Also note that

$$\frac{\partial \bar{X}^k}{\partial X^j}\frac{\partial X^j}{\partial \bar{X}^p} = \delta_p^{\ k}.$$

Therefore

$$(3\text{-}5.12) \qquad \frac{\partial \bar{X}^k}{\partial X^j} = \frac{\text{cofactor of } \partial X^j/\partial \bar{X}^k \text{ in } |\partial X/\partial \bar{X}|}{|\partial X/\partial \bar{X}|}.$$

Theorem 3-5.4. Let X^j and $\bar{X}^j$ represent rectangular Cartesian and general coordinates, respectively. Then

$$(3\text{-}5.13) \qquad \frac{\partial \bar{X}^k}{\partial X^j}\frac{\partial^2 X^j}{\partial \bar{X}^k \partial \bar{X}^p} = \frac{1}{\sqrt{\bar{g}}}\frac{\partial \sqrt{\bar{g}}}{\partial \bar{X}^p}.$$

PROOF. According to (3-5.7)

$$\sqrt{\bar{g}} = \left|\frac{\partial X}{\partial \bar{X}}\right|.$$

By straightforward partial differentiation

$$(3\text{-}5.14\text{a}) \qquad \frac{\partial \sqrt{\bar{g}}}{\partial \bar{X}^k} = \frac{\partial\,|\partial X/\partial \bar{X}|}{\partial \bar{X}^k} = \frac{\partial\,|\partial X/\partial \bar{X}|}{\partial(\partial X^p/\partial \bar{X}^q)}\frac{\partial^2 X^p}{\partial \bar{X}^k \partial \bar{X}^q}.$$

According to (3-5.10a,b) and (3-5.12),

$$(3\text{-}5.14\text{b}) \qquad \frac{\partial\,|\partial X/\partial \bar{X}|}{\partial(\partial X^p/\partial \bar{X}^q)} = \left|\frac{\partial X}{\partial \bar{X}}\right|\frac{\partial \bar{X}^q}{\partial X_p}.$$

Again using (3-5.7), we obtain

$$(3\text{-}5.14\text{c}) \qquad \frac{\partial\,|\partial X/\partial \bar{X}|}{\partial(\partial X^p/\partial \bar{X}^q)} = \sqrt{\bar{g}}\,\frac{\partial \bar{X}^q}{\partial X^p}.$$

The proof is completed by substituting (3-5.14c) into (3-5.14a).

With the appropriate algebraic tools at hand, it becomes a simple matter to obtain curvilinear coordinate expressions for divergence and curl.

Theorem 3-5.5. Let **W** be a differentiable vector field. Then

$$(3\text{-}5.15) \qquad \boxed{\bar{\nabla}\cdot\bar{\mathbf{W}} = \frac{1}{\sqrt{\bar{g}}}\frac{\partial \sqrt{\bar{g}}\,\bar{W}^p}{\partial \bar{X}^p},}$$

where the $\bar{W}^p$ are contravariant components of the vector.

PROOF. The method of derivation consists of starting with the rectangular Cartesian expression for the divergence of **W** and then performing the transformation to a general system of curvilinear coordinates. We have

$$\nabla \cdot \mathbf{W} = \frac{\partial W^j}{\partial X^j} = \frac{\partial \bar{X}^k}{\partial X^j} \frac{\partial(\partial X^j/\partial \bar{X}^p)\bar{W}^p}{\partial \bar{X}^k} \tag{3-5.16a}$$

$$= \frac{\partial \bar{X}^k}{\partial X^j} \frac{\partial^2 X^j}{\partial \bar{X}^k \partial \bar{X}^p} \bar{W}^p + \frac{\partial X^j}{\partial \bar{X}^p} \frac{\partial \bar{W}^p}{\partial \bar{X}^k}$$

By employing (3-5.13) we can put this relation in the form

$$\bar{\nabla} \cdot \bar{\mathbf{W}} = \frac{1}{\sqrt{\bar{g}}} \left(\frac{\partial \sqrt{\bar{g}}}{\partial \bar{X}^p}\right) \bar{W}^p + \frac{\partial \bar{W}^p}{\partial \bar{X}^p} \tag{3-5.16b}$$

Careful examination shows that (3-5.16b) is a differentiated form of (3-5.15). This completes the proof.

If, in an orthogonal curvilinear system, the physical components of **W** are represented by $\bar{w}_j$, then, according to Definition 3-4.8,

$$\bar{W}^j = \frac{1}{\bar{g}_j} \bar{w}_j.$$

Therefore, in terms of physical components,

$$\bar{\nabla} \cdot \bar{\mathbf{W}} = \frac{1}{\sqrt{\bar{g}}} \sum_{p=1}^{3} \frac{\partial[(\sqrt{\bar{g}}/\bar{g}_p)\bar{w}_p]}{\partial \bar{X}^p} . \tag{3-5.17a}$$

In orthogonal systems

$$\sqrt{\bar{g}} = \bar{g}_1 \bar{g}_2 \bar{g}_3.$$

Therefore an equivalent representation for (3-5.17a) is

$$\boxed{\bar{\nabla} \cdot \bar{\mathbf{W}} = \frac{1}{\bar{g}_1 \bar{g}_2 \bar{g}_3} \left(\frac{\partial \bar{g}_2 \bar{g}_3 \bar{w}_1}{\partial \bar{X}^1} + \frac{\partial \bar{g}_1 \bar{g}_3 \bar{w}_2}{\partial \bar{X}^2} + \frac{\partial \bar{g}_1 \bar{g}_2 \bar{w}_3}{\partial \bar{X}^3}\right).} \tag{3-5.17b}$$

Example 3-5.2. Again considering the cylindrical coordinate system of the preceding example, in terms of physical components

$$\bar{\nabla} \cdot \bar{\mathbf{W}} = \frac{1}{\rho}\left(\frac{\partial \rho \bar{w}_1}{\partial \rho} + \frac{\partial \bar{w}_2}{\partial \theta} + \frac{\partial \rho \bar{w}_3}{\partial z}\right) \tag{3-5.18}$$

$$= \frac{\partial \bar{w}_1}{\partial \rho} + \frac{1}{\rho} \frac{\partial \bar{w}_2}{\partial \theta} + \frac{\partial \bar{w}_3}{\partial z} + \frac{1}{\rho} \bar{w}_1.$$

It is of some interest to note that the form $\partial W^j/\partial X^j$ is not preserved under general transformations [see (3-5.16a)]; hence it is not of scalar character. The question is asked whether the concept of divergence can be expressed in a way that gives it scalar character. It is answered under the heading of covariant differentiation. (See Chapter 4.)

Example 3-5.3 A special case of the divergence $\bar{\nabla} \cdot \bar{\mathbf{W}}$ is presented when $\bar{\mathbf{W}}$ is a gradient vector; that is

(3-5.19a) $$\bar{\mathbf{W}} = \bar{\nabla}\Phi.$$

In this example the physical components of $\bar{\mathbf{W}}$ are

(3-5.19b) $$\bar{w}_j = \frac{1}{\bar{g}^j}\frac{\partial \Phi}{\partial \bar{X}^j}. \qquad \text{[See (3-5.2).]}$$

Under this circumstance [substitute into (3-5.16b)]

(3-5.19c)
$$\bar{\nabla}\cdot\bar{\nabla}\,\Phi = \frac{1}{\bar{g}_1\bar{g}_2\bar{g}_3}\left[\frac{\partial\left(\dfrac{\bar{g}_2\bar{g}_3}{\bar{g}_1}\dfrac{\partial\Phi}{\partial\bar{X}^1}\right)}{\partial\bar{X}^1} + \frac{\partial\left(\dfrac{\bar{g}_1\bar{g}_3}{\bar{g}_2}\dfrac{\partial\Phi}{\partial\bar{X}^2}\right)}{\partial\bar{X}^2} + \frac{\partial\left(\dfrac{\bar{g}_1\bar{g}_2}{\bar{g}_3}\dfrac{\partial\Phi}{\partial\bar{X}^3}\right)}{\partial\bar{X}^3}\right]$$

This is the general orthogonal coordinate form of the Laplacian $\nabla^2\Phi$ mentioned in Chapter 3, Section 2. It is named for the French mathematician Pierre-Simon, Marquis de Laplace (1749–1827). The spherical and cylindrical forms of Laplace's partial differential equation

$$\nabla^2\Phi = 0$$

are often met in problems involving electric and magnetic phenomena and in other aspects of mathematical physics.

The algebraic groundwork preceding the introduction of the curvilinear representation for divergence also suffices for the development of the curl concept.

Theorem 3-5.6. Let $\bar{\mathbf{W}}$ be a differentiable vector field. Then, in an orthogonal system,

(3-5.20)
$$\bar{\nabla}\times\bar{\mathbf{W}} = \frac{1}{\sqrt{\bar{g}}}\begin{vmatrix} \bar{g}_1\bar{\mathbf{e}}_1 & \bar{g}_2\bar{\mathbf{e}}_2 & \bar{g}_3\bar{\mathbf{e}}_3 \\ \dfrac{\partial}{\partial\bar{X}^1} & \dfrac{\partial}{\partial\bar{X}^2} & \dfrac{\partial}{\partial\bar{X}^3} \\ \bar{g}_1\bar{w}_1 & \bar{g}_2\bar{w}_2 & \bar{g}_3\bar{w}_3 \end{vmatrix},$$

where $(\bar{w}_1, \bar{w}_2, \bar{w}_3)$ are the physical components of $\bar{\mathbf{W}}$.

PROOF. $\bar{\nabla}\times\mathbf{W}$ can be represented in the rectangular Cartesian form $\iota_j E^{jkp}\,\bar{\partial}_k W_p$ where the W_p are covariant components of $\mathbf{W}$ and

$$\bar{\partial}_k = \frac{\partial}{\partial\bar{X}^k}.$$

Then

(3-5.21a)

$$\begin{aligned}\boldsymbol{\iota}_j E^{jkp}\,\partial_k W_p &= \left(\frac{\partial \bar{X}^u}{\partial X^j}\frac{\partial \mathbf{r}}{\partial \bar{X}^u}\right)\bar{E}^{jkp}\frac{\partial \bar{X}^v}{\partial X^k}\frac{\partial}{\partial \bar{X}^t}\left(\frac{\partial \bar{X}^t}{\partial X^p}\,\bar{W}_v\right)\\ &= \frac{\partial \bar{X}^u}{\partial X^j}\frac{\partial \mathbf{r}}{\partial \bar{X}^u}\,\bar{E}^{jkp}\frac{\partial \bar{X}^t}{\partial X^k}\left(\frac{\partial^2 \bar{X}^v}{\partial X^q\,\partial X^p}\frac{\partial X^q}{\partial \bar{X}^t}\,W_v + \frac{\partial \bar{X}^v}{\partial X^p}\,\bar{\partial}_t \bar{W}_v\right)\\ &= \frac{\partial \mathbf{r}}{\partial \bar{X}^u}\left(\bar{E}^{jkp}\frac{\partial \bar{X}^u}{\partial X^j}\,\delta_k^{\ q}\,\frac{\partial^2 \bar{X}^v}{\partial X^q\,\partial X^p}\,\bar{W}_v + \bar{E}^{jkp}\frac{\partial \bar{X}^u}{\partial X^j}\frac{\partial \bar{X}^t}{\partial X^k}\frac{\partial \bar{X}^v}{\partial X^p}\,\bar{\partial}_t \bar{W}_v\right)\\ &= \frac{\partial \mathbf{r}}{\partial \bar{X}^u}\left(\bar{E}^{jkp}\frac{\partial \bar{X}^u}{\partial X^j}\frac{\partial^2 \bar{X}^v}{\partial X^k\,\partial X^p}\,\bar{W}_v + \bar{E}^{utv}\left|\frac{\partial \bar{X}}{\partial X}\right|\bar{\partial}_t \bar{W}_v\right).\end{aligned}$$

Because $\bar{E}^{jkp}$ is skew symmetric in the indices k and p, whereas $\partial^2 \bar{X}^v/\partial X^k\,\partial X^p$ is symmetric in k and p, the product and summation

$$\bar{E}^{jkp}\frac{\partial^2 \bar{X}^v}{\partial X^k\,\partial X^p} = 0.$$

Therefore (3-5.21a) reduces to (recall $|\partial \bar{X}/\partial X| = |\partial X/\partial \bar{X}|^{-1} = 1/\sqrt{\bar{g}}$.

(3-5.21b)
$$\boldsymbol{\iota}_j E^{jkp}\,\partial_k W_p = \frac{1}{\sqrt{\bar{g}}}\frac{\partial \mathbf{r}}{\partial \bar{X}^u}\,\bar{E}^{utv}\,\bar{\partial}_t \bar{W}_v.$$

Since

$$\frac{\partial \mathbf{r}}{\partial \bar{X}^j} = \bar{g}_j \bar{\mathbf{e}}_j, \qquad \bar{W}_v = \bar{g}_v \bar{w}_v,$$

it follows that (3-5.21b) is equivalent to (3-5.20).

Example 3-5.4. Again considering the cylindrical coordinate system of the preceding examples, we have

(3-5.22)

$$\begin{aligned}\bar{\nabla}\times\bar{\mathbf{W}} &= \frac{1}{\rho}\begin{vmatrix}\boldsymbol{\rho} & \rho\boldsymbol{\theta} & \mathbf{k}\\ \dfrac{\partial}{\partial \rho} & \dfrac{\partial}{\partial \theta} & \dfrac{\partial}{\partial z}\\ \bar{w}_1 & \rho\bar{w}_2 & \bar{w}_3\end{vmatrix}\\ &= \frac{1}{\rho}\left[\boldsymbol{\rho}\left(\frac{\partial \bar{w}_3}{\partial \theta} - \frac{\partial \rho\bar{w}_2}{\partial z}\right) + \rho\boldsymbol{\theta}\left(\frac{\partial \bar{w}_1}{\partial z} - \frac{\partial \bar{w}_3}{\partial \rho}\right) + \mathbf{k}\left(\frac{\partial \rho\bar{w}_2}{\partial \rho} - \frac{\partial \bar{w}_1}{\partial \theta}\right)\right].\end{aligned}$$

Problems

1. Show that with respect to the spherical system of Example 3-4.2 ∇ has the form

$$\bar{\nabla} = \bar{\mathbf{e}}_1\frac{\partial}{\partial r} + \bar{\mathbf{e}}_2\frac{1}{r}\frac{\partial}{\partial \theta} + \bar{\mathbf{e}}_3\frac{1}{r\sin\theta}\frac{\partial}{\partial \phi}.$$

2. Compute $\overline{\nabla}\Phi$ with respect to the cylindrical coordinate system of Example 3-4.1 if
   (a) $\Phi = \rho$,
   (b) $\Phi = \theta$,
   (c) $\Phi = \rho\theta$,
   (d) $\Phi = \rho \sin \theta$.
3. (a) Compute $\bar{g}$ for the cylindrical system of Example 3-4.1.
   (b) Compute $\bar{g}$ for the spherical system of Example 3-4.2.
4. Determine the form of $\overline{\nabla} \cdot \overline{\mathbf{W}}$ for the spherical coordinate system of Example 3-4.2.
5. For the cylindrical system of Example 3-4.1 compute $\overline{\nabla} \cdot \overline{\mathbf{W}}$ in each of the following:
   (a) $\overline{\mathbf{W}} = \boldsymbol{\rho}$,
   (b) $\overline{\mathbf{W}} = \boldsymbol{\theta}$,
   (c) $\overline{\mathbf{W}} = \sin\theta\boldsymbol{\rho} + \cos\theta\boldsymbol{\theta}$.
6. Show that for orthogonal curvilinear coordinate systems
$$\overline{\nabla}\bar{X}^1 \cdot \overline{\nabla}\bar{X}^2 \times \overline{\nabla}\bar{X}^3 = \left|\frac{\partial \bar{X}^j}{\partial X^k}\right|.$$
7. What is the form of $\overline{\nabla} \cdot \overline{\nabla}\Phi$ in the spherical coordinate system of Example 3-4.2?
8. What is the form of $\overline{\nabla} \times \overline{\mathbf{W}}$ in the spherical coordinate system of Example 3-4.2
9. Show that Definition 3-5.1 implies (3-5.7).
10. Show that the natural representation of $\overline{\nabla}\Phi$ is in terms of covariant components by showing that
$$\boldsymbol{\iota}^j\, \partial_j\Phi = \bar{\mathbf{r}}^p\, \bar{\partial}_p\Phi.$$

6. *Maxima and Minima of Functions of Two Variables*

The cylindrical coordinate system representation of ∇ offers an opportunity to handle certain maxima and minima problems in an appealing way. Since the method utilizes the results concerning maxima and minima of functions of one variable, it is worthwhile to review the fundamental facts of this theory.

Suppose that f were a differentiable function on a domain $a \leq X^1 \leq b$. Let c be an interior point of the closed interval $[a, b]$. Furthermore, assume that the second derivative of f exists at c. If $f'(c) = 0$, then c is called a critical value of f. The nature of f at a critical value is explored by examination of the second derivative. In particular,

(a)[11] if $f''(c) > 0$, f has a relative minimum at $X^1 = c$,

[11] f has a relative maximum at c is equivalent to the statement: there exists $\delta > 0$ such that $f(X^1) \leq f(c)$ whenever $|X^1 - c| < \delta$. A similar definition holds for relative minimum.

(b) if $f''(c) < 0$, f has a relative maximum at $X^1 = c$,

(c) if $f''(c) = 0$, no conclusion can be drawn.

When $f''(c) = 0$, f may have a relative maximum, a relative minimum, or $(c, f(c))$ may be coordinates of a point of inflection.

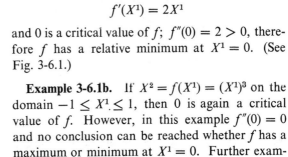

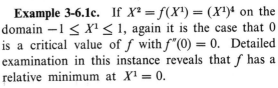

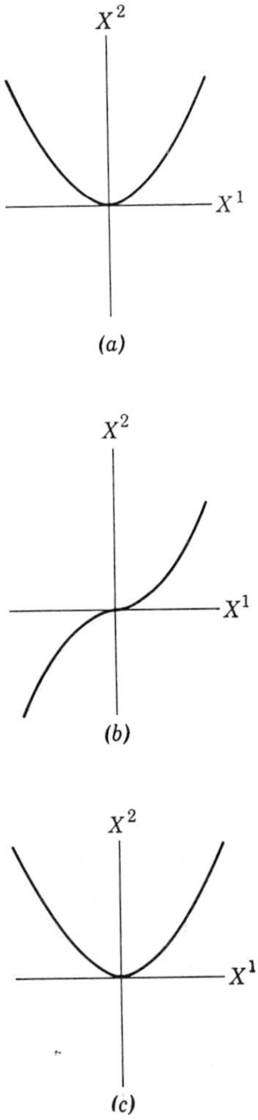

Fig. 3-6.1

Example 3-6.1a. If $X^2 = f(X^1) = (X^1)^2$ on the domain $-1 \leq X^1 \leq 1$, then

$$f'(X^1) = 2X^1$$

and 0 is a critical value of f; $f''(0) = 2 > 0$, therefore f has a relative minimum at $X^1 = 0$. (See Fig. 3-6.1.)

Example 3-6.1b. If $X^2 = f(X^1) = (X^1)^3$ on the domain $-1 \leq X^1 \leq 1$, then 0 is again a critical value of f. However, in this example $f''(0) = 0$ and no conclusion can be reached whether f has a maximum or minimum at $X^1 = 0$. Further examination would show that $(0, f(0))$ are coordinates of a point of inflection. [See Fig. 3-6.1b.)

Example 3-6.1c. If $X^2 = f(X^1) = (X^1)^4$ on the domain $-1 \leq X^1 \leq 1$, again it is the case that 0 is a critical value of f with $f''(0) = 0$. Detailed examination in this instance reveals that f has a relative minimum at $X^1 = 0$.

Having recalled the major results concerning relative maxima and minima of functions of one variable, let us consider functions of two variables. In the remarks that follow it is assumed that f has continuous first partial derivatives on a given plane region and that (X_0^1, X_0^2) are coordinates of an interior point of this region. Furthermore, it is assumed that the second partial derivatives of f exist at (X_0^1, X_0^2).

Definition 3-6.1. Let $f = f(X^1, X^2)$; f has a relative maximum at (X_0^1, X_0^2) means that there exist $\delta_1 > 0$ and $\delta_2 > 0$ such that $f(X^1, X^2) \leq f(X_0^1, X_0^2)$ whenever $|X^1 - X_0^1| < \delta_1$ and $|X^2 - X_0^2| < \delta_2$.

The definition of relative minimum of f is acquired by simply reversing the inequality relating $f(X^1, X^2)$ and $f(X_0^1, X_0^2)$.

Theorem 3-6.1. If f has a relative maximum at (X_0^1, X_0^2), then $\partial f/\partial X^1 = 0$, $\partial f/\partial X^2 = 0$ at (X_0^1, X_0^2).

PROOF. By definition,

$$\frac{\partial f(X_0^1, X_0^2)}{\partial X^1} = \lim_{\Delta X^1 \to 0} \frac{f(X_0^1 + \Delta X^1, X_0^2) - f(X_0^1, X_0^2)}{\Delta X^1} \tag{3-6.1a}$$

According to the definition of relative maximum,

$$f(X_0^1 + \Delta X^1, X_0^2) - f(X_0^1, X_0^2) \leq 0. \tag{3-6.1b}$$

Therefore the sign of the difference quotient involved in (3-6.1a) depends on ΔX^1. This means that the sign changes at (X_0^1, X_0^2). Since $\partial f/\partial X^1$ is continuous, this can happen only if $\partial f/\partial X^1 = 0$ at (X_0^1, X_0^2). The proof that $\partial f/\partial X^2 = 0$ at the given coordinate pair follows the same pattern.

It is important to realize that the converse to Theorem 3-6.1 does not hold. This fact is illustrated by subsequent examples.

Definition 3-6.2. If at (X_0^1, X_0^2)

$$\frac{\partial f}{\partial X^1} = 0, \qquad \frac{\partial f}{\partial X^2} = 0,$$

then (X_0^1, X_0^2) is called a critical coordinate pair of f, and the point with these coordinates is called a critical point of f.

Example 3-6.2. The theory of maxima and minima of functions of two variables is a great deal more complicated than the corresponding theory of one variable. For instance, consider

$$f(X^1, X^2) = 1 + (X^1)^2 - (X^2)^2$$

at (0, 0). We can write

$$X^3 = f(X^1, X^2) \tag{3-6.2a}$$

and interpret (3-6.2a) as a surface. (See Fig. 3-6.2.) We have

$$\begin{aligned} \frac{\partial f(X_0^1, 0)}{\partial X^1} &= (2X^1)_0 = 0, \\ \frac{\partial f(0, X^2)}{\partial X^2} &= -2(X^2)_0 = 0. \end{aligned} \tag{3-6.2b}$$

Furthermore,

$$\begin{aligned} \frac{\partial^2 f(X^1, 0)}{(\partial X^1)^2} &= 2, \\ \frac{\partial^2 f(0, X^2)}{\partial (X^2)^2} &= -2. \end{aligned} \tag{3-6.2c}$$

From the geometric point of view, putting $X^1 = 0$ corresponds to restricting the considerations to the coordinate plane $X^1 = 0$. Hence we think of studying the curve of intersection of $X^1 = 0$ and $X^3 = 1 + (X^1)^2 - (X^2)^2$

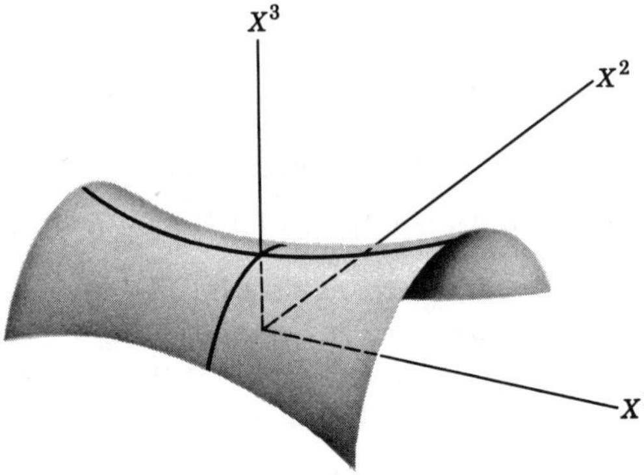

Fig. 3-6.2

in the plane $X^1 = 0$. From the first parts of the relations in (3-6.2b,c) it is seen that this curve has a relative minimum at $(0, 0, 1)$. On the other hand, the second parts of the relations in (3-6.2b,c) imply that the curve $X^2 = 0$, $X^3 = 1 + (X^1)^2 - (X^2)^2$ has a relative maximum at $(0, 0, 1)$. Therefore the function f has neither a relative minimum nor a relative maximum at $(0, 0)$. The critical point presented in this example is often called a saddle point.

The last example clearly indicates that $\partial f/\partial X^1 = \partial f/\partial X^2 = 0$ is a necessary condition that f have a relative maxima or relative minima, but it is not a sufficient condition. We now utilize cylindrical coordinates to find sufficient conditions.

The value of the cylindrical coordinate system in investigating maxima and minima resides in the fact that $\theta =$ constant represents a plane. (See Fig. 3-6.3.) If the critical point lies on the z axis, then for each value of θ we can investigate the curve $z = z(\rho)$, which is the trace of the surface on the plane $\theta =$ constant. These investigations can be carried out in terms of the criteria for a function of one variable.

If the critical point to be investigated does not lie on the z axis, a new coordinate system, obtained from the original by means of a translation, can be introduced.

It is important that the theoretical considerations made in the cylindrical coordinate system be set up in vector form, for the vector representation simplifies the transformation of the derived information into rectangular Cartesian coordinates. The equations of transformation take the form

$$\begin{aligned} X^1 &= X_0^1 + \rho \cos \theta, \\ X^2 &= X_0^2 + \rho \sin \theta, \qquad (3\text{-}6.3) \\ X^3 &= z, \end{aligned}$$

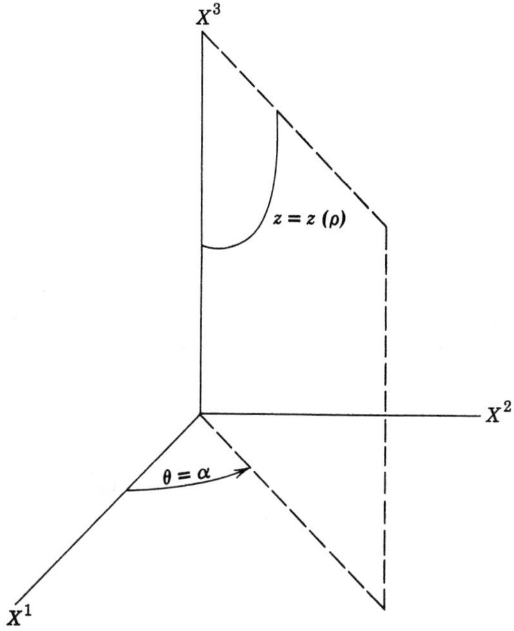

Fig. 3-6.3

where (X_0^1, X_0^2) is a critical coordinate pair of a function f. Note that the inclusion of the terms X_0^1 and X_0^2 simply provides for a translation such that the z axis passes through the point with coordinates (X_0^1, X_0^2) (i.e., $\rho = 0$, and any θ).

Theorem 3-6.2. Let f have continuous first partial derivatives on a given plane region and be such that the second partial derivatives exist at an interior point (X_0^1, X_0^2) of the region. Suppose that $\bar{f}(\rho, \theta) = f[X^1(\rho, \theta), X^2(\rho, \theta)]$. Then

(3-6.4)

$$\text{(a)} \quad \boldsymbol{\rho} \cdot \bar{\boldsymbol{\nabla}}\bar{f} = \frac{\partial \bar{f}}{\partial \rho},$$

$$\text{(b)} \quad \boldsymbol{\rho} \cdot \bar{\boldsymbol{\nabla}}(\boldsymbol{\rho} \cdot \bar{\boldsymbol{\nabla}}\bar{f}) = \frac{\partial^2 \bar{f}}{\partial \rho^2}.$$

PROOF. By expressing the gradient in cylindrical coordinates, we have

$$\bar{\boldsymbol{\nabla}}\bar{f} = \boldsymbol{\rho}\frac{\partial \bar{f}}{\partial \rho} + \frac{1}{\rho}\boldsymbol{\theta}\frac{\partial \bar{f}}{\partial \theta} + \mathbf{k}\frac{\partial \bar{f}}{\partial z}. \tag{3-6.5a}$$

Taking the dot product of (3-6.5a) with $\boldsymbol{\rho}$ produces the result

$$\boldsymbol{\rho} \cdot \overline{\nabla} \bar{f} = \frac{\partial \bar{f}}{\partial \rho}. \tag{3-6.5b}$$

This proves (3-6.4a).

By applying the operator $\overline{\nabla}$ to (3-6.5b), we have

$$\nabla(\boldsymbol{\rho} \cdot \overline{\nabla} \bar{f}) = \boldsymbol{\rho} \frac{\partial^2 \bar{f}}{\partial \rho^2} + \frac{1}{\rho} \boldsymbol{\theta} \frac{\partial^2 \bar{f}}{\partial \theta \, \partial \rho} + \mathbf{k} \frac{\partial^2 \bar{f}}{\partial z \, \partial \rho}. \tag{3-6.5c}$$

Therefore

$$\boldsymbol{\rho} \cdot \overline{\nabla}(\boldsymbol{\rho} \cdot \overline{\nabla} \bar{f}) = \frac{\partial^2 \bar{f}}{\partial \rho^2}. \tag{3-6.5d}$$

This completes the proof.

Quite clearly the criteria for maxima and minima of a function of one variable can be employed in the cylindrical case (with the additional understanding that it must hold for all θ). Therefore the next task is to determine the meaning of this information in terms of rectangular Cartesian representations.

Theorem 3-6.3a. At a domain value (X_0^1, X_0^2) of f, $\partial \bar{f}/\partial \rho = 0$ for all θ imples that $\partial f/\partial X^1 = 0$ and $\partial f/\partial X^2 = 0$.

PROOF. We have

$$\boldsymbol{\rho} = \frac{\partial \mathbf{r}}{\partial \bar{X}^1} = \frac{\partial \mathbf{r}}{\partial X^k} \frac{\partial X^k}{\partial \bar{X}^1} = \cos \theta \boldsymbol{\iota}_1 + \sin \theta \boldsymbol{\iota}_2,$$

$$\nabla f = \boldsymbol{\iota}_1 \frac{\partial f}{\partial X^1} + \boldsymbol{\iota}_2 \frac{\partial f}{\partial X^2} + \boldsymbol{\iota}_3 \frac{\partial f}{\partial X^3}.$$

Therefore

$$0 = \frac{\partial \bar{f}}{\partial \rho} = \boldsymbol{\rho} \cdot \overline{\nabla} \bar{f} = \boldsymbol{\rho} \cdot \nabla f = \cos \theta \frac{\partial f}{\partial X^1} + \sin \frac{\partial f}{\partial X^2}. \tag{3-6.6}$$

Suppose that $\partial f/\partial X^1$ and $\partial f/\partial X^2$ were not individually zero at (X_0^1, X_0^2). Then (3-6.6) determines a discrete set of values for θ. But this is contrary to the assumption that (3-6.6) holds for all θ. The contradiction implies the validity of the theorem.

Theorem 3-6.3b. At a domain value (X_0^1, X_0^2) of f at which the second partial derivatives are continuous,

$$\frac{\partial^2 \bar{f}}{\partial \rho^2} = \cos^2 \theta \frac{\partial^2 f}{(\partial X^1)^2} + 2 \cos \theta \sin \theta \frac{\partial^2 f}{\partial X^1 \, \partial X^2} + \sin^2 \theta \frac{\partial^2 f}{(\partial X^2)^2}. \tag{3-6.7}$$

PROOF. According to (3-6.4b),

$$\text{(3-6.8)}\quad \frac{\partial^2 \bar{f}}{\partial \rho^2} = \boldsymbol{\rho} \cdot \bar{\nabla}(\boldsymbol{\rho} \cdot \bar{\nabla} \bar{f}) = \boldsymbol{\rho} \cdot \nabla(\boldsymbol{\rho} \cdot \nabla f)$$

$$= (\cos\theta \boldsymbol{\iota}_1 + \sin\theta \boldsymbol{\iota}_2) \cdot \left[\boldsymbol{\iota}_1 \frac{\partial(\boldsymbol{\rho} \cdot \nabla f)}{\partial X^1} + \boldsymbol{\iota}_2 \frac{\partial(\boldsymbol{\rho} \cdot \nabla f)}{\partial X^2} + \boldsymbol{\iota}_3 \frac{\partial(\boldsymbol{\rho} \cdot \nabla f)}{\partial X^3} \right].$$

Now $\boldsymbol{\rho} \cdot \bar{\nabla} \bar{f} = \cos\theta\,(\partial f/\partial X^1) + \sin\theta\,(\partial f/\partial X^2)$ where θ is to be thought of as a fixed value (i.e., θ = constant represents the plane in which the investigation of maxima and minima takes place). Therefore the result (3-6.7) is obtained from (3-6.8) by straightforward calculation.

The theorems (3-6.3a,b) along with the theory of relative maxima and minima of one variable lead to the following conclusions:

If $\partial f/\partial X^1 = 0$, $\partial f/\partial X^2 = 0$ at (X_0^1, X_0^2) the second partial derivatives of f are continuous, and (3-6.7),

$$\begin{cases} < 0 \\ > 0 \\ = 0 \end{cases} \text{then} \begin{cases} f \text{ has a relative maximum at } (X_0^1, X_0^2), \\ f \text{ has a relative minimum at } (X_0^1, X_0^2), \\ \text{no conclusion can be reached.} \end{cases}$$

These results can be put in still better form. Let

$$A = \frac{\partial^2 f}{(\partial X^1)^2}, \qquad B = \frac{\partial^2 f}{\partial X^1\, \partial X^2}, \qquad C = \frac{\partial^2 f}{(\partial X^2)^2}.$$

Theorem 3-6.4. Suppose that f has continuous first derivatives on a plane region. Let (X_0^1, X_0^2) represent coordinates of an interior point of this region. Furthermore, suppose the second partials are continuous at (X_0^1, X_0^2). If (X_0^1, X_0^2) is a critical coordinate pair (see Definition 3-6.2), then

(3-6.9)

(a) $B^2 - AC < 0$, $\quad A < 0 \to f$ has a relative maximum at (X_0^1, X_0^2),

(b) $B^2 - AC < 0$, $\quad A > 0 \to f$ has a relative minimum at (X_0^1, X_0^2),

(c) $B^2 - AC > 0 \to f$ has neither a maximum nor a minimum at (X_0^1, X_0^2),

(d) $B^2 - AC = 0 \to$ no conclusion can be drawn.

PROOF [of (a)]. The right-hand member of (3-6.7) can be written in the form (whenever $\cos\theta \neq 0$)

$$\text{(3-6.10)}\qquad P(\theta) = \cos^2\theta[A + 2B\tan\theta + C\tan^2\theta],$$

where we think of $P(\theta)$ as a polynomial in θ. Suppose that (3-6.9a) holds and that $\cos\theta \neq 0$ ($\cos\theta = 0$ is considered separately). $B^2 - AC < 0$ guarantees that $P(\theta)$ will have no zeros, hence is positive for all θ or negative for all θ. Since $P(\theta) = A < 0$ and $P(\theta)$ does not change sign, we have $P(\theta) < 0$ for all θ (excluding θ such that $\cos\theta = 0$). Finally, if $\cos\theta = 0$, it follows from the right-hand member of (3-6.7) that

$$P(\theta) = C.$$

But C must have the same sign as A. Otherwise $B^2 - 4AC < 0$ could not hold. Since $P(\theta) < 0$ implies $\partial^2 f/\partial \rho^2 < 0$, it follows that f has a relative maximum at (X_0^1, X_0^2).

The proof of (3-6.9b) follows the same pattern. In (3-6.9c) the relation $B^2 - AC > 0$ implies that $P(\theta)$ has zeros. [See (3-6.10a).] Therefore the sign of $\partial^2 f/\partial \rho^2$ is dependent on θ, and there is neither a minimum nor a maximum at (X_0^1, X_0^2). If the relation (3-6.9d) holds, then there are values of θ for which $\partial^2 f/\partial \rho^2 = 0$. The theory of one variable indicates that in this instance no conclusion can be drawn.

Example 3-6.3. Suppose that an ellipsoidal surface were given according to the representation of Monge.

$$\text{(3-6.11a)} \quad \Phi(X^1, X^2, X^3) \equiv \frac{(X^1)^2}{4} + \frac{(X^2)^2}{9} + \frac{(X^3)^2}{16} - 1 = 0.$$

Two of the possible functions that can be constructed from Φ have rules:

$$\text{(3-6.11b)} \qquad f_1 \equiv X^3 = 4\left\{1 - \left[\frac{(X^1)^2}{4} + \frac{(X^2)^2}{9}\right]\right\}^{1/2},$$

$$\text{(3.6-11c)} \qquad f_2 \equiv X^3 = -4\left\{1 - \left[\frac{(X^1)^2}{4} + \frac{(X^2)^2}{9}\right]\right\}^{1/2}.$$

Let each of these functions have domain $[(X^1)^2/4] + [(X^2)^2/9] \leq 1$. Then, examining f_1, we have

$$\text{(3-6.11d)} \qquad \frac{\partial f_1}{\partial X^1} = -X^1\left\{1 - \left[\frac{(X^1)^2}{4} + \frac{(X^2)^2}{9}\right]\right\}^{-1/2}$$

$$\text{(3-6.11e)} \qquad \frac{\partial f_1}{\partial X^2} = -\frac{4}{9}X^2\left\{1 - \left[\frac{(X^1)^2}{4} + \frac{(X^2)^2}{9}\right]\right\}^{-1/2}.$$

The critical coordinate pair is (0, 0). This is an interior point of the set; therefore we proceed to investigate its nature.

$$\text{(3-6.11f)} \qquad \frac{\partial^2 f_1}{(\partial X^1)^2} = -\left\{1 - \left[\frac{(X^1)^2}{4} + \frac{(X^2)^2}{9}\right]\right\}^{-1/2} - \frac{1}{4}(X^1)^2\left\{1 - \left[\frac{(X^1)^2}{4} + \frac{(X^2)^2}{9}\right]\right\}^{-3/2}.$$

$$\text{(3-6.11g)} \qquad \frac{\partial^2 f_1}{\partial X^2\,\partial X^1} = -X^1X^2\left\{1 - \left[\frac{(X^1)^2}{4} + \frac{(X^2)^2}{9}\right]\right\}^{-3/2}.$$

$$\text{(3-6.11h)} \qquad \frac{\partial^2 f_1}{(\partial X^2)^2} = -\frac{4}{9}\left\{1 - \left[\frac{(X^1)^2}{4} + \frac{(X^2)^2}{9}\right]^{-1/2}\right. - \frac{4}{81}(X^2)^2\left\{1 - \left[\frac{(X^1)^2}{4} + \frac{(X^2)^2}{9}\right]\right\}^{-3/2}$$

By evaluating at (0, 0), we have

$$A = \left(\frac{\partial^2 f_1}{(\partial X^2)}\right)_{(0,0)} = -1,$$

(3-6.11i)
$$B = \left(\frac{\partial^2 f_1}{\partial X^2\,\partial X^1}\right)_{(0,0)} = 0,$$

$$C = \left(\frac{\partial^2 f_1}{(\partial X^2)^2}\right)_{(0,0)} = -\tfrac{4}{9}$$

Therefore

$$B^2 - AC = -\tfrac{4}{9} < 0, \qquad A = -1 < 0.$$

The function f_1 has a relative maximum at (0, 0). Similarly, it can be shown that f_2 has a relative minimum at (0, 0).

Problems

1. Find the relative maximum and/or relative minimum for the function with domain $(X^1)^2 + (X^2)^2 \leq 1$ and rule

$$f(X^1, X^2) = (X^1)^2 + (X^2)^2 - X^1.$$

2. Find the critical coordinate pairs for each of the following functions and test for maxima and minima:

(a) $f(X^1, X^2) = 4 - (X^1)^2 + (X^2)^2,\quad \begin{matrix} -\infty < X^1 < \infty, \\ -\infty < X^2 < \infty, \end{matrix}$

(b) $f(X^1, X^2) = 9 - (X^1)^2 - (X^2)^2,\quad \begin{matrix} -\infty < X^1 < \infty, \\ -\infty > X^2 < \infty. \end{matrix}$

chapter 4 integration of vectors

This chapter is concerned with the integration of scalar and vector fields over curves, surfaces, and space regions. It is assumed that the reader is familiar with the fundamental ideas of integration of continuous functions. To attempt to go into detail concerning the foundations of the integration concept would involve a long digression into the field of analysis. Therefore the discussion is restricted for the most part to extending the ideas of integration over line and plane region to integration over curve and surface element. Some attention is given to volume integrals. Naturally, the use of vector language is emphasized.

Whether the integration involves a curve, a surface, or a space region, it is defined in terms of a summation process. When vector fields are concerned, the use of a summation process presents a difficulty that is worthy of comment. Arithmetic addition of vectors depends on a composition of corresponding components. Since the discussion is restricted to euclidean space, an orthogonal basis of constant elements $\boldsymbol{\iota}_1, \boldsymbol{\iota}_2, \boldsymbol{\iota}_3$ always exists and therefore this addition can be carried out even though the vectors involved are not associated with the same point of space. (Note that the existence of the constant basis depends on the space being euclidean.) However, if a given integrand is expressed in terms of a non-constant basis, we cannot remove the basis elements inside the integration sign. Such variable elements present computational difficulties. The immediately important aspects of these remarks may be summarized. The various integrals to be discussed have meaning in Euclidean three-space regardless of the nature of the basis used. However, more often than not, actual evaluation of an integral will depend on a constant basis $\boldsymbol{\iota}_1, \boldsymbol{\iota}_2, \boldsymbol{\iota}_3$.

1. Line Integrals

In an elementary calculus course we consider the integral $\int_a^b f(x)\,dx$ where f is a continuous function defined on the closed interval $[a, b]$ of the line $y = 0$. (See Fig. 4-1.1.) The concept of line integral extends this idea by replacing f by a scalar function Φ or vector function $\mathbf{U}$ defined on a space curve $X^j = X^j(t)$, $t_0 \leq t \leq t_n$. The term line integral is a little misleading, since geometric emphasis is placed on a curve over which the integration takes place. "Curve integral" might be more appropriate. However, to avoid confusion, we shall stick to the standard terminology.

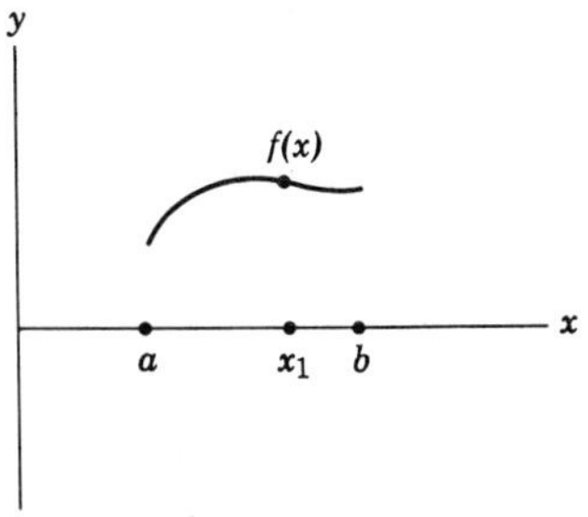

Fig. 4-1.1

Suppose that functions $X^j = X^j(t)$, with continuous derivatives that are not simultaneously zero on the domain $t_0 \leq t \leq t_n$, determine a space curve C. Then the concept of line integral can be expressed as follows.

Definition 4-1.1. Let $\Phi(X^j)$ be a continuous scalar field defined on C; then

$$\int_{t_0}^{t_n} \Phi[X^j(t)]\,dt = \lim_{|\Delta t| \to 0} \sum_{i=1}^{n} \Phi[X^j(\tau_i)]\,\Delta t_i \tag{4-1.1a}$$

is said to be a scalar line integral on C.

Definition 4-1.1b. Let $\mathbf{G}(X^j)$ be a continuous vector field on C; then

$$\int_{t_0}^{t_1} \mathbf{G}[X^j(t)]\,dt = \lim_{|\Delta t| \to 0} \sum_{i=1}^{n} \mathbf{G}[X^j(\tau_i)]\,\Delta t_i \tag{4-1.1b}$$

is said to be a vector line integral on C.

The symbolism used in (4-1.1a,b) is the usual. In particular,

$$\Delta t_i = t_i - t_{i-1},$$

and $|\Delta t| \to 0$ means that all subdivisions of the refinements $(t_0, \cdots, t_n)$ of the partition approaches zero as $n \to \infty$. The symbols τ_i represent intermediate values in the intervals Δt_i. (See Fig. 4-1.2.)

It may often be convenient to write $\int_C$ rather than $\int_{t_0}^{t_n}$.

For purposes of analysis we can think of the parameter t as representing the domain values of Φ and of $\mathbf{G}$. The continuity of Φ and of $\mathbf{G}$ implies

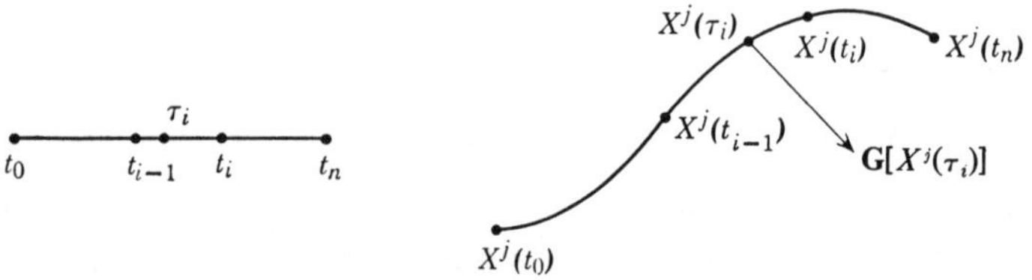

Fig. 4-1.2

the existence of the integrals. (This is an assumption on our part. The proof that continuity of a function on a closed interval is sufficient for the existence of the integral is usually given in an elementary calculus course.) The evaluation of the vector line integral is usually simplest when the basis elements are constant. For a basis $\boldsymbol{\iota}_1$, $\boldsymbol{\iota}_2$, $\boldsymbol{\iota}_3$ we have

$$\int_C \mathbf{G}\,dt = \lim_{|\Delta t| \to 0} \sum_{i=1}^{n} \mathbf{G}(\tau_i)\,\Delta t_i = \lim_{|\Delta t|} \sum_{i=1}^{n} G^j(\tau_i)\boldsymbol{\iota}_j\,\Delta t_i.$$

Because $\boldsymbol{\iota}_1$, $\boldsymbol{\iota}_2$, $\boldsymbol{\iota}_3$ are constant and therefore do not depend on the limit process,

$$\begin{aligned}\int_C \mathbf{G}\,dt &= \boldsymbol{\iota}_1 \lim_{|\Delta t| \to 0} \sum_{i=1}^{n} G^1(\tau_i)\,\Delta t_i + \boldsymbol{\iota}_2 \lim_{|\Delta t| \to 0} \sum_{i=1}^{n} G^2(\tau_i)\,\Delta t_i \\ &\qquad + \boldsymbol{\iota}_3 \lim_{|\Delta t| \to 0} \sum_{i=1}^{n} G^3(\tau_i)\,\Delta t \\ &= \boldsymbol{\iota}_1 \int_{t_0}^{t_n} G^1\,dt + \boldsymbol{\iota}_2 \int_{t_0}^{t_n} G^2\,dt + \boldsymbol{\iota}_3 \int_{t_0}^{t_n} G^3\,dt.\end{aligned}$$

Therefore the evaluation depends on a component-by-component integration. Difficulties that may exist are independent of the vector process.

For the most part we are concerned with the scalar line integral in which Φ has the form

$$\Phi = \mathbf{G} \cdot \frac{d\mathbf{r}}{dt}. \tag{4-1.2}$$

Therefore Φ is dependent on the curve. For this reason we demand that in general the curve be smooth. As defined in Chapter 2, Section 1, a smooth curve is characterized by continuous derivatives dX^1/dt, dX^2/dt, and dX^3/dt, which are not simultaneously zero. The geometric implication of these conditions is simply that the curve has a continuously turning tangent.

Our conditions on the curve are slightly relaxed by the following definition.

Definition 4-1.2a. A curve is said to be sectionally smooth if it is composed of a finite number of smooth curves attached sequentially end to end.

Definition 4-1.2b. The line integral over a sectionally smooth curve is the sum of the line integrals over the corresponding smooth pieces.

The preceding definitions make it possible to include such an intuitively simple curve as a broken line segment in our considerations.

Geometric interpretation is often facilitated by representing a curve in terms of a parameter s, which denotes arc length. The tangent vector field $d\mathbf{r}/ds$ is a unit field and the line integral

$$\int_C \frac{d\mathbf{r}}{ds} \cdot \frac{d\mathbf{r}}{ds}\, ds = L, \tag{4-1.3}$$

where L is the length of C. More generally, the scalar $\mathbf{G} \cdot (d\mathbf{r}/ds)$ can be interpreted as the projection (with appropriate sign) of $\mathbf{G}$ onto the tangential direction at each point of C. If arc length is not used in the parametric representation of the curve, then the preceding statement must be modified to include the magnitude of $d\mathbf{r}/dt$.

The $\int_C \mathbf{G} \cdot (d\mathbf{r}/dt)\, dt$ can be evaluated in at least three different ways. These methods are presented in the following examples.

Example 4-1.1. As previously indicated, the parametric equations

$$\begin{aligned} X^1 &= \cos t, \\ X^2 &= \sin t, \qquad t \geq 0, \\ X^3 &= t. \end{aligned} \tag{4-1.4a}$$

represent a circular helix. The position vector $\mathbf{r}$ has the form

$$r = \cos t\boldsymbol{\iota}_1 + \sin t\boldsymbol{\iota}_2 + t\boldsymbol{\iota}_3 \tag{4-1.4b}$$

and

$$\frac{d\mathbf{r}}{dt} = -\sin t\boldsymbol{\iota}_1 + \cos t\boldsymbol{\iota}_2 + \boldsymbol{\iota}_3. \tag{4-1.4c}$$

Suppose that a vector field

$$\mathbf{G} = \sin t\boldsymbol{\iota}_1 + \cos t\boldsymbol{\iota}_2 + \boldsymbol{\iota}_3 \tag{4-1.4d}$$

is defined along the curve; that is, $\mathbf{r}$ and $\mathbf{G}$ have a common domain such that if t_0 gives rise to $\mathbf{r}_0$ then $\mathbf{G}$ is associated with the curve point with coordinates X_0^j (i.e., with the end point of $\mathbf{r}_0$). Then, as a consequence of (4-1.4c,d),

$$\begin{aligned} \int_0^{t_0} \mathbf{G} \cdot \frac{d\mathbf{r}}{dt}\, dt &= \int_0^{t_0} (-\sin^2 t + \cos^2 t + 1)\, dt, \\ &= \int_0^{t_0} (1 + \cos 2t)\, dt = \left(t + \tfrac{1}{2}\sin 2t\right)\Big|_0^{t_0}, \\ &= t_0 + \frac{\sin 2t_0}{2}. \end{aligned}$$

In this example both the curve and the vector field are represented in terms of the parameter t and the integration was straightforward. The next example presents a variation from this method.

Example 4-1.2. The $\int_C \mathbf{G} \cdot (d\mathbf{r}/dt)\, dt$ also can be expressed in the form

$$\int_C \mathbf{G} \cdot d\mathbf{r} = \int_C G^1\, dX^1 + \int_C G^2\, dX^2 + \int_C G^3\, dX^3. \tag{4-1.5}$$

This representation depends on a change of the variable of integration from t to X^1, X^2, X^3, respectively. There are certain advantages to this procedure. For example, if the curve is given by

$$\begin{aligned} X^2 &= (X^1)^2, \\ X^3 &= 0, \end{aligned} \qquad 0 \leq X^1 \leq 2. \tag{4-1.6a}$$

and

$$\mathbf{G} = X^2\boldsymbol{\iota}_1 + (X^1)^2\boldsymbol{\iota}_2, \tag{4-1.6b}$$

then

$$\mathbf{r} = X^1\boldsymbol{\iota}_1 + (X^1)^2\boldsymbol{\iota}_2, \tag{4-1.6c}$$

and

$$d\mathbf{r} = dX^1\boldsymbol{\iota}_1 + 2X^1\, dX^1\boldsymbol{\iota}_2. \tag{4-1.6d}$$

(See Fig. 4-1.3a.) Therefore

$$\int_{(0,0)}^{(2,4)} \mathbf{G} \cdot d\mathbf{r} = \int_{(0,0)}^{(2,4)} (X^2\, dX^1 + 2(X^1)^3\, dX^1). \tag{4-1.6e}$$

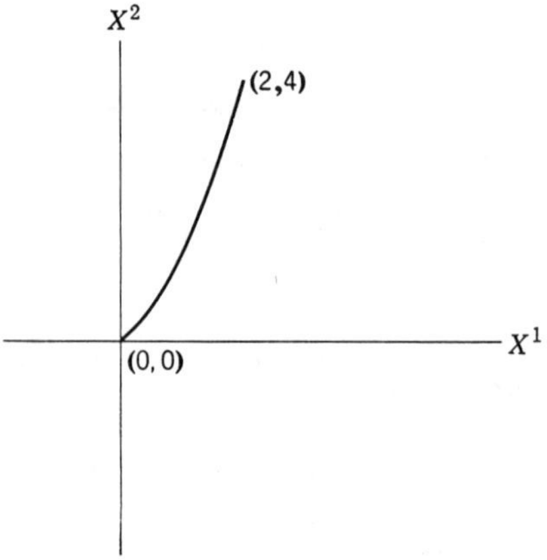

Fig. 4-1.3a

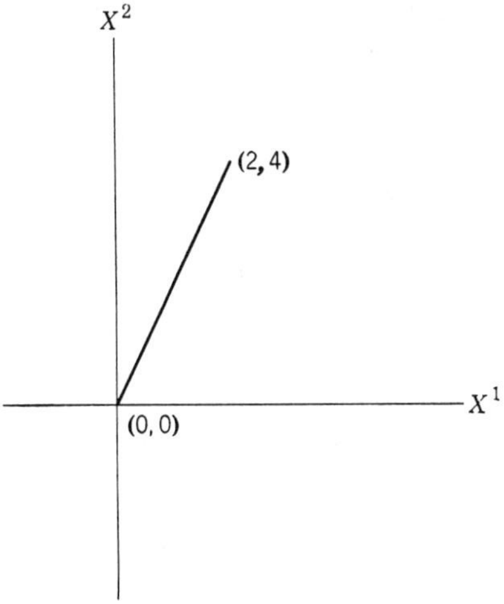

Fig. 4-1.3b

Again, using the curve representation, the right member of (4-1.6e) can be put entirely in terms of X^1 and dX^1; that is,

$$\int_{(0,0)}^{(2,4)} \mathbf{G} \cdot d\mathbf{r} = \int_0^2 [(X^1)^2 + 2(X^1)^3]\, dX^1 = \tfrac{8}{3} + 8 = \tfrac{32}{3}. \tag{4-1.6f}$$

Note that if the vector field $\mathbf{G}$ of this example is integrated over

$$\begin{aligned} X^2 &= 2X^1, \\ X^3 &= 0, \end{aligned} \tag{4-1.7a}$$

then

$$d\mathbf{r} = dX^1\boldsymbol{\iota}_1 + 2dX^1\boldsymbol{\iota}_2 \tag{4-1.7b}$$

and

$$\begin{aligned} \int_{(0,0)}^{(2,4)} \mathbf{G} \cdot d\mathbf{r} &= \int_{(0,0)}^{(2,4)} X^2\, dX^1 + 2(X^1)^2\, dX^1 = \int_0^2 (2X^1 + 2(X^1)^2)\, dX^1 \\ &= \left[(X^1)^2 + \frac{2(X^1)^3}{3}\right]_0^2 = 4 + \tfrac{16}{3} = \tfrac{28}{3}. \end{aligned} \tag{4-1.7c}$$

(See Fig. 4-1.3b.) Thus, comparing (4-1.6f) and (4-1.7c), it is clear that the value of the integral depends on the path joining (0, 0) and (2, 4).

The next example illustrates a case in which the path of integration is of no concern.

Example 4-1.3. Let

$$\mathbf{G} = X^2\boldsymbol{\iota}_1 + X^1\boldsymbol{\iota}_2. \tag{4-1.8a}$$

Consider any plane path given by differentiable equations which joins (0, 0) and (2, 8). For example,

(4-1.8b)
$$X^2 = (X^1)^3, \quad X^3 = 0, \qquad dX^2 = 3(X^1)^2\,dX^1,$$

or

(4-1.8c)
$$X^2 = 4X^1, \quad X^3 = 0, \qquad dX^2 = 4dX^1.$$

In (4-1.8b) and (4-1.8c), respectively, we obtain

(4-1.8d)
$$\int_{(0,0)}^{(2,8)} \mathbf{G}\cdot d\mathbf{r} = \int_{(0,0)}^{(2,8)} X^2\,dX^1 + 3(X^1)^3\,dX^1 = \int_{(0,0)}^{(2,8)} [(X^1)^3 + 3(X^1)^3]\,dX^1 = 16.$$

(4-1.8e)
$$\int_{(0,0)}^{(2,8)} \mathbf{G}\cdot d\mathbf{r} = \int_{(0,0)}^{(2,8)} X^2\,dX^1 + 4X^1\,dX^1 = \int_0^2 8X^1\,dX^1 = 16.$$

Choice of other paths joining (0, 0) and (2, 8) leads to this same numerical value of 16; therefore we are led to doubt the importance of any particular path in evaluating the integral. Closer examination shows that

$$\int_C \mathbf{G}\cdot d\mathbf{r} = \int_C X^2\,dX^1 + X^1\,dX^2 = \int_{(0,0)}^{(2,8)} d(X^2X^1) = 16;$$

that is, the integrand can be written as a perfect differential, and choice of path does not matter. This question is examined more fully later in the section.

Example 4-1.4. Consider the curve C whose parameterization is

$$\begin{array}{llll} X^1 = t, & X^2 = t^2, & X^3 = 0, & 0 \le t < 1 \\ X^1 = 1, & X^2 = 1, & X^3 = 1 - t, & 1 \le t < 2, \\ X^1 = 3 - t, & X^2 = 3 - t, & X^3 = 1, & 2 \le t \le 3. \end{array}$$

Let $\mathbf{G} = X^1X^2\boldsymbol{\iota}_1 + X^2\boldsymbol{\iota}_2 + X^3\boldsymbol{\iota}_3$. (See Fig. 4-1.4.) This curve is sectionally smooth. Therefore the line integral can be written as a sum of three line integrals.

$$\int_C \mathbf{G}\cdot\frac{d\mathbf{r}}{dt}\,dt = \int_0^1 (t^3 + 2t^3)\,dt + \int_1^2 -(1-t)\,dt + \int_2^3 [-(3-t)^2 - (3-t)]\,dt$$
$$= \tfrac{3}{4} + \tfrac{1}{2} - \tfrac{5}{6} = \tfrac{5}{12}.$$

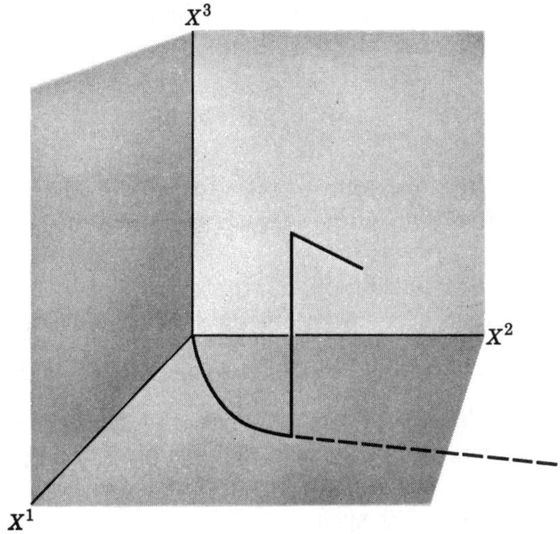

Fig. 4-1.4

The next example introduces an application of the concept of line integral to physics.

Example 4-1.5. In its most elementary form the concept of work[1] is defined as the product of a constant force by the straight line distance over which it acts. Thus, if F represents the measure of the constant force, Δr, the distance, and W symbolizes the measure of work, then

(4-1.9a) $$W = F\,\Delta r.$$

A slight generalization of (4-1.9a) consists in representing the constant force by means of a vector $\mathbf{F}$ and allowing the vector to have a direction different from that of the line segment over which it acts. Then

(4-1.9b) $$W = F\,\Delta r \cos\theta;$$

that is, the component of force in the direction of action is multiplied by the distance through which it acts.

These two elementary definitions of work (the second includes the first) were supplemented after the introduction of the integral calculus by

(4-1.9a) $$W = \int_{x_0}^{x_1} F(x)\,dx,$$

where the force is variable along the line of action.

[1] The term work was used by G. G. Coriolis (1792–1843), and later by J. V. Poncelet (1789–1867, French). These men were among the first to promote reforms in the teaching of rational mechanics. (See Florian Cajori, *A History of Physics*, Dover, 1962, p. 59.)

All these representations are included in the more general definition given in terms of the line integral; that is,

$$W = \int_C \mathbf{F} \cdot d\mathbf{r}. \tag{4-1.9d}$$

There will be further occasion to refer to (4-1.9d) after a more detailed development of certain ideas associated with the concept of line integral.

The next theorem introduces a property of the line integral that is useful in applications in geometry and physics. The function $\mathbf{G}$ of Example 4-1.3 led to a line integral $\int_C \mathbf{G} \cdot d\mathbf{r}$, the value of which did not depend on the path C. The theorem gives a first answer to the question of what form $\mathbf{G}$ must take in general in order that the line integral may be independent of path. Independence of path means that there is a scalar function, which we shall call $W(X^j)$, such that

$$W(X^j) = \int_{P_0(X_0{}^j)}^{P(X^j)} \mathbf{G} \cdot d\mathbf{r}. \tag{4-1.10}$$

In other words, the value of the integral depends only on the end points P_0 and P, where P is variable and P_0 is fixed.

Theorem 4-1.1. Let $\mathbf{G}$ be a vector field that is continuous on R, a preassigned region of space. Suppose that C were any smooth curve lying in R. Denote the end points of C by P_0 and P_1. Then

$$\left\{\begin{array}{l}\text{the value of } \int_{P_0}^{P} \mathbf{G} \cdot d\mathbf{r} \text{ is in-}\\ \text{dependent of the path } C\\ \text{joining } P_0 \text{ and } P\end{array}\right\} \leftrightarrow \left\{\begin{array}{l}\text{there exists a differentiable}\\ \text{scalar field } \Phi \text{ such that}\\ \mathbf{G} = \nabla\Phi.\end{array}\right\} \tag{4-1.11}$$

PROOF. Assume that $\mathbf{G}$ can be represented as the gradient of a differentiable scalar field Φ. Because Φ is differentiable, a change of variables from X^1, X^2, X^3 to Φ is possible. This change takes the form of the chain rule of differentiation and $\nabla\Phi \cdot d\mathbf{r}$ is replaced by $d\Phi$. Thus

(4-1.12a)

$$\int_C \mathbf{G} \cdot d\mathbf{r} = \int_C \nabla\Phi \cdot d\mathbf{r} = \int_{P_0(X_0{}^j)}^{P_1(X_1{}^j)} d\Phi = \Phi(X_1{}^j) - \Phi(X_0{}^j).$$

It follows that $\int_C \mathbf{G} \cdot d\mathbf{r}$ depends on the coordinates $X_0{}^j$ and $X_1{}^j$ of the end points of C but does not depend on C.

Conversely, assume that the line integral is independent of path. From (4-1.10) we can derive the expression

(4-1.12b)

$$\frac{W(X^1+\Delta X^1, X^2, X^3) - W(X^1, X^2, X^3)}{\Delta X^1}$$
$$= \frac{1}{\Delta X^1}\left(\int_{P_0}^{X+\Delta X^1, X^2, X^3} \mathbf{G}\cdot d\mathbf{r} - \int_{P_0}^{X^1, X^2, X^3} \mathbf{G}\cdot d\mathbf{r}\right)$$
$$= \frac{1}{\Delta X^1}\int_{X^1, X^2\, X^3}^{X^1+\Delta X^1, X,^2 X^3} \mathbf{G}\cdot d\mathbf{r}.$$

The path determined by the limits of integration is such that $X^2 =$ constant and $X^3 =$ constant; therefore

$$\frac{W(X^1+\Delta X^1, X^2, X^3) - W(X^1, X^2, X^3)}{\Delta X^1} = \frac{1}{\Delta X^1}\int_{X^1, X^2, X^3}^{X^1+\Delta X^1, X^2, X^3} G^1\, dX^1.$$

According to the mean value theorem of integral calculus, there exists Z^1 in $(X^1, X^1 + \Delta X^1)$ such that

$$\frac{1}{\Delta X^1}\int_{X^1, X^2, X^3}^{X^1+\Delta X^1, X^2, X^3} G^1\, dX^1 = G^1(Z^1, X^2, X^3).$$

In turn, $G^1(Z^1, X^2, X^3)$ can be replaced by $G^1(X^1, X^2, X^3) + \epsilon^1$ for a sufficiently small value of ΔX^1 because G^1 is a continuous function. Therefore

(4-1.12c)
$$\frac{\partial W}{\partial X^1} = \lim_{|\Delta X^1|\to 0} \frac{W(X^1+\Delta X^1, X^2, X^3) - W(X^1, X^2, X^3)}{\Delta X^1}$$
$$= \lim_{\epsilon^1\to 0} G^1(X^1, X^2, X^3) + \epsilon^1 = G^1.$$

It can be shown in a similar manner that

$$\frac{\partial W}{\partial X^2} = G^2, \qquad \frac{\partial W}{\partial X^3} = G^3.$$

Clearly, the function W is the required scalar field and $\mathbf{G}$ can be expressed as a gradient. This completes the proof.

Example 4-1.6. Suppose that a vector field

$$\mathbf{G} = (X^2X^3 + 2)\boldsymbol{\iota}_1 + (X^1X^3 + 5)\boldsymbol{\iota}_2 + X^1X^2\boldsymbol{\iota}_3$$

is given. If the vector field is representable as a gradient field, an appropriate scalar function Φ can be found by making the identification

$$\mathbf{G} = \nabla\Phi.$$

Then

$$\frac{\partial\Phi}{\partial X^1} = X^2X^3 + 2, \qquad \frac{\partial\Phi}{\partial X^2} = X^1X^3 + 5, \qquad \frac{\partial\Phi}{\partial X^3} = X^1X^2.$$

From the first of these relations we find on integrating that

$$\Phi = X^1X^2X^3 + 2X^1 + f(X^2, X^3).$$

Therefore

$$X^1X^3 + 5 = \frac{\partial\Phi}{\partial X^2} = X^1X^3 + \frac{\partial f}{\partial X^2}$$

From this relation we ascertain that

$$f(X^2, X^3) = 5X^2 + g(X^3);$$

thus

$$\Phi = X^1X^2X^3 + 2X^1 + 5X^2 + g.$$

The given expression for $\partial\Phi/\partial X^3$ leads to the result

$$X^1X^2 = \frac{\partial\Phi}{\partial X^3} = X^1X^2 + \frac{\partial g(X^3)}{\partial X^3}.$$

Consequently

$$g(X^3) = \text{constant}.$$

By the foregoing process of partial integration we find that

$$\Phi = X^1X^2X^3 + 2X^1 + 5X^2 + \text{constant}.$$

The flaw in the method of this example is that it depends on **G** being a gradient field. If **G** is not a gradient vector field, the method will fail, but doubt may linger in our minds whether the failure was actually owing to the nonexistence of a function Φ or to personal inadequacy. Fortunately, a more powerful test will become available later in the book.

In the next theorem the symbol $\oint$ is used to indicate integration over a closed path; that is, the initial and end points of the path coincide.

Theorem 4-1.2. Let **G** be continuous on a region R. Suppose P_0 and P_1 are points in R and both C_1 and C_2 are smooth paths in R joining these points. Then

$$\text{(4-1.13)} \qquad \left\{\begin{matrix}\text{The line integral } \int_{P_0}^{P_1} \mathbf{G}\cdot d\mathbf{r} \\ \text{is independent of the path} \\ \text{joining } P_0 \text{ to } P_1.\end{matrix}\right\} \leftrightarrow \oint \mathbf{G}\cdot d\mathbf{r} = 0.$$

PROOF. To prove the equivalence of (4-1.13), start with the implication from left to right. According to the preceding theorem, $\int_C \mathbf{G}\cdot d\mathbf{r}$ independent of path implies that there is a differentiable function Φ such that $\mathbf{G} = \boldsymbol{\nabla}\Phi$. Therefore

$$\oint_C \mathbf{G} \cdot d\mathbf{r} = \int_{C_1} \mathbf{G} \cdot d\mathbf{r} + \int_{C_2} \mathbf{G} \cdot d\mathbf{r} = \int_{P_0}^{P_1} \mathbf{G} \cdot d\mathbf{r} + \int_{P_1}^{P_0} \mathbf{G} \cdot d\mathbf{r}$$

$$= \int_{P_0}^{P_1} d\Phi + \int_{P_1}^{P_0} d\Phi$$

$$= \int_{P_0}^{P_1} d\Phi - \int_{P_0}^{P_1} d\Phi = 0.$$

Conversely, if $\oint \mathbf{G} \cdot d\mathbf{r} = 0$, then

$$\oint \mathbf{G} \cdot d\mathbf{r} = \int_{C_1} \mathbf{G} \cdot d\mathbf{r} + \int_{C_2} \mathbf{G} \cdot d\mathbf{r} = 0.$$

Consequently,

(4-1.14a) $$\int_{C_1} \mathbf{G} \cdot d\mathbf{r} = -\int_{C_2} \mathbf{G} \cdot d\mathbf{r}.$$

The relation (4-1.14a) can also be written in the form

(4-1.14b) $$\int_{C_1} \mathbf{G} \cdot d\mathbf{r} = \int_{-C_2} \mathbf{G} \cdot d\mathbf{r},$$

where $-C_2$ is oppositely oriented to C_2. Therefore the integration over path C_1 or $-C_2$ (from P_0 to P_1, see Fig. 4-1.5) produces the same result. Since C_1 and $-C_2$ are arbitrary curves joining P_0 to P_1 (they must meet the restrictions of the hypothesis), this completes the proof.

Example 4-1.7. Disregarding part of the hypothesis of a theorem can have disastrous results. The function

$$\Phi = \tan^{-1}\left(\frac{X^2}{X^1}\right)$$

provides an opportunity for such neglect. This function can be expressed in the form

$$\mathbf{G} = \boldsymbol{\nabla}\Phi = \frac{-X^2\boldsymbol{\iota}_1 + X^1\boldsymbol{\iota}_2}{(X^1)^2 + (X^2)^2}.$$

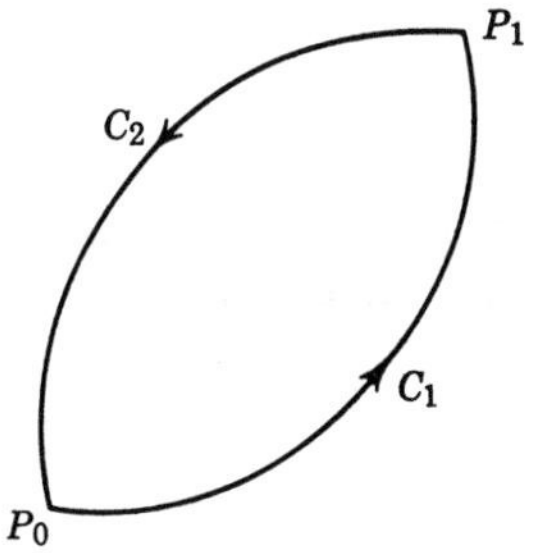

Fig. 4-1.5

Therefore we might expect the line integral around a closed path to have the value zero. However, consideration of the circle

$$X^1 = \cos\theta, \qquad X^2 = \sin\theta, \qquad X^3 = 0$$

leads to the result

$$\oint_C \mathbf{G} \cdot d\mathbf{r} = \int_0^{2\pi} \frac{\cos^2\theta + \sin^2\theta}{1} = \int_0^{2\pi} d\theta = 2\pi.$$

A careful analysis of Φ solves the dilemma. The function Φ is not defined for $X^1 = 0$, that is, anywhere along the X^2 axis. Therefore the statement on the right-hand side of (4-1.11) is not satisfied and the conclusion cannot be drawn.

Fields of force which can be represented as the gradient of a scalar field (i.e., by $\mathbf{F} = -\nabla\Phi$, where the minus sign is introduced for future convenience) play an important role in mathematical physics. Such fields of force are called conservative and the scalar function is a potential function. Among the physical phenomena that can be so represented are gravitational, electrostatic, and magnetostatic force fields. It follows at once from Theorem 4-1.1 that the work done by a conservative field of force in moving an object from P_0 to P_1 does not depend on the path of motion.

Classical Newtonian mechanics makes much use of the concept of energy. It is assumed that there are two kinds of energy, potential (a measure of the capacity to do work expressed by means of the potential function) and kinetic (defined as $\frac{1}{2}mV^2$, where m is the mass of the particle and V is the magnitude of its instantaneous velocity). The next example illustrates that for conservative force fields the total energy (i.e., the sum of potential and kinetic energy) does not vary. This is one of the three famous conservation principles of classical mechanics.

Example 4-1.8. In Newtonian mechanics the force field associated with a given path of motion is described in terms of the rate of change of the momentum field; that is,

$$\mathbf{F} = \frac{dm\mathbf{V}}{dt}, \tag{4-1.15a}$$

where m represents mass (which is constant according to Newtonian mechanics) and $\mathbf{V}$ represents the velocity vector field along the path of motion.

When $\mathbf{F}$ is conservative (i.e., $\mathbf{F} = -\nabla\Phi$),

(4-1.15b)

$$\int_{P_0}^{P} \mathbf{F} \cdot d\mathbf{r} = \int_{P_0}^{P} -\nabla\Phi \cdot d\mathbf{r} = -\int_{P_0}^{P} d\Phi = \Phi(P_0) - \Phi(P).$$

From the point of view of relation (4-1.15a),

$$\int_{P_0}^{P} \mathbf{F} \cdot d\mathbf{r} = \int_{t_0}^{t} \frac{dm\mathbf{V}}{dt} \cdot \frac{d\mathbf{r}}{dt}\, dt = \frac{m}{2} \int_{t_0}^{t} \frac{d\mathbf{V} \cdot \mathbf{V}}{dt}\, dt \tag{4-1.15c}$$

$$= \frac{m}{2} \int dV^2 = \frac{m}{2} [V^2(P) - V^2(P_0)].$$

Equating the results in (4-1.15b) and (4-1.15c) we obtain

(4-1.15d) $$\Phi(P_0) + \frac{m}{2} V^2(P_0) = \Phi(P) + \frac{m}{2} V^2(P).$$

Therefore the total energy is the same at any point P (or equivalently at any time t) as it was at the initial position of the particle. Hence it is said to be conserved.

Example 4-1.9. As a consequence of the conservation of energy principle, illustrated in the preceding example, it is possible to predict the initial velocity needed by a particle in order that it may escape from the earth's gravitational field.

It is assumed that as the distance from the earth's surface $\rightarrow \infty$, the total energy tends toward being all potential energy. On the other hand, at the surface of the earth the energy is all kinetic.

According to Newton's universal law of gravitation,

(4-1.16a) $$\mathbf{F} = \frac{\gamma m M \mathbf{r}}{r^3} = -m\,\nabla\Phi,$$

where

(4-1.6b) $$\Phi = \frac{\gamma M}{r},$$

R = radius of earth,
M = mass of earth,
m = mass of body,
γ = gravitational constant.

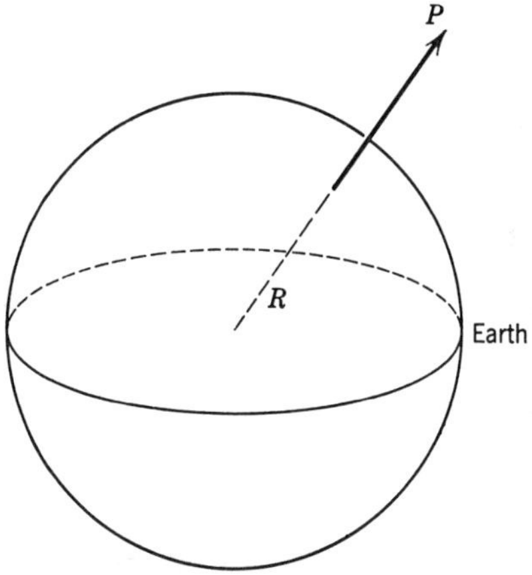

Fig. 4-1.6

Since work is the negative of potential energy, the fact that the total energy at the earth's surface (i.e., all kinetic energy) equals the total energy at ∞ (i.e., all potential energy) produces the result

$$\tfrac{1}{2}mV^2 = \lim_{P\to\infty}\int_R^P \mathbf{F}\cdot d\mathbf{r} = \lim_{P\to\infty} -m\int_R^P \nabla\Phi\cdot d\mathbf{r} = \lim_{P\to\infty} -m\int_R^P d\Phi$$

$$= -\lim_{P\to\infty} \frac{m\gamma M}{r}\bigg]_R^P = -\lim_{P\to\infty}\left(\frac{m\gamma M}{P} - \frac{m\gamma M}{R}\right) = \frac{m\gamma M}{R}.$$

Therefore

$$V = \sqrt{\frac{2\gamma M}{R}}. \tag{4-1.16c}$$

Using the values

$$R = (6.37)10^8 \text{ cm}, \qquad \gamma = (6.67)10^{-8} \text{ dyne/gm}^2\text{ cm}^2,$$
$$M = (6.1)10^{27} \text{ gm},$$

we find that

$$V = (1.13)10^6 \text{ cm/sec},$$

or about 7 mi./sec.

Problems

1. Find the values of the following line integrals:

(a) $\displaystyle\int_c \frac{d\mathbf{r}}{dt}\cdot\frac{d\mathbf{r}}{dt}\,dt \begin{cases} X^1 = a\cos t, \\ X^2 = a\sin t, \\ X^3 = 0. \end{cases} \quad 0 \le t \le 2.$

(b) $\displaystyle\int_C \frac{d\mathbf{r}}{ds}\cdot\frac{d\mathbf{r}}{ds}\,ds \begin{cases} X^1 = a\cos s/a, \\ X^2 = a\sin s/a, \\ X^3 = 0. \end{cases} \quad 0 \le s \le 2\pi.$

2. Find the value of $\displaystyle\int_C \mathbf{G}\cdot(d\mathbf{r}/dt)\,dt$ in each of the following cases:

(a) $\mathbf{G} = -X^2\iota_1 + X^1\iota_2 \begin{cases} X^1 = t^2, \\ X^2 = 2t, \end{cases} \quad 0 \le t \le 2.$

(b) $\mathbf{G} = (X^2)^2\iota_1 + (X^1)^2\iota_2, \quad X^1 = \sqrt{1-(X^2)^2},$ from $(0, -1)$ to $(0, 1)$.

(c) $\mathbf{G} = -\sin t\,\iota_1 + \cos t\,\iota_2 + \iota_3 \quad \begin{cases} X^1 = \cos t, \\ X^2 = \sin t, \\ X^3 = t, \end{cases} \quad 0 \le t \le \pi.$

(d) $\mathbf{G} = -5X^2\iota_1 + 3X^1\iota_2 + 2X^3\iota_3 \begin{cases} X^1 = t, \\ X^2 = \sqrt{1-t^2}, \\ X^3 = 1, \end{cases} \quad -1 \le t \le 1.$

3. If $\mathbf{G} = 2X^1X^2\iota_1 + (X^1)^2\iota_2$, evaluate $\int_C \mathbf{G} \cdot d\mathbf{r}$ over each of the following paths:

(a) $X^2 = \sqrt{1 + (X^1)^2}$, $X^3 = 0$, from $(0, 1)$ to $(2, \sqrt{5})$.

(b) $X^2 = \dfrac{\sqrt{5} - 1}{2} X^1 + 1$, $X^3 = 0$, from $(0, 0)$ to $(2, \sqrt{5} - 1)$.

(c) The line path:

$$\begin{aligned} X^1 &= t, & X^1 &= 2, \\ X^2 &= 0, \ 0 \le t \le 2, \qquad & X^2 &= t - 2, \qquad 2 \le t \le \sqrt{5} + 2. \\ X^3 &= 0, & X^3 &= 0, \end{aligned}$$

(d) Is $\int_C \mathbf{G} \cdot d\mathbf{r}$ independent of path? Prove that your answer is valid.

4. Suppose that for all t a vector force field $\mathbf{F}$ is perpendicular to the tangent vector $d\mathbf{r}/dt$ along the path of motion. Show that the work done by $\mathbf{F}$ is zero.

5. Suppose that a force vector field $\mathbf{F} = \mathbf{r}$ is defined on

$$\begin{aligned} X^1 &= h + a\cos\theta, \\ X^2 &= k + a\sin\theta, \qquad 0 \le \theta \le \pi. \\ X^3 &= 0, \end{aligned}$$

(a) Show that the work done is $-2ah$.
(b) What is the work done over the interval $0 \le \theta \le 2\pi$?
(c) Let $k = 0$, $h = a$. Then calculate

$$\int_0^\pi \mathbf{G} \cdot \frac{d\mathbf{r}}{d\theta}\, d\theta.$$

6. Show that $\int_C \mathbf{G} \cdot d\mathbf{r}$ is independent of path for each of the following vector fields:

(a) $\mathbf{G} = \dfrac{X^2}{(X^1)^2}\iota_1 - \dfrac{X^1}{(X^1)^2}\iota_2, \qquad X^1 \ne 0.$

(b) $\mathbf{G} = \mathbf{r}$.

(c) $\mathbf{G} = \dfrac{\mathbf{r}}{r^n}, \qquad n \ge 0$ and an integer, $r \ne 0$.

Hint: We can show that $\mathbf{G} \cdot d\mathbf{r}$ is a perfect differential or we can produce Φ such that $\mathbf{G} = +\nabla\Phi$.

7. Prove that $\int_C \mathbf{G} \cdot d\mathbf{r}$ is not independent of path if

$$\mathbf{G} = -X^2\iota_1 + X^1\iota_2.$$

2. *Surface Integrals*

The fundamental ideas of this section are based on Definition 3-1.2c of a surface. Analysis is restricted to surface sections, every point of which is regular; that is, surface sections which have a parameterization

$$X^j = X^j(v^\beta)$$

such that at each point of the section

(4-2.1)
(a) the X^j are continuous functions,
(b) the first partial derivatives are continuous,
(c) the matrix $(\partial X^j/\partial v^\beta)$ has rank 2 (i.e., $\partial \mathbf{r}/\partial v^1 \times \partial \mathbf{r}/\partial v^2 \neq \mathbf{0}$).

By referring to Definition 3-1.2 we observe that these conditions guarantee the coordinate pairs (v^1, v^2) to be in D.

Definition 4-2.1. A surface section satisfying conditions (4-2.1) is said to be smooth.

The restrictions in (4-2.1) are severe. They eliminate from consideration sharp points, such as the vertex of a cone, nonorientable[2] surfaces, such as the Möbius strip,[3] and degenerate representations, such as

$$X^1 = v^1 - v^2, \qquad X^2 = (v^1 - v^2)^2, \qquad X^3 = (v^1 - v^2)^3. \tag{4-2.2}$$

The equations in (4-2.2) obviously can be expressed in terms of a single parameter $u = v^1 - v^2$. Therefore they represent a space curve rather than a surface.

Unfortunately, the restrictions go further than our intuition might desire. Most so-called surfaces met in analytic geometry and calculus conform to the conditions in (4-2.1), except perhaps at isolated points. In particular, quadratic surfaces are among those subjects to our considerations. But such intuitively simple configurations as cubes, parallelepipeds, and polyhedra in general have edges that present difficulties. The problem can be overcome by introducing the concept of sectionally smooth surfaces in analogy to sectionally smooth curves. Then, for example, an integral over a sectionally smooth surface can be defined as a sum of integrals over the sections that compose the surface.

The major difficulty involved in defining sectional smoothness has to do with the process of joining. For example, we cannot allow the definition

[2] A surface is orientable if and only if there is a continuous unit normal at every point.

[3] A Möbius strip is a one-sided surface, constructed by half twisting a strip of paper and then sticking the ends together.

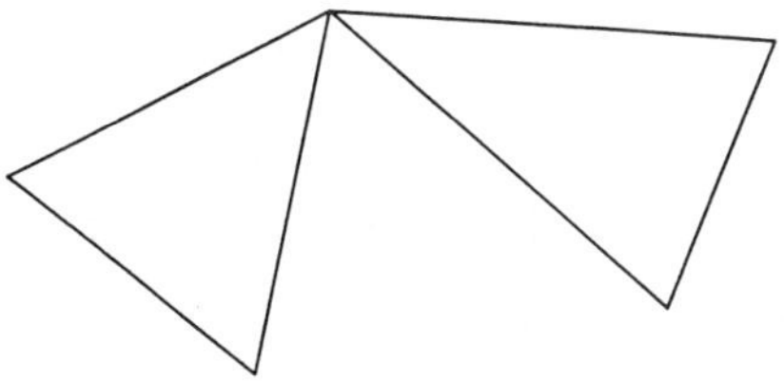

Fig. 4-2.1

of a sectionally smooth surface to include a configuration composed of two sections with a single point in common. (See Fig. 4-2.1.)

Definition 4-2.2. A surface is said to be sectionally smooth if and only if the smooth sections that compose the surface satisfy the following conditions:

(4-2.3)

(a) Any two sections have no points, one point, or an edge in common.
(b) If two sections have exactly one point P in common, then there exists a chain of sections with P in common and such that each consecutive pair has a common edge.
(c) No three sections have more than one point in common.

The preceding definitions and remarks enable us to continue with some hope of intelligibility. The reader who is interested in pursuing a detailed treatment of the theory of surfaces may begin with *Advanced Calculus* by John Olmsted (Appleton-Century-Crofts, 1961), in which a very nice presentation, as well as further references, is given.

The parameters v^j, in terms of which the surface equations are expressed, are subject to two geometric interpretations. On the one hand, the set (v^1, v^2) can be designated as coordinates of the points of a plane region. From this viewpoint $X^j = X^j(v^\beta)$ can be thought of as equations of transformation relating the plane coordinates to the space coordinates. We say that the points of the plane region are mapped onto the surface region by means of the transformation. This concept of "mapping" has proved fruitful in modern mathematics. The idea is illustrated by the following example.

Example 4-2.1. As we saw in Chapter 3, Section 1, the parametric equations

$$X^1 = v^1,$$

$$X^2 = v^2, \qquad (v^1)^2 + (v^2)^2 < 1. \tag{4-2.4}$$

$$X^3 = [1 - (v^1)^2 - (v^2)^2]^{1/2},$$

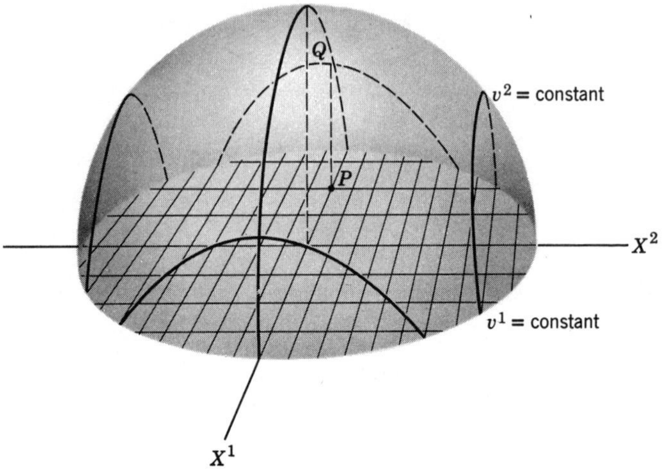

Fig. 4-2.2

are associated with the upper hemisphere of a sphere. From the standpoint of mappings, we think of the plane point with coordinates $(v^1, v^2) = (X^1, X^2)$ as being mapped into the space point (X^1, X^2, X^3). (In this case the inverse mapping would be an orthogonal projection of the hemisphere onto the plane region $(X^1)^2 + (X^2)^2 < 1$.) In Fig. 4-2.2 point P is thought of as being mapped into point Q. In particular, $P(\frac{1}{2}, \frac{1}{2})$ is sent into

$$Q\left(\frac{1}{2}, \frac{1}{2}, \frac{1}{\sqrt{2}}\right).$$

On the other hand, the families of parametric curves, v^1 = constant, v^2 = constant, introduce a coordinate net on the surface; that is, each surface point is the intersection of a unique pair of curves, one from each family. The parameters (v^1, v^2) can be interpreted as surface coordinates of a point. This approach, due to Gauss, leads to the development of the intrinsic geometry of surfaces, that is, the study of the geometry of surfaces dissociated from an embedded three-space.[4]

Example 4-2.2. The planes, X^1 = constant, X^2 = constant, intersect the hemisphere

$$X^1 = v^1, \qquad X^2 = v^2, \qquad X^3 = [1 - (v_1)^2 - (v_2)^2]^{\frac{1}{2}}$$

[4] Bernhard Riemann followed up Gauss's work on curved surfaces by setting the foundations for the development of n-dimensional geometry which included Euclidean geometry as a special case. This gave impetus to the introduction by Beltrami and others of surface models of non-Euclidean geometries. Furthermore, the studies of Riemann have played an important part in modern physical theory and philosophy.

in two sets of curves. By denoting these sets by $v^1 =$ constant and $v^2 =$ constant (see Fig. 4-2.2), we obtain a coordinate net on the surface that is analogous to the coordinate net determined in the plane by a rectangular Cartesian coordinate system. From the viewpoint of Gauss, we forget the correspondence which we have just used to obtain the coordinate net on the surface. The intrinsic geometry of the surface is developed by simply assuming the existence of some such net (and all those that can be introduced by surface coordinate transformations) and then proceeding with an analytical development of the surface geometry.

In the development of the integral concepts we feel free to use the geometrical viewpoint concerning surfaces that best fits the immediate need.

Integrals with either a scalar or a vector integrand are considered. The symbol **N** is used to denote a unit surface normal (i.e., **N** is, in general, a unit vector field defined over the surface region), and dA is symbolic of the element of surface area. Before stating a definition of surface integral, it is convenient to obtain representations for both **N** and dA. The following theorems lead to this end. As in previous considerations of surfaces, Greek letters have the range 1, 2.

Theorem 4-2.1. The square of the differential element of arc on a surface has the form

$$\text{(4-2.5a)} \quad ds^2 = g_{\alpha\beta}\, dv^\alpha\, dv^\beta = g_{11}(dv^1)^2 + 2g_{12}\, dv^1\, dv^2 + g_{22}(dv^2)^2,$$

where

$$\text{(4-2.5b)} \quad g_{\alpha\beta} = \frac{\partial \mathbf{r}}{\partial v^\alpha} \cdot \frac{\partial \mathbf{r}}{\partial v^\beta}.$$

PROOF. We have

$$ds^2 = d\mathbf{r} \cdot d\mathbf{r} = \frac{\partial \mathbf{r}}{\partial v^\alpha}\, dv^\alpha \cdot \frac{\partial \mathbf{r}}{\partial v^\beta}\, dv^\beta = g_{\alpha\beta}\, dv^\alpha\, dv^\beta,$$

as was to be shown.

Definition 4-2.3. The form (4-2.5a) is called the fundamental metric form of the surface.

The determination of an expression for the differential element of surface area is, of course, the basic step in defining that measurement, associated with the surface, which is called area. The problems involved in giving a meaning, consistent with intuition, to the concept of surface area are very deep. In fact, it is of some interest to realize that a completely general definition of surface area is of concern to modern research mathematics. The obvious approach in defining the surface area as the limiting value of

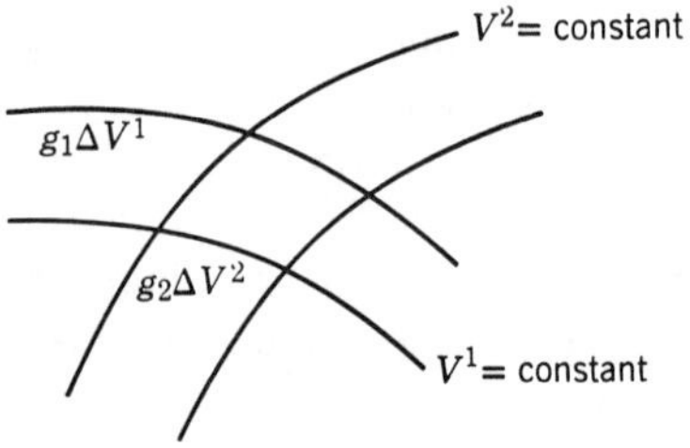

Fig. 4-2.3

areas of inscribed polyhedra (this approach would be in direct analogy to that of determining curve length as a limiting value of lengths of inscribed polygons) leads to inconsistent results. A simple example, called the Schwarz example, which demonstrates this fact, can be found on page 616 of *Advanced Calculus* by John Olmsted. We restrict ourselves here to the presentation of an intuitional background and a definition valid for smooth surface sections. We should expect the forthcoming definition of surface area to meet two fundamental requirements. First of all, the formula should reduce to the usual thing if the surface is a plane. Second, the representation of surface area should be invariant with respect to surface coordinate transformation.

The fundamental metric form is an aid in the determination of an appropriate representation for dA. Let ds_1 and ds_2, respectively, denote differential elements of arc corresponding to $v^2 =$ constant and $v^1 =$ constant. Then, according to the form (4-2.5a),

(4-2.6a) $$ds_1 = g_1\, dv^1, \qquad ds_2 = g_2\, dv^2,$$

where

(4-2.6b) $$g_\beta = \sqrt{g_{\beta\beta}}.$$

The property of magnitude cannot be associated with the differentials. However, if dv^1, dv^2 are momentarily replaced by Δv^1 and Δv^2, then the area of a parameter parallelogram is approximated by

(4-2.6c) $$\Delta A = g_1 g_2\, \Delta v^1\, \Delta v^2 \sin\theta, \qquad 0 < \theta < \pi,$$

where θ represents the angle determined by the positive senses of parameter curves (i.e., the angle determined by $\partial\mathbf{r}/\partial v^1$ and $\partial\mathbf{r}/\partial v^2$). The representation in (4-2.6c) motivates the following definition. (See Fig. 4-2.3.)

Definition 4-2.4. The differential element of surface area is

(4-2.7) $$dA = g_1 g_2 \sin\theta\, dv^1\, dv^2.$$

The next two theorems introduce a surface unit normal **N**.

Theorem 4-2.2. The partial derivatives $\partial \mathbf{r}/\partial v^1$ and $\partial \mathbf{r}/\partial v^2$ are tangent vector fields along the parameter curves $v^2 = \text{constant}$ and $v^1 = \text{constant}$, respectively.

PROOF. Suppose that the parameter curve $v^2 = \text{constant}$ is represented in terms of its arc length s. Then $(d\mathbf{r}/ds)_1$ is a unit tangent vector field along the curve. The subscript of 1 simply indicates the particular curve. Since $v^2 = \text{constant}$, we have

$$\left(\frac{d\mathbf{r}}{ds}\right)_1 = \frac{\partial \mathbf{r}}{\partial v^\beta}\frac{dv^\beta}{ds} = \frac{\partial \mathbf{r}}{\partial v^1}\frac{dv^1}{ds}. \tag{4-2.8}$$

Therefore $\partial \mathbf{r}/\partial v^1$ is proportional to the unit tangent field $(d\mathbf{r}/ds)_1$. By letting $v^1 = \text{constant}$, we can show that $\partial \mathbf{r}/\partial v^2$ is proportional to $(d\mathbf{r}/ds)_2$.

Theorem 4-2.3. A unit normal vector field associated with a smooth surface section is

$$\mathbf{N} = \frac{\partial \mathbf{r}/\partial v^1 \times \partial \mathbf{r}/\partial v^2}{|\partial \mathbf{r}/\partial v^1 \times \partial \mathbf{r}/\partial v^2|}. \tag{4-2.9}$$

PROOF. The fact that $(\partial X^j/\partial v^\beta)$ has rank 2 is equivalent to $\partial \mathbf{r}/\partial v^1 \times \partial \mathbf{r}/\partial v^2 \neq 0$. According to the preceding theorem $\partial \mathbf{r}/\partial v^1$ and $\partial \mathbf{r}/\partial v^2$ are, at a point of the surface, tangent vectors to parametric curves, hence determine the tangential plane at the point. It follows that the cross product is normal to the surface and, when divided by its magnitude, is a unit vector.

The relations in (4-2.7) and (4-2.9) together lead to a vector element of surface area. Since

$$\left|\frac{\partial \mathbf{r}}{\partial v^1} \times \frac{\partial \mathbf{r}}{\partial v^2}\right| = \left|\frac{\partial \mathbf{r}}{\partial v^1}\right|\left|\frac{\partial \mathbf{r}}{\partial v^2}\right| \sin\theta = g_1 g_2 \sin\theta, \qquad 0 < \theta < \pi, \tag{4-2.10}$$

we obtain

$$\mathbf{N}\, dA = \frac{\partial \mathbf{r}}{\partial v^1} \times \frac{\partial \mathbf{r}}{\partial v^2}\, dv^1\, dv^2. \tag{4-2.11}$$

This expression forms the basis for the definition of the integral of a scalar over a surface.

Definition 4-2.5a. Let $\mathbf{W}$ be a continuous vector field defined on a smooth surface sector S. The scalar surface integral of $\mathbf{W}$ on S is

$$\int_S \mathbf{W} \cdot \mathbf{N}\, dA = \lim_{|\Delta A| \to 0} \sum_{j=1}^{n} \sum_{k=1}^{m} \Phi(v_j^{\,1}, v_k^{\,2})\, \Delta A_{jk}, \tag{4-2.12a}$$

where $\Phi = \mathbf{W} \cdot \mathbf{N}$.

Definition 4-2.5b. If **W** is defined on a sectionally smooth surface S composed of surface sectors $S_1 \cdots S_j$, then

$$\int_S \mathbf{W} \cdot \mathbf{N}\, dA = \sum_{i=1}^{j} \int_{S_i} \mathbf{W} \cdot \mathbf{N}\, dA. \tag{4-2.12b}$$

From the analytic point of view, the integral (4-2.12a) is precisely the same as that studied in the calculus. Its evaluation depends on the fact that the double integral can be represented as an iterated integral. By using (4-2.11), we obtain the formula

$$\boxed{\int \mathbf{G} \cdot \mathbf{N}\, dA = \int_{v_0{}^2}^{v_1{}^2} \int_{v_0{}^1}^{v_1{}^1} \mathbf{G} \cdot \frac{\partial \mathbf{r}}{\partial v^1} \times \frac{\partial \mathbf{r}}{\partial v^2}\, dv^1\, dv^2.} \tag{4-2.13}$$

Example 4-2.3. (See Fig. 4-2.4). Let $X^3 = (X^1)^2 + (X^2)^2$, where

$$X^1 = v^1, \qquad X^2 = v^2, \qquad X^3 = (v^1)^2 + (v^2)^2, \qquad \mathbf{G} = v^1\boldsymbol{\iota}_1 + 2v^2\boldsymbol{\iota}_2;$$

then

$$\frac{\partial \mathbf{r}}{\partial v^1} \times \frac{\partial \mathbf{r}}{\partial v^2} = -2v^1\boldsymbol{\iota}_1 - 2v^2\boldsymbol{\iota}_2 + \boldsymbol{\iota}_3,$$

and

$$\int_S \mathbf{G} \cdot \mathbf{N}\, dA = \int_{v_0{}^2}^{v_1{}^2} \int_{v_0{}^1}^{v_1{}^1} [-2(v^1)^2 - 4(v^2)^2]\, dv^1\, dv^2.$$

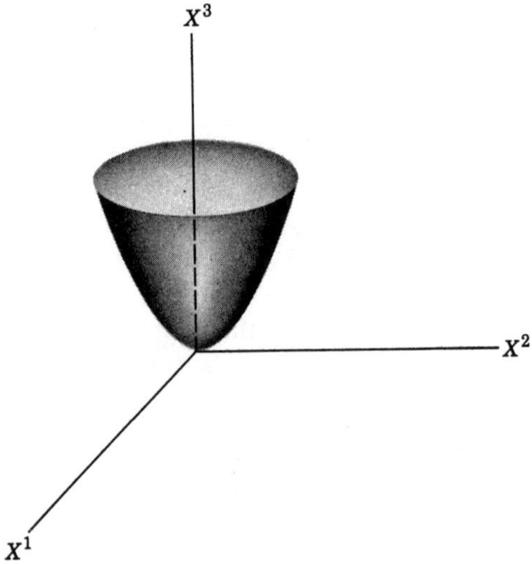

Fig. 4-2.4

Therefore the integral over the surface region in the first quadrant and otherwise bounded by $X^1 = v^1 = 3$ and $X^2 = v^2 = 4$ is

$$-2\int_0^4\int_0^3 [(v^1)^2 + 2(v^2)^2]\, dv^1\, dv^2 = -432.$$

The concept of surface area is defined in a manner consistent with (4-2.12). In fact, if we let $\mathbf{W} = \mathbf{N}$ and recall that the expression in (4-2.7) for dA is equivalent to $|\partial\mathbf{r}/\partial v^1 \times \partial\mathbf{r}/\partial v^2|$, then (4-2.12) reduces to the relation stated in the following definition.

Definition 4-2.6. The area of a smooth surface is given by the following integral:

$$A = \int_S dA = \int_{v_0{}^2}^{v_1{}^2}\int_{v_0{}^1}^{v_1{}^1} \left|\frac{\partial\mathbf{r}}{\partial v^1} \times \frac{\partial\mathbf{r}}{\partial v^2}\right| dv^1\, dv^2. \tag{4-2.14}$$

Example 4-2.4. With respect to the surface region considered in Example 4-2.3, we have

$$\int_S dA = \int_0^4\int_0^3 \sqrt{4[(v^1)^2 + (v^2)^2] + 1}\, dv^1\, dv^2.$$

It is interesting to note that the area integral can also be represented in terms of the determinant g of the surface metric tensor components $g_{\alpha\beta}$.

Theorem 4-2.4. We have

$$\int_S dA = \int_{v_0{}^2}^{v_1{}^2}\int_{v_0{}^1}^{v_1{}^1} \sqrt{g}\, dv^1\, dv^2, \tag{4-2.15a}$$

where

$$g = \det(g_{\alpha\beta}). \tag{4-2.15b}$$

PROOF. Recall Lagrange's identity (Chapter 1, Section 7, Problem 18)

$$(\mathbf{A} \times \mathbf{B}) \cdot (\mathbf{C} \times \mathbf{D}) = (\mathbf{A} \cdot \mathbf{C})(\mathbf{B} \cdot \mathbf{D}) - (\mathbf{A} \cdot \mathbf{D})(\mathbf{B} \cdot \mathbf{C}).$$

We have

$$\left|\frac{\partial\mathbf{r}}{\partial v^1} \times \frac{\partial\mathbf{r}}{\partial v^2}\right|^2 = \left(\frac{\partial\mathbf{r}}{\partial v^1} \times \frac{\partial\mathbf{r}}{\partial v^2}\right) \cdot \left(\frac{\partial\mathbf{r}}{\partial v^1} \times \frac{\partial\mathbf{r}}{\partial v^2}\right)$$

$$= \left(\frac{\partial\mathbf{r}}{\partial v^1} \cdot \frac{\partial\mathbf{r}}{\partial v^1}\right)\left(\frac{\partial\mathbf{r}}{\partial v^2} \cdot \frac{\partial\mathbf{r}}{\partial v^2}\right) - \left(\frac{\partial\mathbf{r}}{\partial v^1} \cdot \frac{\partial\mathbf{r}}{\partial v^2}\right)^2.$$

By making the identifications

$$g_{11} = \left|\frac{\partial\mathbf{r}}{\partial v^1}\right|^2, \qquad g_{22} = \left|\frac{\partial\mathbf{r}}{\partial v^2}\right|^2, \qquad g_{12} = g_{21} = \frac{\partial\mathbf{r}}{\partial v^1} \cdot \frac{\partial\mathbf{r}}{\partial v^2},$$

we obtain

$$\left|\frac{\partial \mathbf{r}}{\partial v^1} \times \frac{\partial \mathbf{r}}{\partial v^2}\right|^2 = \begin{vmatrix} g_{11} & g_{12} \\ g_{21} & g_{22} \end{vmatrix} = g.$$

The proof is completed by substituting this result in (4-2.14).

Example 4-2.5. Suppose the surface region S is, in fact, a region of the X^1, X^2 plane. (See Fig. 4-2.5.)

$$X^1 = v^1, \qquad X^2 = v^2, \qquad X^3 = 0,$$
$$a \leq v^1 \leq b, \qquad v_0^2(v^1) \leq v^2 \leq v_1^2(v^1).$$

Then

$$(g_{\alpha\beta}) = \begin{pmatrix} 1 & 0 \\ 0 & 1 \end{pmatrix}, \qquad g = 1.$$

The area formula (4-2.15a) reduces to

$$\int_a^b \int_{v_0^2}^{v_1^2} dv^2\, dv^1.$$

In other words, we obtain the usual result. This confirms that the definition of surface area is consistent with that of plane area. The second criterion we ask of the definition, that of invariance, is considered in Section 2*.

Example 4-2.6. Suppose that a smooth surface is given by the explicit representation

$$X^3 = X^3(X^1, X^2).$$

A parameterization for the surface is

$$X^1 = v^1, \qquad X^2 = v^2, \qquad X^3 = X^3(v^1, v^2).$$

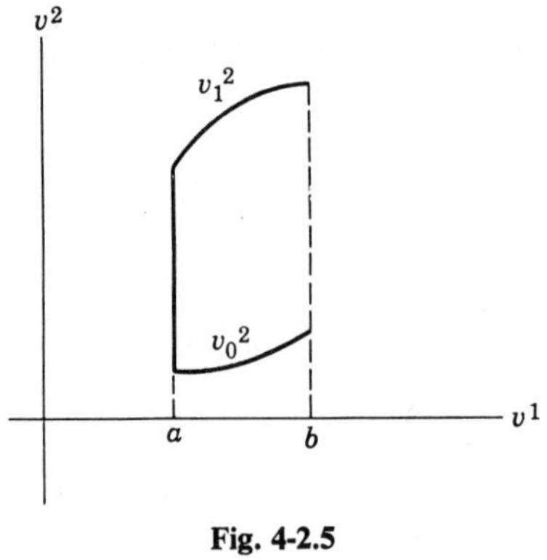

Fig. 4-2.5

Then

$$\frac{\partial \mathbf{r}}{\partial v^1} \times \frac{\partial \mathbf{r}}{\partial v^2} = \begin{vmatrix} \iota_1 & \iota_2 & \iota_3 \\ 1 & 0 & \dfrac{\partial X^3}{\partial v^1} \\ 0 & 1 & \dfrac{\partial X^3}{\partial v^2} \end{vmatrix} = -\frac{\partial X^3}{\partial v^1}\iota_1 - \frac{\partial X^3}{\partial v^2}\iota_2 + \iota_3.$$

Therefore

$$\left|\frac{\partial \mathbf{r}}{\partial v^1} \times \frac{\partial \mathbf{r}}{\partial v^2}\right| = \left[1 + \left(\frac{\partial X^3}{\partial v^1}\right)^2 + \left(\frac{\partial X^3}{\partial v^2}\right)^2\right]^{1/2}$$

By substituting this last relation into (4-2.14) we obtain the formula

$$A = \int_{v_0{}^1}^{v_1{}^1} \int_{v_0{}^2}^{v_1{}^2} \left[1 + \left(\frac{\partial X^3}{\partial v^1}\right)^2 + \left(\frac{\partial X^3}{\partial v^2}\right)^2\right]^{1/2} dv^2\, dv^1. \tag{4-2.16}$$

The surface integral of a vector field is also a useful idea. This concept is defined as follows.

Definition 4-2.7. Let **W** be a continuous vector field defined on a smooth surface. The surface integral of **W** is

$$\int_S \mathbf{W}\, dA = \lim_{|\Delta A| \to 0} \sum_{j,k} \mathbf{W}(v_j{}^1, v_k{}^2)\, \Delta A_{jk}. \tag{4-2.17}$$

In order to evaluate surface vector integrals, it is, in general, necessary to represent **W** in terms of a constant basis. Therefore it is convenient to think of the factors of the integrand as initially defined in terms of rectangular Cartesian coordinates. The surface equations $X^j = X^j(v^\beta)$ can be construed in the sense of a change of variables made for the purpose of evaluating the integral. This procedure is illustrated in the example that follows the next theorem.

Theorem 4-2.5. The integral of Definition 4-2.7 can be expressed in the form

$$\int_S \mathbf{W}\, dA = \int_{v_0{}^1}^{v_1{}^1} \int_{v_0{}^2}^{v_1{}^2} \sqrt{g}\mathbf{W}\, dv^2\, dv^1. \tag{4-2.18}$$

Proof. The proof follows from the fact that the relation (4-2.7) for dA is equivalent to

$$\left|\frac{\partial \mathbf{r}}{\partial v^1} \times \frac{\partial \mathbf{r}}{\partial v^2}\right| dv^1\, dv^2$$

which in turn is equivalent to $\sqrt{g}\, dv^1\, dv^2$.

Example 4-2.7. Suppose that

$$\mathbf{W} = X^1\iota_1 + X^2\iota_2$$

and the region of definition is

$$(X^1)^2 + (X^2)^2 + (X^3)^2 = 1,$$
$$X^3 > 0, \qquad X^2 > 0, \qquad X^1 > 0,$$

that is, that part of the unit sphere in the first quadrant. In order to evaluate $\int_S \mathbf{W}\,dA$, we can make use of the usual parametric equations

$$X^1 = \sin v^1 \cos v^2,$$
$$X^2 = \sin v^1 \sin v^2,$$
$$X^3 = \cos v^1.$$

In this example

$$\sqrt{g} = \sin v^1.$$

Therefore

$$\begin{aligned}\int_S \mathbf{W}\,dA &= \iota_1 \int_0^{\pi/2}\int_0^{\pi/2} \sin^2 v^1 \cos v^2\,dv^1\,dv^2 + \iota_2 \int_0^{\pi/2}\int_0^{\pi/2} \sin v^1 \sin v^2\,dv^1\,dv^2 \\ &= \frac{\pi}{4}\iota_1 \int_0^{\pi/2} \cos v^2\,dv^2 + \iota_2 \int_0^{\pi/2} \sin^2 dv^2 \\ &= \frac{\pi}{4}(\iota_1 + \iota_2).\end{aligned}$$

Problems

1. Given

$$X^1 = \sin v^1 \cos v^2,$$
$$X^2 = \sin v^1 \sin v^2,$$
$$X^3 = \cos v^1,$$

show that

$$\frac{\partial \mathbf{r}}{\partial v^1} \times \frac{\partial \mathbf{r}}{\partial v^2} = \sin v^1 \mathbf{r}.$$

2. If $v^\beta = v^\beta(\bar{v}^\alpha)$ and $\left|\dfrac{\partial v^\beta}{\partial \bar{v}^\alpha}\right| \neq 0$, show that

(a)
$$\frac{\partial \mathbf{r}}{\partial v^1} \times \frac{\partial \mathbf{r}}{\partial v^2} = \left|\frac{\partial \bar{v}^\alpha}{\partial v^\beta}\right| \frac{\partial \mathbf{r}}{\partial \bar{v}^1} \times \frac{\partial \mathbf{r}}{\partial \bar{v}^2}.$$

Hint: Start with $\varepsilon_{ijk}(\partial X^j/\partial v^1)(\partial X^k/\partial v^2)$ and transform to the bar system.

(b) Parameterize $X^3 - (X^1)^2 - (X^2)^2 = 0$ in each of the following ways:

(1) $X^1 = v^1, \qquad X^2 = v^2, \qquad X^3 = (v^1)^2 + (v^2)^2,$

(2) $X^1 = \bar{v}^1 \cos \bar{v}^2, \qquad X^2 = \bar{v}^1 \sin \bar{v}^2, \qquad X^3 = (\bar{v}^1)^2.$

Show that at $X^1 = \sqrt{2}$, $X^2 = \sqrt{2}$, $X^3 = 4$.

$$\frac{\partial \mathbf{r}}{\partial v^1} \times \frac{\partial \mathbf{r}}{\partial v^2} = \frac{1}{2} \frac{\partial \mathbf{r}}{\partial \bar{v}^1} \times \frac{\partial \mathbf{r}}{\partial \bar{v}^2}.$$

3. Suppose that a given surface can be represented by either $f(X^1, X^2, X^3) = 0$ or $X^j = X^j(v^\beta)$. Then $\partial \mathbf{r}/\partial v^1 \times \partial \mathbf{r}/\partial v^2$ and ∇f would be normal to the surface, hence proportional to one another.

(a) Show that if

$$f(X^1, X^2, X^3) = X^3 + g(X^1, X^2)$$

and

$$X^1 = v^1, \qquad X^2 = v^2, \qquad X^3 = X^3(v^1, v^2),$$

then

$$\nabla f = \frac{\partial \mathbf{r}}{\partial v^1} \times \frac{\partial \mathbf{r}}{\partial v^2}.$$

(b) Show that in general

$$\frac{\partial \mathbf{r}}{\partial v^1} \times \frac{\partial \mathbf{r}}{\partial v^2} = \sqrt{g}\, \frac{\nabla f}{|\nabla f|}.$$

4. Let θ be the angle determined by the positive senses of $\partial \mathbf{r}/\partial v^1$ and $\partial \mathbf{r}/\partial v^2$. Show that

$$\sin \theta = \frac{\sqrt{g}}{\sqrt{g_{11}}\sqrt{g_{22}}}.$$

Hint: Use the trigonometric identity $\sin^2 \theta = 1 - \cos^2 \theta$.

5. Compute each of the following scalar surface integrals $\int_S \mathbf{W} \cdot \mathbf{N}\, dA$.

(a) $$\mathbf{W} = \mathbf{r} \begin{cases} X^1 = v^1, & v^1 \geq 0, \\ X^2 = v^2, & v^2 \geq 0, \\ X^3 = 1 - v^1 - v^2, & v^1 + v^2 \leq 1 \end{cases}$$

(i.e., the surface is a first octant plane area with vertices (1, 0, 0), (0, 1, 0), (0, 0, 1).

(b) $\mathbf{W} = X^1\boldsymbol{\iota}_1 + 2X^2\boldsymbol{\iota}_2$, S is that part of the surface $X^1 + X^2 = 1$ which lies in the first octant and is bounded by $X^3 = 0$, $X^3 = 2$, $X^1 = 0$, $X^2 = 0$. Determine your own parameterization such that $\mathbf{N}$ points away from the origin.

(c) $$\mathbf{W} = (X^2)^2\boldsymbol{\iota}_1 + X^1\boldsymbol{\iota}_2, \begin{cases} X^1 = \cos v^1, \\ X^2 = \sin v^1, & 0 \leq v^1 < 2\pi \qquad 0 \leq v^2 \leq 3. \\ X^3 = v^2, \end{cases}$$

6. Find the surface area in each part of (5) by letting $\mathbf{W} = \mathbf{N}$.

7. Compute each of the vector surface integrals $\int_S \mathbf{W}\, dA$:

(a) $$\mathbf{W} = X^1\iota_1 - X^2\iota_2, \quad \begin{cases} X^1 = \cos v^1, \\ X^2 = \sin v^1, \\ X^3 = v^2, \end{cases} \quad 0 \le v^1 \le \pi \quad 0 \le v^2 \le 2;$$

(b) $$\mathbf{W} = \mathbf{r}, \quad \begin{cases} X^1 = \sin v^1 \cos v^2, \\ X^2 = \sin v^1 \sin v^2, \\ X^3 = \cos v^2, \end{cases} \quad \begin{matrix} 0 \le v^1 \le \pi, \\ 0 \le v^2 \le 2\pi. \end{matrix}$$

2*. *An Introduction to Surface Tensors and Surface Invariants*

The purpose of this section is to point out some of the elementary aspects of the intrinsic geometry of a surface and to establish the invariance of the area integral with respect to allowable surface coordinate transformations. Throughout the discussion it is assumed that a smooth surface section (see Definition 4-2.1) is under consideration.

If the surface has the vector equation

(4-2*.1a) $$\mathbf{r} = \mathbf{r}(v^\beta),$$

then

(4-2*.1b) $$d\mathbf{r} = \frac{\partial \mathbf{r}}{\partial v^\beta}\, dv^\beta, \qquad ds^2 = g_{\alpha\beta}\, dv^\alpha\, dv^\beta$$

where

(4-2*.1c) $$g_{\alpha\beta} = \frac{\partial \mathbf{r}}{\partial v^\alpha} \cdot \frac{\partial \mathbf{r}}{\partial v^\beta}.$$

In the preceding section we made use of the quantities $\partial \mathbf{r}/\partial v^\beta$ and $g_{\alpha\beta}$ to develop the desired integral formulas. Now we emphasize their geometric nature and their properties under transformation of surface coordinates.

At a point P of the surface S the vectors $\partial \mathbf{r}/\partial v^1$, $\partial \mathbf{r}/\partial v^2$ determine the tangential plane. (See Theorem 4-2.2.) Furthermore $\partial \mathbf{r}/\partial v^1 \times \partial \mathbf{r}/\partial v^2 \neq \mathbf{0}$ implies that $\partial \mathbf{r}/\partial v^1$, $\partial \mathbf{r}/\partial v^2$ are linearly independent. Therefore these two vectors determine a basis in the tangential plane, $T(P)$, to S at P. The building block of the intrinsic geometry of a surface is the geometry of the tangential plane.

Let $\mathbf{W}$ be a Cartesian vector defined at a point P of S and lying in $T(P)$. Then

(4-2*.2a) $$\mathbf{W} = W^1\mathbf{r}_1 + W^2\mathbf{r}_2 = W^\beta\mathbf{r}_\beta$$

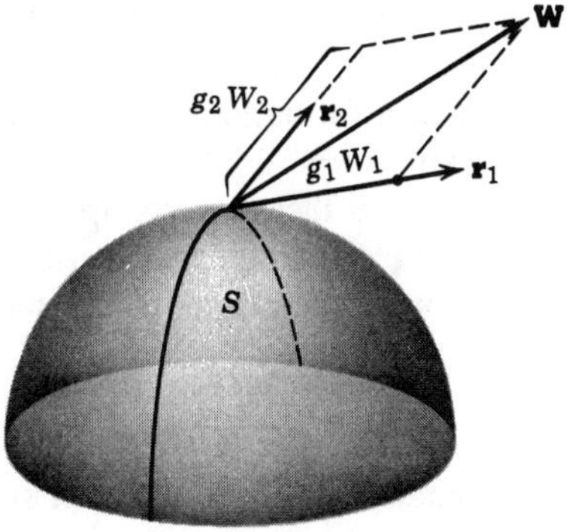

Fig. 4-2*.1

where

(4-2*.2b)
$$\mathbf{r}_\beta = \frac{\partial \mathbf{r}}{\partial v^\beta}.$$

(See Fig. 4-2*.1.) The Cartesian vectors $\mathbf{r}_\beta$ are not unit vectors in general; rather they have magnitudes

$$g_\beta = \sqrt{g_{\beta\beta}} = \sqrt{\mathbf{r}_\beta \cdot \mathbf{r}_\beta}.$$

If we introduce unit vectors $\mathbf{e}_\beta$ with the direction and sense of $\mathbf{r}$, then

(4-2*.2c)
$$\mathbf{r} = g_\beta \mathbf{e}_\beta.$$

The vector $\mathbf{W}$ can now be expressed in the form

(4-2*.3)
$$\mathbf{W} = W^1 g_1 \mathbf{e}_1 + W^2 g_2 \mathbf{e}_2;$$

and $W^1 g_1$, $W^2 g_2$ are, according to the parallelogram law of addition, the parallel projections of $\mathbf{W}$ onto the directions of $\mathbf{r}_1$ and $\mathbf{r}_2$, respectively. If it is desired, we can introduce a two-dimensional Cartesian coordinate system into $T(P)$ with the coordinate axes along $\mathbf{r}_1$ and $\mathbf{r}_2$. Hence the geometry of $T(P)$ is local Cartesian.

Example 4-2*.1. Suppose that the equations

$$X^j = X^j[v^\beta(t)]$$

represented a curve on a surface S. The tangent vector

(4-2*.4)
$$\frac{d\mathbf{r}}{dt} = \frac{\partial \mathbf{r}}{\partial v^\beta}\frac{dv^\beta}{dt} = \frac{dv^\beta}{dt}\mathbf{r}_\beta$$

is a concrete illustration of a vector at a point P in $T(P)$. From the viewpoint of the three-dimensional space in which the surface is embedded, $d\mathbf{r}/dt$, evaluated at P, is a free vector. In passing over to the intrinsic geometry of the surface it is necessary to consider the vector as bound to P.

Before commenting in a precise way on the vector concepts employed in the intrinsic geometry of a surface, it is necessary to secure information about surface coordinate transformations. It has been demonstrated that a surface parameterization $X^j = X^j(v^\beta)$ is not unique; that is, other parameterizations of a given surface, for example $X^j = X^j(\bar{v}^\beta)$, also exist. Now it is clear that any one of these parameterizations must satisfy the requirements designated by the term smooth surface section. In establishing the allowable transformations,

$$v^\beta = v^\beta(\bar{v}^\alpha), \tag{4-2*.5}$$

we guarantee that the requirements of a smooth surface section will be met by making the following assumptions.

Agreement 4-2*.1. The surface coordinate transformations of the form in (4-2*.5) satisfy these properties:

(4-2*.6)

(a) The functions in (4-2*.5) are defined on a domain D' of $\bar{v}^\beta$ values such that the range values v^β include D. [See (4-2.1) and (3-1.2).]

(b) At least the first partial derivatives $\partial v^\beta/\partial \bar{v}^\alpha$ are continuous everywhere in D'.

(c) The Jacobian of transformation, $|\partial v/\partial \bar{v}|$, is different from zero everywhere in D'.

The coordinate transformations satisfying agreement (4-2*.1) are called allowable coordinate transformations.

The preceding agreement establishes the nature of the allowable coordinate transformations; therefore we can proceed with some of the elementary aspects of the theory of surface coordinate transformations.

Example 4-2*.2. If $v^\beta = v^\beta(\bar{v}^\alpha)$, then

$$\frac{dv^\beta}{dt} = \frac{\partial v^\beta}{\partial \bar{v}^\alpha}\frac{d\bar{v}^\alpha}{dt}. \tag{4-2*.7a}$$

Therefore the element of the set $\{dv^\beta/dt\}$ are related by a rule previously designated contravariant. The existence of the class of components $\{dv^\beta/dt\}$ motivates the following definition.

Definition 4-2*.1. The class $\{W^\beta\}$, whose elements satisfy the transformation rule

$$W^\beta = \frac{\partial v^\beta}{\partial \bar{v}^\alpha}\overline{W}^\alpha \tag{4-2*.7b}$$

is said to be a contravariant vector on the surface S with respect to the allowable transformation group. (See Fig. 4-2.*1.)

At this stage of development it is appropriate to dissociate from the embedding three-dimensional space and to concentrate on the intrinsic geometry of the surface. In terms of the geometry of the surface, which is in general non-Euclidean, a contravariant vector cannot be endowed with precise geometric meaning. The intuitional properties we can employ are inferred by the expression "locally Euclidean"; that is, at a point P of S, the linear algebraic relations to which the components of a vector are subjected are precisely the same as those employed in the Cartesian geometry of a plane. Therefore the surface can be thought of as Euclidean in a small neighborhood of a point, or, in other words, the algebraic relations can be interpreted in the tangential plane at the point. Clearly, there are many difficulties awaiting discussion. For example, the tangential plane $T(P)$ is a function of P; therefore the linear geometry previously mentioned is valid only at a point of S. The way in which we make comparisons of vectors at different points of P is discussed in Chapter 5.

A second basis can be introduced into $T(P)$ in analogy to the developments of Chapter 1, Section 5*, and Chapter 3, Section 4. We define contravariant basis arrows by means of the relations

(4-2*.8) $$\mathbf{r}_\gamma \cdot \mathbf{r}^\beta = \delta_\gamma{}^\beta.$$

This sets the stage for a complete algebraic duality.

Just as the components of a tangential vector give rise to the concept of a contravariant vector, so do the components of a Cartesian gradient vector serve as the prototype of a covariant vector.

Example 4-2*.3. Let Φ be a scalar function defined on a smooth surface section S. Let $v^\beta = v^\beta(\bar{v}^\gamma)$ be an allowable coordinate transformation. Then

(4-2*.9a) $$\frac{\partial \Phi}{\partial v^\beta} = \frac{\partial \bar{v}^\gamma}{\partial v^\beta} \frac{\partial \Phi}{\partial \bar{v}^\gamma}.$$

Definition 4-2*.2. The class $\{W_\beta\}$, whose elements satisfy the transformation rule

(4-2*.9b) $$W_\beta = \frac{\partial \bar{v}^\gamma}{\partial v^\beta} \overline{W}_\gamma,$$

is said to be a covariant vector on the surface S with respect to the allowable transformation group.

As long as a metric is associated with a given space, covariant and contravariant vectors are just different algebraic representations of the same geometric entity. The relation between the two types of vector is established by use of the components $g_{\alpha\beta}$. The development of these ideas is precisely the same as the curvilinear coordinate development of Chapter

3, Section 4; therefore we restrict the following to a statement of fundamental facts.

A vector $\mathbf{W}$ has the representations

(4-2*.10a) $$\mathbf{W} = W^\beta \mathbf{r}_\beta = W^\beta g_{\beta\gamma} \mathbf{r}^\gamma = W_\gamma \mathbf{r}^\gamma.$$

In particular, the contravariant and covariant components satisfy the relations

(4-2*.10b) $$W_\gamma = g_{\gamma\beta} W^\beta, \qquad W^\beta = g^{\beta\gamma} W_\gamma,$$

where the $g^{\beta\gamma}$ satisfy the defining equations

(4-2*.10c) $$g_{\alpha\beta} g^{\alpha\gamma} = \delta_\beta{}^\gamma.$$

Definition 4-2*.3. The class $\{g_{\alpha\beta}\}$, whose elements satisfy the transformation rule

(4-2*.11a) $$g_{\alpha\beta} = \frac{\partial \bar{v}^\lambda}{\partial v^\alpha} \frac{\partial \bar{v}^\gamma}{\partial v^\beta} \bar{g}_{\lambda\gamma},$$

is called the fundamental metric tensor of the surface and is a tensor of covariant order 2.

It is shown (as in Chapter 3, Section 4) that $g^{\alpha\beta}$ satisfies the contravariant rule of transformation

(4-2*.11b) $$g^{\alpha\beta} = \frac{\partial v^\alpha}{\partial \bar{v}^\lambda} \frac{\partial v^\beta}{\partial \bar{v}^\gamma} \bar{g}^{\lambda\gamma}.$$

Definition 4-2*.4. The class $\{g^{\alpha\beta}\}$ is called the associated fundamental metric tensor of the surface. It is a tensor of contravariant order 2.

The determinant of the components of the fundamental metric tensor will prove useful in the sequel. Its rule of transformation, as well as that of the determinant of the contravariant metric tensor, is indicated by the next theorem.

Theorem 4-2*.1. The classes of elements $\{|g_{\alpha\beta}|\}$ and $\{|g^{\alpha\beta}|\}$ are scalars of weights -2 and $+2$, respectively, with regard to the allowable group of surface transformations.

PROOF. The proofs follow by noting that the determinants of the right-hand sides of (4-2*.11a,b), respectively, can be expressed as products of three determinants. Thus

(4-2*.12)
$$\text{(a)} \quad |g_{\alpha\beta}| = \left|\frac{\partial \bar{v}}{\partial v}\right|^2 |\bar{g}_{\lambda\gamma}|.$$
$$\text{(b)} \quad |g^{\alpha\beta}| = \left|\frac{\partial v}{\partial \bar{v}}\right|^2 |\bar{g}^{\lambda\gamma}| = \left|\frac{\partial \bar{v}}{\partial v}\right|^{-2} |\bar{g}^{\lambda\gamma}|.$$

According to preceding convention, the factors $|\partial v/\partial \bar{v}|^{-2}$, $|\partial v/\partial \bar{v}|^{+2}$ represent weight factors of -2 and $+2$, respectively.

Let us investigate the invariance of the representation for surface area under allowable surface coordinate transformations. (See Fig. 4-2*.2.)

In order to attack this question, we introduce a local plane Cartesian coordinate system[5] in the tangential plane at a point P of a surface S. Suppose the Cartesian system with coordinates X^j, $j = 1, 2$ and origin at P is rectangular. Let **U** and **W** be two surface vectors at P [hence in $T(P)$]. If the Cartesian coordinates are related to the surface parameters by means of the equations

(4-2*.13a) $$X^j = X^j(v^\beta), \qquad j = 1, 2$$

and $\mathcal{E}_{jk}$, $\mathcal{E}_{\alpha\beta}$ satisfy the transformation rule

(4-2*.13b) $$\mathcal{E}_{jk} = \left|\frac{\partial X}{\partial v}\right| \frac{\partial v^\alpha}{\partial X^j}\frac{\partial v^\beta}{\partial X^k}\mathcal{E}_{\alpha\beta}, \qquad j, k = 1, 2$$

(see Chapter 1, Section 6, for the definition of $\mathcal{E}_{jk}$), the area of the parallelogram determined by **U** and **W** is

(4-2*.14a) $$\mathcal{E}_{jk}U^jW^k = \left|\frac{\partial X}{\partial v}\right| \mathcal{E}_{\alpha\beta}U^\alpha W^\beta.$$

That the preceding relation represents the indicated area can be made clear by considering the left-hand member as the third component, $\mathcal{E}_{3jk}U^jW^k$, of the cross product of **U** and **W**. This statement is made under the supposition that a three-dimensional Cartesian system can be introduced by taking X^3 perpendicular to the plane of X^1, X^2.

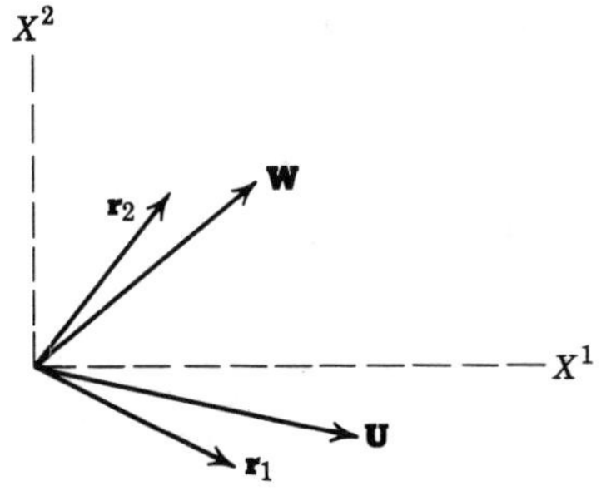

Fig. 4-2*.2

[5] The existence of such a local system can be demonstrated. See Synge and Schild, *Tensor Calculus*, University of Toronto Press, 1956, p. 59.

According to Theorem 3-5.2,

$$\left|\frac{\partial X}{\partial v}\right| = \sqrt{g}.$$

Therefore the generalized expression for the area of a parallelogram, as indicated by the right-hand member of (4-2*.14a), is

$$\sqrt{g}\,\mathcal{E}_{\alpha\beta}U^{\alpha}W^{\beta}. \tag{4-2*.14b}$$

It is this expression that gives rise to a general definition of the element of surface area.

The procedure established in this development can be generalized to elements of volume in a straightforward manner.

Definition 4-2*.5. Let $\delta_1 v^{\alpha} = (\delta_1 v^1, \delta_1 v^2)$ and $\delta_2 v^{\alpha} = (\delta_2 v^1, \delta_2 v^2)$ be differential representations of independent directions in $T(P)$ at a point P of a smooth surface section S. The differential element of area determined by $\delta_1 v$ and $\delta_2 v$ is

$$\sqrt{g}\,\mathcal{E}_{\alpha\beta}\,\delta_1 v^{\alpha}\,\delta_2 v^{\beta}. \tag{4-2*.15}$$

Since $\sqrt{g}$ is a scalar density of weight -1, $\mathcal{E}_{\alpha\beta}$ is a covariant density of order 2 and weight $+1$ and the differentials are contravariant vectors of weight zero, it follows that (4-2*.15) is an invariant. Furthermore, if

$$\begin{aligned} \delta_1 v^1 &= dv^1, \qquad & \delta_1 v^2 &= 0, \\ \delta_2 v^1 &= 0, \qquad & \delta_2 v^2 &= dv^2; \end{aligned} \tag{4-2*.16}$$

that is, if the differential vectors have the directions of the parameter curves (4-2*.15) reduces to the expression

$$\sqrt{g}\,dv^1\,dv^2. \tag{4-2*.16b}$$

This is precisely the area element introduced in the last section.

Problem

1. Carry out the proof of the scalar character of $\sqrt{g}\,\mathcal{E}_{\alpha\beta}\,\delta_1 v^{\alpha}\,\delta_2 v^{\beta}$.

3. *Volume Integrals*

This section deals with integrals of the types

(a) $\displaystyle\int_V \Phi(\bar{X}^1, \bar{X}^2, \bar{X}^3)\,dV,$

(b) $\displaystyle\int_V \mathbf{G}\,dV = \boldsymbol{\iota}_1\int_V G^1\,dV + \boldsymbol{\iota}_2\int_V G^2\,dV + \boldsymbol{\iota}_3\int_V G^3\,dV.$

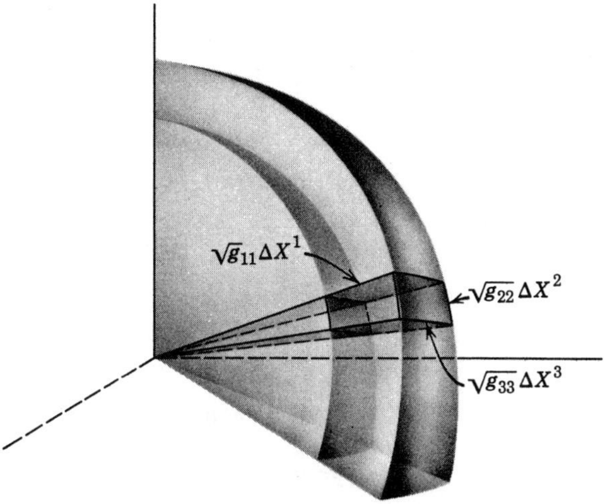

Fig. 4-3.1

In both cases the integration theory discussed in the usual calculus course applies. The standard method for evaluation of the volume integral depends on its equivalence to an iterated triple integral.

A problem which might arise is that of determining the element of volume associated with a given coordinate system. The key to this determination is again the fundamental metric form

$$ds^2 = \bar{g}_{jk}\, d\bar{X}^j\, d\bar{X}^k.$$

Because orthogonal curvilinear coordinate systems are most often of practical value, this discussion is restricted to them. (For an introduction to a more general procedure see Section 2*.)

For $n = 3$, each of the equations, $\bar{X}^1 =$ constant, $\bar{X}^2 =$ constant, $\bar{X}^3 =$ constant represents a coordinate surface. In rectangular Cartesian coordinates each of these surfaces is a plane, whereas in cylindrical coordinates $\rho =$ constant is a circular cylinder and $\theta =$ constant and $z =$ constant represent planes. The intersection of two coordinate surfaces determines a coordinate curve. Therefore three families of coordinate curves are associated with each system. It is in terms of the set of three coordinate curves associated with each point P (see Fig. 4-3.1) that we offer the following intuitive argument leading to a definition of an orthogonal differential element of volume. The pattern follows that established for surface elements of area.

Let $(ds)_1$, $(ds)_2$, $(ds)_3$, respectively, represent the differentials obtained from the fundamental metric form by holding $\bar{X}^2$, $\bar{X}^3$ constant, $\bar{X}^1$, $\bar{X}^3$ constant, and $\bar{X}^1$, $\bar{X}^2$ constant.

Theorem 4-3.1. For orthogonal curvilinear systems, we have

$$(ds)_1(ds)_2(ds)_3 = \sqrt{\bar{g}}\, d\bar{X}^1\, d\bar{X}^2\, d\bar{X}^3. \tag{4-3.1}$$

PROOF. Holding $\bar{X}^2$ and $\bar{X}^3$ constant reduces ds^2 to

$$(ds^2)_1 = \bar{g}_{11}(d\bar{X}^1)^2.$$

Therefore

$$ds_1 = \sqrt{\bar{g}_{11}}\, d\bar{X}^1.$$

Similarly

$$ds_2 = \sqrt{\bar{g}_{22}}\, d\bar{X}^2, \qquad ds_3 = \sqrt{\bar{g}_{33}}\, d\bar{X}^3.$$

Since the system is orthogonal $\bar{g}_{jk}$ is diagonal,

$$\bar{g} = \bar{g}_{11}\bar{g}_{22}\bar{g}_{33}.$$

This completes the proof.

Because magnitude cannot be associated with the quantities $(ds)_j$, it is necessary to follow the pattern of intuitive development in Chapter 4, Section 2; that is, we consider the finite quantities

$$\sqrt{\bar{g}_{11}}\, \Delta\bar{X}^1, \qquad \sqrt{\bar{g}_{22}}\, \Delta\bar{X}^2, \qquad \sqrt{\bar{g}_{33}}\, \Delta\bar{X}^3.$$

Consideration of the finite volume $(\sqrt{\bar{g}_{11}}\, \Delta\bar{X}^1)(\sqrt{\bar{g}_{22}}\, \Delta\bar{X}^2)(\sqrt{\bar{g}_{33}}\, \Delta\bar{X}^3)$ inspires the following definition.

Definition 4-3.1. The differential element of volume associated with an orthogonal curvilinear coordinate system $\bar{X}^j$ is

$$dV = \sqrt{\bar{g}}\, d\bar{X}^1\, d\bar{X}^2\, d\bar{X}^3. \tag{4-3.2}$$

Example 4-3.1. The differential volume element for rectangular Cartesian coordinates is

$$dV = dX^1\, dX^2\, dX^3, \tag{4-3.3a}$$

whereas for cylindrical coordinates

$$dV = \rho\, d\rho\, d\theta\, dz, \tag{4-3.3b}$$

and for spherical coordinates

$$dV = r^2 \sin\theta\, dr\, d\theta\, d\phi. \tag{4-3.3c}$$

The reader may recall having used these forms in a course in calculus.

Problems

1. According to (4-3.3a), the rectangular Cartesian differential volume element is $dX^1\,dX^2\,dX^3$. Let $\delta_1 X^j$, $\delta_2 X^j$, $\delta_3 X^j$ be defined such that

$$(\delta_1 X^1, \delta_1 X^2, \delta_1 X^3) = (dX^1, 0, 0),$$
$$(\delta_2 X^1, \delta_2 X^2, \delta_2 X^3) = (0, dX^2, 0),$$
$$(\delta_3 X^1, \delta_3 X^2, \delta_3 X^3) = (0, 0, dX^3).$$

Assume a corresponding definition in any other system. Define the $\delta_k X^j$ to transform as contravariant vector components. Then show that

(a) $dX^1\,dX^2\,dX^3 = \mathcal{E}_{jkp}\,\delta_1 X^j\,\delta_2 X^k\,\delta_3 X^p$,

(b) $dX^1\,dX^2\,dX^3 = \sqrt{\bar{g}}\,d\bar{X}^1\,d\bar{X}^2\,d\bar{X}^3$.

2. Show that $\sqrt{\bar{g}}\bar{\mathcal{E}}_{ijk}\,\delta\bar{X}^i\,\delta\bar{X}^j\,\delta\bar{X}^k$ is a scalar of weight zero with respect to the allowable transformation group.

3. Find the element of volume in a system of conical coordinates related to rectangular Cartesian coordinates by means of the transformation equations

$$X^1 = (s\sin\beta + r\cos\beta)\cos\theta, \qquad \bar{X}^1 = r,$$
$$X^2 = (s\sin\beta + r\cos\beta)\sin\theta, \qquad \bar{X}^2 = \theta,$$
$$X^3 = s\cos\beta - r\sin\beta, \qquad \bar{X}^3 = s,$$

where β is a constant.

4. *Integral Theorems*

Certain theorems relating integrals of different types (line, surface, volume) have proved to be of value in the development of theoretical physics. The purpose of this section is to investigate some of the best known. In particular, Stokes's theorem and the divergence theorem are discussed.

The first theorem to be considered is the so-called Green theorem, named for the English mathematician G. Green (1793–1841). The theorem appeared in his "Essay on the application of mathematical analysis to the theory of electricity and magnetism," a fundamental work in the mathematical formulation of electric and magnetic phenomena. From the viewpoint of our present development this theorem serves as an aid in obtaining the more general Stokes theorem.

Definition 4-4.1. A closed curve is said to be convex if any line in the plane cuts it in, at most, two points.

Theorem 4-4.1. (Green's theorem in the plane). Let R be a plane region bounded by a convex, sectionally smooth curve C. Suppose that $P^1(X^1, X^2)$, $P^2(X^1, X^2)$, $\partial P^1/\partial X^2$, $\partial P^2/\partial X^1$ are continuous on a domain containing C, hence R. Then

$$\int_R \left(\frac{\partial P^2}{\partial X^1} - \frac{\partial P^1}{\partial X^2}\right) dX^1\, dX^2 = \oint_C (P^1\, dX^1 + P^2\, dX^2). \tag{4-4.1}$$

(C is oriented so that an observer moving along the curve always has the inside to his left.)

PROOF. The functions P^1 and P^2 are independent of each other. Hence we must show that

$$\text{(a)} \quad \int_R \frac{\partial P^2}{\partial X^1}\, dX^1\, dX^2 = \oint_C P^2\, dX^2,$$

(4-4.2)

$$\text{(b)} \quad \int_R -\frac{\partial P^1}{\partial X^2}\, dX^1\, dX^2 = \oint_C P^1\, dX^1,$$

where C is a composite of the smooth sections C_1, C_2 and C_3; C_3 represents the possibility of vertical joins $X^1 = \text{constant}$. Suppose we consider (4-4.2b). The left-hand integral can be written as an iterated integral. (See Fig. 4.4.1.) Therefore

$$
\begin{aligned}
\text{(4-4.3)} \quad -\int_R \frac{\partial P^1}{\partial X^2}\, dX^1\, dX^2 &= -\int_{X_0}^{X_1}\left(\int_{Y_1}^{Y_2} \frac{\partial P^1}{\partial X^2}\, dX^2\right) dX^1 \\
&= -\int_{X_0}^{X_1} dX^1 [P^1(X^1; Y_2) - P^1(X^1; Y_1)] \\
&= \int_{X_0}^{X_1} P^1(X^1, Y_1)\, dX^1 + \int_{X_1}^{X_0} P^1(X^1, Y_2)\, dX^1 \\
&= \int_{C_1} P^1\, dX^1 + \int_{C_2} P^1\, dX^1 + \int_{C_3} P^1\, dX^1 \\
&= \oint_C P^1\, dX^1.
\end{aligned}
$$

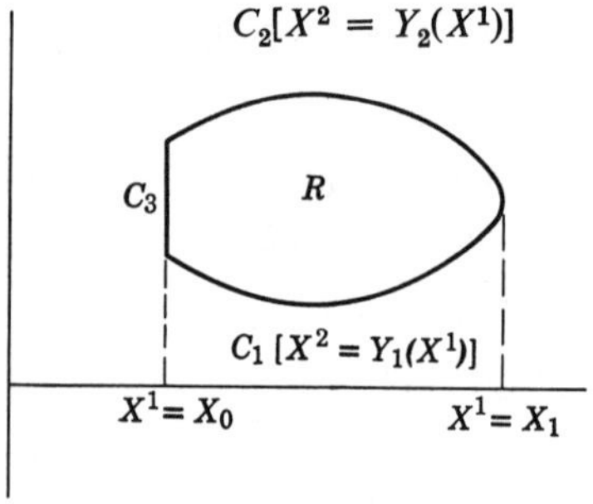

Fig. 4-4.1

Since $X^1 =$ constant on C_3, it follows that $dX^1 = 0$. Therefore $\int_{C_3} P^1\, dX^1 = 0$ and is included only for completeness. By the same process we can show the validity of (4-4.2a). The conclusion of the theorem follows by putting these results together.

It is left as an exercise for the reader to show that Green's theorem in the plane can be put in the vector form

$$\int_R \mathbf{N} \cdot \nabla \times \mathbf{P}\, dA = \oint_C \mathbf{P} \cdot d\mathbf{r}, \tag{4-4.4}$$

where $\mathbf{N} = \boldsymbol{\iota}_3$.

Example 4-4.1. Let $\mathbf{P} = -X^2\boldsymbol{\iota}_1 + X^1\boldsymbol{\iota}_2$.
According to Green's theorem (note $P^1 = -X^2, P^2 = X^1$),

$$2\int_R dA = \int_R \left(\frac{\partial X^1}{\partial X^1} - \frac{\partial(-X^2)}{\partial X^2}\right) dX^1\, dX^2 = \oint_C - X^2\, dX^1 + X^1\, dX^2;$$

that is

$$\boxed{A = \frac{1}{2}\oint_C - X^2\, dX^1 + X^1\, dX^2.} \tag{4-4.5}$$

In particular, the area of an ellipse can be calculated by taking as parametric equations

$$\begin{aligned} X^1 &= a \cos t, \\ X^2 &= b \sin t, \end{aligned} \qquad 0 \le t < 2\pi.$$

Then

$$\begin{aligned} A &= \frac{1}{2}\int_0^{2\pi} [-b \sin t(-a \sin t) + a \cos t(b \cos t)]\, dt \\ &= \frac{1}{2}\int_0^{2\pi} ab\, dt = \pi ab. \end{aligned}$$

G. G. Stokes (1819–1903, English), a contemporary of Green, was one of the leading developers of the English school of mathematical physics. The following theorem which bears his name plays a significant role in the mathematical development of physical theories such as electromagnetism. In fact, as previously hinted, the mathematical expression of physical phenomena gave rise to many of the integral ideas being studied in this section.

Recall that the term surface implies that the parametric equations $\mathbf{r} = \mathbf{r}(v^\beta)$ have continuous partials of at least order 1.

Theorem 4-4.2 (Stokes's theorem).[6] Suppose that a closed curve C bounds a smooth surface section S. If the component functions of $\mathbf{r} = \mathbf{r}(v^\beta)$ have continuous mixed partials and the image of C on the v^1, v^2 plane is convex, then for a vector field $\mathbf{G}$ with continuous partial derivatives on S we have

$$\boxed{\int_S \mathbf{N}\cdot\nabla\times\mathbf{G}\,dA = \oint_C \mathbf{G}\cdot d\mathbf{r}.} \tag{4-4.6}$$

Proof. The method of proof consists in expressing the left-hand member of (4-4.6) as an iterated integral and applying Green's theorem in the v^1, v^2 plane. According to (4-2.11),

$$\int_S \mathbf{N}\cdot\nabla\times\mathbf{G}\,dA = \int_{v_0{}^2}^{v_1{}^2}\int_{v^1=v_0{}^1(v^2)}^{v^1=v_1{}^1(v^2)} \left(\frac{\partial\mathbf{r}}{\partial v^1}\times\frac{\partial\mathbf{r}}{\partial v^2}\cdot\nabla\times\mathbf{G}\right) dv^1\,dv^2. \tag{4-4.7a}$$

Recall the Lagrange identity

$$\mathbf{A}\times\mathbf{B}\cdot\mathbf{C}\times\mathbf{D} = (\mathbf{A}\cdot\mathbf{C})(\mathbf{B}\cdot\mathbf{D}) - (\mathbf{A}\cdot\mathbf{D})(\mathbf{B}\cdot\mathbf{C}).$$

By applying this relation to the right member of (4-4.7a), we can express the parenthetic factor of the integrand

$$\frac{\partial\mathbf{r}}{\partial v^1}\times\frac{\partial\mathbf{r}}{\partial v^2}\cdot\nabla\times\mathbf{G} = \left(\frac{\partial\mathbf{r}}{\partial v^1}\cdot\nabla_G\right)\left(\mathbf{G}\cdot\frac{\partial\mathbf{r}}{\partial v^2}\right) - \left(\frac{\partial\mathbf{r}}{\partial v^2}\cdot\nabla_G\right)\left(\mathbf{G}\cdot\frac{\partial\mathbf{r}}{\partial v^1}\right). \tag{4-4.7b}$$

Since

$$\frac{\partial\mathbf{r}}{\partial v^\beta}\cdot\nabla = \frac{\partial X^j}{\partial v^\beta}\boldsymbol{\iota}_j\cdot\sum_k \boldsymbol{\iota}_k\frac{\partial}{\partial X^k} = \frac{\partial X^j}{\partial v^\beta}\delta_j^k\frac{\partial}{\partial X^k} = \frac{\partial}{\partial v^\beta}, \tag{4-4.7c}$$

the relation (4-4.7b) assumes the form

$$\begin{aligned}\frac{\partial\mathbf{r}}{\partial v^1}\times\frac{\partial\mathbf{r}}{\partial v^2}\cdot\nabla\times\mathbf{G} &= \frac{\partial\mathbf{G}}{\partial v^1}\cdot\frac{\partial\mathbf{r}}{\partial v^2} - \frac{\partial\mathbf{G}}{\partial v^2}\cdot\frac{\partial\mathbf{r}}{\partial v^1}\\ &= \frac{\partial[\mathbf{G}\cdot(\partial\mathbf{r}/\partial v^2)]}{\partial v^1} - \frac{\partial[\mathbf{G}\cdot(\partial\mathbf{r}/\partial v^1)]}{\partial v^2}.\end{aligned} \tag{4-4.7d}$$

By putting this result into (4-4.7a), we have

$$\begin{aligned}\int_S \mathbf{N}\cdot\nabla\times\mathbf{G}\,dA = \int_{v_0{}^2}^{v_1{}^2}\int_{v_0{}^1}^{v_1{}^1}\Bigg\{&\frac{\partial[\mathbf{G}\cdot(\partial\mathbf{r}/\partial v^2)]}{\partial v^1}\\ &- \frac{\partial[\mathbf{G}\cdot(\partial\mathbf{r}/\partial v^1)]}{\partial v^2}\Bigg\}\,dv^1\,dv^2.\end{aligned} \tag{4-4.7e}$$

[6] This theorem first appeared as a problem in a Cambridge examination paper in 1854.

Now if we identify P^1 with $\mathbf{G} \cdot (\partial\mathbf{r}/\partial v^1)$ and P^2 with $\mathbf{G} \cdot (\partial\mathbf{r}/\partial v^2)$, it can be seen that the hypothesis of Green's theorem is satisfied, hence the conclusion may be employed; that is,

$$\int_S \mathbf{N} \cdot \nabla \times \mathbf{G}\, dA = \oint_C \left(\mathbf{G} \cdot \frac{\partial \mathbf{r}}{\partial v^1} dv^1 + \mathbf{G} \cdot \frac{\partial \mathbf{r}}{\partial v^2} dv^2 \right) = \oint_C \mathbf{G} \cdot d\mathbf{r}.$$

This completes the proof.

Example 4-4.2. It has been shown (Chapter 3, Section 1) that the independence of path of a line integral $\int_C \mathbf{F} \cdot d\mathbf{r}$ was equivalent to the field $\mathbf{F}$ being conservative, that is, having a representative $\mathbf{F} = \nabla\Phi$. To show that there is Φ such that $\mathbf{F}$ is the gradient of Φ can be a difficult task. With the help of Stoke's theorem, we can show that

$$\nabla \times \mathbf{F} = \mathbf{0}$$

is a necessary and sufficient condition for the field $\mathbf{F}$ to be conservative. In particular, if $\mathbf{F}$ is conservative,

$$\nabla \times \mathbf{F} = \nabla \times \nabla\Phi = \mathbf{0}.$$

On the other hand, if $\nabla \times \mathbf{F} = 0$,

$$\oint_C \mathbf{F} \cdot d\mathbf{r}^7 = \int_S \nabla \times \mathbf{F} \cdot \mathbf{N}\, dA = 0.$$

Since the left-hand member of this expression represents a way of expressing the integral's independence of path, we can conclude that $\mathbf{F}$ is a conservative vector field.

It should be mentioned that some of the restrictions on the bounding curve C can be removed without destroying the validity of the theorem. However, easing of the conditions on C makes the proof quite a bit more difficult. It is also the case that the theorem can be extended to a surface that is a composite of smooth sections by a simple process of addition.

Stoke's theorem is not restricted to the scalar form in (4-4.6). This statement is verified by the following two theorems.

Theorem 4-4.3a. Let the vector field $\mathbf{G}$ of Theorem 4-4.2 be replaced according to the relation $\mathbf{G} = f(X^i)\mathbf{a}$ where f is a scalar function of

---

[7] A vector field $\mathbf{F}$ with the property $\oint_C \mathbf{F} \cdot d\mathbf{r} = 0$ often is said to be irrotational. The term arises from a physical interpretation of $\mathbf{F}$ as a velocity vector in a fluid. The relation $\nabla \times \mathbf{F} = \mathbf{0}$ can be shown to correlate with the lack of rotational velocity of the particles of the fluid.

position and **a** is an arbitrary constant vector. Then

$$\boxed{\int_S (\mathbf{N} \times \nabla f)\, dA = \oint_C f\, d\mathbf{r}.} \tag{4-4.8}$$

PROOF. We have

$$\nabla \times f\mathbf{a} = (\nabla f) \times \mathbf{a} = -\mathbf{a} \times \nabla f. \tag{4-4.9a}$$

Therefore

$$\mathbf{N} \cdot \nabla \times \mathbf{G} = -\mathbf{N} \cdot \mathbf{a} \times \nabla f = \mathbf{a} \cdot \mathbf{N} \times \nabla f. \tag{4-4.9b}$$

By employing (4-4.9b) we can put the relation in (4-4.6) in the form

$$\mathbf{a} \cdot \left[\int_S (N \times \nabla f)\, dA - \oint f\, d\mathbf{r}\right] = 0. \tag{4-4.9c}$$

The validity of removing **a** from the integral sign is a consequence of its constancy. Since **a** is arbitrary, the parenthetic expression must be zero. This completes the proof.

Theorem 4-4.3b. Let the vector field **G** of Theorem 4-4.2 be replaced by

$$\mathbf{G} = \mathbf{F} \times \mathbf{a},$$

where **F** is a differentiable vector field on S and **a** is an arbitrary vector constant. Then

$$\boxed{\int_S (\mathbf{N} \times \nabla) \times \mathbf{F}\, dA = \oint_C d\mathbf{r} \times \mathbf{F}.} \tag{4-4.10}$$

PROOF. We have

$$\begin{aligned} \mathbf{N} \cdot \nabla \times \mathbf{G} &= \mathbf{N} \cdot \nabla \times (\mathbf{F} \times \mathbf{a}) = \mathbf{N} \times \nabla \cdot (\mathbf{F} \times \mathbf{a}) \\ &= (\mathbf{N} \times \nabla) \times \mathbf{F} \cdot \mathbf{a} \end{aligned} \tag{4-4.11a}$$

and

$$\mathbf{G} \cdot d\mathbf{r} = \mathbf{F} \times \mathbf{a} \cdot d\mathbf{r} = d\mathbf{r} \cdot \mathbf{F} \times \mathbf{a} = d\mathbf{r} \times \mathbf{F} \cdot \mathbf{a}. \tag{4-4.11b}$$

Therefore Stoke's formula (4-4.6) can be written

$$\mathbf{a} \cdot \int_S (\mathbf{N} \times \nabla) \times \mathbf{F}\, dA = \mathbf{a} \cdot \oint_C d\mathbf{r} \times \mathbf{F}, \tag{4-4.11c}$$

where again the constancy of **a** makes it possible to pull that vector out of the integral. Relation (4-4.11c) is equivalent to

$$\mathbf{a} \cdot \left[\int_S (\mathbf{N} \times \nabla) \times \mathbf{F}\, dA - \oint_C d\mathbf{r} \times \mathbf{F}\right] = 0. \tag{4-4.11d}$$

Since **a** is arbitrary, the relation (4-4.10) follows.

Example 4-4.3. Let

$$\mathbf{G} = -X^2\boldsymbol{\iota}_1 + X^1\boldsymbol{\iota}_2 + X^3\boldsymbol{\iota}_3.$$

Suppose $\mathbf{G}$ is defined on $(X^1)^2 + (X^2)^2 + (X^3)^2 = r^2$, $X^3 \geq 0$. Then, according to Stoke's theorem [see (4-4.6)],

$$\text{(4-4.12a)} \qquad \int_S \frac{\mathbf{r}}{r} \cdot (2\iota_3)\, dA = \oint_C -X^2\, dX^1 + X^1\, dX^2 + X^3\, dX^3.$$

The bounding curve C is a circle in the plane $X^3 = 0$; therefore it follows from (4-4.5) (or by direct use of the parameterization $X^1 = r\cos\phi$, $X^2 = r\sin\phi$, $X^3 = 0$) that the right-hand line integral has the value $2\pi r^2$. Hence from (4-4.12a)

$$\text{(4-4.12b)} \qquad \int_S \frac{2X^3}{r}\, dX^1\, dX^2\, dX^3 = 2\pi r^2.$$

The next integral theorem to be considered is the so-called divergence theorem, often called Gauss's theorem. It states a relationship between an integral over a closed surface and an integral over the enclosed volume. As with Stoke's theorem, we cut down on the generality of the theorem, in this case by assuming that the surface is oval-shaped. (Any line in space cuts the surface in, at most, two points.) This makes possible the presentation of a relatively simple proof.

Theorem 4-4.4. Let $\mathbf{G}$ be a vector field with continuous derivatives on a closed space region R. Suppose that S were a smooth oval-shaped surface bounding R. Then

$$\text{(4-4.13)} \qquad \boxed{\int_V \nabla \cdot \mathbf{G}\, dV = \oint_S \mathbf{G} \cdot \mathbf{n}\, dA,}$$

where $\mathbf{n}$ is an outwardly drawn normal.

PROOF. Assume that the quantities involved in (4-4.13) are referred to a rectangular Cartesian coordinate system. In orthogonally projecting the surface S onto a coordinate plane (say the X^1, X^2 plane) each point P of the projection (see Fig. 4-4.2) corresponds to two surface points. Suppose $\mathbf{n}_2$ and $\mathbf{n}_1$ represent the outward drawn normals on the upper portion S_2 and lower portion S_1 of the surface, respectively. The vector element of surface area is (assume parameterizations $X^1 = v^1$, $X^2 = v^2$, $X^3 = z_\beta(v^1, v^2)$, $\beta = 1, 2$, respectively, for the surfaces S_1 and S_2)

$$\text{(4-4.14a)} \qquad \mathbf{N}\, dA = \frac{\partial \mathbf{r}}{\partial v^1} \times \frac{\partial \mathbf{r}}{\partial v^2}\, dv^1\, dv^2$$

$$= \left(\iota_1 + \frac{\partial X^3}{\partial v^1}\iota_3\right) \times \left(\iota_2 + \frac{\partial X^3}{\partial v^2}\iota_3\right) dv^1\, dv^2$$

$$= \left(\iota_3 - \frac{\partial X^3}{\partial v^2}\iota_2 - \frac{\partial X^3}{\partial v^1}\iota_1\right) dv^1\, dv^2.$$

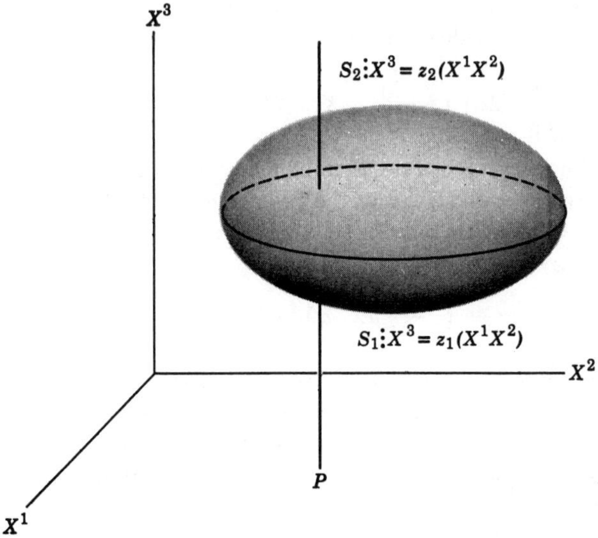

Fig. 4-4.2

Hence

(4-4.14b) $$\iota_3 \cdot \mathbf{N}\, dA = dv^1\, dv^2 = dX^1\, dX^2.$$

According to (4-4.14b), $\mathbf{N}$ has the orientation of ι_3. Therefore

(4-4.14c) $$\begin{aligned} \iota_3 \cdot \mathbf{n}_2\, dA &= dX^1\, dX^2, \\ \iota_3 \cdot \mathbf{n}_1\, dA &= -dX^1\, dX^2. \end{aligned}$$

This result will prove useful in the sequel.

The next step in the proof consists of performing one of the integrations indicated by the left-hand member of (4-4.13). We have

(4-4.15a) $$\int_V \nabla \cdot \mathbf{G}\, dV = \iiint\limits_V \left(\frac{\partial G^1}{\partial X^1} + \frac{\partial G^2}{\partial X^2} + \frac{\partial G^3}{\partial X^3} \right) dX^1\, dX^2\, dX^3.$$

Consider the third of these integrals; that is,

(4-4.15b) $$\iiint\limits_V \frac{\partial G^3}{\partial X^3}\, dX^3\, dX^2\, dX^1 = \int_{x_1}^{x_2} \int_{y_1}^{y_2} dX^2\, dX^1 \int_{z_1}^{z_2} \frac{\partial G^3}{\partial X^3}\, dX^3,$$

where

$$x_1 \leq X^1 \leq x_2, \qquad y_1(X^1) \leq X^2 \leq y_2(X^1), \qquad z_1(X^1, X^2) \leq X^3 \leq z_2(X^1, X^2).$$

By performing the indicated integration, we obtain

$$\iiint_V \frac{\partial G^3}{\partial X^3} dX^3 dX^2 dX^1 = \int_{x_1}^{x_2}\int_{y_1}^{y_2} [G^3(X^1, X^2, z_2) - G^3(X^1, X^2, z_1)]\, dX^2 dX^1$$

$$= \int_{S_2} G^3 \mathbf{\iota}_3 \cdot \mathbf{n}_2\, dA - \int_{S_1} G^3 \mathbf{\iota}_3 \cdot (-\mathbf{n}_1)\, dA$$

$$= \oint_S G^3 \mathbf{\iota}_3 \cdot \mathbf{n}\, dA.$$

In a similar manner, the remaining two integrals of (4-4.15a) can be expressed in the forms

(4-4.16)
$$\iiint_V \frac{\partial G^2}{\partial X^2} dV = \oint_S G^2 \mathbf{\iota}_2 \cdot \mathbf{n}\, dA,$$
$$\iiint_V \frac{\partial G^1}{\partial X^1} dV = \oint_S G^1 \mathbf{\iota}_1 \cdot \mathbf{n}\, dA.$$

Together these results produce the conclusion (4-4.13).

Among the integral relations developed during the English upsurge of interest in applied mathematics (nineteenth century) were the so-called Green identities. The formulas, stated in the next two theorems, are named after the same George Green previously mentioned in this section.

Theorem 4-4.5a (Green's first identity). Suppose $U(X^j)$, $W(X^j)$ are scalar functions with continuous second partial derivatives on a space region R. Suppose R is such that the bounding surface S meets the conditions of the divergence theorem. Then

(4-4.17) $$\int_V [U\, \nabla^2 W + (\nabla U) \cdot (\nabla W)]\, dV = \oint_S U \mathbf{n} \cdot (\nabla W)\, dA, \quad \nabla^2 = \nabla \cdot \nabla$$

PROOF. Let

(4-4.18a) $$\mathbf{G} = U\, \nabla W;$$

then

(4-4.18b) $$\nabla \cdot \mathbf{G} = \nabla U \cdot \nabla W + U\, \nabla^2 W.$$

By making the substitutions (4-4.18a,b) in (4-4.13) we obtain the conclusion of the theorem.

Theorem 4-4.5b (Green's second identity). Under the hypothesis in Theorem 4-4.5a

(4-4.19) $$\int_V (U\, \nabla^2 W - W\, \nabla^2 U)\, dV = \oint_S \mathbf{n} \cdot (U\, \nabla W - W\, \nabla U)\, dA.$$

PROOF. Let

(4-4.20a) $$\mathbf{G}_1 = U\,\nabla W, \qquad \mathbf{G}_2 = W\,\nabla U.$$

Form

(4-4.20b) $$\mathbf{G} = \mathbf{G}_1 - \mathbf{G}_2.$$

The result (4-4.19) is obtained by replacing $\mathbf{G}$ in (4-4.13) according to (4-4.20b).

Example 4-4.4. Maxwell's equations, which serve as a foundation for electromagnetic theory, can be expressed in the form (in Gaussian units)

(4-4.21)
$$\text{(a)}\quad \nabla \times \mathbf{E} = -\frac{1}{c}\frac{\partial \mathbf{B}}{\partial t}, \qquad \text{(b)}\quad \nabla \cdot \mathbf{D} = 4\pi\rho,$$
$$\text{(c)}\quad \nabla \times \mathbf{H} = \frac{1}{c}\frac{\partial \mathbf{D}}{\partial t} + \frac{4\pi\mathbf{j}}{c}, \qquad \text{(d)}\quad \nabla \cdot \mathbf{B} = 0,$$ [8]

where $\mathbf{E}$ and $\mathbf{H}$ represent electric and magnetic field strengths, respectively, whereas $\mathbf{B}$ denotes magnetic induction, $(1/4\pi)\mathbf{D}$ signifies displacement current, and $\mathbf{j}$ is an electric current vector. The constant c represents the velocity of light. For the most part these laws were derived from the experiments of Oersted, Faraday, and others. (See Chapter 1, Section 6.) However, the form of (4-4.21c), which resulted from experimentation, did not contain the term $1/c\,(\partial\mathbf{D}/\partial t)$. This term was added by Maxwell on the basis of intuition supported by mathematical consistency. Maxwell noticed that the divergence of the left-hand side of (4-4.21c) was identically zero, whereas that of the right-hand side, in the absence of the partial derivative term, was not. With this inconsistency in mind, he considered the flux of electric current through an arbitrary area, $\int_S \mathbf{j} \cdot \mathbf{n}\, dA$. According to the divergence theorem,

(4-4.22a) $$\int_S \mathbf{j} \cdot \mathbf{n}\, dA = \int_V \nabla \cdot \mathbf{j}\, dV.$$

Furthermore, Maxwell assumed the conservation of charge, that is,

(4-4.22b) $$\int_V \nabla \cdot \mathbf{j}\, dV = -\frac{\partial Q}{\partial t}.$$

where Q represents the charge inside the volume at any moment, that is,

(4-4.22c) $$Q = \int_V \rho\, dV.$$

---

[8] A vector field $\mathbf{G}$ such that $\oint_S \mathbf{G} \cdot \mathbf{n}\, dA = 0$ is said to be solenoidal. $\mathbf{B}$ is clearly a solenoidal vector field.

From (4-4.22b) and (4-4.22c) it results that

(4-4.22d) $$\int_V \left(\nabla \cdot \mathbf{j} + \frac{\partial \rho}{\partial t}\right) dV = 0.$$

Since this relation is independent of the volume V, it follows that the integrand is equal to zero.

(4-4.22e) $$\nabla \cdot \mathbf{j} + \frac{\partial \rho}{\partial t} = 0.$$

Maxwell assumed the validity of (4-4.21b), which, when solved for ρ, has the form

$$\rho = \frac{1}{4\pi} \nabla \cdot \mathbf{D}.$$

From this it follows that

(4-4.22f) $$\frac{\partial \rho}{\partial t} = \frac{1}{4\pi} \nabla \cdot \frac{\partial \mathbf{D}}{\partial t}.$$

Putting this result in (4-4.22e), we have

$$0 = \nabla \cdot \mathbf{j} + \frac{1}{4\pi} \nabla \cdot \frac{\partial \mathbf{D}}{\partial t} = \nabla \cdot \left(\mathbf{j} + \frac{1}{4\pi} \frac{\partial \mathbf{D}}{\partial t}\right).$$

In other words, the divergence of $\mathbf{j}$ is not, in general, zero, but the divergence of $\mathbf{j} + 1/4\pi(\partial \mathbf{D}/\partial t)$ is equal to zero. Therefore Maxwell hypothesized that

$$\nabla \times \mathbf{H} = \frac{4\pi}{c}\left(\mathbf{j} + \frac{1}{4\pi} \frac{\partial \mathbf{D}}{\partial t}\right).$$

The existence of the displacement current, $1/4\pi(\partial \mathbf{D}/\partial t)$, was confirmed later by experiments carried out by Hertz.

Problems

1. Use Green's theorem to evaluate the integral of $\mathbf{P}$ over the closed path C:

(a) $\mathbf{P} = X^2 \iota_1$, $\quad C: \begin{cases} X^2 = (X^1)^2 + 4 \\ X^2 = -[(X^1)^2 - 8] \end{cases}$ counterclockwise.

(b) $\mathbf{P} = (X^2)^2 \iota_1 + X^1 X^2 \iota_2$, $\quad C$: the unit square with origin as center.

(c) $\mathbf{P} = X^1 (X^2)^2 \iota_1 + 2(X^1)^2 X^2 \iota_2$, $\quad C$: counterclockwise around the ellipse

$$\frac{(X^1)^2}{9} + \frac{(X^2)^2}{4} = 1.$$

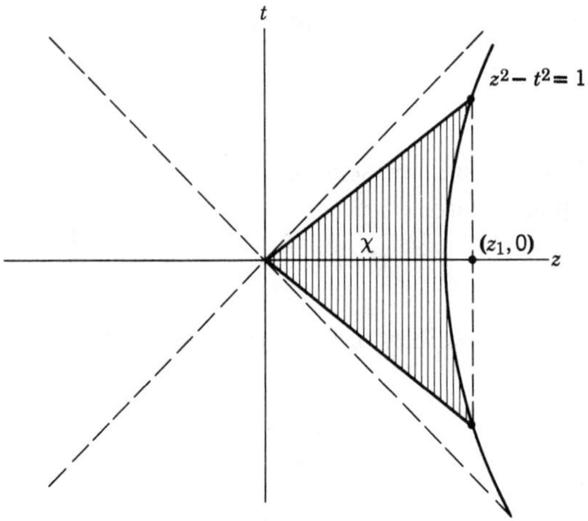

2. Use Green's theorem and the method of Example 4-4.1 to show that the area $\chi = \cosh^{-1} z_1$. (See the accompanying diagram.)
3. Let $\mathbf{P} = U^2\iota_1 - U^1\iota_2$. Show that (4-4.4) can be put in the form

$$\int_S \nabla \cdot \mathbf{U}\, dA = \oint_C (\mathbf{U} \times d\mathbf{r}) \cdot \iota_3.$$

4. Use Stoke's theorem to evaluate the integral of **G** over the closed path C bounding the region R.

$$\mathbf{G} = X^2\iota_1 + X^3\iota_2 + X^1\iota_3, \qquad R: \begin{cases} X^1 = v^1, \ -1 \le v^1 \le 1, \\ X^2 = (v^1)^2, \\ X^3 = v^2, \ 0 \le v^2 \le 4. \end{cases}$$

$$C: \begin{cases} X^1 = v^1, & X^2 = (v^1)^2, & X^3 = 0, & -1 \le v^1 \le 1, \\ X^1 = 1, & X^2 = 1, & X^3 = v^2, & 0 \le v^2 \le 4, \\ X^1 = -v^1, & X^2 = (v^1)^2, & X^3 = 4, & -1 \le v^1 \le 1, \\ X^1 = -1, & X^2 = 1, & X^3 = 4 - v^2, & 0 \le v^2 < 4. \end{cases}$$

5. Show that the integral of $\mathbf{G} = -X^2\iota_1 + X^1\iota_2 + X^3\iota_3$ over any surface satisfying the properties of Stokes's theorem and bounded by a convex curve in the plane $X^3 = 0$ is equal to the integral of **G** over the plane region bounded by C.
6. If $\mathbf{G} = X^1\iota_1$, find the $\int_S \mathbf{G} \cdot \mathbf{n}\, dA$, where S is the surface $(X^1)^2 + (X^2)^2 + (X^3)^2 = 1$ by (a) direct integration, (b) the divergence theorem.
7. Find $\int_S \mathbf{G} \cdot \mathbf{n}\, dA$ over $(X^1)^2 + (X^2)^2 + 2X^3$ by using the divergence theorem, where **G** is defined in Problem 6.

8. Suppose that W satisfied Laplace's equation

$$\nabla^2 W = 0.$$

Show that Green's first identity reduces to

$$\int_V \nabla U \cdot \nabla W \, dV = \oint_S U\mathbf{n} \cdot \nabla W \, dA.$$

9. (a) Let $\mathbf{P} = X^2 \iota_1$ and show that $A = -\oint_C X^2 \, dX^1$.

(b) Let $\mathbf{P} = X^1 \iota_2$ and show that $A = \oint_C X^1 \, dX^2$.

(c) Show that the results of (a) and (b) imply the area formula of Example 4-4.1.

(d) Use the formula of (a) to obtain the area of the ellipse of Example 4-4.1.

10. The experimental laws from which the Maxwell equations arose can be regained from these equations.

(a) Obtain Faraday's law of electromagnetic induction

$$\oint_C \mathbf{E} \cdot d\mathbf{r} = -\frac{1}{c}\frac{\partial}{\partial t}\int_S \mathbf{B} \cdot \mathbf{N} \, dA$$

from (4-4.21a) and the use of Stokes's theorem.

(b) Obtain Oersted's law

$$\int_C \mathbf{H} \cdot d\mathbf{r} = \frac{4\pi}{c}\int_S \mathbf{j} \cdot \mathbf{N} \, dA$$

from

$$\nabla \times \mathbf{H} = \frac{4\pi}{c}\mathbf{j}.$$

(c) Show that the total magnetic flux (i.e., $\int_S \mathbf{B} \cdot \mathbf{n} \, dA$) through a closed oval-shaped surface is zero.

11. Suppose that a continuous vector function $\mathbf{F}$ with continuous first partial derivations over a region including a smooth surface section S is identically zero on S. Show that $\nabla \times \mathbf{F}$ is either tangential to S or equal to $\mathbf{0}$.

chapter 5 tensor algebra and analysis

Tensor calculus came into prominence with the development of the general theory of relativity by Einstein in 1916. It evolved into a cohesive body of knowledge through the efforts of Ricci and Levi-Civita, who published a paper[1] in 1900 illustrating the application of the tensor methods to geometry and mathematical physics. Tensor calculus comes near to being a universal language in mathematical physics. It allows for compact expression of equations and also provides a guide to the selection of physical laws.

The role of tensor analysis is that of a theoretic tool. In general we can expect qualitative results rather than quantitative; that is, relationships can be found and examined but numerical answers are not so likely to be obtained.

From the mathematical point of view, we shall be studying geometric invariants and the modes of transformation of their coordinate system representations. The intuitive background for the formal development of the ensuing pages can be found in Chapters 1, 2, and 3. In fact, much of the present task is simply that of extending the ideas previously developed in terms of Euclidean two- or three-space to a generalized space of n dimensions. This is precisely the process carried out by Riemann in extending the surface theory developed by Gauss to a space of n dimensions. In carrying out a process of generalization, the results must contain the initial information as a special case or as an approximation. Furthermore, it is hoped that the generalizations will lead to new results and new applications. This hope has been realized for the ideas presented in this chapter

[1] Méthodes de calcul différential et leurs applications, *Math. Ann.*, **54,** 1901.

by the development of general relativity and numerous other applications to mathematical physics.

1. *Fundamental Notions in n-space*

The term "space" has been used consistently as an undefined concept. The nature of man's reality is such that the terms "euclidean two- or three-space" have a meaning common to a large group of individuals. It is usual (a practice on which our previous work depends) to assume a one-to-one correspondence between the points of these spaces and the sets of all real-number pairs or real-number triples, respectively. In fact, it is not uncommon to say, for example, that the set of all real-number triples is the three-space. I prefer to preserve the geometric identity of the term "space," as well as the language of geometry, as an intermediate between the physical and the algebraic. Furthermore, surfaces provide examples of non-Euclidean two-spaces, that is, spaces for which the pythagorean theorem does not intrinsically determine the metric. For a surface it is interesting to note that the set of real number pairs put in one-to-one correspondence with the points may be a restricted set. For example, in the sphere with parameterization $X^1 = \sin\theta\cos\phi$, $X^2 = \sin\theta\sin\phi$, $X^3 = \cos\theta$, the pairs (θ, ϕ) satisfy the restrictions $0 \leq \theta \leq \pi$, $0 \leq \phi < 2\pi$.

It is assumed that there is a one-to-one correspondence between the points of a general n-space[2] and the ordered set of n-tuples of real numbers $\{X^1, \cdots, X^n\}$. The real-number domain of any one of the coordinates $X^1, \cdots, X^n$ may be restricted to a finite interval. Whether the n-space can be attributed a physical existence is of philosophic importance but need not concern us. The usefulness of an n-dimensional structure is pointed out by the many physical problems involving more than three independent variables; for example, the problem of describing the behavior of a particle whose velocity does not depend on the position. In this case there are six independent variables and the physicist deals with a so-called six-dimensional configuration space.

As a first step toward the development of the tensor algebra associated with an n-dimensional space, let us fix our attention on a particular coordinate system. Rules of operation for the following entities are introduced.

Definition 5-1.1. Sets of n^k elements are called systems of order k.

---

[2] The words "hyperspace," "variety," or "manifold" are often used for n-space in order to distinguish it from the usual three-space.

Example 5-1.1. The linear homogeneous function

$$a_j X^j, \qquad j = 1, \cdots, n$$

is a system of order zero, whereas the sets $(a_1, \cdots, a_n)$ and $(X^1, \cdots, X^n)$ are systems of order 1. The Kronecker delta

$$(\delta_j^{\ k}) = \begin{pmatrix} 1 & 0 & \cdots & 0 \\ 0 & 1 & \cdots & 0 \\ \cdot & \cdot & \cdot\cdot\cdot & \cdot \\ 0 & 0 & \cdots & 1 \end{pmatrix} \tag{5-1.1}$$

is a system of order 2.

Definition 5-1.2. A system, $T_{j_1 \cdots j_p}^{k_1 \cdots k_q}$, expressed by means of q superscripts and p subscripts is said to have contravariant valence q and covariant valence p.

The Kronecker delta has a contravariant valence 1 and covariant valence 1.

The terms covariant and contravariant were first discussed in Chapter 1 Section 5*. They refer to modes of transformation, and, of course, it is our present purpose to set up an algebraic structure that will be consistent with a theory of tensor transformations.

The systems introduced by Definition 5-1.1 are subject to the following rules of operation.

Definition 5-1.3a. If two systems have the same covariant valences and the same contravariant valences, the set of elements obtained by component addition is a system of the same type (i.e., the covariant and contravariant valences agree with those of the original systems).

Example 5-1.2a. If $a_j^{\ k}$, $b_j^{\ k}$ are given systems, then

$$d_j^{\ k} = a_j^{\ k} + b_j^{\ k} \tag{5-1.2}$$

is a system of the same covariant and contravariant valences. The components of $d_j^{\ k}$ are n^2 in number. For example, if $n = 5$.

$$d_1^{\ 1} = a_1^{\ 1} + b_1^{\ 1}$$
$$d_2^{\ 5} = a_2^{\ 5} + b_2^{\ 5}, \text{ etc.}$$

Clearly, the addition process can be extended to any number of systems, all of which have the same covariant and contravariant valences.

Definition 5-1.3b. Let $T_{j_1 \cdots j_p}^{k_1 \cdots k_q}$ and $R_{l_1 \cdots l_r}^{m_1 \cdots m_s}$ be arbitrary systems. Then the system

$$S_{j_1 \cdots j_p l_1 \cdots l_r}^{k_1 \cdots k_q m_1 \cdots m_s} = T_{j_1 \cdots j_p}^{k_1 \cdots k_q} R_{l_1 \cdots l_r}^{m_1 \cdots m_s} \tag{5-1.3a}$$

of contravariant valence $q + s$ and covariant valence $p + r$ is said to be the outer product of the two original systems. The number of components of the outer product is $n^{p+q+r+s}$.

Example 5-1.2b. If P^j and Q^k are components of n-space n-tuples, then P^jQ^k is a system of contravariant valence 2. The number of components of this system is n^2.

Definition 5-1.3c. Let $T^{k_1\cdots k_q}_{j_1\cdots j_p}$ be a system of mixed order; that is, $p > 0, q > 0$. The system

(5-1.3b)
$$T^{k_1\cdots k_{b-1}k_{b+1}\cdots k_q}_{j_1\cdots j_{a-1}j_{a+1}\cdots j_p} = T^{k_1\cdots k_{b-1}ck_{b+1}\cdots k_q}_{j_1\cdots j_{a-1}cj_{a+1}\cdots j_p}$$

is said to be obtained from the original system by means of contraction. The summation on $j_a = k_b = c$ reduces both the covariant and the contravariant valence of the system by 1. Hence the order of the system is reduced by 2.

Example 5-1.2c. If the Kronecker delta is contracted, we obtain a scalar; that is,

(5-1.4)
$$\delta_j{}^j = n.$$

Definition 5-1.3d. Let $T^{k_1\cdots k_q}_{j_1\cdots j_p}$ and $R^{m_1\cdots m_s}_{l_1\cdots l_r}$ be arbitrary systems. The system

(5-1.5)
$$S^{k_1\cdots k_q m_1\cdots m_{b-1}m_{b+1}\cdots m_s}_{j_1\cdots j_{a-1}j_{a+1}\cdots j_p l_1\cdots l_r} = T^{k_1\cdots k_q}_{j_1\cdots j_{a-1}cj_{a+1}\cdots j_p}R^{m_1\cdots m_{b-1}cm_{b+1}\cdots m_q}_{l_1\cdots l_r}$$

obtained by summing on j_a and m_b is said to result from transvection of the original systems. The summation could just as well be with respect to a contravariant index of T and a covariant index of R.

The process of transvection can be thought of as a combination of multiplication followed by contraction.

Example 5-1.2d. Suppose that P^j and Q^k are components of three-space n-tuples. The cross product

(5-1.6)
$$R_p = \varepsilon_{pjk}P^jQ^k$$

is a system of order 1 obtained by a double application of the process of transvection.

The concepts of symmetry and skew symmetry for double indexed systems have been discussed; that is, if

(5-1.7a)
$$A_{jk} = A_{kj},$$

the system A_{jk} is said to be symmetric, whereas

(5-1.7b) $$A_{jk} = -A_{kj}$$

symbolizes the property of skew symmetry. It is advantageous to extend the terminology to systems of higher order.

Definition 5-1.4a. A system $A_{j_1 \cdots j_k}$, which satisfies the property

(5-1.7c) $$A_{j_1 \cdots j_p j_q \cdots j_k} = A_{j_1 \cdots j_q j_p \cdots j_k}$$

for all pairs $p, q = 1, 2;\ 2, 3, \cdots, k-1, k$ is said to be completely symmetric.

Definition 5-1.4b. A system $A_{j_1 \cdots j_k}$, which satisfies the property

(5-1.7d) $$A_{j_1 \cdots j_p j_q \cdots j_k} = -A_{j_1 \cdots j_q j_p \cdots j_k}$$

for any two adjacent indices is said to be completely skew symmetric.

Definitions 5-1.4a,b apply to contravariant systems as well. However, they are not to be extended to mixed indices.

Example 5-1.3. The $\mathcal{E}$-system $\mathcal{E}_{ijk}$ is completely skew symmetric; that is,

$$\mathcal{E}_{123} = \mathcal{E}_{231} = \mathcal{E}_{312} = -\mathcal{E}_{213} = -\mathcal{E}_{132} = -\mathcal{E}_{321}.$$

A system A^{ijk} such that

$$A^{ijk} = A^{jki} = A^{kij} = A^{jik} = A^{ikj} = A^{kji}$$

illustrates a completely symmetric set of components. The six possible ways of expressing the three index systems found in this example are easily obtained by a process of cycling. The procedure can be correlated to that of interchanging an even or odd number of adjacent pairs, that is, an even or odd transposition.

In Chapter 1, Section 8, equality of matrices A, B was defined as $A_{jk} = B_{jk}$ for all pairs j, k. Therefore the transpose A' of a matrix A is a new entity unless, of course, the matrix is symmetric. Just as a given matrix gives rise to another matrix (i.e., the transpose), systems give rise to other systems. This idea, and the meaning of equality of two systems, is expounded in the following definition.

Definition 5-1.5. Two systems $T^{k_1 \cdots k_q}_{j_1 \cdots j_p}$ and $R^{m_1 \cdots m_q}_{l_1 \cdots l_p}$ are equal if and only if

(5-1.8) $$R^{k_1 \cdots k_q}_{j_1 \cdots j_p} = T^{k_1 \cdots k_q}_{j_1 \cdots j_p}.$$

A given system, for example $A_{j_1 \cdots j_p}$ gives rise to $p!$ systems by reordering of the elements of the set. Since an interchange of indices is equivalent to either an even number or an odd number of transpositions, it is possible to classify the new systems.

Definition 5-1.6. A system $T_{j_1 \cdots j_p}$ arising from $T_{k_1 \cdots k_p}$ by means of a reordering of components is said to be an isomer of the original system. If $k_1 \cdots k_p$ is an even permutation of $j_1 \cdots j_p$, the isomer is said to be even. If $k_1 \cdots k_p$ is an odd permutation of $j_1 \cdots j_p$, the isomer is said to be odd.

Since all isomers of a given system are of the same order and valence, we can carry the process of construction a step further and combine isomers. This leads to the interesting consequence of being able to construct completely symmetric or completely skew symmetric systems from a given system.

Theorem 5-1.1. The systems $A_{(j_1 \cdots j_p)}$ and $A_{[j_1 \cdots j_p]}$, defined by

$$\text{(5-1.9)} \quad \begin{aligned} &\text{(a)} \quad A_{(j_1 \cdots j_p)} = \frac{1}{p!} \begin{array}{l}(\text{sum of all even isomers } + \\ \text{sum of all odd isomers of} \\ A_{j_1 \cdots j_p}),\end{array} \\ &\text{(b)} \quad A_{[j_1 \cdots j_p]} = \frac{1}{p!} \begin{array}{l}(\text{sum of all even isomers } - \\ \text{sum of all odd isomers of} \\ A_{j_1 \cdots j_p}),\end{array} \end{aligned}$$

are completely symmetric and completely skew symmetric, respectively. Corresponding statements hold for contravariant systems.

PROOF. Consider (5-1.9a). If an interchange of adjacent indices is made, the odd isomers become even and vice versa. Since a plus sign joins the two groupings, the resulting expression is the same as the original. In (5-1.9b) the two groupings are connected by a minus sign. This accounts for the skew symmetry.

Example 5-1.4a. Let A_{jk} be a given system. Then the completely symmetric systems constructed from A_{jk} are, respectively,

$$\text{(5-1.10)} \quad \begin{aligned} &\text{(a)} \quad A_{(jk)} = \tfrac{1}{2}(A_{jk} + A_{kj}), \\ &\text{(b)} \quad A_{[jk]} = \tfrac{1}{2}(A_{jk} - A_{kj}). \end{aligned}$$

Suppose that A_{jk} is a symmetric system. Then from (5-1.10a)

$$A_{(jk)} = \tfrac{1}{2}(A_{jk} + A_{kj}) = \tfrac{1}{2}2A_{jk} = A_{jk}.$$

Therefore the factor of $\frac{1}{2}$, which is not necessary for the symmetry of the expression, is needed for conformity.

Example 5-1.4b. Let A_{jkp} be a given system. Then

$$\text{(c)}\quad A_{(jkp)} = \frac{1}{3!}\,[(A_{jkp} + A_{kpj} + A_{pjk}) + (A_{kjp} + A_{jpk} + A_{pkj})]$$

(5-1.10)

$$\text{(d)}\quad A_{[jkp]} = \frac{1}{3!}\,[(A_{jkp} + A_{kpj} + A_{pjk}) - (A_{kjp} + A_{jpk} + A_{pkj})].$$

The differential form

$$ds^2 = g_{\alpha\beta}\, dX^\alpha\, dX^\beta$$

plays a part in later considerations concerning Riemannian geometry. The form is symmetric in $dX^\alpha\, dX^\beta$. The next two theorems point out that $g_{\alpha\beta}$ might as well be chosen as a symmetric system.

Theorem 5-1.2a. A quadratic system A_{jk} can be expressed as a sum of its symmetric and skew symmetric parts. The same statement holds for a contravariant system.

PROOF. We have

$$\text{(5-1.11a)}\quad A_{(jk)} + A_{[jk]} = \tfrac{1}{2}(A_{jk} + A_{kj}) + \tfrac{1}{2}(A_{jk} - A_{kj}) = A_{jk}.$$

In the same way, we determine that

$$\text{(5-1.11b)}\qquad A^{jk} = A^{(jk)} + A^{[jk]}.$$

Theorem 5-1.2b. Suppose that B^{jk} is symmetric and A_{jk} is skew symmetric. Then

$$\text{(5-1.12)}\qquad A_{jk}B^{jk} = 0.$$

PROOF. First we use skew symmetry and symmetry, respectively, to interchange indices and then replace dummy indices in order to switch back. Symbolically, we have

$$A_{jk}B^{jk} = -A_{kj}B^{kj} = -A_{jk}B^{jk}.$$

Therefore

$$2A_{jk}B^{jk} = 0.$$

This completes the proof.

The task of applying Theorems 5-1.2a,b to the differential form is left to the reader. (See Problem 2.)

We next consider the extension of $\mathcal{E}$ systems and related entities to n-space.

Definition 5-1.7. Let

(5-1.13)

$$\mathcal{E}_{j_1 \cdots j_n} = E^{j_1 \cdots j_n} = \begin{cases} 1 \\ -1 \text{ if } j_1 \cdots j_n \text{ is an} \\ 0 \end{cases} \begin{cases} \text{even permutation} \\ \text{odd permutation of } 1 \cdots n. \\ \text{otherwise} \end{cases}$$

This definition differs from that in Chapter 1, Section 6, only in extent. The meaning of the Kronecker delta is unchanged; that is,

$$\delta_j^{\ k} = \begin{cases} 1 \\ 0 \end{cases} \text{ if } \begin{cases} j = k, \\ j \neq k. \end{cases}$$

It is convenient to have available the generalized Kronecker delta introduced in the next definition.

Definition 5-1.8. Let

$$\delta_{j_1 \cdots j_q}^{k_1 \cdots k_q} = \delta_{j_1}^{k_1} \delta_{j_2}^{k_2} \cdots \delta_{j_q}^{k_q}. \tag{5-1.14}$$

The relation between the covariant and contravariant $\mathcal{E}$ systems can now be stated as follows.

Theorem 5-1.3. We have

$$\mathcal{E}_{j_1 \cdots j_p k_{p+1} \cdots k_n} E^{k_1 \cdots k_n} = p!\,(n-p)!\,\delta_{j_1 \cdots j_p}^{[k_1 \cdots k_p]}, \tag{5-1.15}$$

where the bracket notation has the meaning indicated in (5-1.9b).

PROOF. Consider the left-hand member of (5-1.15). When the summation on k_n is carried out, we obtain at most $n - p$ nonzero terms. The number of nonzero terms is exactly $n - p$ if $j_1 \cdots j_p$ is a distinct set chosen from $1 \cdots n$, and $k_1 \cdots k_p$ is some permutation of this set. [If $j_1 \cdots j_p$ and $k_1 \cdots k_p$ are otherwise, (5-1.15) reduces to the identity $0 = 0$.] Under this circumstance, next sum on k_{n-1}. The resulting number of nonzero terms is $(n-p)(n-p-1)$. Continuing this process of writing out the indicated summations, we obtain

$$(n-p)(n-p-1) \cdots 1 = (n-p)!$$

nonzero terms. One typical term would be

$$\mathcal{E}_{j_1 \cdots j_p p+1 \cdots n} E^{k_1 \cdots k_p p+1 \cdots n}.$$

Since the last $n - p$ sub- and superscripts agree in each term, the value $(0, 1, -1)$ of any term is dependent on the subscripts $j_1 \cdots j_p$ and the superscripts $k_1 \cdots k_p$. If the one set is an even permutation of the other, the product is $+1$; if it is an odd permutation of the other, the product is -1. This result is represented exactly by the right-hand side of (5-1.15). The $p!$ compensates for its reciprocal in the bracket notation.

Special examples of products of the type in (5-1.15) are listed in Chapter 1, Section 6.

The fundamental ideas concerning determinants were presented in Chapter 1, Section 6. It would serve no purpose to repeat the details of that development. However, the results were expressed for determinants of order 3. Therefore statements of generalization may serve to clarify and summarize the conclusions.

Definition 5-1.9. A *p*th order determinant symbolized by

(5-1.16a)
$$|a_j{}^k| = \begin{vmatrix} a_1{}^1 & \cdots & a_1{}^p \\ \cdot & & \cdot \\ \cdot & & \cdot \\ \cdot & & \cdot \\ a_p{}^1 & \cdots & a_p{}^p \end{vmatrix}$$

has the numerical value a which is obtained in either of the following ways:

(5-1.16)
$$\text{(b)}\quad \mathcal{E}_{j_1 \cdots j_p} a = \mathcal{E}_{k_1 \cdots k_p} a^{k_1}_{j_1} \cdots a^{k_p}_{j_p},$$
$$\text{(c)}\quad E^{j_1 \cdots j_p} a = E^{k_1 \cdots k_p} a^{j_1}_{k_1} \cdots a^{j_p}_{k_p}.$$

The expressions (5-1.16b,c) are not independent. One can be obtained from the other, as indicated in Chapter 1, Section 6.

Theorem 5-1.4. The numerical value a of the determinant $|a_j{}^k|$ is given by

(5-1.17)
$$a = \frac{1}{p!} E^{j_1 \cdots j_p} \mathcal{E}_{k_1 \cdots k_p} a^{k_1}_{j_1} \cdots a^{k_p}_{j_p}$$

The proof of (5-1.17) is left to the reader. The forms in (5-1.16b,c) and (5-1.17) make trivial the proofs of the elementary properties of determinants. (See Chapter 1, Section 6.)

Definition 5-1.10. A determinant $|c_j{}^k|$ is said to be the product of $|a_j{}^k|$ and $|b_j{}^k|$ if and only if

(5-1.18a)
$$c_j{}^k = a_j{}^p b_p{}^k.$$

Theorem 5-1.5. If c is the value of $|a_j{}^p b_p{}^k|$ and $|a_j{}^p|$, $|b_p{}^k|$ have the values a and b, respectively, then

$$c = ab.$$

Again the proof is left to the reader. It differs only in generality from that of Chapter 1, Section 6.

The multiplication in (5-1.18a) is not commutative. If

(5-1.18b)
$$\bar{c}_j{}^k = b_j{}^p a_p{}^k,$$

then, in general,

$$\bar{c}_j{}^k \neq c_j{}^k.$$

However, it is true that

$$\bar{c} = c.$$

The cofactor of an element $a_j{}^k$ in $|a_j{}^k|$ has played and will continue to play a part in solving systems of linear equations. By the cofactor of $a_j{}^k$ we mean the determinant of order $p - 1$ obtained by striking out the jth row and kth column of $|a_j{}^k|$ and prefixed with the sign $(-1)^{j+k}$. According to our notation, the cofactor $A_k{}^j$ of $a_j{}^k$ is represented by

$$(5\text{-}1.19) \qquad \mathcal{E}_{ji_2\cdots i_p} A_k{}^j = \mathcal{E}_{kj_2\cdots j_p} a_{i_2}^{j_2} \cdots a_{i_p}^{j_p}.$$

Theorem 5-1.6. The cofactor $A_k{}^j$ of $a_j{}^k$ in $|a_p{}^q|$ can be expressed in the form

$$(5\text{-}1.20a) \qquad A_k{}^j = p\, \delta_{kq_2\cdots q_p}^{[ji_2\cdots i_p]} a_{i_2}^{q_2} \cdots a_{i_p}^{q_p},$$

where, according to (5-1.15),

$$(5\text{-}1.20b) \qquad \delta_{kq_2\cdots q_p}^{[ji_2\cdots i_p]} = \frac{1}{p!} E^{ji_2\cdots i_p} \mathcal{E}_{kq_2\cdots q_p}.$$

The proof is left to the reader.

For $n = 3$, especially in relation to the development of transformation concepts, we used the facts

$$(5\text{-}1.21) \qquad a_j{}^p A_p{}^k = \delta_j{}^k a, \qquad a_p{}^k A_j{}^p = \delta_j{}^k a, \qquad a \neq 0.$$

It was assumed that these expressions were consequences of the definition of cofactor. This can be demonstrated rigorously by obtaining (5-1.21) from (5-1.20a). (See Problem 7 at the end of the section.)

The preceding facts concerning determinants are of an algebraic nature. The following result concerns a determinant whose elements are functions.

Theorem 5-1.7. Let $|a_j{}^i|$ be a determinant whose elements are independent functions defined in some region of space. Suppose that at least the first partial derivatives were continuous on the domain of definition. Then

$$(5\text{-}1.22) \qquad \frac{\partial |a_j{}^i|}{\partial a_p{}^q} = A_q{}^p.$$

Since this theorem imitates Theorem 3-5.3, except for generality, the proof is left to the reader.

Problems

1. Suppose that A_{jk} were skew symmetric. Evaluate $A_{(jk)}$.
2. Show that $g_{\alpha\beta}\, dX^\alpha\, dX^\beta$ reduces to $g_{(\alpha\beta)}\, dX^\alpha\, dX^\beta$.

3. If the components of an Euclidean three-space n-tuple A_j are defined by

$$A_j = \mathcal{E}_{jkl} B^k C^l,$$

show that B^k and C^l are orthogonal to A_j.

4. (a) Prove Theorem 5-1.4.
(b) State a corresponding theorem for $|a^{jk}|$.

5. Prove Theorem 5-1.5.

6. Prove Theorem 5-1.6.

7. (a) Show that $a_j{}^p A_p{}^k = \delta_j{}^k a$ is a consequence of (5-1.20a).
(b) Do the same for the right-hand member of (5-1.21).
(c) Show that $|A_p{}^k| = a^{n-1}$.

8. Use (5-1.21) to establish Kramer's rule for solving simultaneous linear equations; that is, show that the solution of the system of p equations in p unknowns,

$$a_j{}^k X^j = b^k, \qquad a \neq 0,$$

is

$$X^j = \frac{b^k A_k{}^j}{a}$$

9. Prove Theorem 5-1.7.

10. Prove that the determinant of a skew symmetric system A_{jk} is zero whenever its order is odd.

11. If the system A_{jk} satisfies the relation $\alpha A_{jk} + \beta A_{kj} = 0$, prove that either $\alpha = -\beta$ and A_{jk} is symmetric or $\alpha = \beta$ and A_{jk} is skew symmetric.

12. Put the Maxwell equations (referred to rectangular Cartesian coordinates)

$$\text{(a) } \nabla \times \mathbf{E} = -\frac{1}{c}\frac{\partial \mathbf{B}}{\partial t}, \qquad \text{(b) } \nabla \cdot \mathbf{D} = 4\pi\rho,$$

$$\text{(c) } \nabla \times \mathbf{H} = \frac{1}{c}\frac{\partial \mathbf{D}}{\partial t} + \frac{4\pi}{c}\mathbf{j}, \qquad \text{(d) } \nabla \cdot \mathbf{B} = 0,$$

in component form.
Answer:

$$\text{(a) } E^{ijk}\, \partial_j E_k = -\frac{1}{c}\frac{\partial B^i}{\partial t}, \qquad \text{(b) } \delta^{ij}\, \partial_i D_j = 4\pi\rho,$$

$$\text{(c) } E^{ijk}\, \partial_j H_k = \frac{1}{c}\frac{\partial D^i}{\partial t} + \frac{4\pi}{c} j^i, \qquad \text{(d) } \delta^{ij}\, \partial_i B_j = 0.$$

13. Make the identifications

$$B^i = \mu H^i, \qquad D^i = \epsilon E^i, \qquad j^i = \sigma E^i, \qquad \mu, \epsilon, \sigma \text{ constants};$$

then show that both the systems H^i and E^i satisfy the wave equations

$$2\partial^q\, \partial_{[p} E_{q]} + \frac{\mu\epsilon}{c^2}\frac{\partial^2 E_p}{\partial t^2} + \frac{4\pi\mu\sigma}{c^2}\frac{\partial E_p}{\partial t} = 0,$$

$$2\partial^q\, \partial_{[p} H_{q]} + \frac{\mu\epsilon}{c^2}\frac{\partial^2 H_p}{\partial t^2} + \frac{4\pi\mu\sigma}{c^2}\frac{\partial H_p}{\partial t} = 0.$$

2. *Transformations and Tensors*

The major objective of tensor analysis is to determine algebraic representations for physical or geometric relations in a form independent of coordinate system, that is, we look for the algebraic and geometric invariants of a given transformation group.

We suppose that a set of coordinates $X^1 \cdots X^n$ can be introduced into the given n-space and that other coordinate systems are introduced by means of functions

(5-2.1a) $$\bar{X}^j = \bar{X}^j(X^k)$$

with unique inverses

(5-2.1b) $$X^j = X^j(\bar{X}^k).$$

The allowable transformations, in other words, are introduced by Definition 3-4.1. (The term triple must be replaced by n-tuple.) Since this definition guarantees the continuity of the partial derivatives $\partial X^j/\partial \bar{X}^k$, $\partial \bar{X}^k/\partial X^j$, the transformations (5-2.1a,b) give rise to the sets of linear transformations

(5-2.2) $$dX^j = \frac{\partial X^j}{\partial \bar{X}^k}\, d\bar{X}^k, \qquad d\bar{X}^k = \frac{\partial \bar{X}^k}{\partial X^j}\, dX^j.$$

These linear transformations relating the differentials dX^j, $d\bar{X}^j$ form a group derived from the general group. This derived group lies at the foundation of tensor analysis. In particular, the laws of tensor transformation are expressed in terms of the coefficients of the linear transformations.

We suppose that the physical or geometric entities under discussion can be represented by a set of one or more functions in every allowable coordinate system. Such an entity is called a tensor if its coordinate representations satisfy the following definition.

Definition 5-2.1. The class $\{T^{k_1 \cdots k_q}_{j_1 \cdots j_p}\}$ is said to be a tensor of contravariant valence q and covariant valence p with respect to the allowable transformation group if and only if the elements satisfy the transformation law,

(5-2.3) $$T^{k_1 \cdots k_q}_{j_1 \cdots j_p} = \frac{\partial X^{k_1}}{\partial \bar{X}^{r_1}} \cdots \frac{\partial X^{k_q}}{\partial \bar{X}^{r_q}} \frac{\partial \bar{X}^{s_1}}{\partial X^{j_1}} \cdots \frac{\partial \bar{X}^{s_p}}{\partial X^{j_p}} \bar{T}^{r_1 \cdots r_q}_{s_1 \cdots s_p}.$$

The elements, $T^{k_1 \cdots k_q}_{j_1 \cdots j_p}$, representing the tensor in a particular coordinate system, are called the components of the tensor with respect to that system. The tensor is said to be of order $p + q$.

It is convenient to identify the physical or geometric entity or concept represented by the class $\{T^{k_1 \cdots k_q}_{j_1 \cdots j_p}\}$ with the class and refer to it as a tensor.

As long as the distinction between the algebraic class and the physical or geometric object is understood, the aforementioned identification is valuable in that it enables us to express thoughts concisely.

It should be clearly noted that in general the components $T^{k_1 \cdots k_q}_{j_1 \cdots j_p}$ are functions of the space coordinates X^j.

Definition 5-2.2a. A tensor of contravariant valence 1 and covariant valence zero is said to be a contravariant vector with respect to the allowable transformation group. According to (5-2.3), the transformation law for the components of a contravariant vector is

$$A^j = \frac{\partial X^j}{\partial \bar{X}^k} \bar{A}^k. \tag{5-2.4a}$$

Definition 5-2.2b. A tensor of covariant valence 1 and contravariant valence zero is said to be a covariant vector with respect to the allowable transformation group. According to (5-2.3), the transformation law for the components of a covariant vector is

$$A_j = \frac{\partial \bar{X}^k}{\partial X^j} \bar{A}_k. \tag{5-2.4b}$$

Definition 5-2.2c. A tensor of order zero is said to be a scalar with respect to the allowable transformation group. According to (5-2.3), the transformation law for scalar representatives is

$$\Phi = \bar{\Phi}. \tag{5-2.4c}$$

Example 5-2.1. Contravariant and covariant vectors have rather familiar prototypes. In Euclidean three-space a space curve is represented by parametric equations

$$X^j = X^j(t)$$

with continuous first derivatives that do not simultaneously vanish. Let us extend this representation of space curve to an arbitrary n-space. Then, with respect to a coordinate transformation,

$$X^j = X^j[\bar{X}^k(t)],$$

we have

$$\frac{dX^j}{dt} = \frac{\partial X^j}{\partial \bar{X}^k} \frac{d\bar{X}^k}{dt}; \tag{5-2.5a}$$

that is, the class $\{dX^j/dt\}$ is a contravariant vector. Its geometric correspondence is defined as a tangent field, and a possible physical analogue is a velocity field.

Next consider a scalar function Φ. Then

$$\frac{\partial \Phi}{\partial X^j} = \frac{\partial \bar{X}^k}{\partial X^j}\frac{\partial \Phi}{\partial \bar{X}^k}. \tag{5-2.5b}$$

Hence the class of gradients of a function Φ is a covariant vector. In structuring an n-dimensional geometry, we use the three-space analogy and visualize the class $\{\partial\Phi/\partial X^j\}$ as the normal field to a surface.[3]

$$\Phi(X^j) = 0.$$

The covariant and contravariant vectors (5-2.5a,b) give rise to a scalar. The preceding fact is algebraically determined by the computation

$$\frac{\partial \Phi}{\partial \bar{X}^j}\frac{dX^j}{dt} = \left(\frac{\partial \bar{X}^k}{\partial X^j}\frac{\partial \Phi}{\partial \bar{X}^k}\right)\left(\frac{\partial X^j}{\partial \bar{X}^p}\frac{d\bar{X}^p}{dt}\right) = \bar{\delta}_p^{\ k}\frac{\partial \Phi}{\partial \bar{X}^k}\frac{d\bar{X}^p}{dt} = \frac{\partial \bar{\Phi}}{\partial \bar{X}^k}\frac{d\bar{X}^k}{dt}$$

Example 5-2.2. An endless quest of science in investigating the world around us is the determination of universal laws. Newton's resolution of his laws of motion was an outstanding achievement in this direction. The second of his three laws states that the change of momentum of an object is proportional to the applied force. The symbolic formulation of the law is

$$F^j = \frac{dmV^j}{dt}, \tag{5-2.6}$$

where the measure of mass m is considered to be a constant. Associated with the Newtonian laws is a principle of relativity of motion which implies that if the formulation (5-2.6) is valid with respect to a given frame of reference then it is valid in any frame in uniform rectilinear motion with respect to the original. We verify this by subjecting (5-2.6) to transformations relating rectangular Cartesian systems of the form

$$X^j = a_k^{\ j}\bar{X}^k + b^j t. \tag{5-2.7a}$$

The $a_k^{\ j}$ and b^j are constants. It is assumed that F^j is a contravariant vector and

$$V^j = \frac{dX^j}{dt}.$$

We have

$$V^j = \frac{dX^j}{dt} = a_k^{\ j}\frac{d\bar{X}^k}{dt} + b^j = a_k^{\ j}\bar{V}^k + b^j.$$

[3] In n-space there is the question whether a surface should be defined in terms of two or $n - 1$ independent parameters. Either generalization from three-space is possible; $n - 1$ parameters, hence the form $\Phi = 0$, wins out, since the set of points satisfying the relation divides the space into two sets, those points for which $\Phi > 0$ and those for which $\Phi < 0$. In order to distinguish $n > 3$ from $n = 3$, the term hypersurface is often employed.

Consequently,

$$\frac{dV^j}{dt} = a_k{}^j \frac{d\bar{V}^k}{dt}. \tag{5-2.7b}$$

By using this result and noting that $\partial X^j/\partial \bar{X}^k = a_k{}^j$, we have

$$0 = F^j - m\frac{dV^j}{dt} = \frac{\partial X^j}{\partial \bar{X}^k}\left(\bar{F}^k - m\frac{d\bar{V}^k}{dt}\right).$$

By multiplying and summing the relation

$$\frac{\partial X^j}{\partial \bar{X}^k}\left(\bar{F}^k - \bar{m}\frac{d\bar{V}^k}{dt}\right) = 0$$

with $\partial \bar{X}^p/\partial X^j$, we obtain the result

$$\bar{\delta}_k{}^p\left(\bar{F}^k - \bar{m}\frac{d\bar{V}^k}{dt}\right) = 0. \tag{5-2.7c}$$

From this it immediately follows that the form (5-2.6) holds in the barred system.

Example 5-2.3. We have encountered other tensors than those already mentioned. For example, in Chapter 3, Section 4, the fundamental metric tensor with components

$$\bar{g}_{jk} = \sum_{i=1}^{3} \frac{\partial X^i}{\partial \bar{X}^j}\frac{\partial X^i}{\partial \bar{X}^k} \tag{5-2.8a}$$

was introduced. In this instance the X^j coordinates were rectangular Cartesian. If the n-dimensional space is euclidean, the fundamental metric tensor can be reintroduced by means of 5-2.8a. More generally, in the manner of Riemann, we can start with a fundamental metric tensor and develop the notions concerning the space from it.

The class $\{\delta_j{}^k\}$ is a tensor with respect to an allowable transformation group since

$$\delta_j{}^k = \frac{\partial X^k}{\partial \bar{X}^p}\frac{\partial \bar{X}^q}{\partial X^j}\bar{\delta}_q{}^p \tag{5-2.8b}$$

is an identity.

The $\mathcal{E}$ systems $\mathcal{E}_{ijk}$ and E^{ijk} are tensor components with respect to the orthogonal Cartesian group of transformations, but, as we saw in Chapter 1, Section 5*, these components transform in terms of more general transformation groups with density factors of $|\partial X/\partial \bar{X}|$ and $|\partial \bar{X}/\partial X|$, respectively.

In order that they may be included in our theory, we make the following definition.

Definition 5-2.3. The class $T^{k_1 \cdots k_q}_{j_1 \cdots j_p}$ is said to be a tensor density of contravariant valence q, covariant valence p, and weight r if and only if the elements satisfy the transformation law

$$(5\text{-}2.9)\quad T^{k_1 \cdots k_q}_{j_1 \cdots j_p} = \left|\frac{\partial X}{\partial \bar{X}}\right|^r \frac{\partial X^{k_1}}{\partial \bar{X}^{s_1}} \cdots \frac{\partial X^{k_q}}{\partial \bar{X}^{s_q}} \frac{\partial \bar{X}^{t_1}}{\partial X^{j_1}} \cdots \frac{\partial \bar{X}^{t_p}}{\partial X^{j_p}} \bar{T}^{s_1 \cdots s_q}_{t_1 \cdots t_p}.$$

Theorem 5-2.1. The $\mathcal{E}$ systems defined according to Definition 5-1.7 in each allowable coordinate system satisfy the transformation laws

$$(5\text{-}2.10)\quad \begin{aligned} &\text{(a)}\quad \mathcal{E}_{j_1 \cdots j_n} = \left|\frac{\partial X}{\partial \bar{X}}\right| \frac{\partial \bar{X}^{k_1}}{\partial X^{j_1}} \cdots \frac{\partial \bar{X}^{k_n}}{\partial X^{j_n}} \bar{\mathcal{E}}_{k_1 \cdots k_n}, \\ &\text{(b)}\quad E^{j_1 \cdots j_1} = \left|\frac{\partial \bar{X}}{\partial X}\right| \frac{X^{j_1}}{\bar{X}^{k_1}} \cdots \frac{X^{j_n}}{\bar{X}^{k_n}} \bar{E}^{k_1 \cdots k_n}, \end{aligned}$$

hence are components of tensors of weight $+1$ and -1, respectively.

The proof of this theorem is a consequence of the fact that (5-2.10a,b) are identities.

Example 5-2.4. The classification of an entity depends on the underlying transformation group. This is clearly pointed out by considering the components

$$\bar{g}_{jk} = \frac{\partial \mathbf{r}}{\partial \bar{X}^j} \cdot \frac{\partial \mathbf{r}}{\partial \bar{X}^k}$$

in a Euclidean space. With respect to rectangular Cartesian transformations, each component is a scalar. Hence we have a set of scalars. With respect to more general transformations, the $\bar{g}_{jk}$ are interpreted as components of a tensor of covariant valence 2.

The algebra of tensors is based on the algebra of systems, as introduced in Section 1. The present objective is to formalize the fact that such operations with tensors result in tensors. It is understood that the terms addition or multiplication applied to tensors is simply shorthand for addition or multiplication of their components in each coordinate system.

Theorem 5-2.2a (Addition). The sum of two tensors of the same kind (i.e., with common contravariant valences and common covariant valences) is a tensor of the original kind.

Theorem 5-2.2b (Outer product). The outer product of a tensor of covariant valence p and contravariant valence q with a tensor of covariant and contravariant valences r and s, respectively, is a tensor of valences $p + r$ and $q + s$, respectively.

Theorem 5-2.2c (Contraction). Given a tensor with component representation $T^{k_1 \cdots k_q}_{j_1 \cdots j_p}$, if any contravariant index is set equal to a covariant index, thereby implying a sum, the resulting entity is a tensor of one less contravariant valence and one less covariant valence.

The proof of Theorem 5-2.2c follows. Proofs of the other two theorems are similar and are left to the reader.

In order to prove Theorem 5-2.2c, it must be shown that the contracted systems satisfy the transformation law (5-2.3). We have

$$\begin{aligned} T^{k_1\cdots k_{q-1}t}_{j_1\cdots j_{p-1}t} &= \left(\frac{\partial X^{k_1}}{\partial \bar{X}^{r_1}}\cdots\frac{\partial X^{k_{q-1}}}{\partial \bar{X}^{r_{q-1}}}\frac{\partial X^{t}}{\partial \bar{X}^{r_q}}\right)\left(\frac{\partial \bar{X}^{s_1}}{\partial X^{j_1}}\cdots\frac{\partial \bar{X}^{s_{p-1}}}{\partial X^{j_{p-1}}}\frac{\partial \bar{X}^{s_p}}{\partial X^{t}}\right)\bar{T}^{r_1\cdots r_q}_{s_1\cdots s_p} \\ &= \left(\frac{\partial X^{k_1}}{\partial \bar{X}^{r_1}}\cdots\frac{\partial X^{k_{q-1}}}{\partial \bar{X}^{r_{q-1}}}\right)\left(\frac{\partial \bar{X}^{s_1}}{\partial X^{j_1}}\cdots\frac{\partial \bar{X}^{s_{p-1}}}{\partial X^{j_{p-1}}}\right)\delta^{s_p}_{r_q}\bar{T}^{r_1\cdots r_q}_{s_1\cdots s_p} \\ &= \left(\frac{\partial X^{k_1}}{\partial \bar{X}^{r_1}}\cdots\frac{\partial X^{k_{q-1}}}{\partial \bar{X}^{r_{q-1}}}\right)\left(\frac{\partial \bar{X}^{s_1}}{\partial X^{j_1}}\cdots\frac{\partial \bar{X}^{s_{p-1}}}{\partial X^{j_{p-1}}}\right)\bar{T}^{r_1\cdots r_{q-1}t}_{s_1\cdots s_{p-1}t}. \end{aligned}$$

Since the components satisfy the appropriate transformation law, the theorem is proved.

Other tensor properties of importance are expressed by the following theorems.

Theorem 5-2.3a. If the components $T^{j_1\cdots j_p\cdots j_q\cdots j_r}$ of a tensor in some one coordinate system are completely $\begin{cases}\text{symmetric}\\ \text{skew symmetric}\end{cases}$ in $j_p\cdots j_q$, then the components in every coordinate system are completely $\begin{cases}\text{symmetric}\\ \text{skew symmetric}\end{cases}$ in the corresponding indices. A corresponding statement holds for covariant components.

PROOF. Suppose the components are symmetric in $j_p\cdots j_q$ $(1 \le p < q \le r)$. Then

$$\bar{T}^{k_1\cdots k_p\cdots k_q\cdots k_r} = \frac{\partial \bar{X}^{k_1}}{\partial X^{j_1}}\cdots\frac{\partial \bar{X}^{k_p}}{\partial X^{j_p}}\cdots\frac{\partial \bar{X}^{k_q}}{\partial X^{j_q}}\cdots\frac{\partial \bar{X}^{k_r}}{\partial X^{j_r}}T^{j_1\cdots j_p\cdots j_q\cdots j_r}.$$

Let $k_{p_1}\cdots k_{q_1}$ be an arbitrary permutation of $k_p\cdots k_q$. Then

$$\bar{T}^{k_1\cdots k_{p_1}\cdots k_{q_1}\cdots k_r} = \frac{\partial \bar{X}^{k_1}}{\partial X^{j_1}}\cdots\frac{\partial \bar{X}^{k_{p_1}}}{\partial X^{j_{p_1}}}\cdots\frac{\partial \bar{X}^{k_{q_1}}}{\partial X^{j_{q_1}}}\cdots\frac{\partial \bar{X}^{k_r}}{\partial X^{j_r}}T^{j_1\cdots j_{p_1}\cdots j_{q_1}\cdots j_r}.$$

The partial derivatives $\partial\bar{X}^{k_{p_1}}/\partial X^{j_{p_1}}\cdots\partial\bar{X}^{k_{q_1}}/\partial X^{j_{q_1}}$ are commutative and can be put in the order $k_p\cdots k_q$. Furthermore, the components $T^{j_1\cdots j_{p_1}\cdots j_{q_1}\cdots j_r}$ by hypothesis are completely symmetric in $j_{p_1}\cdots j_{q_1}$, hence can be replaced by $T^{j_1\cdots j_p\cdots j_q\cdots j_r}$. Therefore the right-hand sides of the foregoing expressions are equal and the equality of the left-hand sides follows. The proof concerning skew symmetry is carried out in the same manner.

Theorem 5-2.3b. If all components of a tensor are zero in some one coordinate system, they are zero in every allowable coordinate system.

PROOF. Suppose that

$$T^{k_1\cdots k_q}_{j_1\cdots j_p} = 0.$$

Then

$$0 = T^{s_1\cdots s_m}_{r_1\cdots r_l} = \frac{\partial \bar{X}^{s_1}}{\partial X^{a_1}} \cdots \frac{\partial \bar{X}^{s_m}}{\partial X^{a_m}} \frac{\partial X^{b_1}}{\partial \bar{X}^{r_1}} \cdots \frac{\partial X^{b_l}}{\partial \bar{X}^{r_l}} \bar{T}^{a_1\cdots a_m}_{b_1\cdots b_l}.$$

Now we can multiply and sum with the successive sets of partial derivatives and arrive at the desired conclusion.

The preceding theorem plays an important part in determining physical laws of an invariant character. The invariance of the Newtonian second law illustrated in Example 5-2.2 is a classical instance of its use.

Suppose the sum of products of tensor components $A_{jk}{}^{i}$ and B^{jk} is formed in each allowable coordinate system. Theorems (5-2.2b,c) assure us that the resultant

$$C^i = A^i{}_{jk}B^{jk}$$

comprises tensor components. But what of the converse process? This question concerning division is answered in part by the following theorem.

Theorem 5-2.4. (Quotient law of tensors). Suppose that systems of the form $A\binom{j_1\cdots j_p}{k_1\cdots k_q}$ are defined in each allowable coordinate system. Let $B^{k_1\cdots k_q\cdots k_s}_{j_1\cdots j_p\cdots j_r}$ be components of an arbitrary tensor. If

$$A\begin{pmatrix} j_1 \cdots j_p \\ k_1 \cdots k_q \end{pmatrix} B^{k_1\cdots k_q\cdots k_s}_{j_1\cdots j_p\cdots j_r} = C^{k_{q+1}\cdots k_s}_{j_{p+1}\cdots j_r}, \tag{5-2.11}$$

where the $C^{k_{q+1}\cdots k_s}_{j_{p+1}\cdots j_r}$ are tensor components, then the systems $A\binom{j_1\cdots j_p}{k_1\cdots k_q}$ determine a tensor of covariant valence q and contravariant valence p.

PROOF. In order to reduce the complexity of the notation, an example exhibiting the method of proof rather than the general proof is presented. Consider the systems $A\begin{pmatrix} i \\ jk \end{pmatrix}$, the arbitrary tensor components B^{jk}, and the contravariant vector components C^i. If (5-2.11) holds, we can write

$$0 = A\begin{pmatrix} i \\ jk \end{pmatrix} B^{jk} - C^i = \frac{\partial X^j}{\partial \bar{X}^p}\frac{\partial X^k}{\partial \bar{X}^q} A\begin{pmatrix} i \\ jk \end{pmatrix} \bar{B}^{pq} - \frac{\partial X^i}{\partial \bar{X}^s}\bar{C}^s.$$

Since (5-2.11) holds in each coordinate system, $\bar{C}^p$ can be replaced by making use of the relation

$$\bar{A}\begin{pmatrix} s \\ pq \end{pmatrix} \bar{B}^{pq} = \bar{C}^s.$$

Then

$$0 = \frac{\partial X^j}{\partial \bar{X}^p}\frac{\partial X^k}{\partial \bar{X}^q} A\binom{i}{jk}\bar{B}^{pq} - \frac{\partial X^i}{\partial \bar{X}^s}\bar{A}\binom{s}{pq}\bar{B}^{pq}$$

$$= \left[\frac{\partial X^j}{\partial \bar{X}^p}\frac{\partial X^k}{\partial \bar{X}^q} A\binom{i}{jk} - \frac{\partial X^i}{\partial \bar{X}^s}\bar{A}\binom{s}{pq}\right]\bar{B}^{pq}$$

Since the tensor $\{B^{pq}\}$ is arbitrary, we can impose various sets of values on the barred system components. For example,

$$(\bar{B}^{pq}) = \begin{pmatrix} 1 & 0 & \cdots & 0 \\ 0 & 0 & \cdots & 0 \\ \cdot & \cdot & & \cdot \\ \cdot & \cdot & & \cdot \\ \cdot & \cdot & & \cdot \\ 0 & 0 & \cdots & 0 \end{pmatrix}$$

By making n^2 such choices, we exhibit the fact that the parenthetic components must be zero; that is,

$$\text{(5-2.12a)} \qquad \frac{\partial X^j}{\partial \bar{X}^p}\frac{\partial X^k}{\partial \bar{X}^q} A\binom{i}{jk} - \frac{\partial X^i}{\partial \bar{X}^s}\bar{A}\binom{s}{pq} = 0.$$

Upon multiplication by $\partial \bar{X}^t/\partial X^i$,

$$\text{(5-2.12b)} \qquad \frac{\partial X^j}{\partial \bar{X}^p}\frac{\partial X^k}{\partial \bar{X}^q}\frac{\partial \bar{X}^t}{\partial X^i} A\binom{i}{jk} = \bar{A}\binom{t}{pq}.$$

Therefore the systems $A\binom{i}{jk}$, etc., are elements of a tensor.

Example 5-2.5. A certain caution must be used in the employment of Theorem 5-2.4. Suppose that

$$A_{jk}W^jW^k = \Phi$$

held in each coordinate system in which W^j are components of an arbitrary contravariant vector and Φ is a scalar representative. By transforming the vector components and replacing $\Phi = \bar{\Phi}$ by its barred system representation, we have

$$A_{jk}\frac{\partial X^j}{\partial \bar{X}^p}\frac{\partial X^k}{\partial \bar{X}^q} W^pW^q = \bar{A}_{pq}\bar{W}^p\bar{W}^q$$

or

$$\text{(5-2.13a)} \qquad \left(A_{jk}\frac{\partial X^j}{\partial \bar{X}^p}\frac{\partial X^k}{\partial \bar{X}^q} - \bar{A}_{pq}\right)\bar{W}^p\bar{W}^q = 0.$$

In this example the tensor with components

$$\bar{B}^{pq} = \overline{W}^p\overline{W}^q$$

is not completely arbitrary. It has the property of symmetry in p and q. If, for example, we choose $\bar{B}^{12} = 1$, then $\bar{B}^{21} = 1$ also. If all other $\bar{B}^{pq} = 0$, the expression (5-2.13a) becomes

$$\text{(5-2.13b)} \quad \left(A_{jk}\frac{\partial X^j}{\partial \overline{X}^1}\frac{\partial X^k}{\partial \overline{X}^2} - \bar{A}_{12}\right) + \left(A_{jk}\frac{\partial X^j}{\partial \overline{X}^2}\frac{\partial X^k}{\partial \overline{X}^1} - \bar{A}_{21}\right) = 0.$$

By interchanging dummy indices in the second term, we can put (5-2.13b) in the form

$$\text{(5-2.13c)} \qquad (A_{jk} + A_{kj})\frac{\partial X^j}{\partial \overline{X}^1}\frac{\partial X^k}{\partial \overline{X}^2} - (\bar{A}_{12} + \bar{A}_{21}) = 0.$$

The procedure indicated in (5-2.13a,b,c) can be carried out with each pair of indices. The consequence of this development is that it cannot be concluded that the systems A_{jk} determine a tensor unless they are symmetric in j, k. However, the symmetric systems $A_{jk} + A_{kj}$ do determine a tensor.

Problems

1. Show that 5-2.4a and 5-2.4b imply
   (a) The components of a covariant vector satisfy the relations
$$\bar{A}_j = \frac{\partial X^k}{\partial \overline{X}^j} A_k.$$
   (b) The components of a contravariant vector satisfy
$$\bar{A}^j = \frac{\partial \overline{X}^j}{\partial X^k} A^k.$$
2. Prove Theorems 5-2.2a,b.
3. Prove that if the contravariant components of a tensor are completely skew symmetric in one system they are so in every coordinate system.
4. Suppose the coordinates X^j are rectangular Cartesian coordinates in an n-dimensional Euclidean space.
   (a) What geometric configuration is represented by $\Phi \equiv A_jX^j = \text{constant}$ if $n = 2$, $n = 3$, $n > 3$?
   (b) What does the set $\partial\Phi/\partial X^j$ represent geometrically in each case of (a)?
5. Systems $A(pqrs)$ are defined in each allowable coordinate system. Their rule of transformation is
$$A(p, q, r, s) = \frac{\partial X^p}{\partial \overline{X}^a}\frac{\partial \overline{X}^b}{\partial X^q}\frac{\partial \overline{X}^c}{\partial X^r}\frac{\partial \overline{X}^d}{\partial X^s}\bar{A}(a, b, c, d).$$
   Is the class $A(p, q, r, s)$ a tensor? If so, determine the covariant and contravariant valences.

6. Suppose $A_{jk}B^{jk} = \Phi$, where B^{jk} are components of an arbitrary skew symmetric tensor and Φ is a scalar of weight 0. Show that $A_{jk} - A_{kj}$ are tensor components.
7. Show that the set $\partial_j A^k$ of partial derivatives of contravariant vector components A^k are not tensor components in general. Under what circumstance are they tensor components?
8. In vacuo, and with the velocity of light $c = 1$, the Maxwell equations have the form

$$\frac{\partial E^3}{\partial X^2} - \frac{\partial E^2}{\partial X^3} + \frac{\partial H^1}{\partial t} = 0, \qquad \frac{\partial H^3}{\partial X^2} - \frac{\partial H^2}{\partial X^3} - \frac{\partial E^1}{\partial t} = \rho_0 V^1,$$

$$\frac{\partial E^1}{\partial X^3} - \frac{\partial E^3}{\partial X^1} + \frac{\partial H^2}{\partial t} = 0, \qquad \frac{\partial H^1}{\partial X^3} - \frac{\partial H^3}{\partial X^1} - \frac{\partial E^2}{\partial t} = \rho_0 V^2,$$

$$\frac{\partial E^2}{\partial X^1} - \frac{\partial E^1}{\partial X^2} + \frac{\partial H^3}{\partial t} = 0, \qquad \frac{\partial H^2}{\partial X^1} - \frac{\partial H^1}{\partial X^2} - \frac{\partial E^3}{\partial t} = \rho_0 V^3,$$

$$\frac{\partial H^1}{\partial X^1} + \frac{\partial H^2}{\partial X^2} + \frac{\partial H^3}{\partial X^3} = 0, \qquad \frac{\partial E^1}{\partial X^1} + \frac{\partial E^2}{\partial X^2} + \frac{\partial E^3}{\partial X^3} = \rho_0.$$

(These equations result from those in Chapter 5, Section 1, Problem 12, by putting $\mu = \epsilon = c = 1$, $\sigma = 1$, $\mathbf{j} = (1/4\pi)\mathbf{V}$, and $\rho = (1/4\pi)\rho_0$. Make the identifications

$$E^1 = F^{41}, \quad E^2 = F^{42}, \quad E^3 = F^{43}, \quad F^{\lambda\gamma} = -F^{\gamma\lambda}.$$
$$H^1 = F^{23}, \quad H^2 = F^{31}, \quad H^3 = F^{12}.$$

and show that with respect to the metric

$$(g_{\alpha\beta}) = \begin{pmatrix} -1 & 0 & 0 & 0 \\ 0 & -1 & 0 & 0 \\ 0 & 0 & -1 & 0 \\ 0 & 0 & 0 & 1 \end{pmatrix}$$

the equations can be put in the form

$$\partial_{[\nu}F_{\beta\gamma]} = 0, \qquad \partial_\beta F^{\alpha\beta} = \rho_0 V^\alpha.$$

3. *Riemannian Geometry*

Definition 5-3.1a. An n-space endowed with a covariant tensor g_{jk} of second order, which is symmetric, is said to be a Riemannian space. The geometry of the space is said to be Riemannian.

Definition 5-3.1b. A quadratic differential form

$$ds^2 = g_{jk}\, dX^j\, dX^k \tag{5-3.1}$$

can be associated with the tensor g_{jk}. It is called the fundamental metric form of the space and g_{jk} is said to be the fundamental metric tensor.

We first met with the concept of a metric tensor in Chapter 1, Section 5*, again in Chapter 3, Section 4, and then in Chapter 4, Section 2. In a Euclidean space it is possible to introduce rectangular Cartesian coordinate systems in which the metric tensor has the form

$$g_{jk} = \delta_{jk} \tag{5-3.2a}$$

and in which the form (5-3.1) reduces to

$$ds^2 = \delta_{jk}\, dX^j\, dX^k = \sum_{j=1}^{n} (dX^j)^2. \tag{5-3.2b}$$

If Cartesian coordinates other than rectangular are used, $g_{jk} \neq \delta_{jk}$, but the components are still constant. Finally, if generalized coordinates, such as three-space spherical coordinates, are introduced, the g_{jk} become functions of position.

In Euclidean three-space the g_{jk} enable us to express length, angle, and volume in forms independent of coordinate systems. In considering surface representations, we saw, at least in part, that the $g_{\alpha\beta}$ were fundamental to the intrinsic determination of length, angle, and area on the surface.

With the introduction of generalized coordinates, we could clearly perceive the dependence of the metric tensor on the space coordinates. Surface representation illustrated the use of the $g_{\alpha\beta}$ in an essentially non-Euclidean situation. This is the background on which Riemannian geometry was developed.

Riemann, a pupil of Gauss, was well aware of Gauss's important and fundamental contributions to the development of the intrinsic geometry of surfaces. The use of parametric curve nets on the surface in analogy to coordinate nets in the plane not only served Gauss's immediate purposes but it also established a firm basis for generalization of his ideas to n-space. The task of generalization was undertaken by Riemann. His resulting theory, as we would expect, included the motivating three-space Euclidean analytic geometry and the intrinsic geometry of surfaces. But it went further. It tied together the theory of non-Euclidean geometries and it set a foundation for the development of the general theory of relativity. In fact, Riemannian geometry has played an important role in the construction of modern physical theories in general.

In his original paper, "On the Hypotheses which Lie at the Foundation of Geometry," Riemann assumed that the fundamental quadratic form $g_{jk}\, dX^j\, dX^k$ was positive definite (i.e., $g_{jk}\, dX^j\, dX^k > 0$). This assumption guaranteed that $ds = \sqrt{g_{jk}\, dX^j\, dX^k}$ was real. We have not included this assumption in Definition 5-3.1, since indefinite forms such as

$$ds^2 = -(dX^1)^2 - (dX^2)^2 - (dX^3)^2 + (dX^4)^2$$

are quite essential in relativity theory and other aspects of geometry and physics. Because of this relaxation of the conditions on $g_{\alpha\beta}$, we shall have to make judicious use of absolute value signs.

It must be clearly understood that the $\{g_{\alpha\beta}\}$ is assumed given and that the geometric structure of a Riemannian space is built up around it by definitions that are based on surface or three-space analogies.

The development of the analytic geometry of a Riemannian space is expedited by the introduction of an associated metric tensor $\{g^{jk}\}$. It is assumed that

(5-3.3) $$|g_{jk}| \neq 0.$$

Definition 5-3.2. In each allowable coordinate system

(5-3.4) $$g^{jk} = \frac{\text{cofactor of } g_{kj} \text{ in } |g_{jk}|}{|g_{jk}|}.$$

According to (5-3.4), the components g^{jk} satisfy the properties (these properties result from determinant expansion by row and by column)

(5-3.5a) $$g_{jk}g^{kp} = \delta_j^{\ p},$$

(5-3.5b) $$g_{kj}g^{pk} = \delta_j^{\ p},$$

where the inner index represents row and the outer index represents column.

Theorem 5-3.1. The system g^{jk} is symmetric; that is,

(5-3.6a) $$g^{jk} = g^{kj}.$$

PROOF. By multiplying and summing each side of (5-3.5a) with g^{jq}, we have

(5-3.6b) $$(g^{jq}g_{jk})g^{kp} = g^{jq}\,\delta_j^{\ p}.$$

According to Definition 5-3.1a, g_{jk} is symmetric. If we use this fact and then employ (5-3.5a) on the parenthetic factors of (5-3.6b) we are led to the expression

$$\delta_k^{\ q}g^{kp} = g^{jq}\,\delta_j^{\ p}.$$

Application of the Kronecker deltas produces the result

(5-3.6c) $$g^{qp} = g^{pq}.$$

This completes the proof.

Theorem 5-3.2. The class of systems $\{g^{jk}\}$ is a second-order tensor of contravariant valence 2.

PROOF. We make use of the fact that the class of systems $\{\delta_j^{\ k}\}$ is a tensor. (See Example (5-2.3.) We have

$$g_{jk}g^{kp} = \delta_j^{\ p}.$$

Since g_{jk} and $\delta_j^{\ p}$ are tensor components,

$$\frac{\partial \bar{X}^q}{\partial X^j}\frac{\partial \bar{X}^r}{\partial X^k}\bar{g}_{qr}g^{kp} = \frac{\partial X^p}{\partial \bar{X}^q}\frac{\partial \bar{X}^r}{\partial X^j}\bar{\delta}_r^{\ q}. \tag{5-3.7a}$$

After multiplication by $\partial X^j/\partial \bar{X}^s$, the relation (5-3.7a) has the form

$$\frac{\partial \bar{X}^r}{\partial X^k}\bar{g}_{sr}g^{kp} = \frac{\partial X^p}{\partial \bar{X}^q}\delta_s^{\ q}. \tag{5-3.7b}$$

Application of $\bar{g}^{st}$ results in

$$\frac{\partial \bar{X}^t}{\partial X^k}g^{kp} = \bar{g}^{qt}\frac{\partial X^p}{\partial \bar{X}^q}. \tag{5-3.7c}$$

When this expression is summed with $\partial X^b/\partial \bar{X}^t$, the following appropriate law of transformation is obtained.

$$g^{bp} = \frac{\partial X^b}{\partial \bar{X}^t}\frac{\partial X^p}{\partial \bar{X}^q}\bar{g}^{tq}. \tag{5-3.7d}$$

This completes the proof.

Definition 5-3.3. $\{g^{jk}\}$ is called the conjugate or associated metric tensor.

The metric tensor and its conjugates play an algebraic role denoted by the terms "lowering of indices" and "raising of indices."[4] For example, if T^{ij} are the components of a given tensor, then

$$T_k^{\ j} = g_{ki}T^{ij}, \qquad T_{kp} = g_{ki}g_{pj}T^{ij}.$$

Note that the same kernel letter T is used in each instance. This notational device emphasizes the fact that, although the tensors $\{T^{ij}\}$, $\{T_{ij}\}$, $\{T_i^{\ j}\}$, $\{T^i_{\ j}\}$ are distinct from the algebraic point of view, each represents the same geometric or physical entity. This sort of situation was illustrated in Chapter 3, Section 9, where W_j and $W^k = g^{kj}W_j$ were components of the same Cartesian vector, although referred to reciprocal bases.[5] We find it convenient to speak of a vector in a Riemannian space and to represent the vector by means of either covariant or contravariant components. The associated metric tensors and the processes of raising and lowering

---

[4] See Chapter 1, Section 5*.

[5] If a space is not metric, that is, if no tensor g_{jk} is given, these remarks do not apply. For example, in a space in which only incidence relations are considered W_j, W^j have distinct geometric interpretations.

indices endow tensor algebra with a duality that has both beauty and utility.

How should the concepts of magnitude and angle be introduced? In other words, what demands should be met? We make two requirements.

1. The definitions introduced in the Riemannian space should reduce to the usual ones of solid analytic geometry if the space is taken as Euclidean.
2. Definitions must satisfy an invariance property; that is, the tensor form should not depend on coordinate system.

Magnitude and angle are introduced as follows.

Definition 5-3.4. Let W^j be the components of a vector. The magnitude of $\mathbf{W}$ is

(5-3.8) $$|\mathbf{W}| = (|g_{jk}W^jW^k|)^{1/2} = (|g^{jk}W_jW_k|)^{1/2} = (|W^jW_j|)^{1/2}.$$

Definition 5-3.5. Let V^j and W^k be contravariant components of a pair of vectors. The angle θ made by the two vectors satisfies the relation

(5-3.9) $$\cos\theta = \frac{g_{jk}V^jW^k}{(|g_{ab}V^aV^b|)^{1/2}(|g_{pq}W^pW^q|)^{1/2}}.$$

Euclidean space is classified by the following definition.

Definition 5-3.6. A Riemannian space is isometric to an Euclidean space if and only if there is an allowable coordinate system such that

(5-3.10) $$g_{jk} = \delta_{jk}\,*$$

throughout the space.

It is easily seen that (5-3.8) and (5-3.9) reduce to the usual formulas of analytic geometry when the space is Euclidean. The invariance of each relation is likewise quickly deduced.

The concept of orthogonality is abstracted directly from the Euclidean form.

Definition 5-3.7a. Two vectors with contravariant components V^j and W^k are said to be orthogonal if and only if

(5-3.11) $$g_{jk}V^jW^k = 0.$$

Example 5-3.1. If

$$g_{jk} = \delta_{jk}, \qquad j, k = 1, 2, 3,$$

then (5-3.11) reduces to

(5-3.12a) $$\sum_{j=1}^{3} V^jW^j = 0.$$

* Two spaces are isometric if measurements of corresponding distances and angles are the same in them. For example a circular cylinder is isometric to a portion of a plane. See any text on differential geometry for more detail.

This is the usual expression from solid analytic geometry. On the other hand, suppose

(5-3.12b) $$(g_{jk}) = \begin{pmatrix} -1 & 0 \\ 0 & 1 \end{pmatrix}.$$

Then the orthogonality condition reduces to

(5-3.12c) $$-V^1W^1 + V^2W^2 = 0.$$

Definition 5-3.7b. **V** is a unit vector if and only if

(5-3.13) $$|g_{jk}V^jV^k| = 1.$$

Example 5-3.2. If g_{jk} is a positive definite metric tensor, that is, if

(5-3.12a) $$g_{jk}U^jU^k \geq 0$$

for all real sets of components U^j and the equality holds if and only if $U^j = 0$ for each j, then

(5-3.14b) $$|\cos \theta| \leq 1.$$

The assertion of the example may be proved by starting with the expression

(5-3.15a) $$g_{jk}(\alpha V^j + \beta W^j)(\alpha V^k + \beta W^k) \geq 0,$$

where α and β are arbitrary parameters not both zero and **V** and **W** are unit vectors. By multiplying out this expression we obtain

$$\alpha^2 + 2(g_{jk}V^jW^k)\alpha\beta + \beta^2 \geq 0.$$

Let

$$\lambda = \frac{\alpha}{\beta}, \qquad \beta \neq 0.$$

Then

(5-3.15b) $$\lambda^2 + 2\cos\theta\lambda + 1 \geq 0.$$

If the inequality holds, the polynomial

$$P(\lambda) = \lambda^2 + 2\cos\theta\lambda + 1$$

has no zeros; hence the discriminant of the right-hand side is negative; that is

$$\cos^2\theta - 1 < 0.$$

Therefore

$$|\cos\theta| < 1.$$

If the equality holds in (5-3.15a,b), then, according to the definition of a positive definite metric,

$$\alpha V^j + \beta W^j = 0.$$

Consequently, by solving for W^j and substituting into (5-3.9), we have

$$\cos\theta = \frac{g_{\gamma\mu}V^\gamma(\lambda V^\mu)}{(|g_{\alpha\beta}V^\alpha V^\beta|)^{1/2}(\lambda^2\,|g_{\eta\delta}V^\eta V^\delta|)^{1/2}} = \frac{\lambda}{|\lambda|}$$

and

$$|\cos\theta| = 1.$$

Combination of this fact with the preceding result produces the relation (5-3.14b).

Problems

1. Show that the transformation relating the fundamental metric forms

$$(dX^1)^2 + (dX^2)^2, \qquad -(d\bar{X}^1)^2 + (d\bar{X}^2)^2$$

is not real. Hence g_{jk} does not reduce to δ_{jk} in a space which admits the right-hand metric.

2. Compute the components of the fundamental metric tensor and the conjugate metric tensor associated with each of the following surfaces:

(a) A sphere related to rectangular Cartesian coordinates by means of of the transformation equations

$$X^1 = r\sin v^1\cos v^2,\ X^2 = r\sin v^1\sin v^2,\ X^3 = r\cos v^1,$$

where r is a constant.

(b) A cylinder related to a rectangular Cartesian system by means of

$$X^1 = \rho\cos v^1, \qquad X^2 = \rho\sin v^1, \qquad X^3 = v^2,$$

where ρ is a constant.

3. Suppose that a Riemannian n-space is referred to the coordinates $X^1\cdots X^n$. The equations

$$\begin{aligned} X^1 &= \text{constant.}\\ &\ \vdots\\ X^{j-1} &= \text{constant.}\\ X^j &= t\\ X^{j+1} &= \text{constant.}\\ &\ \vdots\\ X^n &= \text{constant.}\end{aligned}$$

represent a coordinate curve. Show that the angle θ_{jk} made by the coordinate curves, along which X^j and X^k vary, respectively, is given by

$$\cos\theta_{jk} = \frac{g_{jk}}{(|g_{jj}g_{kk}|)^{1/2}}.$$

4. Prove that if the Riemannian space is referred to an orthogonal coordinate system

$$g^{xx} = \frac{1}{g_{xx}}.$$

4. *Tensor Processes of Differentiation*

In considering the processes of differentiation in a Riemannian space, the concept of invariance should be kept in the foreground. The general form of a geometric or physical law is independent of coordinate system when expressed entirely in terms of tensors. This, of course, is one of the principal values of the tensor representation. We must ask the following question: do the processes of differentiation and partial differentiation transform a tensor into a tensor? In general, the answer is no. Therefore what processes should take the place of partial and ordinary differentiation? Historically, a foundation for answering the question was determined by Elwin Bruno Christoffel, Rudolf Lipschitz, and Eugenio Beltrami. Christoffel and Lipschitz wrote papers on quadratic differential forms in 1870; Beltrami expounded a theory of differential operators in the decade preceding that date. The contributions to invariant theory made by these men were fully developed by Gregorio Ricci Curbastro in his absolute differential calculus of 1884. In particular, it was Ricci who introduced the name covariant differentiation. This term, which signifies a kind of partial differentiation that produces from a tensor a tensor of covariant valence one more, is the main topic of this section.

In order to gather information that will form a basis for answering the foregoing question, we start out on familiar ground. Let $V^i(X^j)$ be a vector field defined on a region of Euclidean space. Suppose that the coordinates X^j are rectangular Cartesian and that the vector field had continuous partial derivatives of at least second order. If

$$X^j = X^j(\bar{X}^k)$$

is an allowable coordinate transformation, then

$$V^j = \frac{\partial X^j}{\partial \bar{X}^k} \bar{V}^k \tag{5-4.1a}$$

and

$$\frac{\partial V^j}{\partial X^p} = \frac{\partial^2 X^j}{\partial \bar{X}^q \partial \bar{X}^k} \frac{\partial \bar{X}^q}{\partial X^p} \bar{V}^k + \frac{\partial X^j}{\partial \bar{X}^k} \frac{\partial \bar{X}^q}{\partial X^p} \frac{\partial \bar{V}^k}{\partial \bar{X}^q}. \tag{5-4.1b}$$

It is the relation (5-4.1b) that commands our attention; $\partial V^j/\partial X^p$ transform as tensor components of covariant and contravariant valence 1 each if and only if the first term of the right-hand member is zero. Because the vector field **V** is arbitrary, this is equivalent to

$$\frac{\partial^2 X^j}{\partial \bar{X}^q \, \partial \bar{X}^k} = 0 \tag{5-4.1c}$$

or, in other words, to the allowable transformation group being linear. In any other circumstance we must look for a tensor quantity to replace the set of partial derivatives.

Simply by replacing $\partial^2 X^j/\partial \bar{X}^q \, \partial \bar{X}^k$ with $\delta_t^{\ j} (\partial^2 X^t/\partial \bar{X}^q \, \partial \bar{X}^k)$ and then substituting a product of partials for $\delta_t^{\ j}$, we can put (5-4.1b) in the form

$$\frac{\partial V^j}{\partial X^p} = \frac{\partial \bar{X}^q}{\partial X^p} \frac{\partial X^j}{\partial \bar{X}^k} \left(\frac{\partial \bar{X}^k}{\partial X^t} \frac{\partial^2 X^t}{\partial \bar{X}^q \partial \bar{X}^s} \bar{V}^s + \frac{\partial \bar{V}^k}{\partial \bar{X}^q} \right). \tag{5-4.2}$$

Two facts are of value in examining (5-4.2). First of all, there is an asymmetry about (5-4.2) in that the parenthetic expression on the right contains the term involving second partials. However, we can expect something of this sort, for if the transformation group were the linear orthogonal Cartesian group the second partials would be zero and (5-4.2) would represent the appropriate law of transformation. Second, if we consider a second transformation

$$X^j = X^j(\bar{\bar{X}}^k)$$

and then eliminate the Cartesian components from the expressions of the form (5-4.2),

(5-4.3a)

$$\frac{\partial \bar{X}^q}{\partial X^p} \frac{\partial X^j}{\partial \bar{X}^k} \left(\frac{\partial \bar{X}^k}{\partial X^t} \frac{\partial^2 X^t}{\partial \bar{X}^q \, \partial \bar{X}^s} \bar{V}^s + \frac{\partial \bar{V}^k}{\partial \bar{X}^q} \right) = \frac{\partial \bar{\bar{X}}^q}{\partial X^p} \frac{\partial X^j}{\partial \bar{\bar{X}}^k} \times \left(\frac{\partial \bar{\bar{X}}^k}{\partial X^t} \frac{\partial^2 X^t}{\partial \bar{\bar{X}}^q \, \partial \bar{\bar{X}}^s} \bar{\bar{V}}^s + \frac{\partial \bar{\bar{V}}^k}{\partial \bar{\bar{X}}^q} \right).$$

By multiplying by $(\partial X^p/\partial \bar{X}^a)(\partial \bar{X}^b/\partial X^j)$, we obtain

(5-4.3b)

$$\left(\frac{\partial \bar{X}^b}{\partial X^t} \frac{\partial^2 X^t}{\partial \bar{X}^a \, \partial \bar{X}^s} \bar{V}^s + \frac{\partial \bar{V}^b}{\partial \bar{X}^a} \right) = \frac{\partial \bar{\bar{X}}^q}{\partial \bar{X}^a} \frac{\partial \bar{X}^b}{\partial \bar{\bar{X}}^k} \left(\frac{\partial \bar{\bar{X}}^k}{\partial X^t} \frac{\partial^2 X^t}{\partial \bar{\bar{X}}^q \, \partial \bar{\bar{X}}^s} \bar{\bar{V}}^s + \frac{\partial \bar{\bar{V}}^k}{\partial \bar{\bar{X}}^q} \right).$$

The parenthetic expressions are of the same form and are related in tensor fashion. Therefore in a Euclidean space in which rectangular Cartesian coordinate systems are available we could replace partial derivatives by the parenthetic expressions. However, we want more generality; hence we must go a step farther.

Following Christoffel, we look for a way of replacing the expressions $(\partial \bar{X}^b/\partial X^t)(\partial^2 X^t/\partial \bar{X}^a \, \partial \bar{X}^s)$ (which still contain references to rectangular

Cartesian coordinates) by quantities involving the components of the metric tensor.

Theorem 5-4.1. If the X^j coordinate system is rectangular Cartesian and $X^j = X^j(\bar{X}^k)$ is an allowable transformation, then

$$\frac{\partial \bar{X}^b}{\partial X^t}\frac{\partial^2 X^t}{\partial \bar{X}^a \partial \bar{X}^s} = \frac{1}{2}\bar{g}^{bc}\left(\frac{\partial \bar{g}_{sc}}{\partial \bar{X}^a} + \frac{\partial \bar{g}_{ca}}{\partial \bar{X}^s} - \frac{\partial \bar{g}_{as}}{\partial \bar{X}^c}\right). \tag{5-4.4a}$$

PROOF. We have

$$\bar{g}_{sc} = \frac{\partial X^p}{\partial \bar{X}^s}\frac{\partial X^q}{\partial \bar{X}^c}\delta_{pq}. \tag{5-4.4b}$$

If for notational convenience we let $\partial_a = \partial/\partial \bar{X}^a$, then

$$\partial_a \bar{g}_{sc} = \frac{\partial^2 X^p}{\partial \bar{X}^a \partial \bar{X}^s}\frac{\partial X^q}{\partial \bar{X}^c}\delta_{pq} + \frac{\partial X^p}{\partial \bar{X}^s}\frac{\partial^2 X^q}{\partial \bar{X}^a \partial \bar{X}^c}\delta_{pq}. \tag{5-4.4c}$$

Simply by permuting indices we obtain two additional expressions of the same type, that is,

$$\partial_s \bar{g}_{ca} = \frac{\partial^2 X^p}{\partial \bar{X}^s \partial \bar{X}^c}\frac{\partial X^q}{\partial \bar{X}^a}\delta_{pq} + \frac{\partial X^p}{\partial \bar{X}^c}\frac{\partial^2 X^q}{\partial \bar{X}^s \partial \bar{X}^a}\delta_{pq}, \tag{5-4.4d}$$

$$\partial_c \bar{g}_{as} = \frac{\partial^2 X^p}{\partial \bar{X}^c \partial \bar{X}^a}\frac{X^q}{\partial \bar{X}^s}\delta_{pq} + \frac{\partial X^p}{\partial \bar{X}^a}\frac{\partial^2 X^q}{\partial \bar{X}^c \partial \bar{X}^s}\delta_{pq}. \tag{5-4.4e}$$

Now we form the parenthetic expression on the right-hand side of (5-4.4a) from (5-4.4c,d,e), thereby obtaining

$$(\partial_a \bar{g}_{sc} + \partial_s \bar{g}_{ca} - \partial_c \bar{g}_{as}) = 2\frac{\partial^2 X^p}{\partial \bar{X}^a \partial \bar{X}^s}\frac{\partial X^q}{\partial \bar{X}^c}\delta_{pq}. \tag{5-4.5a}$$

Since

$$\bar{g}^{bc} = \frac{\partial \bar{X}^b}{\partial X^r}\frac{\partial \bar{X}^c}{\partial X^t}\delta^{rt}, \tag{5-4.5b}$$

we can apply the members of this relation to the corresponding members of (5-4.5a), thereby obtaining the desired result. This completes the proof.

Definition 5-4.1. With respect to each allowable coordinate system, let

$$\Gamma_{ij}{}^{k} = \frac{1}{2}g^{kq}\left(\frac{\partial g_{jq}}{\partial X^i} + \frac{\partial g_{qi}}{\partial X^j} - \frac{\partial g_{ij}}{\partial X^q}\right). \tag{5-4.6}$$

The Γ_{ij}^k, which are functions of the g's and their partial derivatives, are called Christoffel symbols.[6]

[6] The Γ_{ij}^k are usually referred to as Christoffel symbols of the second kind. The quantities $g_{kp}\Gamma_{ij}^k$ are designated as Christoffel symbols of the first kind. The mathematical forms were introduced into the mathematical literature in a paper by Christoffel and Lipschitz in 1870.

With the introduction of the Christoffel symbols, (5-4.3b) can be written

$$\left(\frac{\partial \bar{V}^b}{\partial \bar{X}^a} + \bar{\Gamma}_{sa}{}^b \bar{V}^s\right) = \frac{\partial \bar{\bar{X}}^q}{\partial \bar{X}^a}\frac{\partial \bar{X}^b}{\partial \bar{\bar{X}}}\left(\frac{\partial \bar{\bar{V}}^k}{\partial \bar{\bar{X}}^q} + \bar{\bar{\Gamma}}_{sq}{}^k \bar{\bar{V}}^s\right). \tag{5-4.7}$$

The parenthetic expressions transform as tensor components, reduce to partial derivatives in a Cartesian coordinate system,[7] and are intrinsically defined (i.e., reference to a third coordinate system as in (5-4.3c) is not involved). They are certainly prime candidates as a choice of the generalization of the partial derivative concept. For a Euclidean space we can immediately accept them as such. For non-Euclidean spaces some more work is necessary. The relations (5-4.3b) and (5-4.7) were obtained by making use of an intermediate rectangular Cartesian system. In order to verify that relation (5-4.7) is valid when a Cartesian system is not available, we must know the law of transformation of the Christoffel symbols. Indeed, this is an interesting question in itself, for we might be in doubt concerning their tensor character.

Theorem 5-4.2. The Christoffel symbols satisfy the transformation law

$$\Gamma_{jk}{}^i = \frac{\partial X^i}{\partial \bar{X}^p}\left(\frac{\partial \bar{X}^q}{\partial X^j}\frac{\partial \bar{X}^r}{\partial X^k}\bar{\Gamma}_{qr}{}^p + \frac{\partial^2 \bar{X}^p}{\partial X^j \partial X^k}\right). \tag{5-4.8}$$

The proof, which is left to the reader, can be carried out in the same fashion as the proof of Theorem 5-4.1. Note that the Christoffel symbols are not tensor components.

With the rule of transformation of Christoffel symbols established, the introduction of the partial derivative concept into Riemannian spaces follows.

Definition 5-4.2a. In each allowable coordinate system let

$$\nabla_j V^k = \frac{\partial V^k}{\partial X^j} + \Gamma_{pj}{}^k V^p. \tag{5-4.9a}$$

The set of components $\nabla_j V^k$ is referred to as the covariant derivative of the contravariant vector components V^k.

Theorem 5-4.3a. The class $\{\nabla_j V^k\}$ of covariant derivatives of contravariant vector components is a tensor of contravariant valence 1 and of covariant valence 1.

---

[7] The Christoffel symbols are all zero in a Cartesian system since the components of the metric tensor are constants. See Definition 5-4.6.

PROOF. We have

$$\begin{aligned}
\nabla_j V^k &= \frac{\partial V^k}{\partial X^j} + \Gamma_{pj}{}^k V^p \\
&= \frac{\partial}{\partial X^j}\left(\frac{\partial X^k}{\partial \bar{X}^q}\bar{V}^q\right) + \frac{\partial X^k}{\partial \bar{X}^q}\left(\frac{\partial \bar{X}^r}{\partial X^p}\frac{\partial \bar{X}^s}{\partial X^j}\bar{\Gamma}_{rs}{}^q + \frac{\partial^2 \bar{X}^q}{\partial X^p \partial X^j}\right)\frac{\partial X^p}{\partial \bar{X}^t}\bar{V}^t \\
&= \frac{\partial^2 X^k}{\partial \bar{X}^a \partial \bar{X}^q}\frac{\partial \bar{X}^a}{\partial X^j}\bar{V}^q + \frac{\partial X^k}{\partial \bar{X}^q}\frac{\partial \bar{V}^q}{\partial \bar{X}^t}\frac{\partial \bar{X}^t}{\partial X^j} \\
&\quad + \frac{\partial X^k}{\partial \bar{X}^q}\left(\frac{\partial \bar{X}^r}{\partial X^p}\frac{\partial \bar{X}^s}{\partial X^j}\bar{\Gamma}_{rs}{}^q + \frac{\partial^2 \bar{X}^q}{\partial X^p \partial X^j}\right)\frac{\partial X^p}{\partial \bar{X}^t}\bar{V}^t.
\end{aligned}$$

By taking a partial derivative with respect to $\bar{X}^q$ of the expression

$$\frac{\partial X^k}{\partial \bar{X}^a}\frac{\partial \bar{X}^a}{\partial X^j} = \delta_j^{\,k},$$

we determine that the sum of those terms involving second derivatives is zero. Therefore

$$\begin{aligned}
\nabla_j V^k &= \frac{\partial X^k}{\partial \bar{X}^q}\frac{\partial \bar{X}^t}{\partial X^j}\frac{\partial \bar{V}^q}{\partial \bar{X}^t} + \frac{\partial X^k}{\partial \bar{X}^q}\,\bar{\delta}_t^{\,r}\,\frac{\partial \bar{X}^s}{\partial X^j}\bar{\Gamma}_{rs}{}^q V^t \\
&= \frac{\partial X^k}{\partial \bar{X}^q}\frac{\partial \bar{X}^s}{\partial X^j}\left(\frac{\partial \bar{V}^q}{\partial \bar{X}^s} + \bar{\Gamma}_{ts}{}^q \bar{V}^t\right) \\
&= \frac{\partial X^k}{\partial \bar{X}^q}\frac{\partial \bar{X}^s}{\partial X^j}\bar{\nabla}_s \bar{V}^q.
\end{aligned}$$

This completes the proof.

The introduction of covariant differentiation might just as well have been made through consideration of a covariant vector. This procedure leads to the following consequences.

Definition 5-4.2b. In each allowable coordinate system let

(5-4.9b)
$$\nabla_j V_k = \frac{\partial V_k}{\partial X^j} - \Gamma_{kj}{}^p V_p.$$

The sets of components $\nabla_j V_k$ is referred to as the covariant derivative of the covariant vector components V_k.

Theorem 5-4.3b. The class $\{\nabla_j V_k\}$ of covariant derivatives of covariant vector components is a tensor of covariant valence 2.

The proof is made in the same manner as that of the preceding theorem.

The covariant derivative of a scalar field Φ of weight W is defined as follows.

Definition 5-4.2c. In each allowable coordinate system let

(5-4.9c)
$$\nabla_j \Phi = \frac{\partial \Phi}{\partial X^j} + W \Gamma_{pj}{}^{p} \Phi,$$

where Φ is a scalar field satisfying the transformation rule

(5-4.9d)
$$\Phi = \left| \frac{\partial X}{\partial \bar{X}} \right|^{W} \bar{\Phi}.$$

The set of components $\nabla_j \Phi$ is said to be the covariant derivative of the scalar field Φ.

Theorem 5-4.3c. The class $\{\nabla_j \Phi\}$ of covariant derivatives of scalar fields Φ of weight W is a tensor density of covariant valence 1 and of weight W.

PROOF. In this proof we must make use of the fact that the partial derivative of a determinant, with respect to one of its elements, is the cofactor of that element. In particular,

(5-4.10a)
$$\frac{\partial \left| \partial X / \partial \bar{X} \right|}{\partial (\partial X^r / \partial \bar{X}^s)} = \left| \frac{\partial X}{\partial \bar{X}} \right| \frac{\partial \bar{X}^s}{\partial X^r}.$$

(See Chapter 3, Section 5, also Theorem 5-1.7, page 313.) We have

$$\begin{aligned} \nabla_j \Phi &= \frac{\partial \Phi}{\partial X^j} + W \Gamma_{pj}{}^{p} \Phi \\ &= \frac{\partial \left| \partial X / \partial \bar{X} \right|^{W}}{\partial \bar{X}^q} \frac{\partial \bar{X}^q}{\partial X^j} \bar{\Phi} + \left| \frac{\partial X}{\partial \bar{X}} \right|^{W} \frac{\partial \bar{\Phi}}{\partial \bar{X}^q} \frac{\partial \bar{X}^q}{\partial X^j} \\ &\quad + W \left| \frac{\partial X}{\partial \bar{X}} \right|^{W} \frac{\partial X^p}{\partial \bar{X}^q} \left(\frac{\partial \bar{X}^r}{\partial X^p} \frac{\partial \bar{X}^s}{\partial X^j} \bar{\Gamma}_{rs}{}^{q} + \frac{\partial^2 \bar{X}^q}{\partial X^p \partial X^j} \right) \bar{\Phi}. \end{aligned}$$

The first term of the right member of this expression can be written in the form

$$W \left| \frac{\partial X}{\partial \bar{X}} \right|^{W-1} \frac{\partial \left| \partial X / \partial \bar{X} \right|}{\partial (\partial X^r / \partial \bar{X}^s)} \frac{\partial (\partial X^r / \partial \bar{X}^s)}{\partial \bar{X}^q} \frac{\partial \bar{X}^q}{\partial X^j} \bar{\Phi}$$

or, upon using (5-4.10a), as

(5-4.10b)
$$W \left| \frac{\partial X}{\partial \bar{X}} \right|^{W} \frac{\partial \bar{X}^s}{\partial X^r} \frac{\partial^2 X^r}{\partial \bar{X}^q \partial \bar{X}^s} \frac{\partial \bar{X}^q}{\partial X^j} \bar{\Phi}.$$

By returning to our original expression, of which (5-4.10b) is the first term, it can be shown that the terms involving second derivatives add up to zero. (Differentiate $(\partial X^r / \partial \bar{X}^q)(\partial \bar{X}^q / \partial X^p) = \delta_p{}^r$ with respect to X^j.) Therefore

$$\nabla_j \Phi = \left| \frac{\partial X}{\partial \bar{X}} \right|^{W} \frac{\partial \bar{X}^q}{\partial X^j} \left(\frac{\partial \bar{\Phi}}{\partial \bar{X}^q} + W \bar{\Gamma}_{rq}{}^{r} \bar{\Phi} \right) = \left| \frac{\partial X}{\partial \bar{X}} \right|^{W} \frac{\partial \bar{X}^q}{\partial X^j} \bar{\nabla}_q \bar{\Phi}.$$

This is the appropriate law of transformation. Hence the proof is complete.

Through Definitions 5-4.2a,b,c and Theorems 5-4.3a,b,c the covariant derivatives of contravariant vectors, covariant vectors, and scalar densities have been considered. These individual results indicate the form of a type of partial differentiation that preserves tensor character. In the next definition and theorem the preceding results are summarized and extended to tensors of any order.

Definition 5-4.3. In each allowable coordinate system let

$$\begin{aligned}(5\text{-}4.11)\qquad \nabla_k T^{i_1\cdots i_p}_{j_1\cdots j_q} = {} & \frac{\partial}{\partial X^k} T^{i_1\cdots i_p}_{j_1\cdots j_q} + \Gamma^{i_1}_{rk} T^{r i_2\cdots i_p}_{j_1\cdots j_q} + \cdots \\ & + \Gamma^{i_p}_{rk} T^{i_1\cdots i_{p-1} r}_{j_1\cdots j_q} - \Gamma_{j_1 k}{}^{r} T^{i_1\cdots i_p}_{r j_2\cdots j_q} - \cdots \\ & - \Gamma_{j_q k}{}^{r} T^{i_1\cdots i_p}_{j_1\cdots j_{q-1} r} + W\Gamma^{r}_{rk} T^{i_1\cdots i_p}_{j_1\cdots j_q}.\end{aligned}$$

The set of components $\nabla T^{i_1\cdots i_p}_{j_1\cdots j_q}$ is said to be the covariant derivative of the tensor density components.

Theorem 5-4.4. The class $\{\nabla_k T^{i_1\cdots i_p}_{j_1\cdots j_q}\}$ of covariant derivatives of the tensor density components is a tensor of the same contravariant valence, the same density, and covariant valence one more than the original tensor.

The proof can be accomplished by combining the ideas used in Theorems (5-4.3a,b,c). It is tedious but straightforward. (See Problem 6, also Section 8, Problem 4.)

It is useful to be aware that the covariant derivative satisfies the following rules of elementary calculus.

Theorem 5-4.5. We have

$$\begin{aligned}&\text{(a)}\quad \nabla_i (A^{j_1\cdots j_q}_{k_1\cdots k_p} + B^{j_1\cdots j_q}_{k_1\cdots k_p}) = \nabla_i A^{j_1\cdots j_q}_{k_1\cdots k_p} + \nabla_i B^{j_1\cdots j_q}_{k_1\cdots k_p},\\ (5\text{-}4.12)\quad &\text{(b)}\quad \nabla_i A^{k_1\cdots k_q}_{j_1\cdots j_p} B^{m_1\cdots m_s}_{l_1\cdots l_r} = (\nabla_i A^{k_1\cdots k_q}_{j_1\cdots j_p}) B^{m_1\cdots m_s}_{l_1\cdots l_r} \\ &\qquad\qquad + A^{k_1\cdots k_q}_{j_1\cdots j_p} \nabla_i B^{m_1\cdots m_s}_{l_1\cdots l_r}.\end{aligned}$$

The rules are also valid under contractions.

The proofs of properties (5-4.12a,b) are straightforward but long. They are omitted. (See Problem 8.)

The process of covariant differentiation must be handled with care. For example, examination of Definition 5-4.2a brings forth the fact that covariant differentiation of a set of constant vector components V^k does not necessarily produce a zero result; rather

$$\nabla_j V^k = \Gamma_{pj}{}^{k} V^p.$$

Conversely, if $\nabla_j V^k = 0$, we cannot conclude that the components V^k are constants.

Definition 5-4.4. A tensor $\{T^{k_1 \cdots k_q}_{j_1 \cdots j_p}\}$ is said to be a covariant constant if and only if

(5-4.13) $$\nabla_i T^{k_1 \cdots k_q}_{j_1 \cdots j_p} = 0.$$

The fundamental covariant constants are pointed out by the following theorems.

Theorem 5-4.6a (Ricci's lemma). The fundamental metric tensor is a covariant constant; that is,

(5-4.14a) $$\nabla_j g_{kp} = 0.$$

PROOF. We have (let $\partial_j = \partial/\partial X^j$)

$$\begin{aligned}\nabla_j g_{kp} &= \partial_j g_{kp} - \Gamma_{kj}{}^r g_{rp} - \Gamma_{pj}{}^r g_{kr}\\ &= \partial_j g_{kp} - \tfrac{1}{2}\,\delta_p{}^s(\partial_k g_{js} + \partial_j g_{sk} - \partial_s g_{kj})\\ &\quad - \tfrac{1}{2}\,\delta_k{}^s(\partial_p g_{js} + \partial_j g_{sp} - \partial_s g_{pj})\\ &= 0.\end{aligned}$$

Theorem 5-4.6b. We have

(5-4.14b) $$\nabla_j\, \delta_k{}^p = 0.$$

PROOF.

$$\nabla_j\, \delta_k{}^p = \partial_j\, \delta_k{}^p + \Gamma_{qj}{}^p\, \delta_k{}^q - \Gamma_{kj}{}^q\, \delta_q{}^p = 0.$$

This completes the proof.

Theorem 5-4.6c. The associated metric tensor is a covariant constant; that is,

(5-4.14c) $$\nabla_j g^{kp} = 0.$$

PROOF. Let us start with the relation

$$\delta_j^k = g_{jp} g^{pk}.$$

As a consequence of Theorem 5-4.6b and the product rule stated in Theorem 5-4.5, we have

$$0 = \nabla_q g_{jp} g^{pk} = (\nabla_q g_{jp}) g^{pk} + g_{jp} \nabla_q g^{pk}.$$

Since g_{jp} is a covariant constant, the first term of the right-hand member has the value zero. Multiplication by g^{jr} produces the desired result.

The determinant of the metric tensor is also a covariant constant. However, to establish this fact, the following result concerning contracted Christoffel symbols is needed.

Theorem 5-4.7. We have

$$\Gamma_{jk}^{\ \ j} = \frac{\partial \ln (|g_{pq}|)^{1/2}}{\partial X^k}. \tag{5-4.15}$$

PROOF. According to definition,

$$\Gamma_{jk}^{\ \ j} = \tfrac{1}{2}g^{jq}(\partial_j g_{kq} + \partial_k g_{qj} - \partial_q g_{jk}) = \tfrac{1}{2}g^{jq}\,\partial_k g_{qj}. \tag{5-4.16a}$$

$$g^{jq} = \frac{\text{cofactor } g_{qj} \text{ in } |g_{qj}|}{|g_{rs}|} = \frac{1}{|g_{rs}|}\frac{\partial\, |g_{uv}|}{\partial g_{qj}}. \tag{5-4.16b}$$

By combining (5-4.16a,b), we obtain

$$\Gamma_{jk}^{\ \ j} = \frac{1}{2}\frac{1}{|g_{rs}|}\frac{\partial\, |g_{uv}|}{\partial X^k} = \frac{\partial \ln (|g_{uv}|)^{1/2}}{\partial X^k}, \tag{5-4.16c}$$

as was to be shown.

Theorem 5-4.8. The determinant of the metric tensor is a covariant constant; that is,

$$\nabla_j g = 0. \tag{5-4.17}$$

The proof is left to the reader. Note that g is a scalar density of weight -2.

Thus far we have been considering the Riemannian space analogue of the process of partial differentiation. Let us turn our attention to the problem of an appropriate generalization of the ordinary derivative.

Because it is desirable, at least from the geometric point of view, to think of tensor fields expressed in terms of a single parameter as defined along a space curve, the meaning of the term curve with respect to a Riemannian space should be made precise. In actuality, we simply extend Definition 2-1.2.

Definition 5-4.5. Let $X^1, \cdots, X^n$ be continuous functions defined on a common domain with at least the first derivatives continuous and not simultaneously zero. The set of all points with coordinates $X^1(t) \cdots X^n(t)$ is called a space curve. Other parameter representations which develop from this one through allowable parameter transformations[8] designate the same curve.

$$X^j = X^j(t)$$

are called parametric equations of the curve.

Suppose that Φ is a scalar function defined on a space curve with parametric equations $X^j = X^j(t)$. Then

$$\Phi(t) = \overline{\Phi}(t)$$

[8] See Def. 2-1.2b

for all t. Hence

$$\frac{\Phi(t + \Delta t) - \Phi(t)}{\Delta t} = \frac{\overline{\Phi}(t + \Delta t) - \overline{\Phi}(t)}{\Delta t}. \tag{5-4.18a}$$

By taking the limit as $t \to 0$ of each member of (5-4.18a), we obtain the fact that

$$\frac{d\Phi}{dt} = \frac{d\overline{\Phi}}{dt}; \tag{5-4.18b}$$

that is, the derivative of a scalar field is again a scalar field. There is nothing new in (5-4.18b); however, it does do more than simply point up the fact that this much of prior theory correlates with the generalizations to Riemannian space. Relation (5-4.18b) provides a heuristic approach to the introduction of the derivative of a tensor field. In particular, for Euclidean three-space we have

$$\frac{d\Phi}{dt} = \frac{\partial \Phi}{\partial X^j} \frac{dX^j}{dt}.$$

Since the tangent field is a vector field and because the process of covariant differentiation is the Riemannian space analogue to partial differentiation, the following definition is rather natural.

Definition 5-4.6. In any allowable coordinate system let

$$\frac{D}{dt}(T^{k_1 \cdots k_q}_{j_1 \cdots j_p}) = \frac{dX^r}{dt} \nabla_r T^{k_1 \cdots k_q}_{j_1 \cdots j_p}. \tag{5-4.19}$$

The set $(D/dt)T^{k_1 \cdots k_q}_{j_1 \cdots j_p}$ is said to be the absolute[9] derivative of the tensor field components $T^{k_1 \cdots k_q}_{j_1 \cdots j_p}$.

Theorem 5-4.9. The class $\{(D/dt)T^{k_1 \cdots k_p}_{j_1 \cdots j_p}\}$ of absolute derivatives is a tensor field of the same type as the original tensor field.

PROOF. The proof follows immediately from the fact that the tangential field dX^j/dt and the covariant derivative field are contravariant and covariant vector fields, respectively.

It is to be noted that the rules for differentiation of sums and products are extendable to the absolute derivative. It is also true that covariant constants are in the same sense absolute derivative constants.

Problems

1. (a) Compute the Christoffel symbols for (1) cylindrical coordinates (2) spherical coordinates.

[9] The term intrinsic derivative is used quite often with the same meaning.

(b) Prove that the Christoffel symbols are symmetric, that is,

$$\Gamma_{jk}{}^{i} = \Gamma_{kj}{}^{i}.$$

2. Prove Theorem 5-4.2.

3. Starting with (5-4.8), show that

$$\overline{\Gamma}_{qr}{}^{p} = \frac{\partial \overline{X}^{p}}{\partial X^{i}}\left(\frac{\partial X^{j}}{\partial \overline{X}^{q}}\frac{\partial X^{k}}{\partial \overline{X}^{r}}\Gamma_{jk}{}^{i} + \frac{\partial^2 X^{i}}{\partial \overline{X}^{q}\,\partial \overline{X}^{r}}\right).$$

4. Show that $\nabla_j V_k$ transforms as a tensor of covariant valence 2.

5. Show that

$$\frac{\partial^2 X^{j}}{\partial \overline{X}^{k}\,\partial \overline{X}^{q}}\frac{\partial \overline{X}^{q}}{\partial X^{p}} + \frac{\partial X^{j}}{\partial \overline{X}^{q}}\frac{\partial X^{s}}{\partial \overline{X}^{k}}\frac{\partial^2 \overline{X}^{q}}{\partial X^{s}\,\partial X^{p}} = 0.$$

Hint:

$$\frac{\partial X^{j}}{\partial \overline{X}^{q}}\frac{\partial \overline{X}^{q}}{\partial X^{p}} = \delta_{p}{}^{j}.$$

6. Show that the covariant derivative of tensor components $T_j{}^k$ produces tensor components.

7. Prove Theorem 5-4.8.

8. (a) Show that $\nabla_p(T_j{}^k + W_j{}^k) = \nabla_p T_j{}^k + \nabla_p W_j{}^k$.
(b) Show that $\nabla_i T_j{}^k W_p{}^j = (\nabla_i T_j{}^k) W_p{}^j + T_j{}^k \nabla_i W_p{}^j$.

9. Show that the absolute derivative of a scalar of weight zero is equal to its ordinary derivative.

10. Assume that Newton's second law has the general form

$$F^{j} = \frac{DmV^{j}}{dt}.$$

Express this law in cylindrical and spherical coordinates.

11. Compute the Christoffel symbols associated with the cylindrical and spherical surfaces of Chapter 5, Section 3, Problem 2.

12. Show that

$$\nabla_j W^{j} = \frac{1}{\sqrt{g}}\frac{\partial}{\partial X^{j}}(\sqrt{g}\,W^{j})$$

In curvilinear coordinates the right-hand member of this expression is called the divergence of **W**. (See Chapter 3, Section 5.) The same name is often used for $\nabla_j W^j$ in a Riemannian space.

13. Show that

$$\nabla_{[j} W_{k]} = \partial_{[j} W_{k]}.$$

5. *Geodesics*

In Euclidean three-space a line is the shortest distance between two points. The immediate purpose is to generalize the concept of "straight line." In particular, given two points, *A* and *B*, the objective is to find the path of minimum length joining the two points consistent with the space metric. This is a problem in the calculus of variation. It was first considered (for a surface in three-space) by Johann Bernoulli (1667–1748, Swiss) in 1697.

Definition 5-5.1. Suppose that a curve $X^j = X^j(t)$ joins points P_0 and P_1 (the coordinates of P_0 and P_1 are $X_0^j = X^j(t_0)$ and $X_1^j = X^j(t_1)$, respectively); then

$$s = \int_{t_0}^{t_1} f\,dt, \qquad f = \left(\left| g_{jk} \frac{dX^j}{dt} \frac{dX^k}{dt} \right| \right)^{1/2} \tag{5-5.1}$$

is said to be the distance from P_0 to P_1 along the given curve.

There are many difficulties associated with the proposed problem of determining the curve of minimum distance. Although we have tried to point out these difficulties, they are not attacked but rather left for consideration as a part of the study of the calculus of variations.

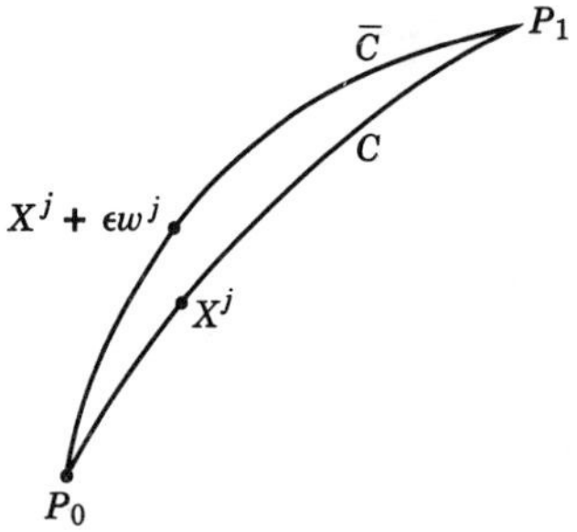

Fig. 5-5.1

First of all, the existence of a path of shortest length is not assured.[10] Second, when such a path does exist, a closed form representation (i.e., a set of parametric equations) more than likely cannot be determined. Therefore, under the assumption that a curve of shortest length does exist (for an indication of the nature of such conditions, see Bolza, pp. 246–247), we proceed to find a set of differential equations which necessarily must be satisfied.

To consider all curves through P_0 and P_1 would defy analysis; therefore the problem is restricted to a band of curves through P_0 and P_1 which contains the curve of minimum length.[11] This hypothesis is valid because of the assumption that such a curve exists. If the parametric equation

$$X^j = X^j(t)$$

are associated with a curve of minimum length C, with end points P_0 and P_1, the coordinates of a neighboring curve $\bar{C}$ are

$$\bar{X}^j = X^j + \epsilon w^j, \tag{5-5.2a}$$

[10] See Oskar Bolza, *Lectures on the Calculus of Variations*, Dover, 1961; University of Chicago Press, 1904.

[11] An intuitive feeling for the analysis can be acquired as follows. Stretch a rubber band tightly over a surface sector, the end points firmly fixed, slide the center of the band from its initial position, and let it return.

where the functions w^j have at least continuous first derivatives and

(5-5.2b) $$w^j(X_0^k) = w^j(X_1^k) = 0.$$

(See Fig. 5-5.1.) The parameter ϵ plays a dual role. On the one hand, it can be thought of as a constant such that $|\epsilon|$ is as small as necessary. On the other hand, the coordinates $\bar{X}^j$ are differentiable functions of ϵ (i.e., assume that a set of functions w^j has been chosen, and there is one-to-one correspondence between values of ϵ and curves $\bar{C}$. The coordinates $\bar{X}^j$ are linear functions of ϵ and are therefore clearly differentiable.)

Theorem 5-5.1a. For a curve C joining P_0 and P_1 to be of minimum length it is necessary that its parametric equations satisfy the Euler-Lagrange equations

(5-5.3a) $$\frac{\partial f}{\partial X^j} - \frac{d(\partial f/\partial \dot{X}^j)}{dt} = 0,$$

where

(5-5.3b) $$f = (|g_{jk}\dot{X}^j\dot{X}^k|)^{1/2}, \qquad \dot{X}^j = \frac{dX^j}{dt}.$$

PROOF. We have

(5-5.4a) $$\bar{s} = \int_{t_0}^{t_1} f(X^j + \epsilon w^j, \dot{X}^k + \epsilon \dot{w}^k)\, dt = \int_{t_0}^{t_1} \bar{f}(\bar{X}^j, \dot{\bar{X}}^k)\, dt.$$

Since $\bar{s}$ is a differentiable function of ϵ,

(5-5.4b) $$\frac{d\bar{s}}{d\epsilon} = \int_{t_0}^{t} \left(\frac{\partial \bar{f}}{\partial \bar{X}^k}\frac{d\bar{X}^k}{d\epsilon} + \frac{\partial \bar{f}}{\partial \dot{\bar{X}}^k}\frac{d\dot{\bar{X}}^k}{d\epsilon}\right) dt.$$

By assumption C is of minimum length.

(5-5.4c) $$0 = \left(\frac{d\bar{s}}{d\epsilon}\right)_{\epsilon=0} = \int_{t_0}^{t_1} \left(\frac{\partial f}{\partial X^k} w^k + \frac{\partial f}{\partial \dot{X}^k} \dot{w}^k\right) dt,$$

where the baf notation does not appear in the right-hand expression because the particular curve C has come under consideration.

(5-5.4d) $$\int_{t_0}^{t_1} \frac{\partial f}{\partial \dot{X}^k} \dot{w}^k\, dt = w^k \frac{\partial f}{\partial \dot{X}^k}\bigg]_{t_0}^{t_1} - \int_{t_0}^{t_1} w^k \frac{d(\partial f/\partial \dot{X}^k)}{dt}\, dt.$$

Since the w^k are zero at t_0 and t_1, the first term on the right-hand side of (5-5.4d) is zero. When the remaining term is substituted into (5-5.4c), we have

(5-5.4e) $$0 = \int_{t_0}^{t_1} \left[\frac{\partial f}{\partial X^k} - \frac{d(\partial f/\partial \dot{X}^k)}{dt}\right] w^k\, dt.$$

Because (5-5.4e) must hold for all sets of functions w^k which vanish at the end points of the arc, it follows that

$$\frac{\partial f}{\partial X^k} - \frac{d(\partial f/\partial \dot{X}^k)}{dt} = 0.$$

This completes the proof.

Theorem 5-5.1b. If f is given by (5-5.3b), that is,

$$f = (|g_{jk}\dot{X}^j\dot{X}^k|)^{1/2}, \qquad f \neq 0,$$

the Euler-Lagrange differential equations (5-5.3a) are equivalent to the set

(5-5.5a) $$\ddot{X}^j + \Gamma_{pq}{}^j \dot{X}^p\dot{X}^q = \frac{d(\ln f)}{dt}\dot{X}^j.$$

Furthermore, if the parameter t represents arc length s, then (5-5.5a) reduces to

(5-5.5b) $$\frac{d^2X^j}{ds^2} + \Gamma_{pq}{}^j \frac{dX^p}{ds}\frac{dX^q}{ds} = 0.$$

PROOF. By straightforward computation of the terms of the Euler-Lagrange equations we obtain the following. (Note that g_{jk} is a function of position, i.e., of the coordinates X^j only and that the $\dot{X}^j$ are independent of the position coordinates X^j; that is, direction at any point may be chosen independently of the coordinates of that point.)

(5-5.6)

(a) $$\frac{\partial f}{\partial \dot{X}^p} = \frac{\epsilon}{2f} g_{jk}(\delta_p{}^j \dot{X}^k + \dot{X}^j \delta_p{}^k) = \frac{\epsilon}{f} g_{pk}\dot{X}^k,$$

(b) $$\frac{d(\partial f/\partial \dot{X}^p)}{dt} = \epsilon\left[\frac{d(1/f)}{dt} g_{pk}\dot{X}^k + \frac{1}{f}\frac{\partial g_{pk}}{\partial X^q}\dot{X}^q\dot{X}^k + \frac{1}{f} g_{pk}\ddot{X}^k\right]$$
$$= \frac{\epsilon}{f}\left[-\frac{d(\ln f)}{dt} g_{pk}\dot{X}^k + \frac{\partial g_{pk}}{\partial X^q}\dot{X}^q\dot{X}^k + g_{pk}\ddot{X}^k\right],$$

where ϵ is the sign of $g_{jk}\dot{X}^j\dot{X}^k$. Furthermore,

(5-5.6c) $$\frac{\partial f}{\partial X^p} = \frac{\epsilon}{2f}\frac{\partial g_{jk}}{\partial X^p}\dot{X}^j\dot{X}^k.$$

By substituting the results (5-5.6b,c) into (5-5.3a), we obtain

(5-5.6d)

$$\frac{1}{2f}\left[2\frac{d(\ln f)}{dt} g_{pk}\dot{X}^k - \left(2\frac{\partial g_{pk}}{\partial X^j} - \frac{\partial g_{jk}}{\partial X^p}\right)\dot{X}^j\dot{X}^k - 2g_{pk}\ddot{X}^k = 0.\right.$$

The Christoffel symbols can be introduced by considering the fact that g_{jk} is a covariant constant. We obtain

(5-5.7a)
$$2\frac{\partial g_{pk}}{\partial X^j} = 2\,\Gamma_{pj}{}^r g_{rk} + 2\,\Gamma_{kj}{}^r g_{pr},$$
$$\frac{\partial g_{jk}}{\partial X^p} = \Gamma_{jp}{}^r g_{rk} + \Gamma_{kp}{}^r g_{jr}.$$

When these relations are plugged into (5-5.6d) and the summation indices, as well as the symmetry of $\dot{X}^j\dot{X}^k$, are taken into account, (5-5.6d) reduces to

(5-5.7b)
$$2\frac{d\ln f}{dt} g_{pk}\dot{X}^k - 2g_{pr}\Gamma_{jk}{}^r\dot{X}^j\dot{X}^k - 2g_{pk}\ddot{X}^k = 0.$$

An alternate form of the set of equations in (5-5.7b) is obtained by multiplying and summing with $\frac{1}{2}g^{pq}$; then

(5-5.7c)
$$\ddot{X}^q + \Gamma_{jk}{}^q\dot{X}^j\dot{X}^k = \frac{d\ln f}{dt}\dot{X}^q,$$

If t represents arc length,

$$\left| g_{jk}\frac{dX^j}{ds}\frac{dX^k}{ds} \right| = 1.$$

Therefore f is a constant and $d\ln f/ds = 0$. Under this assumption (5-5.7c) immediately reduces to the form in (5-5.5b). This completes the proof.

Definition 5-5.2. The curves satisfying the differential equations in (5-5.5a) or (5-5.5b) are said to be the geodesics of the space.[12]

Since the differential equations in (5-5.5a,b) are of second order, a unique solution is determined at a point P_0 when $X_0{}^j$ and $\dot{X}_0{}^j$ are given; that is, a point and a direction uniquely determine a geodesic at a given point.

A geodesic joining two points is not necessarily either unique or the shortest distance between the points. This statement is simply illustrated on the surface of a sphere. It can be shown that great circles (i.e., the intersections of planes through the center of the sphere with the sphere) are the geodesics, and, in fact, the only geodesics, of the surface. Clearly, a geodesic joining the poles is not unique. Furthermore, if we consider

[12] $f \equiv (|g_{jk}\dot{X}^j\dot{X}^k|)^{1/2} = 0$ identically satisfies the Euler-Lagrange equations (5-5.2a). If the direction $\dot{X}^j$ at each point along a curve is such that $f = 0$ and the differential equations in (5-5.4b), expressed in terms of a parameter that does not represent arc length, are satisfied by the equations of the curve, then the curve is said to be a null geodesic. Such geodesics are of importance in the theory of relativity.

two points, other than a diametrically opposite pair, one arc of the geodesic joining them is the shortest distance but the other arc joining them is not. It can be shown that for two points sufficiently near the joining geodesic is both unique and the curve of minimum distance.

The solution of the differential equations of geodesics in a closed form is in general beyond the power of analysis. However, in some simple cases complete solutions can be obtained. The circular cylinder is one such case.

Example 5-5.1. In dealing with geodesics on a surface an alternative form of (5-5.6a) is often useful. If the surface parameters are designated by the symbols v^1, v^2, then (5-5.5a) is equivalent to the pair of equations

(5-5.8a)
$$\frac{d^2v^1}{dt^2} + \Gamma_{pq}{}^1 \frac{dv^p}{dt}\frac{dv^q}{dt} = \frac{d\ln f}{dt}\frac{dv^1}{dt},$$
$$\frac{d^2v^2}{dt^2} + \Gamma_{pq}{}^2 \frac{dv^p}{dt}\frac{dv^q}{dt} = \frac{d\ln f}{dt}\frac{dv^2}{dt}.$$

Either a representation $v^2 = h(v^1)$ or $v^1 = g(v^2)$ can be associated with a surface curve. A parametric representation of the first is

(5-5.8b)
$$v^1 = t,$$
$$v^2 = v^2(t).$$

From (5-5.8b) it follows that

(5-5.8c)
$$\frac{dv^1}{dt} = 1, \qquad \frac{d^2v^1}{dt^2} = 0.$$

By use of these results the equations in (5-5.8a) can be expressed in the form

(5-5.8d)
$$\Gamma_{11}{}^1 + 2\,\Gamma_{12}{}^1 \frac{dv^2}{dv^1} + \Gamma_{22}{}^1 \left(\frac{dv^2}{dv^1}\right)^2 = \frac{d(\ln f)}{dv^1},$$

(5-5.8e)
$$\frac{d^2v^2}{(dv^1)^2} + \Gamma_{11}{}^2 + 2\,\Gamma_{12}{}^2 \frac{dv^2}{dv^1} + \Gamma_{22}{}^2 \left(\frac{dv^2}{dv^1}\right)^2 = \frac{d(\ln f)}{dv^1}\frac{dv^2}{dv^1}.$$

By substituting for $d\ln f/dv^1$ in (5-5.8e) we have

(5-5.8f)
$$\frac{d^2v^2}{(dv^1)^2} + \Gamma_{11}{}^2 + 2\Gamma_{12}{}^2 - \Gamma_{11}{}^1 \frac{dv^2}{dv^1} + (\Gamma_{22}{}^2 - 2\Gamma_{12}{}^1)\left(\frac{dv^2}{dv^1}\right)^2 - \Gamma_{22}{}^1\left(\frac{dv^2}{dv^1}\right)^3 = 0.$$

If we start with the representation $v^1 = g(v^2)$, then

(5-5.8g)
$$\frac{d^2v^1}{(dv^2)^2} + \Gamma_{22}{}^1 + (2\Gamma_{12}{}^1 - \Gamma_{22}{}^2)\frac{dv^1}{dv^2} + (\Gamma_{11}{}^1 - 2\Gamma_{21}{}^2)\left(\frac{dv^1}{dv^2}\right)^2 - \Gamma_{11}{}^2\left(\frac{dv^1}{dv^2}\right)^3 = 0.$$

The two forms (5-5.8f,g) determine the geodesics on a surface.

In the case of a circular cylinder expressed by the parametric equations $X^1 = \rho \cos v^1$, $X^2 = \rho \sin v^1$, $X^3 = v^2$ we have

(5-5.9a) (a) $(g_{jk}) = \begin{pmatrix} \rho^2 & 0 \\ 0 & 1 \end{pmatrix}$, $\rho = \text{constant}$,

(5-5.9b) (b) $\Gamma_{kp}{}^{j} = 0$.

The differential equations (5-5.8f,g) reduce to

(5-5.9)

$$\text{(c)} \quad \frac{d^2v^2}{(dv^1)^2} = 0,$$

$$\text{(d)} \quad \frac{d^2v^1}{(dv^2)^2} = 0.$$

These relations have solutions in the form of

(5-5.9)

$$\text{(e)} \quad v^2 = c_1 v^1 + c_2,$$

$$\text{(f)} \quad v^1 = k_1 v^2 + k_2,$$

where c_1, c_2, k_1, k_2 are constants of integration. If $c_1 = 0$, then $v^2 = c_2$. That the curves $v^2 = \text{constant}$ are circles can be determined by the fact

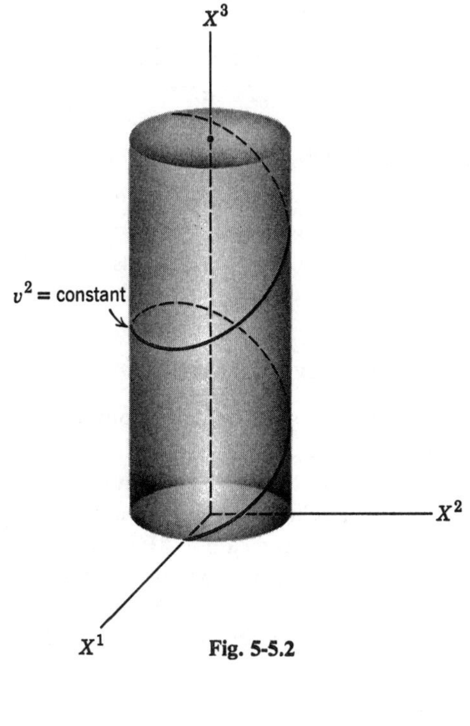

Fig. 5-5.2

that $v^2 =$ constant is equivalent to $X^3 =$ constant. On the other hand, if $c_1 \neq 0$, then

$$X^1 = \rho \cos v^1, \qquad X^2 = \rho \sin v^1, \qquad X^3 = c_1 v^1 + c_2.$$

These are the parametric equations of a circular helix. From (5-5.9f) with $k_1 = 0$, we get the curves $v^1 = k_2$. These are the rulings on the cylinder. If $k_1 \neq 0$, circular helices are obtained again. Hence the geodesics on a circular cylinder are rulings, circles, and helices. (See Fig. 5-5.2.)

Problems

1. Show that the differential equations of geodesics (5-5.5b) imply that the geodesics in a Euclidean space are straight lines.

2. Determine the Christoffel symbols from the Euler-Lagrange equations

$$\frac{\partial f}{\partial X^j} - \frac{d(\partial f/\partial \dot{X}^j)}{ds} = 0, \qquad \dot{X}^j = \frac{dX^j}{ds}$$

for
(a) the cylindrical coordinate system ρ, θ, z,
(b) the spherical coordinate system r, θ, ϕ.
Hints: Note that if one solves (5-5.5b) for d^2X^j/ds^2

$$\frac{d^2X^j}{ds^2} = -\Gamma_{pq}^{\ \ j} \frac{dX^p}{ds}\frac{dX^q}{ds};$$

that is, the Christoffel symbols are coefficients of

$$\frac{dX^p}{ds}\frac{dX^q}{ds}.$$

For cylindrical coordinates

$$f = \left[\left(\frac{d\rho}{ds}\right)^2 + \rho^2\left(\frac{d\theta}{ds}\right)^2 + \left(\frac{dz}{ds}\right)^2\right]^{1/2}$$

Plug this form in the Euler-Lagrange equations and make the appropriate computations. Note that f does not vary with change in s.

6. *The Parallelism of Levi-Civita*

In a Euclidean space there is an unambiguous meaning to the term "parallel vector field." A vector field is parallel if the components of every two members are proportional. In considering surfaces, the only intuitive examples of non-Euclidean spaces, the preceding definition of parallelism of a vector field has no meaning. For, if we are developing the intrinsic geometry of a surface, a member of a surface vector field at a point P must be in the tangential plane to the surface at P. If the components of surface vectors at distinct points P and Q are proportional

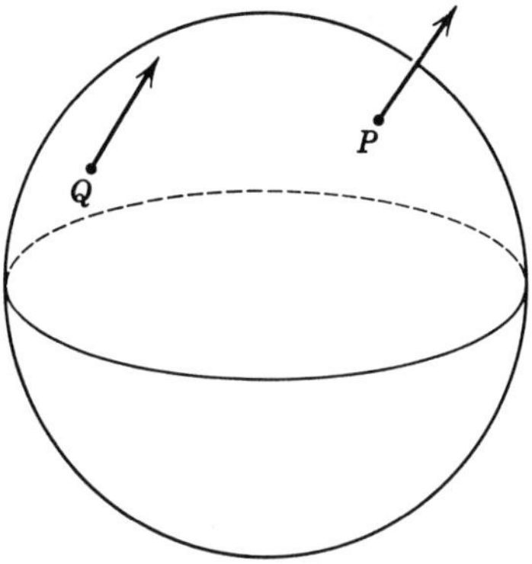

Fig. 5-6.1

with respect to one set of surface parameters, they, in general, will not be with respect to another set. Thus the concept of proportionality of components is not preserved under surface coordinate transformation and cannot serve in more general situations. (See Fig. 5-6.1.)

In casting about for a generalization of the Euclidean concept of parallelism of a vector field, we might again consider the special case of surfaces and attempt to make use of the embedding space. However, this line of thought must be dropped quickly, since a vector at a surface point P parallel in the Euclidean sense to a surface vector at Q would, in general, not be a surface vector.

Note that more often than not comprehensive definitions are induced from special examples. In turn, the general definition must be compatible with elementary cases. In this instance the special examples consist of Euclidean geometry and surface theory. Having stated methods of generalizing the concept of parallelism which might come into our minds but which must be rejected, we now present the accepted generalization.

Although the method of presentation may be quite different from the original, the following ideas are essentially those of the Italian mathematician Tullio Levi-Civita. Suppose we restrict our considerations to a curve C in Euclidean space. Let $V^j = V^j(t)$ be components of a vector field of parallel vectors of constant magnitude along a curve C (parallel in the Euclidean sense); then

$$\frac{dV^j}{dt} = 0. \tag{5-6.1}$$

A definition of a parallel vector field in a Riemannian space is introduced only with respect to a given curve C of that space. The analytic form of the concept is inferred directly from (5-6.1).

Definition 5-6.1. Let **V** be a vector field along a curve C in a Riemannian space. The vector field is said to be a parallel vector field with respect to the curve C if and only if

$$\frac{DV^j}{dt} = 0. \tag{5-6.2}$$

If $V_0{}^j$ is the vector field representative at P_0 of C, then any other vector of the field is said to be obtained from $V_0{}^j$ by means of parallel displacement. It is clear that the general definition (5-6.1) includes the Euclidean case from which it was inferred, since

$$\frac{DV^j}{dt} = \frac{dV^j}{dt} + \Gamma_{pq}{}^j V^p \frac{dX^q}{dt} = 0$$

reduces to the form (5-6.1) in a Cartesian coordinate system where

$$\Gamma_{pq}{}^j = 0.$$

The fundamental properties of parallelism are pointed out in the next theorems.

Theorem 5-6.1. Let **V** be a parallel vector field with respect to a curve C. The magnitude of **V** is constant.

PROOF. Since the ordinary derivative of a scalar coincides with the absolute derivative (Chapter 5, Section 4, Problem 9), we have

$$\frac{d}{dt}(g_{jk}V^jV^k) = \frac{D}{dt}(g_{jk}V^jV^k) = g_{jk}\left(\frac{DV^j}{dt}V^k + V^j\frac{DV^k}{dt}\right)$$

$$= 2g_{jk}V^j\frac{DV^k}{dt} = 0.$$

Since the ordinary derivative of the magnitude is zero, that magnitude is constant.

Theorem 5-6.2. Let **A** and **B** be parallel vector fields along a curve C. The angle determined by **A** and **B** is constant along C.

The proof is analogous to that of the preceding theorem. In fact, that theorem could be considered a special case of this one.

A tangential vector field $\{dX^j/ds\}$ along a geodesic, and referred to arc length, satisfies the relation.

$$\frac{D(dX^j/ds)}{ds} = 0. \tag{5-6.3}$$

Hence this tangential field along a geodesic is a parallel vector field. According to the preceding theorem, any other parallel vector field along the geodesic makes equal angles with the tangential field. This is rather pleasing intuitively, since the geodesics are the lines of the space and the tangential vector field determines the direction of one of the lines at a given point. In other words, the property of a parallel vector field along a geodesic making a fixed angle with that geodesic harmonizes nicely with the Euclidean concept of a parallel field making a constant angle with a line.

It may have occurred to the reader that the relation (5-6.1) from which we generalized to obtain the definition of parallel vector field is somewhat restricted in that V^j was assumed to be of constant magnitude. If this restriction is dropped, we have as a unit field $V^j/(\delta_{pq}V^pV^q)^{1/2}$ along C (assuming that the space is referred to a rectangular Cartesian system). Then

$$\frac{d(V^j/f)}{dt} = 0, \qquad f = (\delta_{pq}V^pV^q)^{1/2}. \tag{5-6.4a}$$

After differentiation, we have

$$\frac{dV^j}{dt} = \left(\frac{d\ln f}{dt}\right)V^j. \tag{5-6.4b}$$

The obvious generalization of (5-6.4b) is

$$\frac{DV^j}{dt} = h(t)V^j. \tag{5-6.4c}$$

Vector fields satisfying (5-6.4c) are referred to in the mathematical literature as "pseudoparallel." If $h(t) = 0$, the German word "pseudoäquipollent" is used. According to Definition 5-6.1, a pseudoäquipollent field is a parallel field.

Parallel vector fields do not satisfy all of the properties that we might intuitively cherish. For example, suppose C is a closed curve and vector components V_0^j are associated with a point P_0 of C. We cannot expect a vector of the field generated by parallel displacement of V_0^j around C to coincide with V_0^j at P_0 again. This fact is illustrated in the next example. In the example Greek indices have the range 1, 2.

Example 5-6.1. In Euclidean three-space and with respect to a rectangular Cartesian coordinate system the equation of a cone is given by

$$(X^3)^2 = (X^1)^2 + (X^2)^2. \tag{5-6.5a}$$

A corresponding parametric representation is

$$\begin{aligned} X^1 &= v^1 \cos v^2, \\ X^2 &= v^1 \sin v^2, \\ X^3 &= v^1, \end{aligned} \tag{5-6.5b}$$

where v^1, v^2 can be thought of as parameters or surface coordinates. To study the intrinsic geometry of the surface of the cone, we begin with the fundamental metric form

$$ds^2 = g_{\lambda\gamma}\, dv^\lambda\, dv^\gamma,$$

where as usual

$$g_{\lambda\gamma} = \sum_{i=1}^{3} \frac{\partial X^i}{\partial v^\lambda} \frac{\partial X^i}{\partial v^\gamma}.$$

By means of straightforward computation, we obtain

$$\text{(a)} \quad (g_{\lambda\gamma}) = \begin{pmatrix} 2 & 0 \\ 0 & (v^1)^2 \end{pmatrix},$$

$$\text{(5-6.6)} \quad \text{(b)} \quad (g^{\lambda\gamma}) = \begin{pmatrix} \dfrac{1}{2} & 0 \\ 0 & \dfrac{1}{(v^1)^2} \end{pmatrix},$$

$$\text{(c)} \quad \Gamma_{12}{}^2 = \Gamma_{21}{}^2 = \frac{1}{v^1}, \qquad \Gamma_{22}{}^1 = -\frac{v^1}{2}, \qquad \Gamma_{\gamma\beta}{}^\lambda = 0 \text{ otherwise.}$$

Consider a surface curve C:

$$\text{(5-6.7a)} \qquad \begin{aligned} v^1 &= c \equiv \text{constant}, \\ v^2 &= t. \end{aligned}$$

Let W be components of a parallel vector field on C. Then

$$\text{(5-6.7b)} \qquad \frac{DW^\lambda}{dt} = 0$$

or

$$\text{(5-6.7c)} \qquad \frac{dW^\lambda}{dt} + \Gamma_{\gamma\beta}{}^\lambda W^\gamma \frac{dv^\beta}{dt} = 0.$$

By making use of the values determined for the Christoffel symbols and the fact that

$$\frac{dv^1}{dt} = 0, \qquad \frac{dv^2}{dt} = 1,$$

we can write out the equations (5-6.7c) as

$$\text{(5-6.7d)} \qquad \begin{aligned} \frac{dW^1}{dt} - \frac{v^1}{2} W^2 &= 0, \\ \frac{dW^2}{dt} + \frac{1}{v^1} W^1 &= 0. \end{aligned}$$

where $V^1 = c$ along the curve of consideration. By differentiating the first of these equations and plugging the second into the result, we obtain

$$\frac{d^2W^1}{dt^2} + \frac{1}{2} W^1 = 0. \tag{5-6.8a}$$

The solution of this differential equation is

$$W^1 = k_1 e^{(1/\sqrt{2})it} + k_2 e^{-(1/\sqrt{2})it} = A \cos \frac{1}{\sqrt{2}} t + B \sin \frac{1}{\sqrt{2}} t. \tag{5-6.8b}$$

From the first of the equations (5-6.7d)

$$W^2 = \frac{\sqrt{2}}{c}\left(-A \sin \frac{1}{\sqrt{2}} t + B \cos \frac{1}{\sqrt{2}} t\right). \tag{5-6.8c}$$

Now suppose that corresponding to $t = 0$ [i.e., at the curve point $(c, 0)$] we have $W^\beta = W_0^{\ \beta}$. Then the constants of integration in (5-6.8b,c) take values

$$A = W_0^{\ 1}, \qquad B = \frac{c}{\sqrt{2}} W_0^{\ 2}. \tag{5-6.8d}$$

At $t = 2\pi$

$$\begin{aligned} W^1 &= W_0^{\ 1} \cos \sqrt{2}\pi + \frac{c}{\sqrt{2}} W_0^{\ 2} \sin \sqrt{2}\pi, \\ W^2 &= -\frac{\sqrt{2}}{c} W_0^{\ 1} \sin \sqrt{2}\pi + W_0^{\ 2} \cos \sqrt{2}\pi. \end{aligned} \tag{5-6.8e}$$

Thus the vector obtained by parallel displacement around the closed curve has a direction different from the original. Of course, the magnitude of the vector is the same as that of $W_0^{\ \beta}$.

Problems

1. To study the intrinsic geometry of a sphere, we begin with

$$ds^2 = g_{\lambda\gamma}\, d\bar{X}^\lambda\, d\bar{X}^\gamma, \qquad \bar{X}^1 = \theta, \qquad \bar{X}^2 = \phi.$$

The fundamental metric tensor components were computed in Chapter 5, Section 3, Problem 2. In Chapter 5, Section 4, Problem 12, the reader found that

$$\Gamma_{22}^{\ \ 1} = -\sin\theta\cos\theta, \qquad \Gamma_{12}^{\ \ 2} = \Gamma_{21}^{\ \ 2} = \cot\theta,$$

and all other Christoffel symbols are zero. Consider the small circle $\theta = \alpha =$ constant for the purpose of parallel propagation of a vector A^j. Show that the equations

$$\frac{DA^\lambda}{d\phi} = 0,$$

where the surface coordinate ϕ is chosen as parameter, reduce to

$$\frac{dA^1}{d\phi} - \cos\alpha \sin\alpha \, A^2 = 0,$$

$$\frac{dA^2}{d\phi} + \cot\alpha \, A^1 = 0.$$

and obtain the solution. Finally, choose $(A^1, A^2) = (1, 0)$ at $\phi = 0$ and then compare with the vector obtained by parallel propagation and corresponding to $\phi = 2\pi$.

7. *The Curvature Tensor*

In studying the calculus of several variables, we find that if second partial derivatives $\partial^2\Phi/\partial X^j\,\partial X^k$, $\partial^2\Phi/\partial X^k\,\partial X^j$ of a function Φ are continuous, they are equal; that is, the second partial derivative is commutative. For simplicity of notation we replace

(5-7.1a)
$$\frac{\partial^2\Phi}{\partial X^j\,\partial X^k} - \frac{\partial^2\Phi}{\partial X^k\,\partial X^j} = 0$$

by

(5-7.1b)
$$\partial_{[j}\partial_{k]}\Phi = 0.$$

In making the generalizations necessary for a proper development of transformation theory in Riemannian spaces, it was seen that covariant differentiation replaces partial differentiation. Therefore it is natural to ask the question whether the covariant derivative satisfies a commutative property. This question is answered in the negative. However, in so doing we introduce a mixed tensor of fourth order. This so-called Riemann Christoffel mixed tensor, $R_{ijk}{}^l$, plays a role in the theory of integration of systems of differential equations and serves to classify the geometric nature of the space. An introduction to these ideas is presented in the following theorem.

Theorem 5-7.1. Let V_i be the components of an arbitrary covariant vector. Suppose that at least the second partial derivatives of the components are continuous on an appropriate region of space. Then

(5-7.2a)
$$\nabla_{[j}\nabla_{k]}V_i = \tfrac{1}{2}R_{jki}{}^l V_l,$$

where

(5-7.2b)
$$R_{jki}{}^l = 2(-\partial_{[j}\Gamma^l_{k]i} + \Gamma^m_{i[j}\Gamma^l_{k]m}).$$

PROOF. We have

(5-7.3a)
$$\nabla_k V_i = \partial_k V_i - \Gamma_{ik}{}^l V_l$$

and

(5-7.3b)
$$\nabla_j(\nabla_k V_i) = \partial_j\,\partial_k V_i - (\partial_j\Gamma_{ik})V_l - \Gamma_{ik}{}^l\,\partial_j V_l - \Gamma_{kj}{}^m\nabla_m V_i - \Gamma_{ij}{}^m\nabla_k V_m.$$

By commuting j and k and subtracting, we obtain

(5-7.3c) $$2\nabla_{[j}\nabla_{k]}V_i = (\partial_k\Gamma_{ji}{}^l - \partial_j\Gamma_{ki}{}^l)\, V_l - (\Gamma_{ik}{}^m\Gamma_{jm}{}^l V_l - \Gamma_{ij}{}^m\Gamma_{km}{}^l V_l).$$

Other terms drop out because of the commutivity of the partial derivatives, the symmetry of the Christoffel symbols, and straightforward cancellation. The relation (5-7.3c) can be written in the form

(5-7.3d) $$2\nabla_{[j}\nabla_{k]}V_i = 2\,\partial_{[k}\Gamma^l_{j]i}V_l - 2\Gamma^m_{i[k}\Gamma^l_{j]m}V_l,$$

or

(5-7.3e) $$2\nabla_{[j}\nabla_{k]}V_i = 2(-\partial_{[j}\Gamma^l_{k]i} + \Gamma^m_{i[j}\Gamma^l_{k]m})V_l.$$

This completes the proof.

Theorem 5-7.2. Suppose that components $R_{jki}{}^l$ are introduced into each allowable coordinate system in accord with (5-7.2b). Then the class $\{R_{jki}{}^l\}$ is a mixed tensor of contravariant valence 1 and covariant valence 3.

PROOF. In examining (5-7.2) it is seen that $\nabla_{[j}\nabla_{k]}$ are components of an operational tensor of covariant order two, whereas V_i are covariant components of an arbitrary tensor. Therefore, according to the quotient law of tensor algebra, $R_{jki}{}^l$ are tensor components of the type indicated.

Definition 5-7.1. The tensor $\{R_{ijk}{}^l\}$ is called the Riemann Christoffel tensor of the second kind. The associated tensor with components

(5-7.4) $$R_{ijkp} = R_{ijk}{}^l g_{lp}$$

is called the Riemann Christoffel tensor of the first kind.

The tensors of the foregoing definition are also referred to as the curvature tensor and the covariant curvature tensor, respectively.

According to (5-7.2a) the covariant derivative is, in general, not commutative. In fact, it is commutative if and only if

(5-7.5) $$R_{ijk}{}^l = 0.$$

We shall investigate the significance of the condition (5-7.5) at a later point.

At present, attention is focused on the concept of parallel displacement. Through this avenue we shall be able to determine, at least in part, the geometric significance of the Riemann-Christoffel tensor. In this development we follow the lead of Levi-Civita.

At a point P of space construct a coordinate parallelogram with vertices P, Q, S, R. Let the parameters for the two coordinate directions be t_1 and t_2. (See Fig. 5-7.1a.) Represent the tangential vectors at P by $dX^j/dt_1 = \underset{1}{dX^j}$, $dX^j/dt_2 = \underset{2}{dX^j}$.

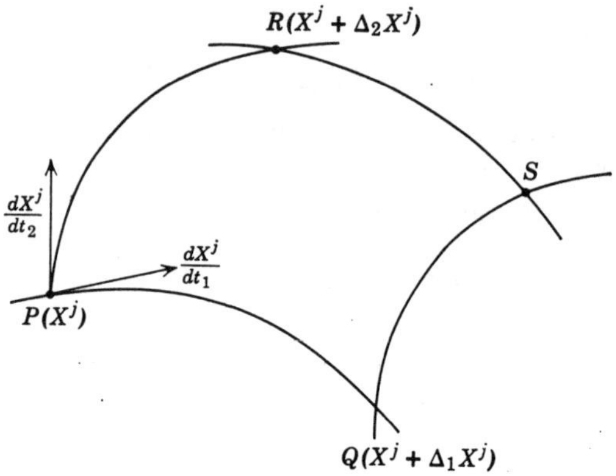

Fig. 5-7.1a

Theorem 5-7.3a. If subjected to parallel displacement the vectors $\underset{1}{dX^j}$, $\underset{2}{dX^j}$ satisfy the parallelogram law of addition.[13]

PROOF. Figure 5-7.1a presents a finite model of the situation at hand. The model is not precise, but it does enable us to describe the algebraic relationships of the derivatives subject to the process of parallel displacement. In terms of Fig. 5-7.1a, our object is to determine the parallel displacement of $\underset{1}{dX^j}$ at R and add this displaced vector to $\underset{2}{dX^j}$. The vector resulting should be the same as that obtained by determining the displaced vector corresponding to $\underset{2}{dX^j}$ at Q and then adding it to $\underset{1}{dX^j}$. A schematic diagram of the process is indicated by Fig. 5-7.1b where the object is to

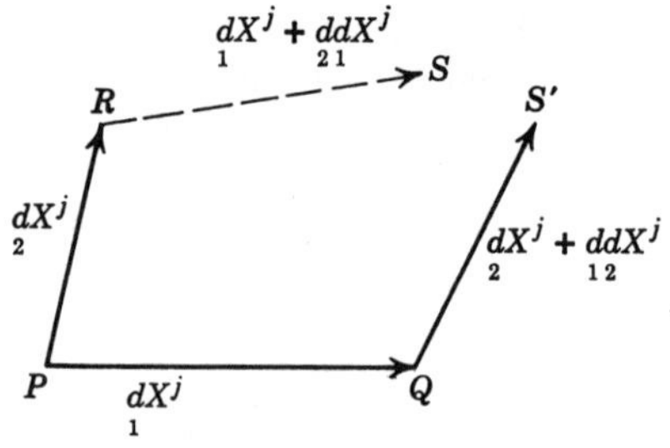

Fig. 5-7.1b

[13] In a Euclidean space two vectors **A** and **B** are added geometrically by displacing **A** such that its initial point coincides with the end point of **B** or vice versa. The fundamental consideration in our present discussion is the nature of such a parallel displacement in Riemannian spaces.

determine that S and S' coincide. The formal steps of the proof are as follows. Since the vectors are subject to parallel displacement,

$$\text{(5-7.6)}\qquad \begin{aligned} &\text{(a)}\quad \underset{2}{D}\,\underset{1}{dX^j} = \underset{21}{ddX^j} + \Gamma_{pq}{}^{j}\,\underset{1}{dX^p}\,\underset{2}{dX^q} = 0,\\ &\text{(b)}\quad \underset{1}{D}\,\underset{2}{dX^j} = \underset{12}{ddX^j} + \Gamma_{pq}{}^{j}\,\underset{2}{dX^p}\,\underset{1}{dX^q} = 0. \end{aligned}$$

Because the Christoffel symbols are symmetric in p and q,

$$\text{(c)}\quad \underset{21}{ddX^j} = \underset{12}{ddX^j},$$

Therefore

$$\underset{2}{dX^j} + (\underset{1}{dX^j} + \underset{21}{ddX^j}) = \underset{1}{dX^j} + (\underset{2}{dX^j} + \underset{12}{ddX^j}),$$

as was to be shown.

Since the symmetry of the Christoffel symbols was essential to the parallelogram law of addition just developed, the law in turn can be construed as the geometric interpretation of the symmetry property of the Christoffel symbols. Indeed, the Christoffel symbol is a special case of the so-called "affine connection." In terms of this more general structure, such a geometric interpretation takes on added meaning. As far as the purpose of this section is concerned, relation (5-7.6c) will be helpful in the next theorem which gets at the significance of the curvature tensor. In this theorem we again consider a coordinate parallelogram and ask what the change is in vector components V^j, defined at P, when displaced over a path P to R to S as compared to displacement over a path P to Q to S. (See Fig. 5-7.2.)

Theorem 5-7.3b. The difference in displacement vectors resulting from parallel displacement along coordinate paths PRS and PQS, respectively, is

$$\text{(5-7.7)}\qquad \underset{12}{ddV^i} - \underset{21}{ddV^i} = R_{jkl}{}^{i}\,\underset{2}{dX^j}\,\underset{1}{dX^k}V^l.$$

Proof. At R we obtain $V^i + \underset{2}{dV^i}$ and at S

$$\text{(5-7.8a)}\qquad V^i + \underset{2}{dV^i} + \underset{1}{d}(V^i + \underset{2}{dV^i}).$$

Following the path of propagation P to Q to S produces the result

$$\text{(5-7.8b)}\qquad V^i + \underset{1}{dV^i} + \underset{2}{d}(V^i + \underset{1}{dV^i}).$$

The difference of (5-7.8a) and (5-7.8b) is represented by

$$\text{(5-7.8c)}\qquad \underset{12}{ddV^i} - \underset{21}{ddV^i} = 2\underset{[12]}{ddV^i}.$$

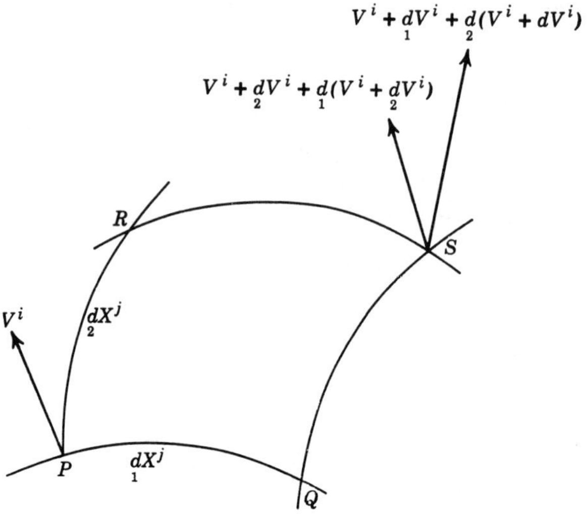

Fig. 5-7.2

For this relation we have, according to the process of parallel displacement,

$$2\underset{[12]}{dd}V^i = -2\,\underset{[1}{d}(\underset{2]}{d}X^k\Gamma_{jk}{}^iV^j)$$
$$= -2[(\underset{[12]}{dd}X^k)\Gamma_{jk}{}^iV^j + \underset{[1}{d}X^p\,\underset{2]}{d}X^k(\partial_p\Gamma_{jk}{}^i)\,V^j + \Gamma_{jk}{}^i(-\Gamma_{qp}{}^jV^q\,\underset{[1}{d}X^p\,\underset{2]}{d}X^k)].$$

Since, according to (5-7.6c),

$$\underset{[12]}{dd}X^j = 0,$$

the last relation reduces to

$$2\underset{[12]}{dd}V^i = -2(\partial_p\Gamma_{kq}{}^i - \Gamma_{jk}{}^i\Gamma_{qp}{}^j)\,\underset{[1}{d}X^p\,\underset{2]}{d}X^kV^q$$
$$= 2(-\partial_{[p}\Gamma_{k]q}{}^i + \Gamma_{q[p}^j\Gamma_{k]j}^i)\,\underset{1}{d}X^p\,\underset{2}{d}X^kV^q$$
$$= R_{pkq}{}^i\,\underset{1}{d}X^p\,\underset{2}{d}X^kV^q.$$

This verifies the conclusion of the theorem.

According to (5-7.7), the tensor $\{R_{jkl}{}^i\}$ classifies the given space at a point P. It does so in the sense that it provides a measure of the deviation of a vector field under parallel displacement with respect to the stated paths around an arbitrarily small coordinate rectangle.

The components $R_{jkl}{}^i$ are defined in terms of the components of the metric tensor g_{jk} and its first and second derivatives. Therefore, if the space is Euclidean, there is a coordinate system in which the g_{jk} are constants and

(5-7.9) $$R_{jkl}{}^i = 0.$$

Of course, if the components of a tensor are all zero in one coordinate system, they are all zero in every allowable coordinate system. Therefore (5-7.9) categorizes a class of spaces. If $n = 2$ these spaces are called developable surfaces.[14] It is because the tensor classifies certain spaces and because in the special case of surfaces it reduces to Gaussian curvature that $R_{jkl}{}^{i}$ is designated by the term "curvature tensor." Of course, the intuitive significance of the term "curvature" becomes clouded when the space dimension is greater than 2.

Definition 5-7.2. If in a given Riemannian space a coordinate system exists in terms of which the components of the fundamental metric tensor are constant, the space is said to be a flat or linear space.

Theorem 5-7.4a. If a space is flat, the Riemann-Christoffel curvature tensor is zero.

The proof is actually stated in the remarks preceding the theorem. The really interesting problem is the converse of Theorem 5-7.4a. However, before tackling it, let us consider the following example.

Example 5-7.1. Parametric equations of a circular cylinder are

$$X^1 = \rho \cos \theta, \qquad X^2 = \rho \sin \theta, \qquad X^3 = z.$$

The metric tensor of the surface has components

$$(g_{jk}) = \begin{pmatrix} \rho^2 & 0 \\ 0 & 1 \end{pmatrix}, \qquad \rho = \text{constant}$$

Therefore, as indicated in previous discussion, the circular cylinder surface is a flat space and its curvature tensor is zero.

If we were given the information that in terms of a certain coordinate system of a space the components of the metric tensor were

$$(g_{jk}) = \begin{pmatrix} 1 & 0 & 0 \\ 0 & r^2 & 0 \\ 0 & 0 & r^2 \sin^2 \theta \end{pmatrix},$$

where r, θ are variables, it would not be clear whether the space were flat. (Of course, in this simple example the representation in terms of spherical coordinates in a Euclidean space is probably easily recognized.) The following theorem, which is the converse of Theorem 5-7.4a, is helpful in such a circumstance.

[14] The developable surfaces are those that can be cut and rolled out on a plane without stretching or tearing. In particular, they are tangential surfaces associated with given curves, cylinders, and cones.

Theorem 5-7.4b. If the curvature tensor of a Riemannian space is zero, then the space is flat.

PROOF. Consider the system of partial differential equations

(5-7.10a)
$$\Gamma_{jk}{}^{i} = \frac{\partial X^i}{\partial \bar{X}^l} \frac{\partial^2 \bar{X}^l}{\partial X^j \partial X^k},$$

or

(5-7.10b)
$$\frac{\partial \bar{X}^p}{\partial X^i} \Gamma_{jk}{}^{i} = \frac{\partial^2 \bar{X}^p}{\partial X^j \partial X^k}.$$

On the one hand, it can be shown by straightforward computation that (5-7.10a) is a particular solution of

$$R_{jkl}{}^{i} = 0.$$

On the other hand, consider the transformation law of the Christoffel symbols

(5-7.10c)
$$\Gamma_{jk}{}^{i} \frac{\partial \bar{X}^p}{\partial X^i} = \frac{\partial \bar{X}^l}{\partial X^j} \frac{\partial \bar{X}^m}{\partial X^k} \bar{\Gamma}_{lm}{}^{p} + \frac{\partial^2 X^p}{\partial X^j \partial X^k}.$$

Employing relation (5-7.10b) to simplify (5-7.10c), we obtain

(5-7.10d)
$$0 = \frac{\partial \bar{X}^l}{\partial X^j} \frac{\partial \bar{X}^m}{\partial X^k} \bar{\Gamma}_{lm}{}^{p},$$

or, on multiplying and summing with $(\partial X^j/\partial \bar{X}^q)(\partial X^k/\partial \bar{X}^s)$

(5-7.10e)
$$\bar{\Gamma}_{qs}{}^{p} = 0;$$

that is, the Christoffel symbols are zero in the barred system. Next consider the covariant derivative of the fundamental metric tensor. We have

(5-7.10f)
$$0 = \bar{\nabla}_j \bar{g}_{kl} = \frac{\partial \bar{g}_{kl}}{\partial \bar{X}^j} - \bar{\Gamma}_{kj}{}^{p} \bar{g}_{pl} - \bar{\Gamma}_{lj}{}^{p} \bar{g}_{kp} = \frac{\partial \bar{g}_{kl}}{\partial \bar{X}^j}.$$

Therefore the components of the metric tensor are constant in the barred system and the space is a flat space.

Often a flat space is simply defined as one for which

$$R_{jkl}{}^{i} = 0.$$[15]

According to Theorems 5-7.4a,b, such a definition is equivalent to Definition 5-7.2.

---

[15] For comments concerning the equations $R_{jkl}{}^{i} = 0$ as integrability conditions for the system (5-7.10b), see Nathaniel Coburn, *Vector and Tensor Analysis*, Macmillan, 1955, p. 184.

Problems

1. Show that

$$\Gamma_{jk}{}^{i} = \frac{\partial X^i}{\partial \bar{X}^l} \frac{\partial^2 \bar{X}^l}{\partial X^j\, \partial X^k}$$

is a particular solution of

$$R_{jkl}{}^{i} = 0.$$

2. Show that a space is flat if and only if there is a coordinate system associated with the space in terms of which the Christoffel symbols are identically zero.

8. *Algebraic Properties of the Curvature Tensor*

The curvature tensor has a number of interesting algebraic properties. We shall begin with a theorem on the decomposition of tensors, which is helpful in investigating the algebraic properties, the main objective of this section.

Theorem 5-8.1. Any tensor can be expressed as a sum of outer products of vectors.

PROOF. An example rather than a formal proof is presented. The method of the example can be extended to the general case without difficulty.

Consider a tensor of contravariant valence 1 and covariant valence 2. Let $T_{ij}{}^{k}$ represent the components in some coordinate system. Define n^2 covariant vectors by means of

$$\underset{(p)}{\overset{(q)}{A}}{}_i = T_{ip}{}^{q}, \tag{5-8.1a}$$

where (p), (q) distinguish between vectors and i indicates the indices of a vector. Also define $2n$ vectors according to the relations

$$\overset{(p)}{B}{}_j = \delta_j{}^{p}, \qquad \underset{(q)}{C}{}^{k} = \delta_q{}^{k}. \tag{5-8.1b}$$

Then

$$T_{ij}{}^{k} = \underset{(p)}{\overset{(q)}{A}}{}_i\, B_j^{(p)} \underset{(q)}{C}{}^{k}. \tag{5-8.1c}$$

The relation (5-8.1c) is of tensor character; hence its validity in one coordinate system implies validity in every allowable coordinate system. This completes the verification of the example.

Now let us investigate the algebraic properties associated with the curvature tensor.

Theorem 5-8.2. If V^i and W_i are arbitrary vector components, then

$$\text{(5-8.2)}\qquad \begin{aligned}&\text{(a)}\quad 2\nabla_{[i}\nabla_{j]}V^k = -R_{ijl}{}^k V^l,\\ &\text{(b)}\quad 2\nabla_{[i}\nabla_{j]}W_k = R_{ijk}{}^l W_l.\end{aligned}$$

PROOF. Relation (5-8.2b) was obtained in Theorem 5-7.1, essentially by straightforward computation. The expression (5-8.2a) can be obtained in the same way.

It follows immediately from either of the relations of Theorem 5-8.2 that the Riemann-Christoffel tensor is skew symmetric in the first two indices. This fact can be expressed by either

$$\text{(5-8.3a)}\qquad R_{ijk}{}^l = -R_{jik}{}^l$$

or

$$\text{(5-8.3b)}\qquad R_{(ij)k}{}^l = 0.$$

Theorem 5-8.3. We have

$$\text{(5-8.4)}\qquad \begin{aligned}&\text{(a)}\quad 2\nabla_{[i}\nabla_{j]}T_k{}^l = R_{ijk}{}^m T_m^l - R_{ijm}{}^l T_k{}^m,\\ &\text{(b)}\quad 2\nabla_{[i}\nabla_{j]}T_{kl} = R_{ijk}{}^m T_{ml} + R_{ijl}{}^m T_{km},\\ &\text{(c)}\quad 2\nabla_{[i}\nabla_{j]}T^{kl} = -R_{ijm}{}^k T^{ml} - R_{ijm}{}^l T^{km}.\end{aligned}$$

PROOF. We consider (5-8.4a). According to Theorem 5-8.1, a vector decomposition can be determined for the components $T_k{}^l$.

$$\text{(5-8.5a)}\qquad T_k{}^l = \overset{(p)}{A}_k \underset{(p)}{B}{}^l$$

where

$$\text{(5-8.5b)}\qquad \overset{(p)}{A}_k = T_k{}^p, \qquad \underset{(p)}{B}{}^l = \delta_p{}^l.$$

Because the covariant derivative possesses the same properties with respect to sums and products as the ordinary or partial derivative

$$\nabla_j T_k{}^l = \nabla_j \overset{(p)}{A}_k \underset{(p)}{B}{}^l = \left(\nabla_j \overset{(p)}{A}_k\right) \underset{(p)}{B}{}^l + \overset{(p)}{A}_k \nabla_j \underset{(p)}{B}{}^l.$$

When the covariant derivative operator ∇_i is applied and the skew symmetric counterpart of the resulting expression is formed, we have

$$\nabla_{[i}\nabla_{j]}T_k{}^l = \left(\nabla_{[i}\nabla_{j]}\overset{(p)}{A}_k\right) \underset{(p)}{B}{}^l + \overset{(p)}{A}_k \nabla_{[i}\nabla_{j]}\underset{(p)}{B}{}^l.$$

Employment of (5-8.2a,b) and (5-8.5a,b) leads to the conclusion

$$2\,\nabla_{[i}\nabla_{j]}T_k{}^l = R_{ijk}{}^m T_m{}^l - R_{ijm}{}^l T_k{}^m.$$

The results (5-8.4b,c) are arrived at in a similar manner.

The two preceding theorems exhibit an operational characteristic of $\nabla_{[i}\nabla_{j]}$ analogous to the form of the covariant derivative itself. This fact makes memorization of relations (5-8.2a,b) and (5-8.4a,b,c) relatively easy. We shall use the results to establish the set of identities satisfied by the curvature tensor.

In particular, the covariant curvature tensor whose components are introduced by the relation

$$R_{ijkl} = g_{lm}R_{ijk}{}^{m} \tag{5-8.6}$$

is considered.

Theorem 5-8.4. We have

$$\begin{aligned} &\text{(a)} \qquad R_{(ij)kl} = 0, \\ \text{(5-8.7)}\quad &\text{(b)} \qquad R_{ij(kl)} = 0, \\ &\text{(c)} \qquad R_{[ijk]l} = 0; \end{aligned}$$

that is, the covariant curvature tensor is skew symmetric in the first pair of indices, skew symmetric in the last two indices, and completely symmetric in the first three indices.

PROOF. Relation (5-8.7a) follows immediately from (5-8.2a). To prove (5-8.7b) we can start with relation (5-8.4b) and let $T_{kl} = g_{kl}$. Since the fundamental metric tensor is a covariant constant,

$$\begin{aligned} 0 = \nabla_{[i}\nabla_{j]}g_{kl} &= R_{ijk}{}^{m}g_{ml} + R_{ijl}{}^{m}g_{km} \\ &= R_{ijkl} + R_{ijlk}. \end{aligned}$$

This completes the proof of (b).

To prove (c), we introduce an arbitrary gradient vector

$$W_l = \nabla_l\Phi. \tag{5-8.8}$$

Then, according to (5-8.2b) and (5-8.7a),

$$\begin{aligned} R_{[ijk]}{}^{l}W_l &= \tfrac{2}{6}(R_{ijk}{}^{l}W_l + R_{jki}{}^{l}W_l + R_{kij}{}^{l}W_l) \\ &= \tfrac{4}{6}(\nabla_{[i}\nabla_{j]}W_k + \nabla_{[j}\nabla_{k]}W_i + \nabla_{[k}\nabla_{i]}W_j). \end{aligned}$$

When W_l is replaced as indicated by (5-8.8), a straightforward evaluation verifies that

$$R_{[ijk]}{}^{l}W_l = 0.$$

Since W_l is arbitrary, it follows that

$$R_{[ijk]}{}^{l} = 0,$$

hence

$$R_{[ijk]p} = g_{pl}R_{[ijk]}{}^{l} = 0.$$

A rather interesting identity, which is a consequence of (5-8.7c), is presented in the following theorem.

Theorem 5-8.5. We have

$$R_{ijkl} = R_{klij}; \tag{5-8.9}$$

that is, the covariant curvature tensor is symmetric in the pairs i, j and k, l.

PROOF. Four equations result from (5-8.7c):

$$\begin{aligned} R_{ijkl} + R_{jkil} + R_{kijl} &= 0. \\ R_{jkli} + R_{klji} + R_{ljki} &= 0. \\ R_{klij} + R_{likj} + R_{iklj} &= 0. \\ R_{lijk} + R_{ijlk} + R_{jlik} &= 0. \end{aligned}$$

By adding these equations and making use of the skew symmetry in the first two indices and the last two indices, we obtain

$$2R_{kijl} - 2R_{jlki} = 0.$$

This relation is equivalent to the desired result.

The identities of Theorems 5-8.4 and 5-8.5 make it clear that the n^4 components of the curvature tensor cannot be chosen independently of one another; that is, if an appropriate basic set of components is chosen, the others are determined by the identities. The next theorem states the number of independent components. The proof of the theorem makes use of the assumption that the set of identities (5-8.7) is "complete;" [16] that is, the assumption that any other identity relating the components can be obtained algebraically from them.

Theorem 5-8.6. The number of independent components of the curvature tensor $\{R_{ijkl}\}$ in any allowable coordinate system is

$$\frac{n^2(n^2 - 1)}{12}. \tag{5-8.10}$$

PROOF. According to (5-8.7a), the first two indices must be different; otherwise the component is zero and among the remaining pairs only half are independent. Therefore the number of distinct choices, disregarding order, of these two indices is $\frac{1}{2}[n(n - 1)]$. A similar number of distinct choices can be made for the second pair of indices, as pointed out by (5-8.7b). Of these $\{[n(n - 1)]/2\}^2$ possibilities, some must be cast out because of the identities in (5-8.9). In particular $[n(n - 1)/2]$ things taken two at a time, that is,

$$\frac{1}{2}\left\{\frac{n(n-1)}{2}\left[\frac{n(n-1)}{2} - 1\right]\right\} \tag{5-8.11a}$$

[16] For further remarks see Tracy Y. Thomas, *Concepts from Tensor Analysis and Differential Geometry*, Academic Press, New York and London, 1961. For proof see Tracy Y. Thomas, *The Differential Invariants of Generalized Spaces*, Cambridge University Press, London and New York, 1934.

can be determined through (5-8.9). The distinct components

$$\left[\frac{n(n-1)}{2}\right]^2 - \frac{1}{2}\left\{\frac{n(n-1)}{2}\left[\frac{n(n-1)}{2} - 1\right]\right\} \tag{5-8.11b}$$

are still subject to the identities (5-8.7c). Since the indices in (5-8.7c) must be distinct, the number of conditions imposed by the relations corresponds to the number of n things taken four at a time; that is

$$\frac{n(n-1)(n-2)(n-3)}{24}, \quad n > 3. \tag{5-8.11c}$$

When (5-8.11b) is reduced by the number of conditions (5-8.11c), we obtain the result

$$\frac{n^2(n^2-1)}{12}.$$

In certain aspects of the general theory of relativity and in other physical and geometric applications we encounter the tensor with components $R_{ijk}{}^i$. This is the so-called Ricci tensor, whose components are represented by the symbols R_{jk}. Since $R_{ijk}{}^j$ and $R_{ijk}{}^k$ are also candidates for the form of representation specified for the Ricci tensor, some clarification is needed. We immediately note that $R_{ijk}{}^j = -R_{jik}{}^j$. The following theorem shows that the components $R_{ijk}{}^k$ are identically zero and that $R_{ijk}{}^i$ is symmetric in j and k.

Theorem 5-8.7. We have

$$\text{(a)} \quad R_{ijk}{}^k = 0, \qquad \text{(b)} \quad R_{i[jk]}{}^i = 0. \tag{5-8.12}$$

PROOF. To show the validity of (a) we start with the defining expressions for the curvature tensor components; that is,

$$R_{ijk}{}^q = 2(-\partial_{[i}\Gamma^q_{j]k} + \Gamma^p_{k[i}\Gamma^q_{j]p}). \tag{5-8.13a}$$

When a summation is imposed on k and q,

$$R_{ijk}{}^k = 2(-\partial_{[i}\Gamma^k_{j]k} + \Gamma^p_{k[i}\Gamma^k_{j]p}). \tag{5-8.13b}$$

For the second term in the right-hand member of (5-1.3b) we have, by interchanging the dummy indices p and k in the second term below,

$$\Gamma_{ki}{}^p\Gamma_{jp}{}^k - \Gamma_{kj}{}^p\Gamma_{ip}{}^k = \Gamma_{ki}{}^p\Gamma_{jp}{}^k - \Gamma_{pj}{}^k\Gamma_{ik}{}^p = 0.$$

Since

$$\Gamma_{jk}{}^k = \partial_j \ln \sqrt{g},$$

it is also the case that

$$-2\partial_{[i}\Gamma^k_{j]k} = -2\partial_{[i}\partial_{j]} \ln \sqrt{g} = 0.$$

This completes the proof of (a).

In order to prove (b), we must first establish the fact that

$$R_{ijk}{}^{i} = R_{jipk}g^{ip}. \tag{5-8.14}$$

We have

$$R_{ijk}{}^{i} = R_{ijkp}g^{pi} = +R_{jipk}g^{ip}.$$

With (5-8.14) in mind, consider

$$2R_{i[jk]}{}^{i} = R_{jipk}g^{ip} - R_{kipj}g^{ip}. \tag{5-8.15a}$$

If we make use of the curvature tensor identities, the second term of (5-8.15a) can be expressed as

$$R_{kipj}g^{ip} = R_{pjki}g^{ip} = R_{jpik}g^{ip}. \tag{5-8.15b}$$

Since p and i are dummy indices and g^{ip} is symmetric, the last member of (5-8.15b) is equivalent to $R_{jipk}g^{ip}$. Substitution of this result into (5-8.15a) leads to the desired consequence.

Since the components $R_{ijk}{}^{k}$ are identically zero and the set $R_{ijk}{}^{i}$ is symmetric in j and k, the following definition is free of ambiguity.

Definition 5-8.1. The Ricci[17] tensor has components

$$R_{jk} = -R_{ijk}{}^{i}. \tag{5-8.16}$$

Since the R_{jk} are symmetric in j and k, the number of independent components is $\frac{1}{2}[n(n+1)]$.

In conjunction with the Ricci tensor, it is convenient to introduce the following scalar.

Definition 5-8.2. The curvature invariant is expressed by means of the relation

$$R = g^{ij}R_{ij}. \tag{5-8.17}$$

The intuitive significance of R can be determined by an examination of the expression in two dimensions. The algebraic part of this task is done in the next theorem.

Theorem 5-8.8. In a two-dimensional Riemannian space

$$R = \frac{2R_{1212}}{g}. \tag{5-8.18}$$

PROOF. We have, according to (5-8.14), (5-8.16), and (5-8.17),

$$\begin{aligned} R &= -g^{jk}R_{jipk}g^{ip}, \\ &= -g^{11}R_{1221}g^{22} - g^{12}R_{1212}g^{21} - g^{21}R_{2121}g^{12} - g^{22}R_{2112}g^{11}, \\ &= 2(-g^{12}g^{21} + g^{11}g^{22})R_{1212} = 2\det(g^{jk})R_{1212}. \end{aligned}$$

[17] Gregorio Ricci Curbastro.

Since det (g^{jk}) and $g = \det(g_{jk})$ are reciprocal values, the proof is complete.

By examination of the last theorem we find that the covariant curvature tensor has one independent component in two-space. Furthermore, the invariant $R/2$ is expressed as the ratio of that component to g. It can be shown[18] that this ratio represents the so-called Gaussian curvature, a fundamental entity in the development of the differential geometry of a surface.

Example 5-8.1. With respect to the surface of a sphere of radius r, we have

$$\frac{R}{2} = \frac{1}{r^2}. \tag{5-8.19}$$

In order to establish this fact, we first observe that the parameterization

$$X^1 = r \sin\theta \cos\phi, \qquad X^2 = r\sin\theta\sin\phi, \qquad X^3 = r\cos\theta$$

leads to the fundamental metric form

$$ds^2 = r^2\, d\theta^2 + r^2 \sin^2\theta\, d\phi^2;$$

that is

$$(g_{jk}) = \begin{pmatrix} r^2 & 0 \\ 0 & r^2\sin^2\theta \end{pmatrix} \tag{5-8.20a}$$

By straightforward computation we find that

$$\Gamma_{22}{}^1 = -\sin\theta\cos\theta, \qquad \Gamma_{12}{}^2 = \Gamma_{21}{}^2 = \operatorname{ctn}\theta. \tag{5-8.20c}$$

Therefore

$$\begin{aligned} R_{1212} &= 2g_{22}(-\partial_{[1}\Gamma^2_{2]1} - \tfrac{1}{2}\Gamma_{12}{}^2\Gamma_{12}{}^2) \\ &= r^2\sin^2\theta\left(-\frac{\partial \operatorname{ctn}\theta}{\partial\theta} - \operatorname{ctn}^2\theta\right) \\ &= r^2\sin^2\theta. \end{aligned} \tag{5-8.20d}$$

From (5-8.20b) and (5-8.20d) we obtain the result

$$\frac{R}{2} = \frac{1}{r^2}. \tag{5-8.20e}$$

This is the expected constant Gaussian curvature of a spherical surface.

A special property of the Ricci tensor in two dimensions is that its components are proportional to the components of the metric tensor. This fact is stated formally in the next theorem.

---

[18] Václav Hlavatý, *Differentialgeometrie der Kurven und Flächen und Tensorrechnung*, P. Noordhoff, 1939, p. 267. Erwin Kreyszig, *Differential Geometry*, University of Toronto Press, 1959, p. 145.

Theorem 5-8.9. In a two-dimension Riemannian space the components of the Ricci tensor are proportional to the components of the metric tensor; that is,

$$R_{jk} = \frac{R_{1212}}{g} g_{jk}, \qquad j, k = 1, 2. \tag{5-8.21}$$

PROOF. We first note that

$$g^{11} = \frac{g_{22}}{g}, \qquad g^{12} = -\frac{g_{21}}{g} = -\frac{g_{12}}{g}, \qquad g^{22} = \frac{g_{11}}{g}, \tag{5-8.22}$$

for

$$g^{jk} = \frac{\text{cofactor } g_{kj} \text{ in det } (g_{kj})}{g}. \tag{5-8.22b}$$

According to (5-8.16),

$$R_{jk} = -R_{ijk}{}^{i} = -R_{ijkp}g^{pi}.$$

Therefore

$$\begin{aligned} R_{11} &= -R_{i11p}g^{pi} = -R_{2112}g^{22} = R_{1212}g^{22}, \\ R_{12} &= R_{21} = -R_{i12p}g^{pi} = -R_{1212}g^{12}, \\ R_{22} &= -R_{i22p}g^{pi} = R_{1212}g^{11}. \end{aligned} \tag{5-8.22c}$$

The proportionalities (5-8.21) are obtained by substituting relations (5-8.22a) into the expressions (5-8.22c). This completes the proof.

The most renowned usage of the tensor calculus is in the development of relativity theory, especially the general theory. A particularly useful tool in its development is the Bianchi identity presented in the next theorem.

Theorem 5-8.10 (Bianchi's[19] identity). We have

$$\nabla_{[i}R_{jk]p}{}^{m} = 0. \tag{5-8.23}$$

PROOF. Let W_p be an arbitrary covariant vector. Then

$$\begin{aligned} 3!\,\nabla_{[i}\nabla_{j}\nabla_{k]}W_p &= 2[\nabla_i(\nabla_{[j}\nabla_{k]}W_p) + \nabla_j(\nabla_{[k}\nabla_{i]}W_p) + \nabla_k(\nabla_{[i}\nabla_{j]}W_p)] \\ &= [\nabla_i(R_{jkp}{}^{q}) + \nabla_j(R_{kip}{}^{q}) + \nabla_k(R_{ijp}{}^{q})]W_q \\ &\quad + R_{jkp}{}^{q}\,\nabla_i W_q + R_{kip}{}^{q}\,\nabla_j W_q + R_{ijp}{}^{q}\,\nabla_k W_q \\ &= 3(\nabla_{[i}R_{jk]p}{}^{q})W_q + 3R_{[jk|p|}{}^{q}\nabla_{i]}W_q. \end{aligned} \tag{5-8.24a}$$

The notation $|p|$ simply indicates that p is not to be included in the bracket symbolism. Note that (5-8.2b) was made use of in this development of a first expression for $3!\,\nabla_{[i}\nabla_{j}\nabla_{k]}W_p$.

[19] Luigi Bianchi (1856–1928, Italian).

On the other hand, we have

(5-8.24b) $$3!\,\nabla_{[i}\nabla_j\nabla_{k]}W_p = 2(\nabla_{[i}\nabla_{j]}\nabla_k W_p + \nabla_{[j}\nabla_{k]}\nabla_i W_p + \nabla_{[k}\nabla_{i]}\nabla_j W_p)$$
$$= (R_{ijk}{}^q\,\nabla_q W_p + R_{ijp}{}^q\,\nabla_k W_q) + (R_{jki}{}^q\,\nabla_q W_p$$
$$+ R_{jkp}{}^q\,\nabla_i W_q) + (R_{kij}{}^q\,\nabla_q W_p + R_{kip}{}^a\,\nabla_j W_a)$$
$$= 3R_{[ijk]}{}^q\,\nabla_q W_p + 3R_{[ij|p|}{}^q\nabla_{k]}W_q.$$

By subtracting (5-8.24b) from (5-8.24a) and employing the identity $R_{[ijk]}{}^q = 0$, we obtain the result

(5-8.24c) $$\nabla_{[i}R_{jk]p}{}^q W_q = 0.$$

Since the preceding relation must hold for all sets of vector components W_q, we obtain

$$\nabla_{[i}R_{jk]p}{}^q = 0,$$

as was to be shown.

The preceding proof of the Bianchi identity does not depend on coordinate system. When studying invariants of a transformation group, as is a major goal of tensor analysis, it is rewarding to find such a proof. However, there are some interesting ideas that can be put forward by making the proof through the expediency of a special coordinate system.

Definition 5-8.3. Let the Riemannian space be referred to a coordinate system X^j. At a point P_0, introduce new coordinates by means of the transformation equations

(5-8.25a) $$\bar{X}_i = X^i - X_0{}^i + \frac{1}{2}(\Gamma_{mn}{}^i)_0(X^m - X_0{}^m)(X^n - X_0{}^n).$$

The coordinates $\bar{X}^i$ are said to be geodesic coordinates at the point P_0.

Since

(5-8.25b) $$\frac{\partial \bar{X}^i}{\partial X^j} = \delta_j{}^i + \frac{1}{2}(\Gamma_{mn}{}^i)_0\,\delta_j{}^m(X^n - X_0{}^n) + \frac{1}{2}(\Gamma_{mn}{}^i)_0(X^m - X_0{}^m)\,\delta_j{}^n$$

and

(5-8.25c) $$\left(\frac{\partial \bar{X}^i}{\partial X^j}\right)_0 = \delta_j{}^i,$$

it follows that the Jacobian of transformation is nonzero at P_0. Hence the transformation is allowable.

The particular advantage to the geodesic coordinate system introduced by the last definition lies in the fact that the Christoffel symbols are all zero at P_0 in that system. The following theorem discusses this idea.

Theorem 5-8.11. In a geodesic coordinate system $\bar{X}^j$ at P_0 we have

(5-8.26) $$(\bar{\Gamma}_{jk}{}^i)_0 = 0.$$

PROOF. According to the rule of transformation of the Christoffel symbols,

$$\bar{\Gamma}_{jk}{}^{i} = \frac{\partial \bar{X}^i}{\partial X^p}\frac{\partial^2 X^p}{\partial \bar{X}^j\,\partial \bar{X}^k} + \frac{\partial \bar{X}^i}{\partial X^p}\frac{\partial X^q}{\partial \bar{X}^j}\frac{\partial X^r}{\partial \bar{X}^k}\Gamma_{qr}{}^{p} \tag{5-8.27a}$$

The set $\partial \bar{X}^i/\partial X^p$ is evaluated at P_0 by (5-8.25c). To evaluate the elements $\partial^2 X^p/\partial \bar{X}^j\,\partial \bar{X}^k$, we first multiply and sum (5-8.25b) with $\partial X^j/\partial \bar{X}^k$, thereby obtaining

$$\delta_k{}^i = \frac{\partial X^i}{\partial \bar{X}^k} + \frac{\partial X^s}{\partial \bar{X}^k}(\Gamma_{sn}{}^{i})_0(X^n - X_0{}^n). \tag{5-8.27b}$$

The result of differentiating this expression with respect to $\bar{X}^j$ is

(5-8.27c)

$$0 = \frac{\partial^2 X^i}{\partial \bar{X}^j\,\partial \bar{X}^k} + \frac{\partial^2 X^s}{\partial \bar{X}^j\,\partial \bar{X}^k}(\Gamma_{sn}{}^{i})_0(X^n - X_0{}^n) + \frac{\partial X^s}{\partial \bar{X}^k}(\Gamma_{sn}{}^{i})_0\frac{\partial X^n}{\partial \bar{X}^j}.$$

When (5-8.27c) is evaluated at P_0, we obtain

$$\left(\frac{\partial^2 X^i}{\partial \bar{X}^j\,\partial \bar{X}^k}\right)_0 = -\left(\frac{\partial X^s}{\partial \bar{X}^k}\right)_0\left(\frac{\partial X^n}{\partial \bar{X}^j}\right)_0(\Gamma_{sn}{}^{i})_0. \tag{5-8.27d}$$

The Christoffel symbols $\bar{\Gamma}_{jk}{}^{i}$ may now be evaluated at P_0 by plugging the results (5-8.25c) and (5-8.25d) into (5-8.27a). We obtain

$$(\bar{\Gamma}_{jk}{}^{i})_0 = -\delta_p{}^i\left(\frac{\partial X^s}{\partial \bar{X}^k}\right)_0\left(\frac{\partial X^n}{\partial \bar{X}^j}\right)_0(\Gamma_{sn}{}^{p})_0 + \delta_p{}^i\left(\frac{\partial X^n}{\partial \bar{X}^j}\right)_0\left(\frac{\partial X^s}{\partial \bar{X}^k}\right)_0(\Gamma_{ns}{}^{p})_0 = 0,$$

as was to be shown.

Example 5-8.2. In a geodesic coordinate system at P_0 we have

$$\nabla_{[i}R_{jk]p}{}^{q} = \partial_{[i}R_{jk]p}{}^{q} = -2\partial_{[i}\partial_j\Gamma^q_{k]p} = 0.$$

The evaluation of the foregoing expression follows from the fact that it is both symmetric and skew symmetric in i, j.

We close this section with the introduction of another tensor relation which plays a part in the development of general relativity theory.

Definition 5-8.4. The components

$$G_{ij} = R_{ij} - \tfrac{1}{2}Rg_{ij} \tag{5-8.28}$$

are components of the so-called Einstein tensor.

Theorem 5-8.12. The components of the Einstein tensor satisfy the properties

$$\nabla^i G_{ij} = 0, \tag{5-8.29}$$

where

$$\nabla^i = g^{ik}\nabla_k.$$

PROOF. The Bianchi identity

$$\nabla_{[i}R_{jk]p}{}^{q} = 0,$$

may be expanded to the form

$$\nabla_i R_{jkp}{}^q + \nabla_j R_{kip}{}^q + \nabla_k R_{ijp}{}^q = 0. \tag{5-8.30a}$$

Summing on i and q, we obtain

$$\nabla_i R_{jkp}{}^i + \nabla_j R_{kp} - \nabla_k R_{jp} = 0. \tag{5-8.30b}$$

Upon multiplication of (5-8.30b) with g^{jp}, it follows that

$$\nabla^i g^{jp} R_{jkpi} + \nabla^p R_{kp} - \nabla_k g^{jp} R_{jp} = 0. \tag{5-8.30c}$$

This relation can also be written in the form

$$\nabla^i R_{ki} + \nabla^i R_{ki} - \nabla^i g_{ik} R = 0. \tag{5-8.30d}$$

Finally (5-8.30d) is equivalent to

$$\nabla^i (R_{ki} - \tfrac{1}{2} g_{ki} R) = 0.$$

This completes the proof.

Problems

1. Decompose $T_{ij}{}^{kl}$ into sums and products of vector components.
2. Prove that relation (5-8.2a) is valid.
3. Prove that relations (5-8.4b,c) hold.
4. Use the decomposition property of tensors to prove that the covariant derivative of a tensor is a tensor with covariant valence one more than that of the original. (See Theorem 5-4.4.)
5. Prove that $R_{ijk}{}^k = 0$ by means of the relation $\nabla_i g^{jk} = 0$.
6. Show that $R = 2R_{1212}/g$ can be written in the tensor form

$$R(g_{ik}g_{jl} - g_{jk}g_{il}) = 2R_{ijkl}$$

7. Show that R is a scalar of weight $+2$ in a Riemannian two space.

9. *An Introduction to the General Theory of Relativity*

In the special theory of relativity (1905) the fundamental laws of mechanics and electromagnetic theory are formulated so that their algebraic forms are invariant with respect to the special Lorentz group of transformations. Physically this means that the inertial systems whose coordinates are related by these transformations are equivalent. A uniform rectilinear motion of any one of the systems is relative to some other.

From the kinematical point of view it is well known that all motion is relative. Therefore the question of producing a structure in which frames of reference in accelerated motion are equivalent for the purpose of formulating physical laws develops in a natural way. Mathematically, this requires algebraic statements of fundamental laws which are invariant under general transformations of coordinates.

In the years following 1905 Einstein considered the problem of generalization stated in the preceding paragraph. He found that the tensor calculus, learned from his friend H. Grossman, met the challenge of generalization. But another problem of major character existed. The special theory of relativity did not include a gravitational theory as part of its structure. The "action at a distance" concept on which the Newtonian gravitational theory was based no longer had meaning because of the relative nature of the simultaneity of events. Einstein met this second problem in the discovery of the equivalence of inertial and gravitational mass. The essence of this discovery is that all motions in a homogeneous gravitational field can just as well be considered as motions in the absence of a gravitational field, but with respect to a uniformly accelerated frame of reference. This viewpoint makes it possible to develop a completely geometric structure as the model for a gravitational theory. The gravitational force is thought of as a pseudo-force, just as centrifugal and coriolis forces are in the classical theory, and is included in the metric structure of a Riemannian space.

It is the purpose of this section to give a brief outline of the geometric structure associated with general relativity. No attempt is made to penetrate into the background of the physical quantities and laws that become involved.

Assume that a fundamental metric tensor, $g_{\lambda\mu}$, and a corresponding differential form,

$$ds^2 = g_{\alpha\beta}\, dX^\alpha\, dX^\beta$$

are given. (Greek indices have a range of $1, \cdots, 4$.) When considering surfaces, we found that the components of the metric tensor could be determined by a parametric representation of the surface valid in the three-dimensional embedding space. In the present situation we have no obvious way of determining the g's, and the precise nature of the Riemannian space is not known. For the moment, we proceed abstractly. The problem of evaluating the g's will arise at a later point.

The geodesics in the space of special relativity are straight lines represented by

$$\frac{d(dX^\lambda/ds)}{ds} = 0. \tag{5-9.1a}$$

We have seen that geodesics in a Riemannian space are signified in general by

(5-9.1b) $$\frac{D(dX^\lambda/ds)}{ds} = 0,$$

or, in more detail,

(5-9.1c) $$\frac{d^2X^\lambda}{ds^2} + \Gamma_{\alpha\beta}{}^\lambda \frac{dX^\alpha}{ds}\frac{dX^\beta}{ds} = 0.$$

The assumption is made that "particles subject to gravitational forces only move along geodesics of the space." This postulate embodies the assertion that inertial and gravitational mass are equivalent, for an inequality of the two would bring about deviations, because of inertial mass, from the geodesics determined by the gravitational field. [With respect to relations (5-9.1a,b,c), note that the dX^λ/ds transform as components of a contravariant vector under general transformations, and therefore they can play the role of velocity components in the general theory.]

The preceding paragraph makes the implication that the gravitational field should somehow determine the components of the fundamental metric tensor. This determination is realized by identifying the components g_{jk} with gravitational potential. Classically gravitational potential is represented by a single function Φ, subject to Poisson's equation

(5-9.2) $$\delta^{jk}\frac{\partial^2\Phi}{\partial X^j\,\partial X^k} = 4\pi\mu,$$

where the function μ represents the density of matter. The discovery of an appropriate generalization of Poisson's equation occupied Einstein for a long period of time. On the basis of physical considerations, he decided that the right-hand side should generalize to a tensor density $\mathrm{T}_{\lambda\gamma}$ of weight -1 representative of the momentum and energy of continuously distributed matter and an associated electromagnetic field.[20] The left-hand side of Poisson's equation is a linear function in the second derivatives of the gravitational potential Φ. Therefore the left-hand side of its generalization is required to be linear in the second partials of the components $g_{\lambda\gamma}$. The tensor on the left-hand side must be geometric in character. Einstein indicates[21] that he spent two years rejecting the Riemann curvature tensor

[20] In particular, $\mathrm{T}^{\lambda\gamma} = M^{\lambda\gamma} - E^{\lambda\gamma}$, where $M^{\lambda\gamma} = \sqrt{|g|}\,u_0(dX^\lambda/ds)(dX^\gamma/ds)$ and $E^{\lambda\gamma} = \sqrt{|g|}\,(F^{\alpha\lambda}F_\alpha{}^\gamma + \frac{1}{4}g^{\gamma\lambda}F^{\alpha\beta}F_{\alpha\beta})$. The skew symmetric tensor $\{F^{\alpha\beta}\}$ represents the electromagnetic field and μ_0 is a scalar.

[21] Einstein, *Essays in Science*, The Wisdom Library, New York, 1934, p. 83.

before deciding that it would do the job. Specifically, he made use of the so-called Einstein tensor density

(5-9.3) $$\mathcal{G}_{\lambda\gamma} = \sqrt{-g}(R_{\lambda\gamma} - \tfrac{1}{2}Rg_{\lambda\gamma})\,\cdot$$

The conditions imposed on the components of the fundamental metric tensor are

(5-9.4) $$\boxed{\mathcal{G}_{\lambda\gamma} = C\mathrm{T}_{\lambda\gamma}.}$$

The underlying space of the general theory of relativity whose metric tensor is defined by (5-9.4) is said to be a Riemannian space.

The equations in (5-9.4) are complicated. A special case of much importance is one in which

(5-9.5) $$\mathrm{T}_{\lambda\gamma} = 0.$$

This assumption corresponds to a physical problem in which matter is represented by the sun. The electromagnetic field is considered to be negligible, as is matter outside the sun (but in a neighborhood confined to the solar system). The best known solution of the field equations in (5-9.4) under the conditions in (5-9.5) is due to K. Schwarzschild.[22] It is essentially this solution that follows. It is presumed that the matter is homogeneous and spherical so that the chosen coordinates r, θ, ϕ, t are essentially spherical coordinates, along with a time variable, at great distances from the matter generating the space. Furthermore, we assume that the field is static and reversible in time. These suppositions lead to a fundamental metric form

(5-9.6a) $$ds^2 = -\xi(r)\,dr^2 - r^2\,d\theta^2 - r^2\sin^2\theta\,d\phi^2 + \eta(r)\,dt^2,$$

where $\xi(r)$ and $\eta(r)$ are positive functions such that

(5-9.6b) $$\xi(r) \to 1, \qquad \eta(r) \to 1,$$

as $r \to \infty$.

Since

$$\mathrm{T}_{\lambda\gamma} = 0,$$

we have

(5-9.7a) $$R_{\lambda\gamma} - \tfrac{1}{2}Rg_{\lambda\gamma} = 0.$$

When the products of this relation with $g^{\lambda\gamma}$ are summed, we obtain

$$R - \frac{4}{2}R = 0$$

or

(5-9.7b) $$R = 0.$$

[22] Berlin Sitzungsberichte, 1916, p. 189.

Therefore (5-9.7a) reduces to

(5-9.7c) $$\boxed{R_{\lambda\gamma} = 0.}$$

It is these conditions that determine the components of the fundamental metric tensor. In order to investigate them further, it is necessary to compute the Christoffel symbols on the basis of the known information; that is,

(5-9.8a) $$(g_{\alpha\beta}) = \begin{pmatrix} -\xi(r) & 0 & 0 & 0 \\ 0 & -r^2 & 0 & 0 \\ 0 & 0 & -r^2 \sin^2\theta & 0 \\ 0 & 0 & 0 & \eta(r) \end{pmatrix}.$$

The computations are left as an exercise for the reader. We obtain

(5-9.8b)

$$\Gamma_{11}{}^1 = \frac{\xi'}{2\xi}, \qquad \Gamma_{22}{}^1 = -\frac{r}{\xi}, \qquad \Gamma_{33}{}^1 = -\frac{r\sin^2\theta}{\xi}, \qquad \Gamma_{44}{}^1 = \frac{\eta'}{2\xi},$$

$$\Gamma_{33}{}^2 = -\sin\theta\cos\theta, \qquad \Gamma_{12}{}^2 = \Gamma_{21}{}^2 = \frac{1}{r},$$

$$\Gamma_{13}{}^3 = \frac{1}{r}, \qquad \Gamma_{23}{}^3 = \Gamma_{32}{}^3 = \operatorname{ctn}\theta,$$

$$\Gamma_{14}{}^4 = \Gamma_{41}{}^4 = \frac{\eta'}{2\eta}, \qquad \text{the remaining } \Gamma_{\lambda\mu}{}^\gamma = 0,$$

where

$$\xi' = \frac{d\xi}{dr}, \qquad \eta' = \frac{d\eta}{dr}.$$

With the Christoffel symbols at hand, we can write out the relations (5-9.7c). It is convenient to express the results in terms of $R_\lambda{}^\nu = g^{\gamma\nu}R_{\lambda\gamma}$. Then

(5-9.9)
$$\begin{aligned} -R_1{}^1 &= \frac{\xi'}{\xi^2 r} + \frac{\xi'\eta'}{4\xi^2\eta} + \frac{(\eta')^2}{4\eta^2\xi} - \frac{\eta''}{2\xi\eta}, \\ -R_2{}^2 = R_3{}^3 &= \frac{\xi'}{2\xi^2 r} + \frac{1}{r^2}\left(1 - \frac{1}{\xi}\right) - \frac{\eta'}{2\xi\eta r}, \\ -R_4{}^4 &= \frac{\xi'\eta'}{4\xi^2\eta} + \frac{(\eta')^2}{4\eta^2\xi} - \frac{\eta''}{2\xi\eta} - \frac{\eta'}{\xi\eta r}, \\ R_\lambda{}^\gamma &= 0, \qquad \lambda \neq \gamma. \end{aligned}$$

The consequence of these computations is set forth in the following theorem

Theorem 5-9.1. The fundamental metric tensor determined by the Schwarzschild solution is represented by the components

$$(5\text{-}9.10\text{a}) \qquad (g_{\lambda\gamma}) = \begin{pmatrix} -\dfrac{1}{\eta} & 0 & 0 & 0 \\ 0 & -r^2 & 0 & 0 \\ 0 & 0 & -r^2 \sin^2\theta & 0 \\ 0 & 0 & 0 & \eta \end{pmatrix},$$

where

$$(5\text{-}9.10\text{b}) \qquad \eta = 1 - \frac{\gamma}{r}$$

and γ is a constant of integration.

PROOF. The equations in (5-9.9) may be solved as follows.

$$R_4^{\ 4} - R_1^{\ 1} = \frac{1}{\xi r}\left(\frac{\eta'}{\eta} + \frac{\xi'}{\xi}\right) = 0.$$

The solution of this separable differential equation is

$$\xi\eta = \text{constant}.$$

By an appropriate choice of the unit of time, this constant can be evaluated as 1; therefore

$$(5\text{-}9.11\text{a}) \qquad \xi\eta = 1.$$

When this result is substituted into the second of the relations (5-9.9), we obtain

$$(5\text{-}9.11\text{b}) \qquad \frac{1}{r^2}[-r\eta' + (1 - \eta)] = 0.$$

The result is an immediate consequence of this relation.

According to the relation (5-9.10b), $r = 0$ is a singular point. We assumed that $\eta > 0$; therefore for a positive constant there is a neighborhood, $r < \gamma$, about $r = 0$, for which the solution is not valid.

The Schwarzschild solution may be summarized as a spherically symmetric, static solution with the metric of special relativity as the limiting metric when r increases indefinitely. Mathematically, the effect of the gravitational field is expressed by taking the components of the Ricci tensor equal to zero. It should be noted that to set the components of the curvature tensor itself equal to zero would characterize a flat space and a physical situation in which the gravitational field had no effect outside the generating matter.

The general theory of relativity has been given a great deal of credence in three experiments concerning the advance of the perihelion of Mercury, the deflection of light rays by the sun, and the shift toward the red end of the spectrum of spectral lines of light originating in dense stars. We examine next the relativistic predictions concerning the first phenomenon.

It is assumed that the mass of a planet is negligible in comparison to that of the sun. Therefore planets can be thought of as moving along geodesics in a spherically symmetric static field. The defining equations of the motion are

$$\text{(5-9.12)}\qquad \begin{aligned} &\text{(a)}\quad \ddot{X}^\lambda + \Gamma_{\alpha\beta}{}^\lambda \dot{X}^\alpha \dot{X}^\beta = 0,\\ &\text{(b)}\quad \frac{\dot{r}^2}{\eta} + r^2\dot{\theta}^2 + r^2\sin^2\theta\dot{\phi}^2 - \eta\dot{t}^2 = -1, \end{aligned}$$

where the · denotes differentiation with respect to the parameter s.

Observations of planetary motions indicate that the Newtonian prediction for their trajectory is not far from wrong. Therefore the equations (5-9.12a,b) should produce certain correspondence with the classical results. One such common fact is determined by the following considerations. The last three of the equations in (5-9.12a) may be written out in the form

$$\text{(5-9.13)}\qquad \begin{aligned} &\text{(a)}\quad \ddot{\theta} + \frac{2}{r}\dot{r}\dot{\theta} - \sin\theta\cos\theta\dot{\phi}^2 = 0,\\ &\text{(b)}\quad \ddot{\phi} + \frac{2}{r}\dot{r}\dot{\phi} + 2\operatorname{ctn}\theta\dot{\theta}\dot{\phi} = 0,\\ &\text{(c)}\quad \ddot{t} + \frac{\eta'}{\eta}\dot{r}\dot{t} = 0. \end{aligned}$$

If we assume the initial conditions

$$\theta_0 = \frac{\pi}{2}, \qquad \dot{\theta}_0 = 0,$$

then from (5-9.13a) it follows that $\ddot{\theta}_0 = 0$. By differentiation of (5-9.13a) and a process of induction we find that

$$\overset{(k)}{\theta_0} = 0, \qquad k = 1, 2, 3\cdots.$$

Next we write a Taylor series expansion for θ. We have

$$\text{(5-9.14)}\qquad \theta = \theta_0 + (s - s_0)\dot{\theta}_0 + \frac{(s-s_0)^2}{2!}\ddot{\theta}_0 + \cdots.$$

Therefore $\theta = \pi/2$ for all values for which the expansion is valid. It is assumed that the expansion can be made at any point along the geodesic; therefore $\theta = \pi/2$ all along it. Consequently, $\dot{\theta} = 0$ along the geodesic. This result reminds us that the classical Newtonian trajectory lies in a plane. The relativistic result just obtained has the same physical interpretation.

The first of the equations in (5-9.13) reduces to an identity as a consequence of the constancy of θ. The other two equations take the form

$$\text{(5-9.15)} \qquad \begin{aligned} &\text{(a)} \quad \ddot{\phi} + \frac{2}{r}\dot{r}\dot{\phi} = 0, \\ &\text{(b)} \quad \ddot{t} + \frac{\eta'}{\eta}\dot{r}\dot{t} = 0. \end{aligned}$$

From these equations we find that

$$\text{(5-9.16)} \qquad \begin{aligned} \dot{\phi}r^2 &= h, \\ \dot{t}\eta &= k, \end{aligned}$$

where h, k are constants of integration. With these results in mind, let us make the substitution

$$\text{(5-9.17)} \qquad U = \frac{1}{r};$$

the differential equation (5-9.12b) can then be put in the more convenient form

$$\text{(5-9.18a)} \qquad \left(\frac{dU}{d\phi}\right)^2 + U^2 = \lambda - \frac{\gamma}{h^2}U + \gamma U^3,$$

where γ, λ, and h are constants such that

$$\text{(5-9.18b)} \qquad \lambda = \frac{k^2 - 1}{h^2}.$$

Recall that γ entered into our considerations in (5-9.10b) as a constant of integration.

In order to investigate relativistic trajectories of planets, the differential equation (5-9.18a) is compared to the corresponding classical equation

$$\text{(5-9.19a)} \qquad \left(\frac{dU}{d\phi}\right)^2 + U^2 = \frac{2M}{H^2}U + \frac{K}{H^2}.$$

In this equation M represents the mass of the sun.

$$\text{(5-9.19b)} \qquad K = \mathbf{V}\cdot\mathbf{V} - \frac{2M}{r}$$

is the total energy and

$$H = \begin{vmatrix} x & y \\ \dfrac{dx}{dt} & \dfrac{dy}{dt} \end{vmatrix} \tag{5-9.19c}$$

is the area swept out by the radius vector per unit time. Since the relativistic orbit differs only slightly from the classical orbit, it is assumed that the distinction in the differential equations is due to the term U^3 and, in particular, that

$$\frac{\gamma}{h^2} U + \frac{k^2 - 1}{h^2} \stackrel{*}{=} \frac{2M}{H^2} U + \frac{K}{H^2},$$

where $\stackrel{*}{=}$ means approximately equal. This assumption can also be implemented by identification of the constants

$$h \stackrel{*}{=} H, \qquad \gamma \stackrel{*}{=} 2M, \qquad k^2 \stackrel{*}{=} K + 1.$$

For all the planets other than Mercury the Newtonian prediction of the trajectory agrees with observation within the range of probable error. In the case of Mercury there is a discrepancy. A slight advance of perihelion (or rotation of the elliptic orbit) of about 42″ of arc per century has been observed and no acceptable explanation on the basis of classical theory has been found.

If in the classical theory we use polar coordinates, as indicated by the differential equation (5-9.19a), and find the extreme values of $U(\phi)$, they will correspond to perihelion and aphelion. Let (U_1, ϕ_1) and (U_2, ϕ_2), respectively, indicate the points of perihelion and aphelion; then $U_2 < U < U_1$. We find that

$$\phi_1 - \phi_2 = \pi.$$

In the relativistic case we make the identifications previously mentioned so that the differential equation under consideration is

$$\left(\frac{dU}{d\phi}\right)^2 + U^2 \stackrel{*}{=} \frac{K}{H^2} + \frac{2M}{H^2} U + 2MU^3. \tag{5-9.20}$$

Theorem 5-9.2. The relativistic prediction of the advance of perihelion is $(6\pi M)/[a(1 - e^2)]$ per revolution of the planet, where a represents the semimajor axis of the elliptic trajectory and e represents its eccentricity.

PROOF. If the term U^2 is put on the right of (5-9.20) and the zeros of the cubic polynomial are denoted by U_1, U_2, and U_3, then (5-9.20) can be expressed in the form

$$\left(\frac{dU}{d\phi}\right)^2 \stackrel{*}{=} 2M(U - U_1)(U - U_2)(U - U_3). \tag{5-9.21a}$$

The relation (5-9.21a) can be written[23]

$$d\phi \overset{*}{=} \frac{dU}{\{(U_1 - U)(U - U_2)[1 - 2M(U + U_1 + U_2)]\}^{1/2}}$$

If it is desired to integrate this expression from U_2 to U_1, the integration is expedited by the substitution

$$\frac{U - U_2}{U_1 - U_2} = \sin^2\beta.$$

We find that

$$\phi_1 - \phi_2 \overset{*}{=} \int_0^{\pi/2} \frac{2d\beta}{[1 - 2M(U_1 + U_2 + U_1 \sin^2 \beta + U_2 \cos^2 \beta)]^{1/2}}$$

By expanding the integrand in a binomial series and dropping the higher degree terms, we have

$$\phi_1 - \phi_2 \overset{*}{=} \int_0^{\pi/2} 2[1 + M(U_1 + U_2 + U_1 \sin^2 \beta + U_2 \cos^2 \beta)]\, d\beta.$$

As a consequence of the half-angle formulas,

$$\sin^2 \beta = \frac{1 - \cos 2\beta}{2}, \qquad \cos^2 \beta = \frac{1 + \cos 2\beta}{2},$$

we obtain the result[24]

$$\begin{aligned}\phi_1 - \phi_2 &\overset{*}{=} \pi + \tfrac{3}{2}M\pi(U_1 + U_2)\\ &= \pi + \tfrac{3}{2}M\pi \frac{2}{a(1 - e^2)}.\end{aligned}$$

Therefore

$$2(\phi_1 - \phi_2) = 2\pi + \frac{6\pi M}{a(1 - e^2)}$$

This concludes the proof.

---

[23] If the polynomial to which $(dU/d\phi)^2$ is equated is set equal to zero, that is,

$$U^3 - \frac{1}{2M}U^2 + \frac{1}{H^2}U + \frac{K}{2MH^2} = 0,$$

the sum of the roots is equal to the negative of the coefficient of U^2, that is,

$$U_1 + U_2 + U_3 = \frac{1}{2M}.$$

[24] If c represents the distance from the center of the ellipse to a focus, then

$$\begin{aligned} r_2 &= a + c = a + ae,\\ r_1 &= a - c = a - ae.\end{aligned}$$

Therefore $r_1 + r_2 = 2a$,

$$r_1 r_2 = a^2(1 - e^2).$$

In the case of the planet Mercury the discrepancy between observed results and the Newtonian prediction is close to $6\pi M/[a(1 - e^2)]$ so that it appears that the relativistic prediction is the better of the two.

Problems

1. (a) The equation $\partial_\lambda[\mu(dX^\lambda/ds)] = 0$ is called the relativistic continuity equation in discussions of special relativity. The generalized form of this equation is obtained by replacing partial differentiation with covariant differentiation. If μ is a scalar density of weight -1, show that $\nabla_\lambda \mu\,(\partial X^\lambda/ds) = 0$ reduces to the partial derivative form.

(b) If

$$\mu = \mu_0\sqrt{|g|} \quad \text{and} \quad m^{\lambda\gamma} = \sqrt{|g|}\, M^{\lambda\gamma} = \sqrt{|g|}\,\mu_0 \frac{dX^\lambda}{ds}\frac{dX^\gamma}{ds},$$

show that

$$\nabla_\lambda M^{\lambda\gamma} = \mu_0 \frac{D(dX^\gamma/ds)}{ds}.$$

2. (a) Compute the Christoffel symbols listed in (5-9.8b).

(b) Derive the equations in (5-9.9).

answers to odd-numbered problems

Chapter 1

SECTION 1

1. (a) 1, 3, 0.

3. $-2, -5, 1$.

5. $\sqrt{14}$, $5\sqrt{3}$, $\sqrt{77}$, respectively.

7. Components 0, 0, 0 magnitude 0.

9. *Hint.* Make use of the corresponding properties of real numbers.

11. $\beta(A^1, \cdots, A^n) = (0 \cdots 0)$. According to the definition of equality of n-tuples, $\beta A^i = 0$, $i = 1 \cdots n$. Since $\beta \neq 0$, it follows that $A^i = 0$ for all i.

13. Since $\mathbf{A} = p\mathbf{C}$, $\mathbf{B} = q\mathbf{D}$, $\mathbf{A} = r\mathbf{B}$, one has $\mathbf{C} = \frac{1}{p}\mathbf{A} = \frac{r}{p}\mathbf{B} = \frac{rq}{p}\mathbf{D}$.

15. If two sides of a triangle with a common vertex are represented by arrows **A** and **B**, the third side is represented by **B** − **A**. Midpoints of the original sides are at the tips of **A**/2 and **B**/2. Hence their join is represented by (**B** − **A**)/2. It has magnitude $\frac{1}{2}(\mathbf{B} - \mathbf{A})$.

17. $\dfrac{320\sqrt{6}}{1+\sqrt{3}}, \dfrac{640}{1+\sqrt{3}}$.

19. (a)

$$X^1 = 1 - t$$
$$X^2 = 5 + 4t$$
$$X^3 = 3 + 6t$$

$$\frac{X^1 - 1}{-1} = \frac{X^2 - 5}{4} = \frac{X^3 - 3}{6}.$$

21. $\mathbf{r}_2 - \mathbf{r}_1 = \mathbf{B}(t_2 - t_1) = \mathbf{B}\,\Delta t$. The magnitude $|\mathbf{r}_2 - \mathbf{r}_1|$ is dependent on Δt but not on particular t_1 and t_2.

SECTION 2

1. $\begin{vmatrix} A^1 & A^2 & A^3 \\ B^1 & B^2 & B^3 \\ C^1 & C^2 & C^3 \end{vmatrix} = \begin{vmatrix} 2 & 1 & -5 \\ 3 & 2 & 1 \\ 1 & 0 & -11 \end{vmatrix} = 0.$ Therefore the n-tuples are linearly dependent.

3. $\begin{vmatrix} 2 & 3 & 3 \\ 2 & 1 & -3 \\ 1 & 5 & -1 \end{vmatrix} \neq 0.$

5. $\begin{vmatrix} 1 & -3 & 5 \\ 2 & 4 & 7 \\ 1 & 6 & 4 \end{vmatrix} \neq 0.$ Therefore the arrows are independent and not coplanar.

7. $X = \frac{1}{2}$, $Y = \frac{5}{2}$.

9. Assume that P_1, P_2, P_3, and P_4 are coplanar points. Let $\mathbf{r}_1$, $\mathbf{r}_2$, $\mathbf{r}_3$, and $\mathbf{r}_4$ be position arrows from an origin outside the plane to these points. Then

$$\mathbf{r}_1 - \mathbf{r}_2 = \alpha(\mathbf{r}_3 - \mathbf{r}_2) + \beta(\mathbf{r}_4 - \mathbf{r}_2).$$

Therefore: $\mathbf{r}_1 = p\mathbf{r}_2 + q\mathbf{r}_3 + s\mathbf{r}_4$, where $p = 1 - \alpha - \beta$, $q = \alpha$, $s = \beta$.

Conversely: $\mathbf{r}_1 = p\mathbf{r}_2 + q\mathbf{r}_3 + s\mathbf{r}_4$, $p + q + s = 1$

implies that $\quad \mathbf{r}_1 = (1 - q - s)\mathbf{r}_2 + q\mathbf{r}_3 + s\mathbf{r}_4$

or $\quad \mathbf{r}_1 - \mathbf{r}_2 = q(\mathbf{r}_3 - \mathbf{r}_2) + s(\mathbf{r}_4 - \mathbf{r}_2).$

This last relation indicates the coplanar nature of the end points of the arrows.

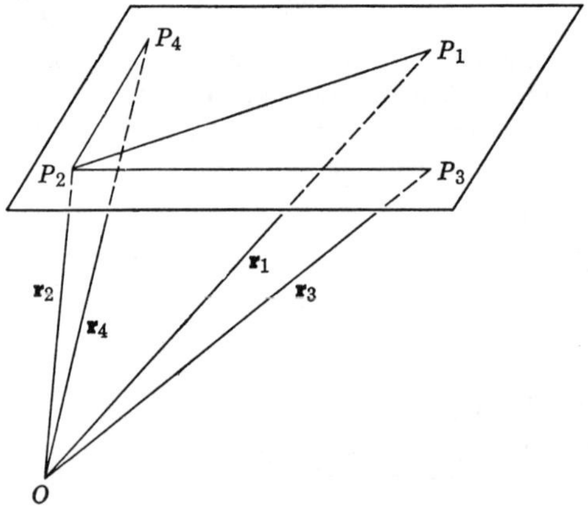

11. $(1, 3, 5) = \alpha(0, 2, 1) + \beta(-2, 1, 4) + \gamma(1, 3, 2)$,

$$\alpha = -\tfrac{7}{3},\ \beta = \tfrac{2}{3},\ \gamma = \tfrac{7}{3}.$$

13. The validity of the relation $(A + B - 3)X + (-5A - 2B - 2) = 0$ on an interval of real numbers leads to the conditions $A + B - 3 = 0$, $-5A - 2B - 2 = 0$ on A and B.

SECTION 3

1. (a) 81, (b) $\frac{1}{2}$, $-\frac{1}{2}$, 0, (c) $\delta_{js}^{[kp]} = \begin{cases} \frac{1}{2} \\ -\frac{1}{2} & \text{if} \\ 0 \end{cases} \begin{cases} j = k,\ s = p,\ j \neq s, \\ j = p,\ s = k,\ j \neq s, \\ \text{otherwise.} \end{cases}$

(d) $\delta_{js}^{[kp]} = \frac{1}{2}[\delta_{js}^{kp} - \delta_{js}^{pk}] = -\frac{1}{2}[\delta_{js}^{pk} - \delta_{js}^{kp}] = -\delta_{js}^{[pk]}$.

(e) $\delta_{js}^{[kp]} = \frac{1}{2}[\delta_j{}^k\delta_s{}^p - \delta_j{}^p\delta_s{}^k] = \frac{1}{2}[\delta_j{}^k\delta_s{}^p - \delta_s{}^k\delta_j{}^p] = \delta_{[js]}^{kp}$.

3. Upon completing squares, we have

$$(X^1 - 3)^2 + (X^2 + 1)^2 + (X^3 - 4)^2 = 1.$$

Therefore $\bar{X}^1 = X^1 - 3,\ \bar{X}^2 = X^2 + 1,\ \bar{X}^3 = X^3 - 4.$

5. See Example 1-3.7.

7. (a) $\left(-\dfrac{1}{\sqrt{2}}, \dfrac{5}{\sqrt{2}}, 0\right)$. (b) $1 - \dfrac{3\sqrt{3}}{2}$, $\sqrt{3} + \frac{3}{2}$.

9. (b) $(A_j{}^k) = \begin{pmatrix} \frac{1}{2} & \sqrt{\frac{3}{2}} & 0 \\ -\sqrt{\frac{3}{2}} & \frac{1}{2} & 0 \\ 0 & 0 & 1 \end{pmatrix}$. *Note.* Use $A_j{}^k = a_k{}^j$.

(c) Comparing with (1-3.23c), we note that $\cos\theta = \frac{1}{2}$, $\sin\theta = -\sqrt{3}/2$. Therefore $\theta = -\pi/3$ and the coefficients represent a clockwise rotation of $\pi/3$ radians of the $\bar{X}^j$ system from the X^j system. The rotation is in the plane $X^3 = 0$.

11. $0 = X^j - X_1{}^j - B^j t = a_k{}^j \bar{X}^k - a_k{}^j \bar{X}_1{}^k - a_k{}^j \bar{B}^k t$, where $\bar{B}^k = \bar{X}_2{}^k - \bar{X}_1{}^k$

Therefore $\quad 0 = a_k{}^j(\bar{X}^k - \bar{X}_1{}^k - \bar{B}^k t).$

Upon multiplying by $A_j{}^p$, we obtain

$$0 = \delta_k{}^p(\bar{X}^k - \bar{X}_1{}^k - \bar{B}^k t) = \bar{X}^p - \bar{X}_1{}^p - \bar{B}^p t.$$

SECTION 4

1. (a) $\quad U^k = a_j{}^k \bar{U}^j = a_1{}^k \bar{U}^1 + a_2{}^k \bar{U}^2 + a_3{}^k \bar{U}^3.$

Therefore $U^1 = 0(\frac{1}{2}) + 1(0) + 0(0) = 0,$

$U^2 = -1(\frac{1}{2}) + 0(0) + 0(0) = -\frac{1}{2},$

$U^3 = 0(\frac{1}{2}) + 0(0) + 1(0) = 0.$

3. (a) $\quad X_2{}^j - X_1{}^j = \bar{X}_2{}^j + \bar{X}_0{}^j - (\bar{X}_1{}^j + \bar{X}_0{}^j) = \bar{X}_2{}^j - \bar{X}_1{}^j.$

Therefore $\quad d^2 = \sum_{j=1}^{3}(X_2{}^j - X_1{}^j)^2 = \sum_{j=1}^{3}(\bar{X}_2{}^j - \bar{X}_1{}^j)^2,$

and it is seen that the algebraic form for d^2 is not dependent on the particular rectangular Cartesian coordinate system.

(b) According to (1-4.1) we have $U^j = \bar{U}^j$. Therefore the components are scalars under translation.

SECTION 5

1. $\mathbf{A} \cdot \mathbf{B} = -28$, $\mathbf{A} \cdot \mathbf{A} = 38$, $\mathbf{B} \cdot \mathbf{B} = 69$.

3. (a) A rotation of $-\pi/6$ radians in the X^1X^2 plane.

(b) $\bar{\mathbf{A}} \cdot \bar{\mathbf{B}} = -14$.

(c) $A^1 = a_k^{\ 1}\bar{A}^k = \frac{\sqrt{3}}{2}(3) - \frac{1}{2}(5), \qquad B^1 = \frac{\sqrt{3}}{2}(2) - \frac{1}{2}(-4),$

$$A^2 = a_k^{\ 2}\bar{A}^k = \frac{1}{2}(3) + \frac{\sqrt{3}}{2}(5), \qquad B^2 = \frac{1}{2}(2) + \frac{\sqrt{3}}{2}(-4).$$

(d) $\mathbf{A} \cdot \mathbf{B} = -14$. This result should have been expected since $\mathbf{A} \cdot \mathbf{B}$ is an invariant with respect to rotations.

5. $\mathbf{A} \cdot \mathbf{B} = 10 - 6 - 4 = 0$, therefore the vectors are orthogonal.

7. $$\mathbf{E}_1 \cdot \mathbf{E}_2 = (\mathbf{A} + \mathbf{B}) \cdot (\mathbf{B} - \mathbf{A}) = \mathbf{B} \cdot \mathbf{B} - \mathbf{A} \cdot \mathbf{A} = 0,$$

Since $\mathbf{A}$ and $\mathbf{B}$ have equal magnitudes.

9. The directions of the lines are determined by $(2, 7, -4)$ and $(-6, 2, \frac{1}{2})$. We have $-12 + 14 - 2 = 0$; therefore the lines are perpendicular.

11. $(3\iota_1 + \iota_2 - 5\iota_3) \cdot [(X^1 + 2)\iota_1 + (X^2 - 4)\iota_2 + (X^3 - 1)\iota_3] = 0,$

i.e., $$3(X^1 + 2) + (X^2 - 4) - 5(X^3 - 1) = 0.$$

13. Their normals are perpendicular.

15. $$d = \left| \frac{2(1 - 3) - 5(2 + 1) + 3(-1 - 4)}{\sqrt{4 + 25 + 9}} \right| = \frac{34}{\sqrt{38}}.$$

17. $$\sin^2\theta = 1 - \cos^2\theta = 1 - \frac{(\mathbf{P} \cdot \mathbf{Q})^2}{(\mathbf{P} \cdot \mathbf{P})(\mathbf{Q} \cdot \mathbf{Q})} = \frac{(\mathbf{P} \cdot \mathbf{P})(\mathbf{Q} \cdot \mathbf{Q}) - (\mathbf{P} \cdot \mathbf{Q})^2}{(\mathbf{P} \cdot \mathbf{P})(\mathbf{Q} \cdot \mathbf{Q})}.$$

SECTION 6

1. Consider the expression $\varepsilon_{str}a_j^{\ s}a_j^{\ t}a_k^{\ r}$. Because s and t are dummy indices, we may replace this expression by either $\varepsilon_{pqr}a_j^{\ p}a_j^{\ q}a_k^{\ r}$ or $\varepsilon_{qpr}a_j^{\ q}a_j^{\ p}a_k^{\ r}$. Considering the second of these and employing the fact that ε_{qpr} is skew symmetric in q and p, we have (also note that $a_j^{\ q}a_j^{\ p} = a_j^{\ p}a_j^{\ q}$)

$$\varepsilon_{qpr}a_j^{\ q}a_j^{\ p}a_k^{\ r} = -\varepsilon_{pqr}a_j^{\ q}a_j^{\ p}a_k^{\ r} = -\varepsilon_{pqr}a_j^{\ p}a_j^{\ q}a_k^{\ r}.$$

Taking into account the opening remark of this proof, we conclude that

$$\varepsilon_{pqr}a_j^{\ p}a_j^{\ q}a_k^{\ r} = -\varepsilon_{pqr}a_j^{\ p}a_j^{\ q}a_k^{\ r},$$

and therefore

$$2\varepsilon_{pqr}a_j^{\ p}a_j^{\ q}a_{kr} = 0.$$

3. Proof of (1-6.11b):

$$E^{ijk}\varepsilon_{ijq} = E^{1jk}\varepsilon_{1jq} + E^{2jk}\varepsilon_{2jq} + E^{3jk}\varepsilon_{3jq}$$
$$= (E^{12k}\varepsilon_{12q} + E^{13k}\varepsilon_{13q}) + (E^{21k}\varepsilon_{21q} + E^{23k}\varepsilon_{23q}) + (E^{31k}\varepsilon_{31q} + E^{32k}\varepsilon_{32q}).$$

Examining the three parenthetic expressions, we see that if $q \neq k$ every term is equal to zero. For $q = k$ there are two nonzero terms and each has the value 1. For example, if $q = k = 3$, then $E^{123}\varepsilon_{123} = E^{213}\varepsilon_{213} = 1$, we conclude that $E^{ijk}\varepsilon_{ijq} = 2\delta_q^{\ k}$. Equation 1-6.11c can be verified by writing out the summations.

5. $\delta^{jp}\delta^{kq}\varepsilon_{pq} = \delta^{j1}\delta^{k1}\varepsilon_{11} + \delta^{j2}\delta^{k1}\varepsilon_{21} + \delta^{j1}\delta^{k2}\varepsilon_{12} + \delta^{j2}\delta^{k2}\varepsilon_{22}$
$= \delta^{j2}\delta^{k1}\varepsilon_{21} + \delta^{j1}\delta^{k2}\varepsilon_{12}$.

For $j, k = 1, 1;\ 1, 2;\ 2, 1;\ 2, 2$, we obtain values 0, 1, -1, 0. This completes the proof, since E^{11}, E^{12}, E^{21}, and E^{22} have values 0, 1, -1, 0 respectively.

7. Express the third-order determinant in the form

$$a = \frac{1}{3!} E^{ijk}E^{pqr}a_{ip}a_{jq}a_{kr}.$$

Use the method of replacing dummy indices indicated in the answer to Problem 1 with the information $a_{uv} = -a_{vu}$ to conclude that $a = -a$, hence $a = 0$.

9. (a) $a = \dfrac{1}{n!}\varepsilon_{i_1\cdots i_n}E^{s_1\cdots s_n}a_{s_1}^{\ i_1}\cdots a_{s_n}^{\ i_n}$

or

$$a = \frac{1}{n!}E^{i_1\cdots i_n}E^{s_1\cdots s_n}a_{i_n s_n}\cdots a_{i_n s_n}.$$

(b) Yes: prove that any skew symmetric determinant of odd order has the value zero.

11. $bB = c$ where c is the numerical value of a determinant with elements

$$b_j^{\ k}B_k^{\ p} = c_j^{\ k}.$$

In this case $c_j^{\ k} = \delta_j^{\ k}$ and $\begin{vmatrix} 1 & 0 & 0 \\ 0 & 1 & 0 \\ 0 & 0 & 1 \end{vmatrix} = 1.$

SECTION 7

1. (a) $\mathbf{P} \times (\mathbf{Q} + \mathbf{R}) = \sum_{j=1}^{3} \varepsilon_{jkr}\iota_j P^k(Q^r + R^r) = \sum_j \varepsilon_{jkr}\iota_j P^k Q^r$
$+ \sum_j \varepsilon_{jkr}\iota_j P^k R^r = (\mathbf{P} \times \mathbf{Q}) + (\mathbf{P} \times \mathbf{R})$.

(b) $\alpha(\mathbf{P} \times \mathbf{Q}) = \alpha \sum_{j=1}^{3} \varepsilon_{jkr}\iota_j P^k Q^r = \sum \varepsilon_{jkr}\iota_j(\alpha P^k)Q^r$.

In both parts (a) and (b) the proofs depend on the corresponding real number properties.

3. Area $= |\mathbf{A} \times \mathbf{B}| = \sqrt{(\mathbf{A} \times \mathbf{B}) \cdot (\mathbf{A} \times \mathbf{B})}$,

$$\mathbf{A} \times \mathbf{B} = \begin{vmatrix} \iota_1 & \iota_2 & \iota_3 \\ 5 & -2 & 3 \\ 1 & -3 & 1 \end{vmatrix} = 7\iota_1 - 2\iota_2 - 13\iota_3,$$

$$|\mathbf{A} \times \mathbf{B}| = \sqrt{222}.$$

5. The plane is spanned by the arrows $\overrightarrow{P_1P_2}$ and $\overrightarrow{P_1P_3}$, hence is orthogonal to their cross product.

$$\overrightarrow{P_1P_2} \times \overrightarrow{P_1P_3} = \begin{vmatrix} \iota_1 & \iota_2 & \iota_3 \\ -1 & -1 & -1 \\ 0 & -2 & -2 \end{vmatrix} = -2\iota_2 + 2\iota_3.$$

The equation of the plane is

$$-2(X^2 - 2) + 2(X^3 - 5) = 0.$$

7. The cross produced is anticommutative: therefore

$$\mathbf{M} = \mathbf{r} \times \mathbf{F} = -\mathbf{F} \times \mathbf{r} = -\mathbf{M}'.$$

9. (a) Compute $\begin{vmatrix} \iota_1 & \iota_2 & \iota_3 \\ 3 & 5 & -2 \\ 7 & 3 & 3 \end{vmatrix} = 21\iota_1 - 23\iota_2 - 26\iota_3$

The parametric equations of the desired line are

$$X^1 = 1 + 21\mu$$
$$X^2 = 5 - 23\mu$$
$$X^3 = 2 - 26\mu$$

(b) The assumption that the two lines have a point in common (say with coordinates X_0^1, X_0^2, X_0^3) leads to the inconsistent set of equations

$$1 + 3t_0 = 7s_0,$$
$$2 + 5t_0 = 4 + 3s_0,$$
$$3 - 2t_0 = 2 + 3s_0.$$

That these equations are inconsistent can be seen by solving two of them for t_0 and s_0 and then showing that the solutions obtained do not satisfy the third equation.

(c) The desired line has direction numbers (21, −23, −26) [see part (a)]. Its parametric equations may be written

$$X^1 = X_0^1 + 21\mu, \qquad X^2 = X_0^2 - 23\mu, \qquad X^3 = X_0^3 - 26\mu.$$

Suppose this line has P^1 in common with the line expressed through the parameter t and that $\mu = 0$ corresponds to the point of the line in coincidence with the second given line. Then we can form the equations

$$\begin{aligned} 1 + 3t_1 &= X_0^{\,1} + 21\mu_1 & 7s_0 &= X_0^{\,1} \\ 2 + 5t_1 &= X_0^{\,2} - 23\mu_1 & 4 + 3s_0 &= X_0^{\,2} \\ 3 - 2t_1 &= X_0^{\,3} - 26\mu_1 & 2 + 3s_0 &= X_0^{\,3} \end{aligned}$$

This is a set of 6 equations in six unknowns. In particular the values obtained for $X_0^{\,1}$, $X_0^{\,2}$, $X_0^{\,3}$ complete the desired information.

11.

$$\mathbf{P}\cdot\mathbf{Q}\times\mathbf{R} = \begin{vmatrix} 2 & -5 & 1 \\ 1 & 3 & 0 \\ 3 & 1 & 1 \end{vmatrix} = 3,$$

$$\mathbf{P}\times\mathbf{Q}\cdot\mathbf{R} = \mathbf{R}\cdot\mathbf{P}\times\mathbf{Q} = \begin{vmatrix} 3 & 1 & 1 \\ 2 & -1 & 5 \\ 1 & 3 & 0 \end{vmatrix}.$$

The two determinants differ by an even number of intercharges of rows, hence have the same numerical value.

SECTION 8

1. The scores from a three-set tennis match might be listed as follows:

$$\begin{pmatrix} 3 & 6 \\ 6 & 4 \\ 2 & 6 \end{pmatrix}.$$

A probability matrix in genetics might have the form

$$\begin{array}{c} \\ D \\ H \\ R \end{array}\begin{array}{c} \begin{array}{ccc} D & H & R \end{array} \\ \begin{pmatrix} 1 & 0 & 0 \\ \frac{1}{4} & \frac{1}{2} & \frac{1}{4} \\ 0 & 0 & 1 \end{pmatrix} \end{array}.$$

The reader should be able to construct many other examples.

3.

$$\begin{pmatrix} \dfrac{\sqrt{2}}{2} & -\dfrac{\sqrt{6}}{4} & \dfrac{\sqrt{2}}{4} \\[2ex] \dfrac{\sqrt{2}}{2} & \dfrac{\sqrt{6}}{4} & -\dfrac{\sqrt{2}}{4} \\[2ex] 0 & \dfrac{1}{2} & \dfrac{\sqrt{3}}{2} \end{pmatrix}$$

5. Yes.

7. $$(X^1, X^2, X^3) = (\bar{X}^1, \bar{X}^2, \bar{X}^3)\begin{pmatrix} a_1^{\ 1} & a_1^{\ 2} & a_1^{\ 3} \\ a_2^{\ 1} & a_2^{\ 2} & a_2^{\ 3} \\ a_3^{\ 1} & a_3^{\ 2} & a_3^{\ 3} \end{pmatrix}$$

13. (a) $$\mathbf{P} \times (\mathbf{Q} \times \mathbf{R}) = \mathbf{P} \times \begin{vmatrix} \iota_1 & \iota_2 & \iota_3 \\ 2 & -3 & 1 \\ 0 & 4 & 2 \end{vmatrix} = \begin{vmatrix} \iota_1 & \iota_2 & \iota_3 \\ 1 & -4 & 4 \\ -10 & -4 & 8 \end{vmatrix}$$
$$= -16\iota_1 - 48\iota_2 - 44\iota_3.$$

(b) According to (1-7.17a),
$$\mathbf{P} \times (\mathbf{Q} \times \mathbf{R}) = \mathbf{Q}(\mathbf{P} \cdot \mathbf{R}) - \mathbf{R}(\mathbf{P} \cdot \mathbf{Q}) = (2\iota_1 - 3\iota_2 + \iota_3)(-8)$$
$$- (4\iota_2 + 2\iota_3)18,$$
$$= -16\iota_1 - 48\iota_2 - 44\iota_3.$$

15. Consider $\iota_1 \times \mathbf{Q} = \iota_3$. Let $\mathbf{Q} = \iota_2$ and $\iota_1 + \iota_2$, in turn.

17. If $\mathbf{A} \cdot \mathbf{B} = \mathbf{A} \cdot \mathbf{C}$ for any $\mathbf{A}$, then $\mathbf{B} = \mathbf{C}$.

19. $$h = |\mathbf{r}_1 - \mathbf{r}_0| \sin\theta = |\mathbf{r}_1 - \mathbf{r}_0| \left| \frac{(\mathbf{r}_1 - \mathbf{r}_0)}{|\mathbf{r}_1 - \mathbf{r}_0|} \times \frac{\mathbf{B}}{|\mathbf{B}|} \right|$$
$$= \left| (\mathbf{r}_1 - \mathbf{r}_0) \times \frac{\mathbf{B}}{|\mathbf{B}|} \right|.$$

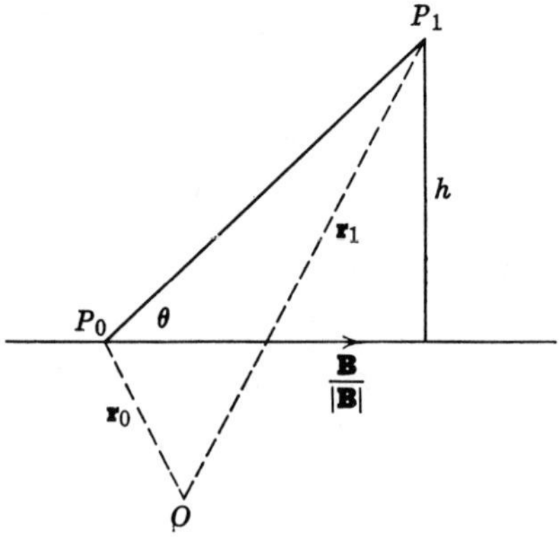

21. If $\mathbf{B} = \alpha\mathbf{A}$, then $$\begin{vmatrix} \mathbf{A}\cdot\mathbf{A} & \mathbf{A}\cdot\mathbf{B} \\ \mathbf{B}\cdot\mathbf{A} & \mathbf{B}\cdot\mathbf{B} \end{vmatrix} = \mathbf{A}\cdot\mathbf{A}\begin{vmatrix} 1 & \alpha \\ \alpha & \alpha^2 \end{vmatrix} = 0.$$

If $\begin{vmatrix} \mathbf{A}\cdot\mathbf{A} & \mathbf{A}\cdot\mathbf{B} \\ \mathbf{B}\cdot\mathbf{A} & \mathbf{B}\cdot\mathbf{B} \end{vmatrix} = 0$, then $(\mathbf{A} \times \mathbf{B}) \cdot (\mathbf{A} \times \mathbf{B}) = 0$ and $\mathbf{A} \times \mathbf{B} = \mathbf{O}$.

If $\mathbf{A}$ or $\mathbf{B}$ is a zero vector, the dependence is immediate. If $\mathbf{A} \neq \mathbf{O}$ and $\mathbf{B} \neq \mathbf{O}$, then $|\mathbf{A} \times \mathbf{B}| = |\mathbf{A}|\,|\mathbf{B}| \sin\theta$. Therefore $\sin\theta = 0$ and $\theta = 0$ or π.

Chapter 2

SECTION 1

1. (a) The orthogonal projection of the curve into the X^1, X^2 plane is that part of the hyperbola $(X^1)^2 - (X^2)^2 = 1$ lying in the first quadrant. The curve makes a constant angle with the X^2 axis and is called a helix.
(b) The orthogonal projection of the curve into the X^1, X^2 plane is the circle $(X^1)^2 + (X^2)^2 = a^2$; X^3 increases exponentially and without bound.
(c) This is a segment of the curve of part (b) determined by the inequality $1/e \leq X^3 \leq e$.

3. (a) $\left(\frac{dX^1}{dt}, \frac{dX^2}{dt}, \frac{dX^3}{dt}\right) = (\sinh t, \cosh t, 1)$

(b) $\left(\frac{dX^1}{dt}, \frac{dX^2}{dt}, \frac{dX^3}{dt}\right) = (-a \sin t, a \cos t, e^t)$

(c) $\left(\frac{dX^1}{dt}, \frac{dX^2}{dt}, \frac{dX^3}{dt}\right) = (-a \sin t, a \cos t, e^{\sin t} \cos t)$

5. Proof of (2-1.5a):

$$\frac{d\mathbf{U} + \mathbf{V}}{dt} = \lim_{h\to 0}\left\{\frac{[\mathbf{U}(t+h) + \mathbf{V}(t+h)] - [\mathbf{U}(t) + \mathbf{V}(t)]}{h}\right\}$$
$$= \lim_{h\to 0}\left[\frac{\mathbf{U}(t+h) - \mathbf{U}(t)}{h} + \frac{\mathbf{V}(t+h) - \mathbf{V}(t)}{h}\right]$$
$$= \lim_{h\to 0}\frac{\mathbf{U}(t+h) - \mathbf{U}(t)}{h} + \lim_{h\to 0}\frac{\mathbf{V}(t+h) - \mathbf{V}(t)}{h} = \frac{d\mathbf{U}}{dt} + \frac{d\mathbf{V}}{dt}$$

The validity of the limit theorem for vector sums is a consequence of the corresponding theorem for sums of functions.
Proof of (2-1.5b):

$$\frac{d(\mathbf{U}\cdot\mathbf{V})}{dt} = \lim_{h\to 0}\frac{\mathbf{U}(t+h)\cdot\mathbf{V}(t+h) - \mathbf{U}(t)\cdot\mathbf{V}(t)}{h}.$$

The proof can be completed by adding and subtracting $\mathbf{U}(t)\cdot\mathbf{V}(t+h)$ in the numerator of the difference quotient and regrouping and employing the limit theorems for products and sums. Proofs of (2-1.5c,d) follow the same pattern.

7. $\mathbf{V}\cdot\mathbf{V} = c$; c is a constant. Therefore

$$\frac{d\mathbf{V}\cdot\mathbf{V}}{dt} = \mathbf{V}\cdot\frac{d\mathbf{V}}{dt} + \frac{d\mathbf{V}}{dt}\cdot\mathbf{V} = 2\mathbf{V}\cdot\frac{d\mathbf{V}}{dt} = 0.$$

Since $\mathbf{V} \neq \mathbf{0}$ and $d\mathbf{V}/dt \neq \mathbf{0}$, they must have perpendicular directions.

9. Parametric equations of the tangent line at $\sqrt{2}, 3/\sqrt{2}, \pi/4$:

$$X^1 = \sqrt{2} - \sqrt{2}u, \qquad X^2 = \frac{3}{\sqrt{2}} + \frac{3}{\sqrt{2}}u, \qquad X^3 = \frac{\pi}{4} + u.$$

Osculating plane at $(\sqrt{2}, 3/\sqrt{2}, \pi/4)$:

$$\begin{vmatrix} X^1 - \sqrt{2} & X^2 - \dfrac{3}{\sqrt{2}} & X^3 - \dfrac{\pi}{4} \\ -\sqrt{2} & \dfrac{3}{\sqrt{2}} & 1 \\ -\sqrt{2} & -\dfrac{3}{\sqrt{2}} & 0 \end{vmatrix} = 0$$

11. $X^1 = u, \quad X^2 = 0, \quad X^3 = 0.$

SECTION 2

3. (a) $\sqrt{2}$, (b) $\sqrt{1 + 4t^2 + 9t^4}$,
(c) 1, (d) $\sqrt{2 + \cos^2 t}$.

5.
$$\frac{d\mathbf{r}}{ds} = \frac{d\mathbf{r}}{dt}\frac{dt}{ds}, \qquad 1 = \frac{d\mathbf{r}}{ds}\cdot\frac{d\mathbf{r}}{ds} = \left(\frac{d\mathbf{r}}{dt}\cdot\frac{d\mathbf{r}}{dt}\right)\left(\frac{dt}{ds}\right)^2;$$

Therefore
$$\left(\frac{dt}{ds}\right)^2 = \left(\frac{d\mathbf{r}}{dt}\cdot\frac{d\mathbf{r}}{dt}\right)^{-1}.$$

These relations and straightforward application of the rules of differentiation lead to the results.

7. (a) $\kappa^2 = 0, \tau = 0$, (b) $\kappa^2 = \dfrac{4(1 + 9t^2 + 9t^4)}{(1 + 4t^2 + 9t^4)^3}$

$$\tau = \frac{3}{1 + 9t^2 + 9t^4}$$

(c) $\kappa^2 = \dfrac{1}{a^2}, \tau = 0$, (d) $\kappa^2 = \dfrac{2\sin^2 t}{(2 + \cos^2 t)^3}$

$$\tau = 0$$

9. Curves a, c, and d have $\tau = 0$, hence are plane curves. This means that each lies in its own osculating plane.

(a) $X^3 = 0$ $\quad \begin{cases} X^1 = (1 + \pi/4) + u, \\ X^2 = \pi/4 + u, \\ X^3 = 0, \end{cases} \quad \begin{cases} X^1 = (1 + \pi/4) - v, \\ X^2 = \pi/4 + v, \\ X^3 = 0. \end{cases}$

(b)
$$\begin{vmatrix} X^1 - \dfrac{\pi}{4} & X^2 - \left(\dfrac{\pi}{4}\right)^2 & X^3 - \left(\dfrac{\pi}{4}\right)^3 \\ 1 & \dfrac{\pi}{2} & 3\left(\dfrac{\pi}{4}\right)^2 \\ 0 & 2 & \dfrac{3\pi}{2} \end{vmatrix} = 0,$$

$$X^1 = \frac{\pi}{4} + u, \qquad X^1 = \frac{\pi}{4},$$

$$X^2 = \left(\frac{\pi}{4}\right)^2 + \frac{\pi}{2} u, \qquad X^2 = \left(\frac{\pi}{4}\right)^2 + 2v,$$

$$X^3 = \left(\frac{\pi}{4}\right)^3 + 3\left(\frac{\pi}{4}\right)^2 u, \qquad X^3 = \left(\frac{\pi}{4}\right)^3 + \frac{3\pi}{2} v.$$

(c) $X^3 = 0 \qquad \begin{cases} X^1 = \frac{1}{2}, \\ X^2 = -u, \\ X^3 = 0, \end{cases} \qquad \begin{cases} X^1 = \frac{1}{2} - 2v, \\ X^2 = 0, \\ X^3 = 0. \end{cases}$

(d) $$\begin{vmatrix} X^1 - \frac{\pi}{4} & X^2 - \frac{\pi}{4} & X^3 - \frac{1}{\sqrt{2}} \\ 1 & 1 & \frac{1}{\sqrt{2}} \\ 0 & 0 & -\frac{1}{\sqrt{2}} \end{vmatrix} = 0$$

$$X^1 = \frac{\pi}{4} + u, \qquad X^1 = \frac{\pi}{4},$$

$$X^2 = \frac{\pi}{4} + u, \qquad X^2 = \frac{\pi}{4},$$

$$X^3 = \frac{1}{\sqrt{2}} + \frac{1}{\sqrt{2}} u, \qquad X^3 = \frac{1}{\sqrt{2}} - \frac{1}{\sqrt{2}} v.$$

SECTION 3

1. (a) $\mathbf{V} = -\sin t\iota_1 + \cos t\iota_2 + \iota_3,$
$\mathbf{a} = -\cos t\iota_1 - \sin t\iota_2.$

(b) $\mathbf{V} \cdot \mathbf{a} = 0$ for all t_3 therefore $\mathbf{V}$ and $\mathbf{a}$ are perpendicular vector fields.

(c) $\tau = -\frac{1}{2}.$

(d) $s = \int_0^t \sqrt{2}\, dt = \sqrt{2}\, t.$

3. (a) $\mathbf{r} = (1 + t^2)\iota_1 + t\iota_2,$
$\mathbf{V} = 2t\iota_1 + \iota_2,$
$\mathbf{a} = 2\iota_1.$

(b) $\mathbf{r} = t\iota_1 + t^2\iota_2 + t^3\iota_3,$
$\mathbf{V} = \iota_1 + 2t\iota_2 + 3t^2\iota_3,$
$\mathbf{a} = 2t\iota_2 + 6t\iota_3.$

The expressions for **r**, **V**, and **a** in parts (c), (d), (e), and (f) are produced by straightforward differentiations following the pattern given above.

5. $$\mathbf{r} = \mathbf{r}_0 + \mathbf{B}t, \qquad \mathbf{V} = \frac{d\mathbf{r}}{dt} = \mathbf{B}, \qquad \mathbf{a} = \frac{d^2\mathbf{r}}{dt^2} = \mathbf{0}$$

7. (a) $\mathbf{r} = 3\cos t\iota_1 + 3\sin t\iota_2 + t\iota_3,$
$\mathbf{V} = -3\sin t\iota_1 + 3\cos t\iota_2 + \iota_3,$
$\mathbf{a} = -3\cos t\iota_1 - 3\sin t\iota_2,$

(b) $\dfrac{\mathbf{r} \times \mathbf{a}}{|\mathbf{r} \times \mathbf{a}|} = \sin t\iota_1 - \cos t\iota_2$

(c) The path of motion is a circular helix.

(d) $\cos\theta = \dfrac{\mathbf{v}\cdot\mathbf{a}}{|\mathbf{v}|\,|\mathbf{a}|} = 0$; therefore **v** and **a** are perpendicular.

9. Take the dot product of $(d^2\mathbf{R}/d\theta^2) = \alpha\mathbf{R} + \beta\mathbf{P}$ and $\mathbf{R}$; since $\mathbf{R}\cdot\mathbf{R} = 1$ and $\mathbf{R}\cdot\mathbf{P} = 0$, it follows that $\alpha = \mathbf{R}\cdot\dfrac{d^2\mathbf{R}}{d\theta^2}$. Since $\mathbf{P}\cdot(d^2\mathbf{R}/d\theta^2) = 0$, a similar procedure produces the result $\beta = 0$.

11. (a) $\mathbf{r} = e^{bt}\mathbf{R}$,
$\mathbf{V} = be^{bt}\mathbf{R} + e^{bt}\mathbf{P}$,
$\mathbf{a} = (b^2 - 1)e^{bt}\mathbf{R} + 2be^{bt}\mathbf{P}$.
Corresponding results are obtained for parts (b) and (c) by making direct substitutions in (2-3.8a) and (2-3.8b).

SECTION 4

1. (a) $\bar{X}^1 = A_j{}^1X^j + \bar{X}_0{}^1 = (2 + 3t) + t = 2 + 4t$,

$\bar{X}^2 = (1 + 4t)\cos\theta + (3 + 5t)\sin\theta + (1 + 2t)$,

$\bar{X}^3 = -(4 + 4t)\sin\theta + (3 + 5t)\cos\theta + (2 + 3t)$.

$$\bar{V}^1 = \frac{dA_k{}^1}{dt}X^k + A_k{}^1\frac{dX^k}{dt} + \frac{d\bar{X}_0{}^1}{dt} = \frac{dX^1}{dt} + \frac{d\bar{X}_0{}^1}{dt} = 3 + 1 = 4,$$

$$\bar{V}^2 = [-(1 + 4t)\sin\theta + (3 + 5t)\cos\theta]\frac{d\theta}{dt} + 4\cos\theta + 5\sin\theta + 2,$$

$$\bar{V}^3 = [-(1 + 4t)\cos\theta - (3 + 5t)\sin\theta]\frac{d\theta}{dt} - 4\sin\theta + 5\cos\theta + 3.$$

$$\bar{a}^1 = \frac{d^2A_k{}^1}{dt^2}X^k + 2\frac{dA_k{}^1}{dt}\frac{dX^k}{dt} + A_k{}^1\frac{d^2X^k}{dt^2} + \frac{d^2\bar{X}_0{}^1}{dt^2} = 0,$$

$$\bar{a}^2 = [-(1 + 4t)\cos\theta - (3 + 5t)\sin\theta]\left(\frac{d\theta}{dt}\right)^2 + 2(-4\sin\theta + 5\cos\theta)\frac{d\theta}{dt}$$
$$+ [-(1 + 4t)\sin\theta + (3 + 5t)\cos\theta]\frac{d^2\theta}{dt^2}$$

$$\bar{a}^3 = [(1 + 4t)\sin\theta - (3 + 5t)\cos\theta]\left(\frac{d\theta}{dt}\right)^2 + 2(-4\cos\theta - 5\sin\theta)\frac{d\theta}{dt}$$
$$+ [(1 + 4t)\cos\theta - (3 + 5t)\sin\theta]\frac{d^2\theta}{dt^2}.$$

$$(\omega_{ik}) = \begin{pmatrix} 0 & 0 & 0 \\ 0 & 0 & 1 \\ 0 & -1 & 0 \end{pmatrix}\frac{d\theta}{dt}.$$

$$\omega^p = \frac{1}{2}E^{pik}\omega_{ik} \quad \therefore \quad (\omega^1, \omega^2, \omega^3) = \left(\frac{d\theta}{dt}, 0, 0\right).$$

The corresponding values for part (b) are obtained by substituting into (2-4.3b), (2-4.3c) (2-4.4a), and (2-4.6a).

3. $\frac{1}{2}\omega^r{}_q\omega^q{}_r = \frac{1}{2}\sum_{r,q}\omega_{rq}\omega_{qr} = \frac{1}{2}\sum_{r,q}\mathcal{E}_{urq}\omega^u\mathcal{E}_{vqr}\omega^v$

$$= -\tfrac{1}{2}\mathcal{E}_{uqr}E^{vqr}\omega^u\omega_v = -\delta^u_v\omega^u\omega_v = -\omega^u\omega_u.$$

5. Substitute into (2-4.13a).

SECTION 4*

1. (a) $\bar{\bar{\omega}}^i{}_k = \sum_p B_i{}^p \frac{dB_k{}^p}{dt} = \sum_p \frac{\partial \bar{X}^p}{\partial \bar{\bar{X}}^i}\frac{d(\partial \bar{X}^p/\partial \bar{\bar{X}}^k)}{dt}$

$$= \sum_p \frac{\partial \bar{X}^p}{\partial X^q}\frac{\partial X^q}{\partial \bar{\bar{X}}^i}\frac{d[(\partial \bar{X}^p/\partial X^r)(\partial X^r/\partial \bar{\bar{X}}^k)]}{dt}$$

$$= \sum_p \frac{\partial \bar{X}^p}{\partial X^q}\frac{\partial X^q}{\partial \bar{\bar{X}}^i}\left[\frac{d(\partial \bar{X}^p/\partial X^r)}{dt}\frac{\partial X^r}{\partial \bar{\bar{X}}^k} + \frac{\partial \bar{X}^p}{\partial X^r}\frac{d(\partial X^r/\partial \bar{\bar{X}}^k)}{dt}\right]$$

$$= \frac{\partial X^q}{\partial \bar{\bar{X}}^i}\frac{\partial X^r}{\partial \bar{\bar{X}}^k}\omega_{qr} + \delta_{qr}\frac{\partial X^q}{\partial \bar{\bar{X}}^i}\frac{d(\partial X^r/\partial X^k)}{dt}.$$

Since the transformations are orthogonal Cartesian rotations, $\partial X^q/\partial \bar{\bar{X}}^i = \partial \bar{\bar{X}}^i/\partial X^q$, and the result can be written in the form (2-4*.7).

(b) The set of equations (2-4*.5) is solved for the $\bar{\bar{\omega}}_p{}^q$ by summing and multiplying with $\partial \bar{\bar{X}}^s/\partial X^i$, then $\partial X^k/\partial \bar{\bar{X}}^t$. Finally we employ the relation obtained by differentiating $(\partial X^k/\partial \bar{\bar{X}}^t)(\partial \bar{\bar{X}}^q/\partial X^k) = \bar{\bar{\delta}}_t{}^q$.

3. The rotational velocity components DX^j/dt have vector character. When U^j is replaced in (2-4*.9) with these components, we obtain

$$\frac{D^2X^j}{dt^2} = \frac{d}{dt}\frac{(DX^j)}{dt} + \omega_i{}^j\frac{DX^i}{dt}.$$

If DX^j/dt is replaced in the right-hand member according to (2-4*.9) we obtain (2-4*.2b).

SECTION 5

1. If $\mathbf{h} = \mathbf{0}$ then the motion is linear

3. $\frac{d\mathbf{L}}{dt} = \mathbf{H} = \mathbf{0}$. Therefore $\mathbf{L}$ is constant.

SECTION 6

1. The galilean transformations have the linear form

$$\bar{X}^j = X^j - v^j t,$$
$$\bar{t} = t.$$

Therefore vector components $\bar{B}^j$ transform according to the law

$$\bar{B}^j = \frac{\partial \bar{X}^j}{\partial X^k}B^k = \delta_k{}^jB^k = B^j,$$

and

$$\bar{X}^j - (\bar{X}_0{}^j + \bar{B}^j\bar{t}) = X^j - v^jt - (X_0{}^j - v^jt + B^jt)$$
$$= X^j - (X_0{}^j + B^jt).$$

3. $-z^2 + t^2 = -(\bar{z} + v\bar{t})^2 + \bar{t}^2 = -(\bar{z})^2 - 2v\bar{z}\bar{t} + (1 - v^2)\bar{t}^2.$

5. If $h_{\alpha\beta}\, dX^\alpha\, dX^\beta = \bar{h}_{\lambda\gamma}\, d\bar{X}^\lambda\, d\bar{X}^\gamma$, then by employing the transformation law of the differential we obtain

$$h_{\alpha\beta}\, dX^\alpha\, dX^\beta = \bar{h}_{\lambda\gamma} \frac{\partial \bar{X}^\lambda}{\partial X^\alpha} \frac{\partial \bar{X}^\gamma}{\partial X^\beta}\, dX^\alpha\, dX^\beta$$

or

$$\left(h_{\alpha\beta} - \frac{\partial \bar{X}^\lambda}{\partial X^\alpha} \frac{\partial \bar{X}^\gamma}{\partial X^\beta}\, \bar{h}^{\lambda\gamma}\right) dX^\alpha\, dX^\beta = 0.$$

Since this expression holds for all sets of differentials $dX^1, \cdots, dX^4$, it follows that the parenthetic expression is identically zero.

7.
$$\frac{\bar{z}}{\bar{t}} = \frac{z - vt}{-vz + t} = \frac{z/t - v}{-vz/t + 1}.$$

If $z/t = 1$, then

$$\frac{\bar{z}}{\bar{t}} = \frac{1 - v}{-v + 1} = 1.$$

9. Differentiation of $h_{\alpha\beta}(dX^\alpha/ds)(dX^\beta/ds) = 1$ with respect to s leads to the form

$$h_{\alpha\beta} \frac{d^2X^\alpha}{ds^2} \frac{dX^\beta}{ds} = 0.$$

When this expression is multiplied by m_0, we obtain according to (2-6.9a)

$$h_{\alpha\beta} \overset{4}{F}{}^\alpha \frac{dX^\beta}{ds} = 0,$$

or

$$\overset{4}{F}{}^4 \frac{dX^4}{ds} = \sum_{a=1}^{3} \overset{4}{F}{}^a \frac{dX^a}{ds} = \sum \frac{F^a}{(1 - v^2)^{1/2}} \frac{dX^a}{dt} \frac{dt}{ds},$$

$$\overset{4}{F}{}^4 = \sum_{a=1}^{3} \frac{F^a(dX^a/dt)}{(1 - v^2)^{1/2}}.$$

Chapter 3

Section 1

1. $X^1 = v^1, \qquad X^2 = (v^2)^3, \qquad X^3 = \frac{1}{A^3}\{A^3X_0^{\ 3} + A^1(X_0^{\ 1} - v^1) + A^2[X_0^{\ 2} - (v)^3]\}.$

$$\left(\frac{\partial X^j}{\partial v^\beta}\right) = \begin{pmatrix} 1 & 0 & \dfrac{-A^1}{A^3} \\ 0 & 3(v^2)^2 & \dfrac{-3A^2}{A^3}(v^2)^2 \end{pmatrix}.$$

The origin is a singular point of this representation.

3. (a) $\begin{cases} X^1 = 3 \sin t \cos 2t \\ X^2 = 3 \sin t \sin 2t \\ X^3 = 3 \cos t. \end{cases}$

$\theta = t$	ϕ
0	0
$\frac{\pi}{8}$	$\frac{\pi}{4}$
$\frac{\pi}{4}$	$\frac{\pi}{2}$
$\frac{3\pi}{8}$	$\frac{3\pi}{4}$
$\frac{\pi}{2}$	π
$\frac{3\pi}{4}$	$\frac{3\pi}{2}$
π	2π

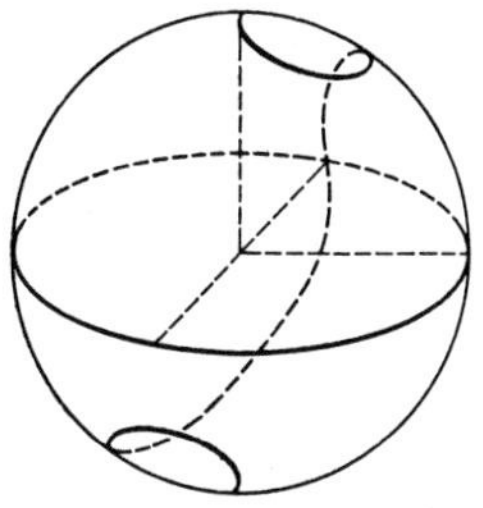

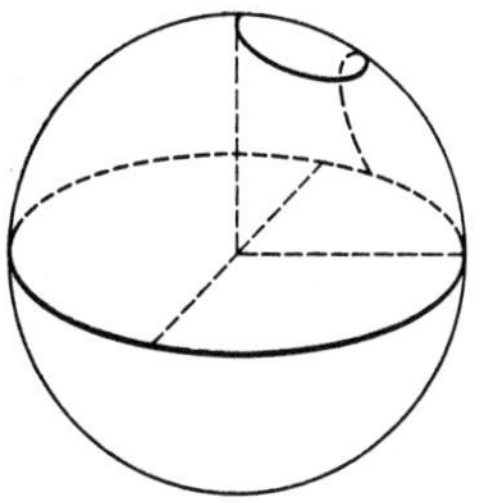

5. Make the identifications $\begin{cases} v^1 = \theta \\ v^2 = \phi\,. \end{cases}$

Meridian $\phi = \phi_0$: $\dfrac{\delta v^1}{\delta \theta} = 1, \qquad \dfrac{\delta v^2}{\delta \theta}\,,$

curve $\phi = \ln \tan \left(\dfrac{\theta}{2} + \dfrac{\pi}{4}\right), \qquad \dfrac{dv^1}{d\theta} = 1, \qquad (dv^2/d\theta) = \sec \theta$

$$\frac{d\mathbf{r}_1/d\theta \cdot d\mathbf{r}_2/d\theta}{|d\mathbf{r}_1/d\theta|\ |d\mathbf{r}_2/d\theta|} = \frac{g_{\alpha\beta}(\delta v^\alpha/d\theta)(\delta v^\beta/d\theta)}{[g_{\lambda\mu}(dv^\lambda/d\theta)(dv^\mu/d\theta)]^{1/2}[g_{\gamma\eta}(\delta v^\gamma/d\theta)(\delta v^\eta/)d\theta]^{1/2}}$$

The result follows upon substitution into this relation.

7. Parameter transformation: $\begin{aligned} {}^\star v^1 &= a \sin v^1 \cos v^2 \\ {}^\star v^2 &= a \sin v^1 \sin v^2\,. \end{aligned}$

$$\left|\frac{\partial {}^\star v^\alpha}{\partial v^\beta}\right| = \begin{vmatrix} a \cos v^1 \cos v^2 & a \cos v^1 \sin v^2 \\ -a \sin v^1 \sin v^2 & a \sin v^1 \cos v^2 \end{vmatrix} = a^2 \sin v^1 \cos v^1.$$

The domain of definition for the transformation involving the upper hemisphere cannot include $v^1 = 0$ or $v^1 = \pi/2$.

SECTION 2

1. (a) $\dfrac{2X^1}{a^2}\iota_1 + \dfrac{2X^2}{b^2}\iota_2 + \dfrac{2X^3}{c^2}\iota_3$, (b) $\dfrac{2X^1}{a^2}\iota_1 + \dfrac{2X^2}{b^2}\iota_2 - \dfrac{2X^3}{c^2}\iota_3$,

(c) $\dfrac{2X^1}{a^2}\iota_1 + \dfrac{2X^2}{b^2}\iota_2 - \dfrac{2X^3}{c^2}\iota_3$, (d) $\dfrac{2X^1}{a^2}\iota_1 - \dfrac{2X^2}{b^2}\iota_2 + 2\iota_3$,

(e) $\dfrac{2X^1}{a^2}\iota_1 + \dfrac{2X^2}{b^2}\iota_2$.

3. $\nabla\Phi = \dfrac{1}{a^2}\mathbf{r}$, $\nabla\Psi = 2\mathbf{r}$; therefore $\nabla\Phi = \dfrac{1}{2a^2}\nabla\Psi$.

5. The tangent plane may be expressed in the form $\nabla\Psi\cdot(\mathbf{r}-\mathbf{r}_0) = 0$, in particular $(\nabla\Psi)_{1,1,\frac{1}{2}} = \frac{1}{2}\iota_1 + \iota_2 + \iota_3$; therefore the equation of the tangent plane is

$$\tfrac{1}{2}(X^1 - 1) + (X^2 - 1) + (X^3 - \tfrac{1}{2}) = 0.$$

7. $\nabla\Phi = 2X^1\iota_1 + 2X^2\iota_2 + 2X^3\iota_3 = 2\mathbf{r}$.

9. $$\nabla f = \frac{\partial f}{\partial X^1}\iota_1 + \frac{\partial f}{\partial X^2}\iota_2 + \frac{\partial f}{\partial X^3}\iota_3 = \left(\frac{\partial f}{\partial u}\frac{\partial u}{\partial X^1} + \frac{\partial f}{\partial v}\frac{\partial v}{\partial X^1}\right)\iota_1$$

$$\left(\frac{\partial f}{\partial u}\frac{\partial u}{\partial X^2} + \frac{\partial f}{\partial v}\frac{\partial v}{\partial X^2}\right)\iota_2 + \left(\frac{\partial f}{\partial u}\frac{\partial u}{\partial X^3} + \frac{\partial f}{\partial v}\frac{\partial v}{\partial X^3}\right)\iota_3$$

$$= \frac{\partial f}{\partial u}\nabla u + \frac{\partial f}{\partial v}\nabla v.$$

11. $$\left(\frac{dx^1}{ds}\frac{dx^2}{ds}\frac{dx^3}{ds}\right) = \left(\cos\theta,\ \sin\theta,\ \frac{\partial\Phi}{\partial X^1}\cos\theta + \frac{\partial\Phi}{\partial X^2}\sin\theta\right).$$

$$\left|\frac{d\mathbf{r}}{ds}\right| = \left[1 + \left(\frac{\partial\Phi}{\partial X^1}\right)^2\cos^2\theta + \left(\frac{\partial\Phi}{\partial X^2}\right)^2\sin^2\theta + \frac{\partial\Phi}{\partial X^1}\frac{\partial\Phi}{\partial X^2}\sin 2\theta\right]^{\frac{1}{2}}$$

SECTION 3

1. We prove (3-3.2a)

$$\nabla\cdot\nabla\times\mathbf{U} = \frac{\partial}{\partial X^i}E^{ijk}\frac{\partial}{\partial X^j}U_k = E^{ijk}\frac{\partial^2}{\partial X^i\,\partial X^j}U_k.$$

The indicator E^{ijk} is skew symmetric in i and j and $\partial^2/\partial X^i\,\partial X^j$ is a symmetric operator with respect to i and j, therefore

$$\nabla\cdot\nabla\times\mathbf{U} = 0.$$

3. $\nabla\cdot(f\mathbf{G}) = \dfrac{\partial fG^j}{\partial X^j} = \dfrac{\partial f}{\partial X^j}G^j + f\dfrac{\partial G^j}{\partial X^j} = \nabla f\cdot\mathbf{G} + f\nabla\cdot\mathbf{G}$.

5. $\nabla\cdot\mathbf{r} = \dfrac{\partial X^1}{\partial X^1} + \dfrac{\partial X^2}{\partial X^2} = 2 \quad (n = 2)$.

7. $\nabla(\mathbf{a}\cdot\mathbf{r}) = \iota_1 \dfrac{\partial \mathbf{a}\cdot\mathbf{r}}{\partial X^1} + \iota_2 \dfrac{\partial \mathbf{a}\cdot\mathbf{r}}{\partial X^2} + \iota_3 \dfrac{\partial \mathbf{a}\cdot\mathbf{r}}{\partial X^3}$

$$= a_1\iota_1 + a_2\iota_2 + a_3\iota_3 = \mathbf{a},$$

since

$$\mathbf{a}\cdot\mathbf{r} = a_1X^1 + a_2X^2 + a_3X^3 \quad \text{and} \quad \frac{\partial a_j X^j}{\partial X^j} = a_j \quad (\text{no sum on } j).$$

9. We prove that the expression of part (a) is of vector character. Consider the rotation equations $X^j = a_k{}^j \bar{X}^k$ where the $a_k{}^j$ are constants. The components of $\nabla \times (\nabla \times \mathbf{U})$ are

$$E^{ijp}\,\partial_j \mathcal{E}_{pqr}\,\partial^q U^r = \bar{E}^{ijp} A_j{}^s\,\bar{\partial}_s \bar{\mathcal{E}}_{pqr} a_t{}^q\,\bar{\partial}t\,a_v{}^r \bar{U}^v$$

$$= \bar{E}^{ijp} A_j{}^s\,\bar{\partial}_s \bar{\mathcal{E}}_{u\,tv} A_p{}^u\,\bar{\partial}^t \bar{U}^v = a_c{}^i \bar{E}^{csu}\,\bar{\partial}_s \bar{\mathcal{E}}_{u\,tv}\,\bar{\partial}^t \bar{U}^v.$$

Note that $\partial_j = \partial/\partial X^j$, ∂^j has the same meaning, and

$$\mathcal{E}_{pqr} a_t{}^q a_v{}^r = \mathcal{E}_{u\,tv} A_p{}^u, \qquad \bar{E}^{ijp} A_j{}^s A_p{}^u = \bar{E}^{csu} a_c{}^i.$$

SECTION 4

1. (a)

$$\left(\frac{\partial X^j}{\partial \bar{X}^k}\right) = \begin{pmatrix} \sin\theta\cos\phi & \sin\theta\sin\phi & \cos\theta \\ r\cos\theta\cos\phi & r\cos\theta\sin\phi & -r\sin\theta \\ -r\sin\theta\sin\phi & r\sin\theta\cos\phi & 0 \end{pmatrix}.$$

(b)

$$\left|\frac{\partial X}{\partial \bar{X}}\right| = r^2 \sin\theta.$$

(c) Those coordinate triples for which $r = 0$ or $\theta = 0$ or $\theta = \pi$.

(d)

$$\left(\frac{\partial \bar{X}^k}{\partial X^j}\right) = \begin{pmatrix} \sin\theta\cos\phi & \dfrac{\cos\theta\cos\phi}{r} & -\dfrac{\sin\phi}{r\sin\theta} \\ \sin\theta\sin\phi & \dfrac{\cos\theta\sin\phi}{r} & \dfrac{\cos\phi}{r\sin\theta} \\ \cos\theta & -\dfrac{\sin\theta}{r} & 0 \end{pmatrix}.$$

(e) Perform the row by column multiplication of $(\partial X^j/\partial \bar{X}^k)(\partial \bar{X}^p/\partial X^j)$ and obtain $(\bar{\delta}_k{}^p)$.

(f) $\bar{\mathbf{r}}_j = \dfrac{\partial \mathbf{r}}{\partial \bar{X}^j} = \dfrac{\partial \mathbf{r}}{\partial X^k}\dfrac{\partial X^k}{\partial \bar{X}^j}$.

Therefore

$$\bar{\mathbf{r}}_1 = \iota_1 \frac{\partial X^1}{\partial \bar{X}^1} + \iota_2 \frac{\partial X^2}{\partial \bar{X}^1} + \iota_3 \frac{\partial X^3}{\partial \bar{X}^1} = \iota_1 \sin\theta\cos\phi + \iota_2 \sin\theta\sin\phi + \iota_3 \cos\theta.$$

$\bar{\mathbf{r}}_2$ and $\bar{\mathbf{r}}_3$ are computed in a similar manner. The expressions $\bar{\mathbf{r}}^j$ result from $\bar{\mathbf{r}}^j = \bar{g}^{jk}\mathbf{r}_k$, whereas $\mathbf{e}_j = \dfrac{\bar{\mathbf{r}}_j}{|\mathbf{r}_j|} = \dfrac{\bar{\mathbf{r}}_j}{\sqrt{\bar{g}_{jj}}}$.

(g) $\bar{\mathbf{r}}_1 = \bar{g}_{1k}\bar{\mathbf{r}}^k = \bar{g}_{11}\bar{\mathbf{r}}^1 = \bar{\mathbf{r}}^1$

$\bar{\mathbf{r}}_2 = r^2\bar{\mathbf{r}}^2, \qquad \bar{\mathbf{r}}_3 = r^2\sin^2\theta\,\bar{\mathbf{r}}^3$

3. Let $(A_j^{\,p})$ be defined so that $A_j^{\,p}(\partial\bar{X}^j/\partial X^k) = \delta_k^{\,p}$; then

$$A_j^{\,p}\,d\bar{X}^j = A_j^{\,p}\frac{\partial\bar{X}^j}{\partial X^k}\,d\bar{\bar{X}}^k = \delta_k^{\,p}\,d\bar{\bar{X}}^k = d\bar{\bar{X}}^p = \frac{\partial\bar{\bar{X}}^p}{\partial\bar{X}^j}\,d\bar{X}^j.$$

From this relation we obtain

$$\left(A_j^{\,p} - \frac{\partial\bar{\bar{X}}^p}{\partial\bar{X}^j}\right) d\bar{X}^j = 0$$

for arbitrary sets $d\bar{X}^j$. Therefore

$$A_j^{\,p} = \frac{\partial\bar{\bar{X}}^p}{\partial\bar{X}^j}\,.$$

5. The properties in (3-4.4a,b) are satisfied, since for the rectangular Cartesian transformations $\partial X^j/\partial\bar{X}^k = a_k^{\,j}$ are constant and $|a_j^{\,k}| = 1$.

7. According to (3-4.18a), $\bar{U}_j = \bar{g}_{jk}\bar{U}^k$, If we multiply and sum this expression with $\bar{g}^{jp}$, then $\bar{g}^{jp}\bar{U}_j = \bar{g}^{jp}\bar{g}_{jk}\bar{U}^k = \delta_k^{\,p}U^k = \bar{U}^p$, as was to be shown.

9. (a) We need to demonstrate that $\bar{\mathbf{r}}^j \cdot \bar{\mathbf{r}}_k = \delta_k^{\,j}$. This can be done by straightforward computation. For example,

$$\bar{\mathbf{r}}_1 \cdot \bar{\mathbf{r}}^1 = \bar{\mathbf{r}}_1 \cdot \frac{\bar{\mathbf{r}}_2 \times \bar{\mathbf{r}}_3}{\bar{\mathbf{r}}_1 \cdot \bar{\mathbf{r}}_2 \times \bar{\mathbf{r}}_3} = 1,$$

whereas

$$\bar{\mathbf{r}}_1 \cdot \bar{\mathbf{r}}^2 = \bar{\mathbf{r}}_1 \cdot \frac{\bar{\mathbf{r}}_3 \times \bar{\mathbf{r}}_1}{\bar{\mathbf{r}}_1 \cdot \bar{\mathbf{r}}_2 \times \bar{\mathbf{r}}_3} = 0.$$

(b) $\mathbf{e}_1 \cdot \mathbf{e}_2 \times \mathbf{e}_3 = 1$.

(c) $\bar{\mathbf{r}}_1 \cdot \bar{\mathbf{r}}_2 \times \bar{\mathbf{r}}_3 = \left|\dfrac{\partial X^j}{\partial\bar{X}^k}\right|$, i.e., the Jacobian of transformation.

11. $\bar{g}_{jk} = \bar{\mathbf{r}}_j \cdot \bar{\mathbf{r}}_k$. The symmetry follows from the commutivity of the dot product. According to (3-4.20a), a similar remark can be made for $\bar{g}^{sk}$.

13. (a) $\bar{U}^j\bar{\mathbf{r}}_j = \dfrac{\partial\bar{X}^j}{\partial\bar{\bar{X}}^k}\bar{\bar{U}}^k\dfrac{\partial\mathbf{r}}{\partial\bar{\bar{X}}^p}\dfrac{\partial\bar{\bar{X}}^p}{\partial\bar{X}^j} = \delta_k^{\,p}\bar{\bar{U}}^k\dfrac{\partial\mathbf{r}}{\partial\bar{\bar{X}}^p} = \bar{\bar{U}}^k\bar{\bar{\mathbf{r}}}_k$.

The proof of part (b) follows the same pattern.

Section 5

1.
$$(\bar{g}_{jk}) = \begin{pmatrix} 1 & 0 & 0 \\ 0 & r^2 & 0 \\ 0 & 0 & r^2\sin^2\theta \end{pmatrix},$$

$$\nabla = \frac{\bar{\mathbf{e}}_1}{\sqrt{\bar{g}_{11}}}\frac{\partial}{\partial\bar{X}^1} + \frac{\bar{\mathbf{e}}_2}{\sqrt{\bar{g}_{22}}}\frac{\partial}{\partial\bar{X}^2} + \frac{\bar{\mathbf{e}}_3}{\sqrt{\bar{g}_{33}}}\frac{\partial}{\partial\bar{X}^3}\,.$$

Therefore the result follows by direct substitution.

3. (a)

$$\bar{g} = \begin{vmatrix} 1 & 0 & 0 \\ 0 & p^2 & 0 \\ 0 & 0 & 1 \end{vmatrix} = p^2.$$

(b)

$$\bar{g} = \begin{vmatrix} 1 & 0 & 0 \\ 0 & r^2 & 0 \\ 0 & 0 & r^2 \sin^2 \theta \end{vmatrix} = r^4 \sin^2 \theta.$$

5. Upon substitution into (3-5.17b), we obtain

(a) $\overline{\nabla} \cdot \overline{\mathbf{W}} = \dfrac{1}{p}$, (b) $\overline{\nabla} \cdot \overline{\mathbf{W}} = 0$,

(c) $\overline{\nabla} \cdot \overline{\mathbf{W}} = -\dfrac{1}{p} \cos \theta + \dfrac{1}{p} \sin \theta.$

7. $$\overline{\nabla} \cdot \overline{\nabla}\Phi = \frac{1}{r^2 \sin \theta} \left\{ \frac{\partial[r^2 \sin \theta (\partial\Phi/\partial r)]}{\partial r} + \frac{\partial[\sin \theta \, (\partial\Phi/\partial\theta)]}{\partial\theta} + \frac{\partial[\frac{1}{\sin \theta} (\partial\Phi/\partial\theta)]}{\partial\theta} \right\}$$

9. According to (3-5.6a)

$$\bar{\mathcal{E}}_{jkq} = \left|\frac{\partial \bar{X}}{\partial \bar{\bar{X}}}\right| \frac{\partial \bar{\bar{X}}^p}{\partial \bar{X}^j} \frac{\partial \bar{\bar{X}}^s}{\partial \bar{X}^k} \frac{\partial \bar{\bar{X}}^t}{\partial \bar{X}^q} \bar{\bar{\mathcal{E}}}_{pst} = \left|\frac{\partial \bar{X}}{\partial \bar{\bar{X}}}\right| \left|\frac{\partial \bar{\bar{X}}}{\partial \bar{X}}\right| \bar{\bar{\mathcal{E}}}_{jkq} = \bar{\bar{\mathcal{E}}}_{jkq}$$

as was to be shown. The verification in part (b) follows the same form.

Section 6

1. $$B^2 - AC = \left(\frac{\partial^2 f}{\partial X^1 \, \partial X^2}\right)^2 - \frac{\partial^2 f}{(\partial X^1)^2} \frac{\partial^2 f}{(\partial X^2)^2}$$

$$= 0 - 2(2) = -4, \qquad A = \frac{\partial^2 f}{\partial X^{12}} = 2 > 0.$$

$$\frac{\partial f}{\partial X^1} = 2X^1 - 1, \qquad \frac{\partial f}{\partial X^2} = 2X^2.$$

$(X^1, X^2) = (\frac{1}{2}, 0)$ is a critical point pair. $B^2 - AC < 0$, $A > 0$ everywhere. Therefore the surface has a relative minimum at this point.

Chapter 4

Section 1

1. (a) $$\left(\frac{dX^1}{dt}, \frac{dX^2}{dt}, \frac{dX^3}{dt}\right) = (-a \sin t, a \cos t, 0),$$

therefore

$$\int_0^2 \frac{d\mathbf{r}}{dt} \cdot \frac{d\mathbf{r}}{dt} \, dt = \int_0^2 a^2 \, dt = 2a^2.$$

(b) $$\left(\frac{dX^1}{ds}, \frac{dX^2}{ds}, \frac{dX^3}{ds}\right) = \left(-\sin \frac{s}{a}, \cos \frac{s}{a}, 0\right);$$

Therefore

$$\int_0^{2\pi} \frac{d\mathbf{r}}{ds} \cdot \frac{d\mathbf{r}}{ds} \, ds = \int_0^{2\pi} ds = 2\pi.$$

3. $\int_C \mathbf{G} \cdot d\mathbf{r} = \int_C (2X^1X^2\,dX^1) + (X^1)^2\,dX^2.$

(a) $X^2 = [1 + (X^1)^2]^{1/2}$; therefore we evaluate

$$\int_0^2 2X^1[1 + (X^1)^2]^{1/2}\,dX^1 + \int_1^{\sqrt{5}} [(X^2)^2 - 1]\,dX^2.$$

We obtain $\frac{2}{3}[5^{3/2} - 1] + [\frac{5}{3}^{3/2} - 5^{1/2} + \frac{2}{3}] = 4\sqrt{5}.$

(b) $X^2 = \dfrac{\sqrt{5} - 1}{2} X^1 + 1, \qquad dX^2 = \dfrac{\sqrt{5} - 1}{2}\,dX^1,$

$$\int_0^2 \left[(\sqrt{5} - 1)(X^1)^2 + 2X^1) + \frac{\sqrt{5} - 1}{2}(X^1)^2\right] dX^1$$

$$= \int_0^2 [\tfrac{3}{2}(\sqrt{5} - 1)(X^1)^2 + 2X^1]\,dX^1 = 4\sqrt{5}.$$

(c) On the portion of the path corresponding to $0 \leq t \leq 2$

$$dX^1 = dt, \qquad dX^2 = 0, \qquad X^2 = 0.$$

On the portion of the path corresponding to $2 \leq t \leq \sqrt{5} + 2$

$$dX^1 = 0, \qquad dX^2 = dt.$$

We have

$$\int_2^{\sqrt{5}+2} 4\,dt = 4\sqrt{5}.$$

(d) $\int \mathbf{G} \cdot d\mathbf{r}$ is independent of path since $\mathbf{G} = \nabla\Phi$,

where $\Phi = (X^1)^2X^2.$

5. (a) $W = \int_C \mathbf{F} \cdot d\mathbf{r} = \int_0^\pi \mathbf{r} \cdot \dfrac{d\mathbf{r}}{d\theta}\,d\theta = \int_0^\pi a[-(h + a\cos\theta)\sin\theta$

$+ (k + a\sin\theta)\cos\theta]\,d\theta,$

$= -2ah.$

(b) 0.

(c) $-2ah$.

7. We only need to show $\oint_C \mathbf{G} \cdot d\mathbf{r} \neq 0.$

Consider C: $\begin{array}{l} X^1 = \cos\theta, \\ X^2 = \sin\theta, \end{array} \quad 0 \leq \theta < 2\pi,$

then

$$\int_0^{2\pi} \left(-X^2\frac{dX^1}{d\theta} + X^1\frac{dX^2}{d\theta}\right) d\theta = 2\pi.$$

Section 2

1.

$$\frac{\partial \mathbf{r}}{\partial v^1} \times \frac{\partial \mathbf{r}}{\partial v^2} = \begin{vmatrix} \iota_1 & \iota_2 & \iota_3 \\ \cos v^1 \cos v^2 & \cos v^1 \sin v^2 & -\sin v^1 \\ -\sin v^1 \sin v^2 & \sin v^1 \cos v^2 & 0 \end{vmatrix}$$

$$= \sin v^1 \mathbf{r}.$$

3. (a) $\nabla f = \dfrac{\partial g}{\partial X^1} \iota_1 + \dfrac{\partial g}{\partial X^2} \iota_2 + \iota_3.$

$$\frac{\partial \mathbf{r}}{\partial v^1} \times \frac{\partial \mathbf{r}}{\partial v^2} = \begin{vmatrix} \iota_1 & \iota_2 & \iota_3 \\ 1 & 0 & \dfrac{\partial X^3}{\partial v^1} \\ 0 & 1 & \dfrac{\partial X^3}{\partial \ \ ^2} \end{vmatrix} = -\frac{\partial X^3}{\partial v^1} \iota_1 - \frac{\partial X^3}{\partial v^2} \iota_2 + \iota_3.$$

Since $X^3 = -g(X^1, X^2)$, we see that $\nabla f = \partial \mathbf{r}/\partial v^1 \times \partial \mathbf{r}/\partial v^2$.

(b) In Theorem 4-2.4 we found that

$$\left| \frac{\partial \mathbf{r}}{\partial v^1} \times \frac{\partial \mathbf{r}}{\partial v^2} \right| = \sqrt{g}.$$

$\dfrac{\partial \mathbf{r}/\partial v^1 \times \partial \mathbf{r}/\partial v^2}{|\partial \mathbf{r}/\partial v^1 \times \partial \mathbf{r}/\partial v^2|}$ and $\dfrac{\nabla f}{|\nabla f|}$ are unit normals to the surface. Therefore the equality expressed in Problem 3b holds except possibly for sign.

5. (a) $\displaystyle\int_s \mathbf{r} \cdot \mathbf{N}\, dA = \int_0^1 \int_0^{1-v^2} dv^1\, dv^2 = \int_0^1 (1 - v^2)\, dv^2 = -\tfrac{1}{2}.$

(b) Let $X^1 = -v^1, \qquad X^2 = 1 + v^1, \qquad X^3 = v^2.$

$$\int_s \mathbf{W} \cdot \mathbf{N}\, dA = \int_0^2 \int_{-1}^0 (2 + v^1)\, dv^1\, dv^2 = \tfrac{3}{2}(2) = 3.$$

(c) $\displaystyle\int_s \mathbf{W} \cdot \mathbf{N}\, dA = \int_0^3 \int_0^{2\pi} (\sin^2 v^1 \cos v^1 + \sin v^1 \cos v^1)\, dv^1\, dv^2$

$= 0.$

7. (a) $\displaystyle\int_0^2 \int_0^{2\pi} \mathbf{W} \left| \frac{\partial \mathbf{r}}{\partial v^1} \times \frac{\partial \mathbf{r}}{\partial v^2} \right| dv^1\, dv^2 = \int_0^2 \int_0^{\pi} (\cos v^1\, \iota_1 - \sin v^1\, \iota_2)\, dv^1\, dv^2$

$= -4\iota_2.$

(b) $\displaystyle\int_0^{2\pi} \int_0^{\pi} (\sin v^1 \cos v^2\, \iota_1 + \sin v^1 \sin v^2 \iota_2 + \cos v^2 \iota_3) \sin v^1\, dv^1\, dv^2$

$$= \int_0^{2\pi} \left(\frac{\pi}{2} \cos v^2 \iota_1 + \frac{\pi}{2} \sin v^2 \iota_2 + 2 \cos v^2 \iota_3 \right) dv^2.$$

$= 0.$

SECTION 2*

$$1.\ \sqrt{g}\,\mathcal{E}_{\alpha\beta}\,\underset{1}{\delta v^{\alpha}}\,\underset{2}{\delta v^{\beta}} = \left|\frac{\partial \bar{v}}{\partial v}\right| \sqrt{\bar{g}} \left|\frac{\partial v}{\partial \bar{v}}\right| \frac{\partial \bar{v}^{\lambda}}{\partial v^{\alpha}}\frac{\partial \bar{v}^{\gamma}}{\partial v^{\beta}}\,\bar{\mathcal{E}}_{\lambda\alpha}\,\frac{\partial v^{\alpha}}{\partial \bar{v}^{\mu}}\frac{\partial v^{\beta}}{\partial \bar{v}^{\gamma}}\,\underset{1}{\delta \bar{v}^{\mu}}\,\underset{2}{\delta \bar{v}^{\nu}}$$

$$= \sqrt{\bar{g}}\,\bar{\delta}_{\mu}{}^{\lambda}\,\bar{\delta}_{\nu}{}^{\gamma}\,\bar{\mathcal{E}}_{\lambda\gamma}\,\underset{1}{\delta \bar{v}^{\mu}}\,\underset{2}{\delta \bar{v}^{\nu}} = \sqrt{\bar{g}}\,\bar{\mathcal{E}}_{\mu\nu}\,\underset{1}{\delta \bar{v}^{\mu}}\,\underset{2}{\delta \bar{v}^{\nu}}.$$

SECTION 3

1. (a) According to the definition of the components of $\underset{1}{\delta X^j}$, $\underset{2}{\delta X^j}$, $\underset{3}{\delta X^j}$, all terms of $\mathcal{E}_{jkp}\,\underset{1}{\delta X^j}\,\underset{2}{\delta X^k}\,\underset{3}{\delta X^p}$ are zero except corresponding to $(j, k, p) = (1, 2, 3)$; that term is $dX^1\,dX^2\,dX^3$.
(b) Following the procedure of Problem 1 in Chapter 4, Section 2*, we see that $\sqrt{g}\mathcal{E}_{jkp}\,\underset{1}{\delta X^j}\,\underset{2}{\delta X^k}\,\underset{3}{\delta X^p}$ is invariant. The result follows from this fact.

$$3.\ (g_{\gamma\beta}) = \begin{pmatrix} \cos\beta\cos\theta & \cos\beta\sin\theta & -\sin\beta \\ -[s\sin\beta + r\cos\beta]\sin\theta & s(\sin\beta + r\cos\beta)\cos\theta & 0 \\ \sin\beta\cos\theta & \sin\beta\sin\theta & \cos\beta \end{pmatrix}.$$

$|g_{\alpha\beta}| = [s\sin\beta + r\cos\beta]$ therefore the element of volume is

$$(s\sin\beta + r\cos\beta)^{1/2}\,dr\,d\theta\,ds.$$

SECTION 4

$$1.\ \text{(a)}\ \oint_C \mathbf{P}\cdot d\mathbf{r} = \int_{-\sqrt{2}}^{\sqrt{2}}\int_{(X^1)^2+4}^{-(X^1)^2+8} -dX^1\,dX^2 = \frac{2^{9/2}}{3}$$

$$\text{(b)}\ \oint_C \mathbf{P}\cdot d\mathbf{r} = \int_0^1\int_0^1 -\,X^2\,dX^2\,dX^1 = -\tfrac{1}{2}$$

$$\text{(c)}\ \oint_C \mathbf{P}\cdot d\mathbf{r} = \int_{-3}^{3}\int_{-2/3[9-(X^1)^2]^{1/2}}^{2/3[9-(X^1)^2]^{1/2}} 2X^1X^2\,dX^2\,dX^1 = 0.$$

3. By straightforward computation we find that

$$\mathbf{N}\cdot\nabla\times\mathbf{P} = -\nabla\cdot\mathbf{U},$$

where

$$\mathbf{U} = U^1\iota_1 + U^2\iota_2;$$

also

$$\mathbf{P}\cdot d\mathbf{r} = U^2\,dX^1 - U^1\,dX^2 = -\mathbf{U}\times d\mathbf{r}.$$

5. If C is in the plane $X^3 = 0$, then

$$\int_S \mathbf{N}\cdot\nabla\times\mathbf{G}\,dA = \oint_C \mathbf{G}\cdot d\mathbf{r} = \oint_C -X^2\,dX^1 + X^1\,dX^2.$$

According to the result in Example 4-4.1, this last integral represents the area of the plane region bound by C.

7. The surface is not closed; hence the theorem is not applicable for any **G**.

9. Substituting into the Green's theorem conclusion, we obtain

(a) $\oint_C X^2\,dX^1 = \int_S -dX^1\,dX^2 = -A.$

(b) $\oint_C X^1\,dX^2 = \int_S dX^1\,dX^2 = A.$

(c) Subtract (a) from (b).

(d) We use the parameterization of Example 4-4.1:

$$-A = \oint_C X^2\,dX^1 = \int_0^{2\pi} b\sin t\,(-a\sin t)\,dt = -ab\pi.$$

Chapter 5

Section 1

1. $A_{(jk)} = \frac{1}{2}(A_{jk} + A_{kj}) = \frac{1}{2}(A_{jk} - A_{jk}) = 0.$

3. $B^jA_j = \mathcal{E}_{jkl}\,B^jB^kC^l = 0.$

Note that $\mathcal{E}_{jkl}$ is skew symmetric in j, k and that B^jB^k is symmetric in these indices.

We show that $C^jA_j = 0$ by the same procedure.

5. Write $\mathcal{E}_{j_1\cdots j_n}b = \mathcal{E}_{i_1\cdots i_n}b_{j_1}{}^{i_1}\cdots b_{j_n}{}^{i_n}$

$E^{j_1\cdots j_n}a = E^{k_1\cdots k_n}a_{k_1}{}^{j_1}\cdots a_{k_n}{}^{j_n}.$

By multiplying and summing these two expressions, we obtain

$$n!\,ab = E^{k_1\cdots k_n}\mathcal{E}_{i_1\cdots i_n}(a_{k_1}{}^{j_1}b_{j_1}{}^{i_1})\cdots(a_{k_n}{}^{j_n}b_{j_n}{}^{i_n})$$

$$n!\,ab = E^{k_1\cdots k_n}\mathcal{E}_{i_1\cdots i_n}c_{k_1}{}^{i_1}\cdots c_{k_n}{}^{i_n}$$

$$= E^{k_n\cdots k_n}\mathcal{E}_{k_1\cdots k_n}c = n!\,c.$$

Therefore

$$ab = c,$$

as was to be shown.

7. (a) If $A_k{}^j = p\,\delta^{[ji_2\cdots i_p]}_{kq_2\cdots q_p}a_{i_2}{}^{q_2}\cdots a_{i_p}{}^{q_p};$

then

$$a_r{}^kA_k{}^j = p\,\delta^{[ji_2\cdots i_p]}_{kq_2\cdots q_p}a_r{}^ka_{i_2}{}^{q_2}\cdots a_{i_p}{}^{q_p}$$

$$= \frac{p}{p!}E^{ji_2\cdots i_p}\mathcal{E}_{kq_2\cdots q_p}a_r{}^ka_{i_2}{}^{q_2}\cdots a_{i_p}{}^{q_p}$$

$$= \frac{p}{p!}E^{ji_2\cdots i_p}\mathcal{E}_{ri_2\cdots i_p}a$$

$$= \frac{p}{p!}(p-1)!\,\delta_r{}^ja = \delta_r{}^ja.$$

(b) Follow the procedure in Part (a)

(c)
$$|a_r{}^kA_k{}^j| = |\delta_r{}^ja|,$$
$$|a_r{}^k|\,|A_k{}^j| = |\delta_r{}^j|\,a^n,$$
$$|A_k{}^j| = |\delta_r{}^j| = a^{n-1}.$$

9. See Theorem 3-5.3.

11. We can write

$$\alpha A_{jk} + \beta A_{kj} = 0,$$
$$\alpha A_{kj} + \beta A_{jk} = 0.$$

When the second of these is subtracted from the first, we obtain

$$\alpha(A_{jk} - A_{kj}) + \beta(A_{kj} - A_{jk}) = 0,$$

that is,

$$(\alpha - \beta)(A_{jk} - A_{kj}) = 0.$$

The result follows from this expression.

13. If we take the curl of relation (a) of Problem 12,

$$\nabla \times (\nabla \times \mathbf{E}) = -\frac{\mu}{c}\frac{\partial}{\partial t}(\nabla \times \mathbf{H}).$$

If we use the vector identity of Chapter 3, Section 3, on the left of this expression and relation (c) of Problem 12 on the right,

$$\nabla(\nabla \cdot \mathbf{E}) - \nabla^2 E = -\frac{\mu}{c}\frac{\partial}{\partial t}\left(\frac{e}{c}\frac{\partial \mathbf{E}}{\partial t} + \frac{4\pi\sigma\mathbf{E}}{c}\right),$$

$$\nabla(\nabla \cdot E) - \nabla^2 E = -\frac{\mu e}{c^2}\frac{\partial^2 \mathbf{E}}{\partial t^2} - \frac{4\pi\mu\sigma}{c^2}\frac{\partial \mathbf{E}}{\partial t},$$

$$\partial^p\,\partial_q E_p - \partial^p\,\partial_p E_q + \frac{\mu e}{c^2}\frac{\partial^2 E_q}{\partial t^2} + \frac{4\pi\mu\sigma}{c^2}\frac{\partial E_q}{\partial t} = 0.$$

This is the desired form. The second relation follows in a corresponding way from Problem 12c.

SECTION 2

1. (a) According to (5-2.4b),

$$A_j = \frac{\partial \bar{X}^k}{\partial X^j}\bar{A}_k$$

If we multiply and sum this relation with $\partial X^j/\partial \bar{X}^p$,

$$\frac{\partial X^j}{\partial \bar{X}^p}A_j = \frac{\partial X^j}{\partial \bar{X}^p}\frac{\partial \bar{X}^k}{\partial X^j}\bar{A}_k = \delta_p{}^k\bar{A}_k = \bar{A}_p,$$

as was to be shown.

(b) This relation follows from (5-2.4a) in a manner similar to that of part (a).

3.

$$\bar{T}^{j_1\cdots j_r j_{r+1}\cdots j_p} = \frac{\partial \bar{X}^{j_1}}{\partial X^{k_1}}\cdots\frac{\partial \bar{X}^{j_r}}{\partial X^{k_r}}\frac{\partial \bar{X}^{j_{r+1}}}{\partial X^{k_{r+1}}}\cdots\frac{\partial \bar{X}^{j_p}}{\partial X^{k_p}}T^{k_1\cdots k_r k_{r+1}\cdots k_p}$$

$$\bar{T}^{j_1\cdots j_{r+1} j_r\cdots j_p} = \frac{\partial \bar{X}^{j_1}}{\partial X^{k_1}}\cdots\frac{\partial \bar{X}^{j_{r+1}}}{\partial X^{k_{r+1}}}\frac{\partial \bar{X}^{j_{r+1}}}{\partial X^{k_{r+1}}}\cdots\frac{\partial \bar{X}^{j_p}}{\partial X^{k_p}}T^{k_1\cdots k_{r+1} k_r\cdots k_p}$$

If we add these expressions,

$$\bar{T}^{j_1\cdots j_r j_{r+1}\cdots j_p} + \bar{T}^{j_1\cdots j_{r+1} j_r\cdots j_p} = \frac{\partial \bar{X}^{j_1}}{\partial X^{k_1}} \cdots \frac{\partial \bar{X}^{j_r}}{\partial X^{k_r}} \frac{\partial \bar{X}^{j_{r+1}}}{\partial X^{k_{r+1}}} \cdots \frac{\partial \bar{X}^{j_p}}{\partial X^{k_p}}$$
$$\times (T^{k_1\cdots k_r k_{r+1}\cdots k_p} + T^{k_1\cdots k_{r+1} k_r\cdots k_p}).$$

Suppose the system $T^{k_1\cdots k_r k_{r+1}\cdots k_p}$ is completely skew symmetric; the right-hand expression is then equal to zero. Therefore the left-hand expression is equal to zero, hence skew symmetric in $j_r j_{r+1}$. The statement is valid for $r = 1 \cdots p - 1$, and the components $\bar{T}^{j_1\cdots j_p}$ are completely skew symmetric.

5. Yes, contravariant valence 1 and covariant valence 3.

7.
$$\frac{\partial A^k}{\partial X^j} = \frac{\partial \bar{X}^p}{\partial X^j} \frac{\partial[\partial X^k/\partial \bar{X}^q)\bar{A}^q]}{\partial \bar{X}^p} = \frac{\partial \bar{X}^p}{\partial X^j}\left(\frac{\partial X^k}{\partial \bar{X}^q}\frac{\partial \bar{A}^q}{\partial \bar{X}^p} + \frac{\partial^2 X^k}{\partial \bar{X}^p\,\partial \bar{X}^q}\bar{A}^q\right)$$

This law of transformation is not tensor in character; it becomes so if and only if $\partial^2 X^k/\partial \bar{X}^p\,\partial X^q = 0$, that is, when the transformations are linear with constant coefficients.

Section 3

1. $X^1 = \sqrt{-1}\bar{X}^1$, $X^2 = \bar{X}^2$ is the required transformation. The allowable transformations of our considerations have real coefficients.

3. In general, the angle made by two tangent vectors dX^j/du, $\delta X^j/dv$ is

$$\cos\theta = \frac{g_{jk}(dX^j/du)(\delta X^k/dv)}{\left[\left|g_{pq}(dX^p/du)\dfrac{dX^q}{du}\right|\right]^{1/2}[|g_{rs}(\delta X^r/dv)(\delta X^s/dv)|]^{1/2}}$$

If we consider the coordinate curves $X^1 = \text{constant}, \cdots X^j = u, \cdots$, $X^n = \text{constant}$ and $X^1 = \text{constant}, \cdots, X^k = v, \cdots, X^n = \text{constant}$. Then only the derivatives of these coordinates are nonzero,

$$\cos\theta_{jk} = \frac{g_{jk}}{\sqrt{|g_{jj}|}\sqrt{|g_{kk}|}}.$$

Section 4

1. (a) The nonzero Christoffel symbols in cylindrical coordinates which are related to rectangular Cartesian coordinates by

$$X^1 = \rho\cos\theta, \qquad X^2 = \rho\sin\theta, \qquad X^3 = z$$

are

$$\Gamma^1_{22} = -\rho, \qquad \Gamma^2_{12} = \Gamma^2_{21} = \frac{1}{\rho}.$$

The nonzero Christoffel symbols in spherical coordinates related to rectangular Cartesian components by

$$X^1 = r\sin\theta\cos\phi, \qquad X^2 = r\sin\theta\sin\phi, \qquad X^3 = r\cos\theta$$

are

$$\Gamma^1_{22} = -r, \qquad \Gamma^1_{33} = -r\sin^2\theta,$$

$$\Gamma^2_{21} = \Gamma^2_{12} = \frac{1}{r}, \qquad \Gamma^2_{33} = -\sin\theta\cos\theta,$$

$$\Gamma^3_{31} = \Gamma^3_{13} = \frac{1}{r}, \qquad \Gamma^3_{32} = \Gamma^3_{23} = \text{ctn}\,\theta.$$

(b) $\Gamma^i_{jk} = \frac{1}{2}g^{iq}(\partial_j g_{kq} + \partial_k g_{qj} - \partial_q g_{jk})$.
The symmetry of the Christoffel symbols follows immediately from the symmetry of the g_{jk}.

3.

$$\Gamma^i_{jk} = \frac{\partial X^i}{\partial \bar{X}^p}\left(\frac{\partial \bar{X}^q}{\partial X^j}\frac{\partial \bar{X}^r}{\partial X^k}\bar{\Gamma}^p_{qr} + \frac{\partial^2 \bar{X}^p}{\partial X^j\,\partial X^k}\right).$$

If we multiply and sum with $(\partial \bar{X}^s/\partial X^i)(\partial X^j/\partial \bar{X}^u)(\partial X^k/\partial \bar{X}^v)$,

$$\frac{\partial \bar{X}^s}{\partial X^i}\frac{\partial X^j}{\partial \bar{X}^u}\frac{\partial X^k}{\partial \bar{X}^v}\Gamma^i_{jk} = \bar{\delta}_p{}^s\left(\bar{\delta}_u{}^q\,\bar{\delta}_v{}^r\,\bar{\Gamma}^p_{qr} + \frac{\partial X^j}{\partial \bar{X}^u}\frac{\partial X^k}{\partial \bar{X}^v}\frac{\partial^2 \bar{X}^p}{\partial X^j\,\partial X^k}\right).$$

The proof can be completed by using the result of Problem 5 of this this section.

5. If

$$\frac{\partial X^j}{\partial \bar{X}^q}\frac{\partial \bar{X}^q}{\partial X^p} = \delta_p{}^j$$

is differentiated partially with respect to $\bar{X}^k$, then

$$\frac{\partial^2 X^j}{\partial \bar{X}^k\,\partial \bar{X}^q}\frac{\partial \bar{X}^q}{\partial X^p} + \frac{\partial X^j}{\partial \bar{X}^q}\frac{\partial^2 \bar{X}^q}{\partial X^r\,\partial X^p}\frac{\partial X^r}{\partial \bar{X}^k} = 0,$$

as was to be shown.

7. Write

$$\bar{g}_{sc} = \frac{\partial X^p}{\partial \bar{X}^s}\frac{\partial X^q}{\partial \bar{X}^c}g_{pq}$$

and follow the procedure of Theorem 5-4.1.

9.

$$\frac{D\Phi}{dt} = \frac{dX^j}{dt}\nabla_j\Phi = \frac{dX^j}{dt}\left(\frac{\partial\Phi}{\partial X^j} + W\Gamma^k_{kj}\Phi\right).$$

If the scalar is of weight zero, $W = 0$ and

$$\frac{D\Phi}{dt} = \frac{dX^j}{dt}\frac{\partial\Phi}{\partial X^j} = \frac{d\Phi}{dt}.$$

11. The nonzero Christoffel symbols associated with the spherical surface of Chapter 5, Section 3, Problem 2, are

$$\Gamma^1_{22} = -\sin\theta\cos\theta, \qquad \Gamma^2_{12} = \Gamma^2_{21} = \text{ctn}\,\theta.$$

13. $\nabla_{[j}W_{k]} = \partial_{[j_k}W_{k]} - \Gamma^q_{[kj]}W_q = \partial_{[j}W_{k]}$.
The last equality results from the fact that the Christoffel symbols are symmetric in the lower indices.

SECTION 5

1. In a Euclidean space there can always be found a coordinate system such that $g_{jk} = \delta_{jk}$, and therefore $\Gamma^k_{ij} = 0$. The differential equations of geodesies $[D(dX^j/ds)]/ds = 0$ then reduce to

$$\frac{d^2X^j}{ds^2} = 0.$$

These equations have the solutions

$$X^j = c_1{}^j s + c_2{}^j,$$

that is, equations of straight lines.

SECTION 6

1.
$$A^1 = \cos [(\cos \alpha)2\pi]$$

at $\phi = 2\pi$

$$A^2 = -\frac{1}{\sin \alpha} \sin [(\cos \alpha)^2\pi].$$

Therefore

$$(A^1, A^2)_{\phi=2\pi} = (A^1, A^2)_{\phi=0} \quad \text{if and only if} \quad \alpha = \frac{\pi}{2}.$$

SECTION 7

1. If

$$\Gamma_{jk}{}^i = \frac{\partial X^i}{\partial \bar{X}^p} \frac{\partial^2 \bar{X}^p}{\partial X^j \, \partial X^k},$$

$$\partial_r \Gamma^i_{jk} = \frac{\partial^2 X^i}{\partial \bar{X}^s \, \partial \bar{X}^p} \frac{\partial \bar{X}^s}{\partial X^r} \frac{\partial^2 \bar{X}^p}{\partial X^j \, \partial X^k} + \frac{\partial X^i}{\partial \bar{X}^p} \frac{\partial^3 \bar{X}^p}{\partial X^r \, \partial X^j \, \partial X^k}$$

$$\text{(a)} \quad \partial_{[r} \Gamma^i_{j]k} = \frac{\partial^2 X^i}{\partial \bar{X}^s \, \partial \bar{X}^p} \frac{\partial \bar{X}^s}{\partial X^{[r}} \frac{\partial^2 \bar{X}^p}{\partial X^{j]} \, \partial X^k}$$

also

$$\text{(b)} \quad \Gamma^m_{k[r} \Gamma^i_{j]m} = \frac{\partial X^m}{\partial \bar{X}^\mu} \frac{\partial^2 \bar{X}^\mu}{\partial X^k \, \partial X^{[r}} \frac{\partial^2 \bar{X}^q}{\partial X^{j]} \, \partial X^m} \frac{\partial X^i}{\partial \bar{X}^q}.$$

Differentiation of $(\partial \bar{X}^s/\partial X^j)(\partial X^i/\partial \bar{X}^s) = \delta_j{}^i$ produces the result

$$\text{(c)} \quad \frac{\partial^2 \bar{X}^s}{\partial X^m \, \partial X^j} \frac{\partial X^i}{\partial \bar{X}^s} = - \frac{\partial \bar{X}^s}{\partial X^j} \frac{\partial^2 X^i}{\partial \bar{X}^p \, \partial \bar{X}^s} \frac{\partial \bar{X}^p}{\partial X^m}.$$

When (c) is applied to (b), we obtain (a); therefore

$$R_{krj}{}^i = 2[-\partial_{[k} \Gamma^i_{r]j} + \Gamma^m_{j[k} \Gamma^i_{r]m}] = 0,$$

as was to be shown.

SECTION 8

1. Define:

$$\underset{(r)}{\overset{(s,t)}{A}} = T_{ir}{}^{st}, \qquad \overset{(r)}{B_j} = \delta_j{}^r,$$

$$\underset{(s)}{C^k} = \delta_s{}^k, \qquad \underset{(t)}{D^l} = \delta_t{}^l,$$

where the symbols in parentheses distinguish vectors and other indices signify components. Then

$$\underset{(r)}{\overset{(s,t)}{A_i}}\ \overset{(r)}{B_j}\ \underset{(s)}{C^k}\ \underset{(t)}{D^l} = T_{ij}{}^{kl}.$$

3. Let the tensor $[T_{kl}]$ have a vector decomposition represented in terms of components by $T_{kl} = A_k B_l$; then

$$2\nabla_{[i}\nabla_{j]}A_k B_l = 2[(\nabla_{[i}\nabla_{j]}A_k)B_l + A_k\nabla_{[i}\nabla_{j}B_{l]}].$$

Relation 5-8.4b follows by employing this relation and (5-8.2b). The expression (5-8.4c) may be verified in the same way.

5. We have

$$0 = 2\nabla_{[r}\nabla_{i]}g^{jk} = -R_{rim}{}^{j}g^{mk} - R_{rim}{}^{k}g^{jm}.$$

If we multiply and sum with g_{pk},

$$R_{rip}{}^{j} = -R_{rim}{}^{k}g_{pk}g^{jm}.$$

If we contract in this expression on p and j,

$$R_{rip}{}^{p} = -R_{rim}{}^{k}\delta_k{}^m = -R_{rik}{}^{k};$$

therefore

$$R_{rip}{}^{p} = 0.$$

7. The result follows immediately from the form of R expressed in Problem 6 and the fact that g is a scalar density of weight -2.

SECTION 9

1. (a) $$\nabla_\lambda\left(\mu\frac{dX^\lambda}{ds}\right) = \frac{\partial[\mu(dX^\lambda/ds)]}{\partial x^\lambda} + \Gamma_{\alpha\lambda}{}^{\lambda}\mu\frac{dX^\alpha}{ds} - \mu\Gamma_{\alpha\lambda}{}^{\gamma}\frac{dX^\lambda}{ds}$$

$$= \frac{\partial[\mu(dX^\lambda/ds)]}{\partial X^\lambda}.$$

(b) $$\nabla_\lambda m^{\lambda\nu} = \nabla_\lambda\left(\mu\frac{dX^\lambda}{ds}\frac{dX^\nu}{ds}\right) = \nabla_\lambda\left(\mu\frac{dX^\lambda}{ds}\right)\frac{dX^\nu}{ds} + \mu\frac{dX^\lambda}{ds}\nabla_\lambda\frac{dX^\nu}{ds}.$$

Since $\sqrt{|g|}$ is a covariant constant and the first term on the right of the foregoing expression corresponds to the relativistic continuity equation, we obtain

$$\sqrt{|g|}\,\nabla_\lambda m^{\lambda\nu} = \sqrt{|g|}\,\mu_0\frac{D(dX^\nu/ds)}{ds}.$$

index

Absolute calculus, 6
Absolute derivative of tensor field, 340
Acceleration, apparent, 160, 161
  arrow form, 157, 160
  centripetal, 161, 174
  components, 160, 163, 169
  Coriolis, 161
  field, 147–149, 152
  linear, 15
  radial, 153
  translational, 161
  vector concept, 127, 155, 157, 167, 170
Adams, 171
Affine connection, 357
Affine transformation group, 66, 67, 113–114, 180
Angle, 42, 60, 65, 68, 82, 328, 329
Angular momentum, 176
Angular velocity, 157, 158, 160, 167
Aphelion, 174, 379
Apolonius, 2
Arc length, 141, 190
Area, 153, 172, 175, 293, 350
Argand diagram, 2
Aristotle, 1
Arrow, 12–20, 31, 52, 76, 77
  basis, 27, 42
  collinear, 24
  components, 161
  coplanar, 24
  dependent, 24, 25
  magnitude, 15
Arrow, parallel, 14
  position, 36
  unit, 25–28, 97
Associated metric tensor, 81, 82, 233, 237, 286, 326
Associativity, cross product, 97, 132
  dot product, 57
  group, 66–67
  n-tuple, 11, 12
Axial vector, 103

Basis, 70
  arrow, 27, 42
  constant, 131, 147, 254
  contravariant, 74, 77, 80, 229, 285
  covariant, 73, 74, 77, 227, 228
  generalized coordinate systems, 225, 234
  linear independence, 32, 33
  nonconstant, 131, 254, 282
  n-tuple, 32, 130, 131
  orthogonal, 28
  physical, 76
  polar coordinate system, 151, 152
  reciprocal, 74, 80, 114
  unit triad, 27, 97, 131, 144, 147, 148
Beltrami, 6, 272, 331
Bernouli, Johann, 342
Bezout, 87
Bianchi's identity, 368, 369, 371
Binary operation, 3, 4, 56, 96
Binormal, 144
Bolza, 342

Bound vector, 283
Brahe, 171

Cartesian coordinates, 8, 9, 10, 19, 35, 38–40, 51, 68–72, 78, 92, 112, 155, 177–179, 227, 325
Cartesian metric tensor, 78, 83, 93
Cauchy, 123
Cayley, 65, 87, 115
Characteristic equation, 94
Christoffel, 6, 331, 333
Christoffel symbols, 333, 345, 348, 369, 375
  contracted, 339
  symmetry, 341, 357
  transformation, 334
Circle, vector equation, 143
Circular functions, 185
Coefficients of transformation, 42, 45, 46, 78, 98, 119, 133
Cofactor, 31, 43, 93, 98, 118
Collinear arrows, 24
Column, index, 37, 88, 89, 115
  interchange, 89
  operation, 89
  zero element, 89
Commutivity, cross product, 97
  divergence, 214
  dot product, 57
  matrix, 116, 117
  n-tuple, 11
Components, 10, 75–77, 156, 160
Composition transformation, 224
Composition of velocities, 180, 189
Cone, 198, 204, 359
  light, 181, 187–188
Conic section, 173
Conjugate metric tensor, *see* Associated metric tensor
Conservation laws, 193, 267, 300
Conservative force, 210, 266
Contravariant, basis, 74, 77
  ε-system, 111
  tensor, 112–113
  transformation, 75, 208
  vector, 75, 76, 81, 169
Coordinate curve, 201–202, 289
  surface, 289
Coordinate system, 2, 7, 43
  curvilinear, 207
Coordinate system, cylindrical, 220, 221, 233, 248
  general Cartesian, 68–72, 78, 112
  inversion of triple scalar product, 106
  left-handed, 9, 100
  polar, 150–152
  rectangular Cartesian, 8, 9, 10, 19, 35, 38–40, 51, 92, 131, 155, 177–179
  right-handed, 9, 10, 100, 144
  spherical, 220, 222, 230
  transformation, 2, 40, 66, 78, 133, 155, 177–178
Copernicus, 171
Coplanar arrows, 24
Coriolis, G. G., 261
Coriolis acceleration, 161
Coulomb, 177
Covariant constant, 338, 340
Covariant derivative, 334–339, 354, 362
  operator, 362
Covariant differentiation, 331
Covariant vector, 75, 81, 112–113
  basis, 73, 74, 77
  transformation, 75, 208
Cramer, 87
Cramer's rule, 43
Critical value, 246, 248
Cross product, 4, 96–114
  algebraic properties, 97–99, 117, 132
  axial vector, 103
  curl, 216, 218
  determinant form, 96
  differentiation, 131, 132
  geometric interpretation of the magnitude as the area of a parallelogram, 100, 101
  linearly independent vectors, 100
  magnitude, 100
  moment of force, 102
  orthogonality, 100
  parallel vectors, 97
  proportional vectors, 100
  reciprocal basis representation, 114
  transformation, 98, 113–114
  of unit arrows, 97
Curl, 216, 238, 243
  transformation, 217
Curvature, 141, 142, 144
  Gaussian, 6, 359, 367

Curvature, invariant, 366
  tensor, 355, 361–371
Curve, arc length, 141, 190
  convex, 291
  latitudinal, 202
  length, 139–140
  longitudinal, 202
  smooth, 125
  surface, 201, 202, 206, 213
  tangent to, 134–137, 141
  vector representation, 125
Curvilinear coordinates, 207, 238, 289
Cylinder, 205, 346, 348, 359

Dependence, linear, 23–33
Derivative, directional, 211, 213
  of scalar field, 128, 206
  vector field, 129–132, 206
  vector sums, 131
  vector triple product, 132
Descartes, 2
Determinant, 42, 85–95, 88
  algebraic properties, 89
  cofactor of, 31, 43, 93, 98, 118, 240,
    313
  expansion of, 43, 89
  of metric tensor, 239, 277, 286, 339
  multiplication of, 91, 240, 312
  order of, 88, 89
  partial derivative of, 313
  product of, 91
  of transformation coefficients, 42, 43,
    49
  triple scalar product as, 103
  value of, 43, 88, 91
Differentiability, 125, 127
  vectors, 128–135, 213
Directional derivative, 211, 213
Direction cosines, 40–42
Direction numbers, 14, 73
Displacement, parallel, 356
Displacement current, 301
Distance, 13, 55, 61, 342
  invariance of, 57
Divergence, 214, 218, 238, 241–243
  differentiation, 214
  noncommutativity, 214
  theorem, 297
  transformation of, 242
Domain, 124

Dot product, 56–65, 131, 206, 214–215
  algebraic properties, 57
Dummy index, 36
Dynamics, 171

Eccentricity, 173–174
Eigenvalue, 94
Einstein, 6, 155, 171, 176, 178, 179,
    193, 304, 372
Einstein convention, 36
Einstein tensor, 370
Electrostatic force field, 177, 266
Ellipse, 127, 174
Ellipsoid, 204
Elliptic cylinder, 205
Energy, 193–194, 266, 267
Equipotential surface, 210, 221
Erlanger Programm, 35, 65, 68, 177
$\mathcal{E}$-systems, 85–95, 111–114, 239, 287,
    311, 318
  permutation, 89
  transformation, 92, 112, 239
Euclidean line, 8
Euclidean metric geometry, 66, 186
Euclidean space, 8, 19, 26, 131, 328
Euler, 66, 123
Euler-Lagrange equations, 343, 344, 345

Faraday, 177, 210, 300
Fermat, 2
Field, conservative, 210, 266, 295
  force, 171, 172, 177, 204, 215, 266
  gravitational, 172, 209, 266, 373
  irrotational, 295
  momentum, 171, 266
  scalar, 127, 128, 133, 203, 210, 215
  tangent, 133, 134, 141, 206, 213
  vector, 127, 133, 141, 147, 152, 204,
    206–207
    parallel, 348–353
    pseudoparallel, 351
Flat space, 359, 360
Föppel, 5
Four-dimensional space, 184
Frame of reference, 147, 184
  motion of, 155–164, 176–179, 187,
    188
Free vector, 283
Frenet-Serret formulas, 145

Function, 124
  two-variable, 246
Fundamental metric form, 6, 190, 273, 324, 372, 374
Fundamental metric tensor, *see* Tensor, metric

Galilean transformations, 177–180, 186
Galileo, 2, 176
Galle, 171
Galois, 68
Gauss, 2, 6, 197, 272, 304, 325
  theorem, 297
Gaussian curvature, 6, 359, 367
Gaussian plane, 2
Geodesic, 345, 346, 348, 372, 373, 377
  coordinates, 369
Geometry, 65
  affine, 66
  Euclidean metric, 66, 186
  hyperbolic, 186
  intrinsic, 284
  non-Euclidean, 325
  projective, 65
  Riemannian, 310
Gibbs, 5, 51, 58
Gradient, 207–209, 215, 218, 231, 249, 262, 263, 266
  transformation, 208
Gram's determinant (Gramian), 111
Grassman, 1, 3, 4, 5, 7, 56, 58, 96
Gravitational constant, 172, 210
Gravitational field, 172, 209, 221
Gravitation law, 172
Green, G., 291
Green's identities, 299
Green's theorem, 291, 293
Grossman, 6, 372
Group, centered affine, 66, 67, 113
  coordinate transformations as, 65, 66, 93, 177, 208, 224
  transformation coefficient matrices as, 119

Hamilton, 1, 3–5, 7, 19, 51, 53, 56, 96, 206, 216
Heaviside, 5
Helix, 126, 150, 348
Henry, 177
Hertz, 301
Hesse's normal form, 61
Hilbert, 7
Hipparchus, 171
Hlavatý, Vácalav, 99, 201, 367
Hyperbola, 173, 204
Hypersurface, 317

Implicit function theorem, 198
Independence, linear, 23–33, 73, 100
Independence of path, 262, 264, 295
Index, column, 88, 89, 115
  dummy, 36
  free, 37
  interchange, 89
  lowering, 80, 112, 231
  permutation, 86, 88
  position, 37
  raising, 80, 112, 231
  row, 88, 115
Indicators in determinant representation, 85–87
Inertial frame, 177
Inertial mass, 193
Inertial system, 19, 371
Inflection point, 246
Inner product, 57
Integral, area, 277
  double, 276
  iterated, 276, 289
  line, 255–269
  surface scalar, 270–282
  surface vector, 270–282
Integration, path of, 259
Invariants, 35, 54, 55, 57, 65, 66, 186, 190, 192, 228, 288, 315
Inverse, 11
  matrix, 118–119
  transformation, 43–45, 66, 67
Inverse square law, 177
Irrotational field, 295
Isomer, 309
Iterated integral, 276

Jacobi, 87, 223
Jacobian, 223, 224, 227, 239, 369

Kepler law, 171–174
Kernel letter, 86, 111
Kinematics, 147–153, 187
Kinetic energy, 193–194, 266, 267

Klein, 35, 54, 65, 68, 87, 177
Kreyszig, Erwin, 125
Kronecker delta, 37, 40, 43, 45, 46, 48, 81, 117, 306, 318
  generalization, 311

Lagrange, 68, 87
  identity, 111, 294
Laplacian, 216
Latitudinal curves, 202
Law of addition, parallelogram, 1, 14, 15
Law of areas, 153, 172, 175
Law of cosines, 59
Leibniz, 87, 123
Length of curve, 139–140
Leverrier, 171
Levi-Civita, 6, 304, 349, 354
Lie, 68
Light cone, 181, 187–188
Linear dependence, 23–33
Linear equation, 7, 8
Linear independence, 23–33, 100
  of basis, 32–33, 226, 229
Linear transformations derived from general transformations, 315
Line, world, 181, 187
Line integral, 255–269
Line representation, 18, 32–33
Line of simultaneity, 181, 187
Liouville, 141
Lipschitz, 6, 331, 333
Longitudinal curves, 202
Lorentz, 178
  transformations, 178, 186, 189–192, 371
Lowering of index, 80, 112, 327
Loxodrome, 205

Magnetic field, 177, 266
Magnitude, 14, 15, 58, 82, 148
  of cross product, 100
  invariance of, 57
Mass, 191, 193, 372
Matrix, 114–122, 229
  equality, 116
  group property, 119
  identity, 117, 118
  inverse, 118, 119
  Kronecker delta, 37
Matrix, multiplication, 115–118
  noncommutativity, 116, 117
  notation, 115
  rank, 28, 29, 197
  sum, 116
  transformation, 42, 45, 119
  transpose, 118, 119
  zero, 116
Maxima, 245–253
Maxwell, 4, 214, 216, 300, 301
  equations, 178, 186, 300, 314
Mercury, 377, 379, 381
Metric tensor, *see* Tensor, metric
Michelson, 179, 182
Minima, 245–253
Minkowski, 184, 186
  metric, 185
  plane, 186–187
  space, 184
Minor, 31
Möbius, 4
  strip, 270
Moment of force, 102, 176
Momentum, 171, 176
Monge, 196
Morley, 179, 182
Motion, planetary, 171–173
Moving frames of reference, 155, 176–179, 184, 187–188

Neptune, 171
Newton, 123, 155, 171, 176
  laws, 18, 19, 148, 172, 191–192, 317, 321
Newtonian orbit, 173, 377, 378
Non-Euclidean geometry, 285
Non-Euclidean space, 19, 20
Nonhomogeneous equations, 32
Normal, principal, 141, 142
  to a surface, 208, 218, 275
Normalized vector, 61
n-space, 6, 7
n-tuple, 10, 11, 20–26, 29–33

Oersted, 177, 300
  law, 303
Olmsted, 271
Operator, 206, 213–216
Orbit, 174
Order of contact, 137

Order of determinant, 88, 89
Order of tensor, 82, 83, 112
Orthogonality conditions, 60, 66, 119–121, 158, 184–185, 328, 329
basis, 25–28, 41, 148
gradient, 208–209
vectors, 40, 45, 60
Orthogonal projection, 78
Orthogonal unit basis triad, 148
Orthogonal vectors, 100
Osculating plane, 136, 137, 142, 148

Parabola, 134–135, 174
Parallel arrows, 14, 15, 29–30
Parallel displacement, 356
Parallelepiped, triple scalar product as representation of volume, 104
Parallelism, 6, 349
Parallelogram, area, 101
law of addition, 1, 14, 15, 76, 356
Parallel projection, 76, 78
Parallel vector, 97
Parallel vector field, 348–353
Parameter transformation, 198
Parametric equations, 18, 125–129, 134, 142, 151, 161, 197, 199–202
Perehelion, 174, 377, 379
Period of revolution, 175
Permutation, 86
Physical basis, 76
Physical components, 76, 235, 236, 237, 243
Plane, 7, 211, 213
Hesse's normal form, 61
Minkowski, 184, 186
of simultaneity, 181
vector equation, 60, 199
Planetary motion, 171–173, 377
Pluto, 171
Poincaré, 179
Point, 7
Poisson's equation, 216, 373
Polar coordinate system, 150–152, 174
Poncelet, 261
Position vector, 36, 149, 170
Positive definite property, 57, 325
Potential, 54, 210, 266, 267
Projection, 76–78, 156, 161, 213
Pseudoparallel vector field, 351
Ptolemy, 171

Quadric cone, 204
Quaternions, 3, 4, 51, 96, 206, 216
Quotient law, 109

Raising index, 80, 112, 327
Range, 17, 124
Rank of matrix, 28, 29, 197
Reciprocal basis, 74, 80, 114
Rectangular Cartesian coordinates, 8, 10, 19, 35, 51, 177–179
Regular point on a surface, 198
Relativity of motion, 155, 177, 317
Relativity theory, 6, 7, 156, 171, 176, 179–194, 304, 325, 368, 371–381
Rest mass, 193
Ricci, 6, 304, 331
lemma, 338
tensor, 365, 366, 367, 376
Riemann, 6, 7, 272, 304, 325
Riemann Christoffel mixed tensor, 354–361, 361–371, 373
Riemannian curvature tensor, 355, 361–374
Rotation, 3, 36, 40, 47, 51, 119–121, 170, 216
orthogonal Cartesian, 75, 119–121
successive plane, 119–122
Rotational derivative, 169–170
Rotational effect, 102
Rotor, 216
Rule of the middle factor, 107

Saddle point, 248
Scalar, 53, 54, 58, 316
density, 104
field, 127, 128, 133, 203, 206, 210, 215
product, 45, 56–65, 131, 206, 214–215
triple product, 103–106, 114
Schwarzschild, K., 374, 376
Sectorial speed, 175
Sense, 14
Serret, 145
Simultaneity, 181, 187
Singular point, 197–198
Skew symmetry, 86, 90, 307, 308, 362
Smooth curve, 125
Solenoidal vector field, 300

Space, 7, 305
  Euclidean, 8, 19, 254
  four-dimensional, 184
  Minkowski, 186–187
  n-, 6, 7
  non-Euclidean, 19, 20
  n-tuple, 10
  Riemannian, 6, 324, 328, 349, 366, 372
  vector, 10
Space curve, 206, 316, 339
Spectral shift, 377
Sphere, 19, 199–202, 210, 367
  vector equation, 62, 63
Stevin, 1
Stokes, G. G., 293
Stokes's theorem, 291, 294, 295, 297
Struik, Dirk, 145
Summation convention, 36, 37
Surface, 270
  area, 273, 274, 275, 287, 288
  continuity conditions, 197
  curvature, 6
  curve, 201–202, 206, 213, 346
  developable, 359
  equipotential, 210
  equivalent representations, 198
  explicit form, 196, 213
  fundamental metric form, 6, 190, 273
  implicit form, 196, 197, 198, 199
  integral, 270–282
  normal, 209–210, 275
  orientable, 270
  parametric representation, 197
  regular point on, 198
  scalar field, 203
  singular point, 197–198, 200
  smooth, 198, 270
  tangential, 359
  tangent to, 211
  transformation, 198, 274, 284
  vector field, 204, 211, 213
Sylvester, 75, 87, 223
Symmetric equations, 18

Tait, 4
Tangent, 133–134, 141, 148, 256
  line, 127, 134, 137, 211
  plane, 209, 282
Tangent, surface, 275
  vector field, 133, 141, 148, 206, 211, 257, 350
Tensor, 5, 82, 315
  addition, 306, 319
  algebra, 6
  analysis, 2, 5, 7, 315
  associated, 81, 82
  calculus, 6, 7, 304, 372
  contraction, 307, 319
  contravariant, 111
  covariant, 111
  covariant curvature, 355, 361–371
  decomposition, 361, 362
  density, 112, 239, 319
  Einstein, 370
  metric, 81, 82, 194, 233, 237, 238, 286, 318, 324, 367, 372
    associated, 81, 82, 233, 237, 286, 326
  mixed, 81, 83
  name, 5
  order, 82, 83, 93, 112, 315
  outer product, 307, 319, 361
  quotient law, 321
  Ricci, 365, 366, 367, 376
  Riemann Christoffel mixed, 354–361, 361–371, 373
  skew symmetry, 307, 308, 320, 362
  symmetry, 307, 308, 320, 322, 326
  systems leading to, 305–314
  transformation law, 315
  transvection, 307
  valence, 306, 319
  weight, 93, 112, 319
Time, 176, 181, 187, 189
Tinseau, 137
Tombaugh, 171
Torque, 176
Torsion, 144
Transformation, 35, 36
  affine, 66, 67, 113–114, 180
  associated metric tensor, 81
  basis arrow, 42
  Cartesian coordinate system, 78, 133, 155, 177–178
  Christoffel symbols, 334
  coefficients, 42–46, 49, 78, 98, 119, 133
  contravariant, 75, 208

Transformation, covariant, 75, 208
  cross product, 98, 114
  curl, 217
  dot product, 67
  ε-systems, 92, 112
  fundamental metric tensor, 81
  general coordinate system, 221–226
  Kronecker delta, 81
  mixed metric tensor, 81
  rotation, 40, 51
  scalar, 54
  similarity, 66
  surface parameter, 198
  translation, 38, 51
  vector, 20, 40, 44, 51
    components, 77
    operator ∇, 206, 213–216, 238
  ω^k components, 167–168
Translation, 36, 38, 51, 52
Transpose of matrix, 118, 119
Transvection, 307
Triple scalar product, 103–106, 114
  dot and cross interchange, 105
  inversion, 106
Triple vector product, 106–108, 114, 132

Universal law of gravitation, 267

Valence, contravariant, 306
  covariant, 306
Vandermonde, 87
Vector, 20, 21
  acceleration, 147, 148, 152, 156–157, 170
  addition, 3, 254
  analysis, 2, 3, 5, 7
  axial, 103
  binormal, 144
  bound, 283
  Cartesian, 51, 52, 76, 151
  components, 55, 75–78, 156–157
  contravariant, 75, 76, 80, 81, 169, 228, 284, 316
  covariant, 75, 80, 81, 113, 231, 316, 385
  density, 113
  division, 108
  field, 127, 129–133, 141, 147, 172, 204, 206, 209–210
  free, 283
Vector, gradient, 207, 215
  magnitude, 58, 82, 148
  operator ∇, 206, 213–216, 238
  origin of term, 3
  orthogonality, 40, 45, 60, 208–209
  parametric equations, 18
  position, 36, 149, 170
  principal normal, 141, 142
  product, 5
    cross, 96
    dot, 56
    triple, 106
  quotient law, 109
  rotational derivative, 169
  solenoidal field, 300
  space, 10
  sum, 131
  symmetric form, 18
  tangential, 133, 141, 148, 206, 211, 213, 228
  transformation law, 20
  triple product, 106–108, 114, 132
  velocity, 127, 147–149, 153, 156, 157, 170, 190
  unit, 211
Velocity, angular, 157, 158, 160, 161
  apparent, 157, 160
  arrow form, 157
  components, 156, 157, 160, 162, 167, 169
  composition, 180, 181
  field, 127, 147–149, 152
  light, 177–179, 182, 183, 186, 188
  linear, 15
  Lorentz transformation equations, 186
  rotation, 161
  translation, 161
Viete, 35
Volume, integral, 288–291
  representation, 105, 290

Wallis, 2
Weight of tensor, 93, 112
Wessel, 2
Wilson, 5
Work, 55, 261, 266
World line, 181, 187

Zero matrix, 116
Zero *n*-tuple, 10